PRINCIPLES OF ISOTOPE GEOLOGY

PRINCIPLES OF ISOTOPE GEOLOGY

by

Gunter Faure

Professor of Geology
The Ohio State University
Columbus, Ohio, 43210

John Wiley & Sons, New York · Santa Barbara · London · Sydney · Toronto

Library of Congress Cataloging in Publication Data:

Faure, Gunter.
   Principles of isotope geology.

   (Smith and Wyllie intermediate geology series)
   Includes bibliographies and index.
   1. Isotope geology. I. Title.
QE501.4.N9F38   550′.28   77-4479
ISBN 0-471-25665-X

Printed in the United States of America

10 9 8 7 6 5 4 3 2 1

For Barbara and our children:

Mary, John, Pamela, and David

PREFACE

Isotope geology has evolved into a highly diversified discipline in the Earth Sciences capable of contributing significantly to the solution of a wide variety of problems. The time has come, therefore, to introduce this subject into the curriculum in order to prepare geologists in all branches of our science to use this source of information. Although the measurements on which isotope geology is based will probably continue to be made by a small but expanding number of experts, the interpretation of the data should be shared increasingly with geologists who are familiar with the complexities of geological problems.

This textbook has been written to stimulate the teaching of isotope geology even at colleges and universities where the facilities for research in this field are not available. Most of the standard journals of geology now publish papers reporting the results of isotopic studies. All geologists should be sufficiently familiar with the principles of isotope geology to be able to make critical evaluations of the data and conclusions these papers contain, or to make their own interpretations. I suggest, therefore, that isotope geology is an important component of the geological curriculum for senior undergraduates or beginning graduate students. It is especially appropriate to teach isotope geology at this level because it touches on all aspects of the Earth Sciences and therefore provides an overview of the entire subject.

The material in this book is presented in three parts. The first part (Chapters 1 through 5) is introductory in nature and presents important background information. Students who are well prepared should be able to bypass these chapters without difficulty. Nevertheless, I have included them, to assure that all students can build on a common base in the more substantive chapters that follow.

The second part (Chapters 6 through 17) presents the interpretation of variations of isotopic compositions of certain chemical elements due to radioactivity. These chapters deal primarily with geochronometry and with the identification of sources and past histories of chemical elements possessing radiogenic isotopes.

The last section of the book (Chapters 18 through 21) contains summaries of the isotopic compositions of hydrogen, oxygen, carbon, and sulfur, whose atoms are fractionated in nature. These elements are among the most abundant in the crust of the Earth and provide useful information on a wide range of geological materials, primarily in terms of isotopic equilibration temperatures and as monitors of certain biological, chemical, and physical processes in nature.

The primary objective of this book is to give a rational exposition of the principles used in the interpretation of isotopic data and to show how such interpretations apply to the solution of geological problems. I have attempted to present the principles of isotope geology by emphasizing the derivation of mathematical equations that are used in the interpretation of isotopic data. In many cases, the geological significance of the conclusions derived from isotopic studies depends on the assumptions that were made in the calculations. I believe that students will better appreciate the limitations of the results of such calculations when they can follow the derivation of the relevant equations step by step and see how various assumptions enter into the process. Isotope geology has no place for handy formulas into which one substitutes data to obtain the magic answer.

All of the substantive chapters include a set of problems that students can use to test their understanding of the principles and to sharpen their skill in making numerical calculations. Most of the problems are from the literature and involve interpretations of data relevant to geology. The answers to many of the problems are given to increase the students' confidence that they are doing it right. In some cases, I have merely given a reference to the literature to let the students sort out for themselves what the "right" answers may be.

Other features of this book that will be helpful are the summaries at the ends of the substantive chapters and the detailed captions of the diagrams. The latter should facilitate browsing and will reinforce the material presented in the text. The references listed at the end of each chapter represent a cross section of the literature up to and including 1975. Although they amount to only a very small fraction of the published literature, the papers I have referred to will enable interested students to pursue any given topic in the library.

This book does not purport to record the current state of the art in the various subdivisions of isotope geology. Consequently, many important studies that have appeared in the recent literature are mentioned only in passing and others may not be cited at all. I do believe, however, that students who have studied this text will be prepared to appreciate the significance of any isotopic study they may find in the literature. Students who intend to specialize in isotope geology may use this book as a stepping-stone toward that goal, but will require much additional study of the theoretical complexities and of the practical problems encountered in the laboratory. This book is intended primarily for the nonspecialists who, nevertheless, should be familiar with the principles of isotope geology.

I am very grateful to my friends and colleagues for their help: G. K. Czamanske, P. Deines, B. R. Doe, H. Faul, R. J. Fleck, E. D. Goldberg, M. J. McSaveney, P. Pushkar, and E. K. Ralph. In addition, I thank R. L. Armstrong, K. Bell, R. D. Dallmeyer, W. G. Deuser, J. D. Obradovich, J. L. Powell, and J. F. Sutter for their suggestions and advice with various aspects of the manuscript. However, all shortcomings of this book and errors of omission or commission are entirely my responsibility. I am grateful to the publishers who gave me permission to use copyrighted material and to the authors who graciously allowed me to do so: P. M. Hurley, R. W. Kistler, E. J. Dasch, G. B. Dalrymple, C. T. Harper, C. H. Stockwell, L. H. Ahrens, G. Turner, G. R. Tilton, S. Epstein, W. Dansgaard, C. Emiliani, R. N. Clayton, H. P. Taylor, Jr., H. Ohmoto, and W. T. Holser. Finally, I thank Kathy Gardlik and Cecilia Weinhoffer for typing the manuscript and C. D. Nardone who helped to proofread it.

Columbus, Ohio

Gunter Faure

CONTENTS

1 THE ROOTS OF ISOTOPE GEOLOGY

1
How Old Is the Earth?

"It is perhaps a little indelicate to ask our Mother Earth her age, but Science acknowledges no shame and from time to time has boldly attempted to wrest from her a secret which is proverbially well guarded."

Arthur Holmes, 1913

The "indelicate" question regarding the age of the planet on which we live has been the cause for scientific controversies for more than 200 years. Although the priests and philosophers of the ancient civilizations had developed some notions regarding the history of the Earth, the subject did not enter the realm of science until geology became established as an independent field of study. Before the middle of the eighteenth century, questions regarding the origin of the Earth and its subsequent history were a matter of theology. Bishop Ussher proclaimed in 1650 that the creation of the world took place in the year 4004 B.C. This statement appeared in most English bibles as a marginal reference and seriously handicapped the thinking of many early geologists who became the victims of theological prejudice. Prior to about 1750 the official view was that all sedimentary rocks were deposited during the Great Flood that befell Noah and his clan and that all other surface features of the Earth resulted from catastrophic events that occurred intermittently.

The rise of geology as a science is commonly associated with the work of James Hutton in Scotland. He emphasized the importance of very slow but continuously acting processes that shape the surface of the Earth. This idea conflicted with Catastrophism and foreshadowed the concept of Uniformitarianism developed by Hutton in his book *Theory of the Earth*, published in 1785. His principal point was that geological processes occurring now have shaped the history of the Earth in the past and would continue to do so in the future. He stated that he could find ". . . no vestige of a beginning—no prospect of an end" for the Earth. The history of the Earth apparently extended over a very long, but not necessarily infinite, amount of time.

Hutton's conclusion regarding the age of the Earth was treated ". . . with righteous horror by the official leaders of the day, most of whom combined the study of theology with that of their favorite science and demanded in the latter a harmonious agreement with the scriptures" (Holmes, 1913, p. 7). However, as time passed, more and more geologists accepted the principle of Uniformitarianism, including the conviction that very long periods of time are required for the deposition of sedimentary rocks whose accumulated thickness amounts to many miles. Geology was emerging as a science, solidly based on the evidence preserved in the rocks, to be interpreted according to the principle of Uniformitarianism. In 1830, Charles Lyell published the first volumes of his *Principles of Geology*. A new generation of geologists gradually replaced the older generations, and with their passing disappeared the requirement that geological theories must agree with the writings of Moses. By the middle of the nineteenth century, geologists seemed to be secure in their conviction that the Earth was indeed very old and that virtually unlimited time was available for the deposition of the great thickness of sedimentary rocks that had been mapped in the field.

The apparent antiquity of the Earth and the principle of Uniformitarianism were unexpectedly attacked by William Thomson, better known as Lord Kelvin (Burchfield, 1975). Thomson was Britain's most prominent physicist during the second half of the nineteenth century. His invasion into geology profoundly influenced geological opinion regarding the age of the Earth for about 50 years. Between 1862 and 1899 Thomson published a number of papers in which he set a series of limits on the possible age of the Earth. His calculations were based on considerations of the luminosity of the sun, the cooling history of the Earth, and the effect of lunar tides on the rate of rotation of the Earth. He initially concluded that the Earth could not be much more than 100 million years old. In subsequent papers the age of the Earth was further reduced. In 1897, Lord Kelvin (he was raised to peerage in 1892) delivered his famous lecture "The Age of the Earth as an Abode Fitted for Life" in which he narrowed the possible age of the Earth to between 20 and 40 million years.

These and earlier estimates of the age of the Earth by Lord Kelvin and others were a serious embarrassment to geologists. Kelvin's arguments seemed to be irrefutable, and yet they were inconsistent with the evidence as interpreted by geologists on the basis of Uniformitarianism. This led to some futile efforts to speed up geological activities in the past in order to squeeze the history of the Earth into the few tens of millions of years that Lord Kelvin would permit. Others clung to the hope that ultimately some flaw would be discovered in his arguments and that the geological evidence would eventually be reconciled with the laws of physics.

Ironically, one year before Lord Kelvin presented his lecture in 1897, the French physicist Henri Becquerel had announced the discovery of radioactivity. Only a few years later it was recognized that the disintegration of radioactive elements is an exothermic process. The natural radioactivity of rocks produces heat, so that the Earth is not merely a cooling body as Lord Kelvin had assumed in his calculation. His conclusions regarding the age of the Earth were therefore invalid. Thus ended a troublesome period in the history of geology. The lesson to be learned from this episode was expressed in these eloquent words:

"The fascinating impressiveness of rigorous mathematical analysis, with its atmosphere of precision and elegance, should not blind us to the defects of the premises that condition the whole process"

T. C. Chamberlin, 1899

2
The Discovery of Radioactivity

The phenomenon of radioactivity was discovered less than 100 years ago, perhaps because we have no sense for detecting it. It cannot be seen, felt, smelled, or heard directly and is detectable only with the aid of mechanical or electronic devices. Its eventual discovery resulted from a series of favorable circumstances that involved experimentation with cathode-ray tubes and studies of luminescence of uranium salts.

Around 1855, the German glassblower Heinrich Geissler succeeded in constructing evacuated glass vessels containing metal electrodes that could be used to study the transmission of electrical charges across a vacuum. The cathode rays that were discovered as a result of experimentation with Geissler's tubes received much attention from physicists during the second half of the nineteenth century. They were explained in 1897 by J. J. Thomson as streams of negatively charged particles (electrons). This was an important milestone in the rapidly developing understanding of the internal structure of atoms for which Thomson received the Nobel Prize for physics in 1906.

The discovery of radioactivity, however, involved cathode-ray tubes only indirectly. Around 1880, Henri Becquerel experimented with the luminescence of uranyl double sulfate crystals caused by exposure to ultraviolet light. Luminescence is the property of certain materials to emit visible light in response to excitation by another source of energy. Physicists working with cathode-ray tubes observed luminescence in the walls of the glass vessels of cathode-ray tubes when they were struck by these rays. The German physicist Wilhelm Konrad Roentgen was studying this phenomenon and discovered that his cathode-ray tube, which was sheathed in black cardboard, caused luminescence in a sheet of paper coated with barium platinocyanide. The luminescence continued as long as the cathode-ray tube was energized, even when he moved the sheet of coated paper into an adjoining room! Apparently the cathode-ray tube emitted a mysterious radiation that could penetrate black paper. The report of his discovery in December 1895 aroused much interest among his colleagues, and several scientists began searching for a connection between Roentgen's x-rays and luminescence.

Roentgen's discovery motivated Henri Becquerel to renew his earlier work with uranium salts. In order to find out if uranium compounds could be made to emit x-rays, he wrapped a photographic plate in black paper, placed some crystals of uranyl sulfate

on it, and exposed the entire package to sunlight. Sure enough, the uranyl sulfate crystals emitted an invisible radiation that penetrated the black paper and fogged the photographic plate. Eventually Becquerel determined that uranium salts, and even uranium-bearing minerals, emitted this radiation continuously without having to be exposed to sunlight. These discoveries, which Becquerel presented to the Academy of Science in Paris in 1896, had far-reaching consequences and led in the next several decades to the development of atomic and nuclear physics and radiochemistry.

3
The Heroic Years

Becquerel's discoveries attracted the attention of several young scientists, among them Marie (Manya) Sklodowska Curie. Manya Sklodowska had come to Paris in 1891 from her native Poland to study at the Sorbonne. On July 25, 1895, she married Pierre Curie, a brilliant physicist working at the Sorbonne. After Becquerel reported his discoveries regarding salts of uranium, Marie Curie decided to devote her doctoral dissertation to a systematic search to determine whether other elements and their compounds emitted similar radiation. Her work was rewarded when she discovered that thorium is also an active emitter of penetrating radiation. Turning to natural uranium and thorium minerals, she noticed that these materials were far more active than the pure salts of these elements. This important observation suggested to her that natural uranium ore, such as pitchblende, should contain more powerful emitters of radiation than uranium. For this reason, Marie and Pierre Curie requested a quantity of uranium ore from the mine of Joachimsthal in Czechoslovakia and in 1898 began a systematic effort to find the powerful emitter whose presence she had postulated. The search eventually led to the discovery of two new active elements which they named polonium and radium. Marie Curie coined the word "radioactivity" on the basis of the emissions of radium. She worked hard to produce pure radium in the metallic form, while her husband concerned himself with the physical properties of the radiation. In 1903 the Curies shared the Nobel Prize for physics with Henri Becquerel for the discovery of radioactivity.

Although Pierre Curie died in 1906 in a traffic accident in Paris, Marie Curie continued her intensive research dealing with the chemistry of radium. She was appointed to the professorship in physics at the Sorbonne left vacant by her husband's death. In 1911, she received the Nobel Prize in chemistry in recognition of her successful efforts to isolate pure radium. In 1914, she founded the Radium Institute which became a center for research in nuclear physics and chemistry. At this institute, Marie Curie continued her work on the chemistry of radioactive elements and on their possible uses in medicine. She died in 1934 of leukemia caused by the effects of radiation emitted by the radioactive elements that she had studied all her life.

The ionizing radiation emitted by radium discovered by the Curies in 1898 aroused the curiosity of Ernest Rutherford who was then working with J. J. Thomson at the Cavendish Laboratory of Cambridge University. He had studied the ionization of gases by x-rays and then took up a study of the nature of the radiation emitted by radium. In the fall of 1898, Rutherford moved to Montreal where he had accepted a professorship in physics at McGill University. A year later he reported that the radiation emitted by radioactive substances consists of three different components which he named alpha, beta, and gamma. The alpha component was

eventually shown to consist of helium nuclei, while the beta rays were identified as electrons. Only the gamma rays proved to be electromagnetic radiation similar to the x-rays discovered by Roentgen.

In 1900, Frederick Soddy came to McGill University as a demonstrator in chemistry. During the two years he spent in Montreal, Soddy worked with Rutherford on the radioactivity of thorium compounds, which led them to formulate the theory of radioactive decay and growth. They suggested that the atoms of radioactive elements disintegrate spontaneously to form atoms of another element. They proposed that the disintegration is accompanied by the emission of alpha and beta particles and that the intensity of the radiation is proportional to the number of radioactive atoms present. Therefore they expressed the rate of disintegration as

$$-\frac{dN}{dt} = \lambda N \qquad (1.1)$$

where λ is the decay constant and represents the probability that an atom will decay in unit time, and N is the number of radioactive atoms present. In 1902, Frederick Soddy went to work with Sir William Ramsay at University College in London. Rutherford left McGill University in 1907 to become Langworthy Professor of physics at Manchester University. He received the Nobel Prize for physics in 1908 in recognition of his work on radioactivity.

However, Rutherford's greatest achievements were yet to come. At Manchester University, he began his famous experiments on the scattering of alpha particles by metal foils. The results of these experiments indicated that atoms have very small, positively charged nuclei and that the nucleus is surrounded by electrons orbiting around it. The positive charge of the nucleus was attributed to the presence of protons, which Rutherford named in 1919. A year later he speculated that the nuclei of atoms might contain a neutral particle. This hypothetical particle, the neutron, was discovered in 1932 on the basis of experiments carried out by W. Bothe, Frederic Joliot, and Sir James Chadwick (Frederic Joliot was the husband of Irene Curie, the eldest daughter of Marie and Pierre Curie).

Rutherford's model of the atom did not endure for long. In 1912, a young Danish physicist, Niels Bohr, came to Manchester as a postdoctoral fellow. He applied the principles of quantum mechanics developed by Max Planck and Albert Einstein, which involved radical departures from classical physics, to the possible energy states of the hydrogen atom. Bohr's model of the hydrogen atom was based on the postulates that the angular momentum of an electron orbiting the nucleus of an atom could only have values that are multiples of $h/2\pi$, where h is Planck's constant, and that the energy of radiation emitted or absorbed by the hydrogen atom results from transitions between different energy states. Bohr's theory was remarkably successful in explaining the spectrum of hydrogen but failed when applied to atoms having many electrons. Bohr's theory was subsequently replaced by wave mechanics developed by Dirac, Heisenberg, and Schrödinger.

In 1919, Rutherford succeeded J. J. Thomson as director of the Cavendish Laboratory of Cambridge University where he had started his scientific career. He had been knighted in 1914 and, in 1931, he was created the first Baron Rutherford of Nelson and Cambridge.

As the radioactive decay series of uranium and thorium were worked out, a confusing problem arose. It was found, for example, that there are several kinds of thorium which decay at different rates. Moreover, careful determinations by the American chemist T. W. Richards showed that the atomic weights of the elements are not whole numbers as had been proposed by William Prout in 1815. Still more disturbing was Richard's report in 1913 that lead produced by the decay of uranium had a different atomic weight than ordinary lead. These problems were solved by a bold suggestion made by Soddy that the place occupied by a particular element in the periodic table could accommodate more than one kind of atom. He named these atoms "isotopes," which in Greek means "same place." Actually William Crookes in 1886 had attempted to explain the failure of Prout's hypothesis by suggesting that the atoms of elements have different whole-number weights and that the atomic weight of an element represents an average weight of the mixture of different atoms. These ideas received direct experimental confirmation when J. J. Thomson observed in 1913 that neon is composed of two kinds of atoms having atomic weights of about 20 and 22. These observations were made by means of a "positive-ray" apparatus built by Thomson.

At that time a young chemist, F. W. Aston, was working in the Cavendish Laboratory. He set out immediately to confirm or disprove J. J. Thomson's claim. After the conclusion of World War I, Aston improved the design of Thomson's positive-ray apparatus and called it a "mass spectrograph." With this instrument he not only confirmed Thomson's earlier work on neon but also discovered a third isotope having a mass of 21. Aston devoted the rest of his life to building increasingly precise mass spectrographs with which he discovered 212 of the 287 naturally occurring isotopes. He also measured the masses of these isotopes and calculated the atomic weights of elements on the basis of the masses and relative abundances of their naturally occurring isotopes. Aston won the Nobel Prize in chemistry in 1922 for his achievements. The design of mass spectrographs was improved in later years by A. J. Dempster, K. T. Bainbridge, A. O. Nier, M. G. Inghram, H. E. Duckworth, and many others. It has evolved into a highly precise and accurate tool for the measurement of isotopic abundances of elements in geological materials. The theory of mass spectrometry is presented in Chapter 5.

4
Impact on Geology

The discovery of radioactivity and the subsequent work of the Curies, Rutherford, Soddy, Thomson, Ramsay, and others had a profound effect on geology. In 1903, Curie and Laborde demonstrated that radioactive decay is an exothermic process. This started a new line of research by geologists to measure the radioactivity of rocks and to calculate the rate of heat production. The first such calculation was made by R. J. Strutt in 1906 on the basis of the radium content of rocks. John Joly recognized in 1907 that pleochroic halos in rocks are caused by the presence of radioactive minerals. He also measured the concentrations of radium and thorium in different kinds of rocks and summarized his important conclusions in 1909 in a book entitled *Radioactivity and Geology*. In it he discussed the origin of pleochroic halos, reported measurements of the radioactivity of rocks, and calculated the resulting heat production. He also speculated that radioactivity may provide the energy required for mountain building. The work of Joly, Strutt, and others regarding the distribution of radioactive elements in the Earth and the resulting heat production has continued to the present. In fact, this subject has received much attention recently as a result of the manned flights to the moon from 1969 to 1972. One of the primary objectives of these flights has been the measurement of the concentrations of radioactive elements in lunar rocks. These measurements have provided the information that is needed to understand the thermal history and present temperature distribution of the moon.

Radioactivity not only causes heat generation in rocks but also provides an accurate method of measuring the ages of rocks and minerals. This possibility was recognized by both Rutherford and B. B. Boltwood around 1905. During a series of lectures at Yale University in 1905, Rutherford proposed that the ages of uranium minerals could be measured by the amount of helium that had accumulated in them. He actually carried out such age determinations on several uranium minerals and obtained dates of about 500 million years. Here was positive evidence that Lord Kelvin's estimate of the age of the Earth was in error, not quite 10 years after he presented his definitive solution of the problem.

In 1904, Rutherford presented a lecture at the Royal Institution on the heat production of the radioactivity of radium and its effect on the prolongation of the heat of the Earth. As he entered the lecture hall, he was made uneasy by the fact that Lord Kelvin was in the audience. However, he avoided a potentially explosive confrontation with the old man by announcing that Lord Kelvin had calculated the age of the Earth from its thermal history, *provided* that no new source of heat was found. Therefore, Lord Kelvin had, in fact, *anticipated* the discovery of radioactivity and the heat produced by this phenomenon! Lord Kelvin, who had slept soundly through the rest of the lecture, beamed with pleasure.

The American chemist Bertram Boltwood reported in 1904 that the U/Ra ratio in most old uranium minerals is constant. In 1905, he speculated that lead is the stable end product of the decay of uranium, based on very careful chemical analyses of uraninite performed by Hillebrandt in 1890 and 1891. In 1907, Boltwood published age determinations of three uraninite specimens based on their U/Pb ratios. His dates ranged from 410 to 535 million years and are in reasonably good agreement with modern age determinations on similar materials from the respective locations. Boltwood's age determinations were made before isotopes had been discovered, before it was known that lead is also produced by decay of thorium, and before the disintegration rate of uranium was known accurately.

The state of the art was lucidly reviewed by Arthur Holmes in his book "The Age of the Earth" which appeared in 1913 when he was 23 years old. In it, Holmes presented an authoritative account of the importance of radioactivity to the question of the age of the Earth and proposed the first geological time scale, based on the thickness of accumulated sedimentary rocks and on the formation of helium and lead in uranium-bearing minerals. Holmes was, from the very beginning, an enthusiastic champion of the importance of radioactivity to geology. However, geologists did not always share this enthusiasm, and Holmes remarked regretfully in 1913 that "the surprises which radioactivity had in store for us have not always been received as hospitably as they deserved." One of the surprises that Holmes referred to was the seemingly excessive length of geologic time indicated by dating of minerals based on radioactivity. This was surely an ironic situation. Hardly 15 years after Lord Kelvin had crushed geologists with his "irrefutable" calculation that the age of the Earth could be no more than 40 million years, geologists were now complaining that age determinations based on radioactivity made the Earth too old! The difficulty arose because dates calculated on the basis of radioactivity were considerably greater than those obtained from rates of erosion, the salt

content of the oceans, and the rates of sedimentation. The discrepancy implied either that sedimentation had been slower in the past than at present or that very large amounts of sediment had been removed by erosion. Neither alternative appealed to some geologists who preferred instead to question the reliability of the radiometric dates. The time scale published by Holmes in 1913 indicated an age of about 1300 million years for Archaean gneisses, but he speculated that the oldest Archaean rocks must be 1600 million years old. Holmes devoted a major portion of his career to the applications of radioactivity in the solution of geological problems. The geological time scale remained a primary concern throughout his life (Appendix II).

We will deal with the subsequent history of geochronometry in the appropriate chapters which describe the several different methods of dating that are available to us now. The study of the radioactivity of rocks and the measurement of geologic time were given formal recognition in the United States in 1923 when the Research Council of the Academy of Sciences formed the Committee on the Measurement of Geologic Time by Atomic Disintegration. This committee sponsored scientific meetings and published periodic reports that have played a vital role in the growth of isotope geology.

5
Fractionation of Stable Isotopes

Before closing this chapter we must mention yet another important discovery. In 1931 Harold C. Urey predicted on theoretical grounds that there should be a difference in the vapor pressures of the isotopes of hydrogen. His interest in hydrogen had been aroused by a suggestion of Birge and Menzel that it may have naturally occurring isotopes. Urey, working with Murphy and

Brickwedde, promptly planned and carried out a definitive experiment to detect ^2_1H and ^3_1H by spectroscopic methods in the residual volume of gas produced by evaporating about six liters of liquid hydrogen. The results immediately confirmed the presence of ^2_1H, but ^3_1H was not found. Urey named the newly discovered isotope "deuterium" because it has nearly twice the mass of hydrogen. The specific reason for this was not yet known because the existence of neutrons was not established until 1932. In 1934, Harold Urey won the Nobel Prize for chemistry for his discovery of deuterium. During World War II, he used his knowledge of isotope fractionation to develop methods for the separation of ^{235}U by gaseous diffusion. When the war ended, he turned his attention to the possibility that the stable isotopes of oxygen may be fractionated by natural processes. He suggested that such fractionation might occur during the formation of calcium carbonate in the oceans and that the extent of fractionation depends on the temperature. Out of these ideas has developed the oxygen isotope method of measuring the temperature of deposition of skeletal calcium carbonate.

The research inspired and led by Harold Urey has evolved into an important branch of isotope geology which deals with the fractionation of the stable isotopes by physical and chemical reactions occurring in nature. The group of elements whose isotopes are especially susceptible to natural isotope fractionation includes hydrogen, carbon, nitrogen, oxygen, and sulfur. These are among the most abundant elements in the Earth, and they are intimately associated with the biosphere, the hydrosphere, and the lithosphere. Consequently, the study of fractionation of their isotopes provides information on a great variety of important geological processes occurring in many different geological environments.

6. BOOKS ON ISOTOPE GEOLOGY

Allegre, C. J., and G. Michard (1974) Introduction to geochemistry. D. Reidel, Dordrecht, Holland, 142 p.

Bandy, O. L., ed. (1970) Radiometric dating and paleontologic zonation. Geol. Soc. Amer. Spec, Paper 124, 247 p.

Cherdyntsev, V. V. (1969) Uranium-234. Atomizdat, Moscow. Translated by J. Schmorak. Israel Program for Scientific Translations, Jerusalem, 1971, 234 p.

Craig, H., S. L. Miller, and G. J. Wasserburg, eds. (1964) Isotopic and cosmic chemistry. North-Holland, Amsterdam, 553 p.

Dalrymple, G. B., and M. A. Lanphere (1969) Potassium-argon dating. Principles, techniques and applications to geochronology. W. H. Freeman, San Francisco, 258 p.

Doe, B. R. (1970) Lead isotopes. Springer-Verlag, Berlin and Heidelberg, 137 p.

Faul, H., ed. (1954) Nuclear geology. John Wiley, New York, 414 p.

Faul, H. (1966) Ages of rocks, planets and stars. McGraw-Hill, New York, 109 p.

Faul, H. (1968) Nuclear clocks. U.S. Atomic Energy Comm., Div. Tech. Information, P.O. Box 62, Oak Ridge, Tenn., 37830, 61 p.

Faure, G., and J. L. Powell (1972) Strontium isotope geology. Springer-Verlag, Berlin and Heidelberg, 188 p.

Fleischer, R. L., P. Buford, and R. M. Walker (1975) Nuclear tracks in solids: Principles and applications. University of California, Berkeley, Los Angeles and London, 605 p.

Geiss, J., and E. D. Goldberg, eds. (1963) Earth Science and meteoritics. North-Holland, Amsterdam, 312 p.

Hamilton, E. I. (1965) Applied geochronology. Academic Press, London, 267 p.

Hamilton, E. I., and R. M. Farquhar, eds. (1968) Radiometric dating for geologists. Interscience, London, 506 p.

Harland, W. B., A. Gilbert Smith, and B. Wilcock, eds. (1964) The Phanerozoic time-scale. Quart. J. Geol. Soc. London, *120s*, 458 p.

Harland, W. B., E. H. Francis, and P. Evans, eds. (1971) The Phanerozoic time-scale. A supplement. Spec. Paper no. 5, Geol. Soc. London, 356 p.

Harper, C. T., ed. (1973) Geochronology: Radiometric dating of rocks and minerals. Dowden, Hutchinson and Ross, Stroudsburg, Pa., 469 p.

Hoefs, J. (1973) Stable isotope geochemistry. Springer-Verlag, Berlin and Heidelberg, 140 p.

Hurley, P. M. (1959) How old is the Earth? Doubleday, 160 p.

Libby, W. F. (1952) Radiocarbon dating, University of Chicago Press, Chicago.

Rankama, K. (1954) Isotope geology. Pergamon Press, London, 535, p.

Rankama, K. (1963) Progress in isotope geology. Interscience, London, 705 p.

Russell, R. D., and R. M. Farquhar (1960) Lead isotopes in geology. Interscience, New York, 243, p.

Schaeffer, O. A., and J. Zähringer, eds. (1966) Potassium argon dating. Springer-Verlag, New York, 234 p.

Shukolyukov, Y. A., I. M. Gorokhov, and O. A. Levchenkov (1974) The graphical methods in isotope geology. Moscow, Nedra, 207 p. (In Russian.)

Van der Merwe, N. J. (1969) The carbon-14 dating of iron. University of Chicago Press, Chicago, 137 p.

Yachenko, M. L., and E. S. Varshavskaya (1971) A brief review of the application of isotopes of strontium and lead in geology. Acad. Sci. U.S.S.R., Institute for Geology and Geochronology of the Precambrian, Leningrad, Nauka, 140 p. (In Russian.)

York, D., and R. M. Farquhar (1972) The Earth's age and geochronology. Pergamon Press, Oxford, 178 p.

REFERENCES

Aston, F. W., (1920) The constitution of atmospheric neon. Phil. Mag., ser. 6, *39*, 449–455.

Becquerel, H. (1896) Sur les radiations invisibles emises par phosphorescence; Sur les radiations invisibles emisses par les corps phosphorescents; Sur les radiations invisibles emisses par les sels d'uranium. Compt. rend., *122*, 420, 501, 689.

Boltwood, B. B. (1907) On the ultimate disintegration products of the radioactive elements. Am. J. Sci. (4), *23*, 77–88.

Burchfield, J. D. (1975) Lord Kelvin and the age of the Earth. Science History Publications, New York, 260 pp.

Chadwick, J. (1932) The existence of a neutron. Proc. Roy. Soc. London, ser A. *136*, 692–708.

Chamberlin, T. C. (1899) Lord Kelvin's address on the age of the earth as an abode fitted for life. Science, *9*, 889–901, *10*, 11–18.

Curie, M. S. (1898) Rayons emis par les composes de l'uranium et du thorium. Compt. rend., *126*, 1101.

Curie, P., and A. Laborde (1903) Sur la chaleur degagee spontanement par les sels de radium. Compt. rend., *136*, 673–675.

Holmes, A. (1911) The association of lead with uranium in rock-minerals, and its application to the measurement of geologic time. Proc. Roy. Soc. (A), *85*, 248–256.

Holmes, A. (1913) The age of the earth. Harper and Brothers, London, 194 p.

Holmes, A. (1947) The construction of a geological time-scale. Trans. Geol. Soc. Glasg., *21*, 117–152.

Holmes, A. (1960) A revised geological time-scale. Trans. Edinburgh Geol. Soc., *17*, 183–216.

Jauncey, G. E. M. (1946) The early years of radioactivity. Am. J. Phys., *14*, 226.

Joly, J. J. (1908) On the radium content of deep-sea sediments. Phil. Mag. (6), *16*, 190–197.

Joly, J. J. (1910) The amount of thorium in sedimentary rocks, II: Arenaceous and argillaceous rocks. Phil. Mag. (6), *20*, 354–357.

Joly, J. J., and J. H. J. Poole (1924) The radioactivity of basalts and other rocks. Phil. Mag. (6), *48*, 819–832.

Knopf, A., C. Schuchert, A. F. Kovarik, A. Holmes, and E. W. Brown (1931) The age of the earth. Nat. Res. Council, Bull. *80*, 487.

Murphy, G. M. (1964) The discovery of deuterium. In Isotopic and Cosmic Chemistry, H. Craig, S. L. Miller, and G. J. Wasserburg, eds., 1–7, North-Holland, Amsterdam, 553 pp.

Richards, T. W., and M. E. Lembert (1914) Atomic weight of lead of radioactive origin. J. Am. Chem. Soc., *36*, 1329–1344.

Rutherford, E., and F. Soddy (1902a) The cause and nature of radioactivity. Pt. I. Phil. Mag., ser. 6, *4*, 370–396.

Rutherford, E., and F. Soddy (1902b) The cause and nature of radioactivity. Pt. II. Phil. Mag., ser. 6, *4*, 569–585.

Rutherford, E., and F. Soddy (1902c) The radioactivity of thorium compounds. I. An investigation of the radioactive emanation. J. Chem. Soc. London, *81*, 321–350.

Rutherford, E., and F. Soddy (1902d) The radioactivity of thorium compounds. II. The cause and nature of radioactivity. J. Chem. Soc. London, *81*, 837–860.

Rutherford, E. (1906) Radioactive transformations. Charles Scribner's Sons, New York, 287 p.

Rutherford, E. (1911) The scattering of α and β particles by matter and the structure of the atom. Phil. Mag., ser. 6, *21*, 669–688.

Soddy, Frederick (1908) Attempts to detect the production of helium from the primary radio-elements. Phil. Mag. (6), *16*, 513–530.

Soddy, Frederick (1914) The chemistry of the radio-elements. Pt. II. Longmans, Green, London, 46 p.

Strutt, R. J. (1906) On the distribution of radium in the Earth's crust, and on the Earth's internal heat. Proc. Roy. Soc. (London), *A77*, 472–488.

Strutt, R. J. (1908) The acculumation of helium in geological time, I. Proc. Roy. Soc. London, *A81*, 272–277.

Thomson, J. J. (1914) Rays of positive electricity. Proc. Roy. Soc. London, ser. S, *89*, 1–20.

Thomson, W. (Lord Kelvin) (1862) On the age of the sun's heat. Popular Lectures and Addresses, vol. 1, p. 349.

Thomson, W. (Lord Kelvin) (1899) The age of the earth as an abode fitted for life. Phil. Mag. (5), *47*, 66–90.

Urey, H. C., Brickwedde, F. G., and G. M. Murphy (1932) An isotope of hydrogen of mass 2 and its concentration. (Abstract) Phys. Rev., *39*, 864.

2 THE INTERNAL STRUCTURE OF ATOMS

Our concept of the internal structure of atoms has evolved dramatically since Rutherford first demonstrated the existence of atomic nuclei. However, it will be sufficient for our purposes to adopt a fairly simple view of atoms. We begin, therefore, by stating that every atom contains a very small, positively charged nucleus in which most of its mass is concentrated. The nucleus is surrounded by a cloud of electrons that are in motion around it. In a neutral atom the negative charges of the electrons exactly balance the total positive charge of the nucleus. The diameters of atoms are of the order of 10^{-8} centimeters (cm) and are conveniently expressed in angstrom units ($1 \text{ Å} = 10^{-8}$ cm). The nuclei of atoms are about 10,000 times smaller than that and have diameters of 10^{-12} cm, or 10^{-4} Å. The density of nuclear matter is about 10^{14} grams per cubic centimeter or 100 million tons per cubic centimeter. Such enormous densities may occur also in neutron stars and "black holes," but are quite outside our realm of experience.

It is now known that the nucleus contains a large number of different elementary particles which interact with each other and which are organized into complex patterns within the nucleus. It will suffice for the time being to introduce only two of these, the proton (p) and the neutron (n), which are collectively referred to as nucleons. Protons and neutrons can be regarded as the main building blocks of the nucleus because they account for its mass and electrical charge. Briefly stated, a proton is a particle having a positive charge that is equal in magnitude but opposite in polarity to the charge of an electron. Neutrons have a very slightly larger mass than protons and carry no electrical charge. Extranuclear neutrons are unstable and decay spontaneously to form protons and electrons with a "half-life" of 10.6 minutes. The only other components of atoms of interest to us now are the electrons that swarm around the nucleus. Electrons at rest have a very small mass (1/1836.1 that of hydrogen atoms) and a negative electrical charge. In a neutral atom the number of extranuclear electrons is equal to the number of protons. The protons in the nucleus of an atom therefore determine how many electrons that atom can have when it is electrically neutral. The number of electrons and their distribution about the nucleus in turn determine the chemical properties of that atom.

1
Nuclear Systematics

The composition of atoms is conveniently described by specifying the number of protons and neutrons that are present in the nucleus. The number of protons (Z) is called the "atomic number" and the number of neutrons (N) is the "neutron number." The atomic number Z also indicates the number of extranuclear electrons in a neutral atom. The sum of protons and neutrons in the nucleus of an atom is the "mass number," (A). We can therefore represent the composition of the nuclei of atoms by means of the simple relationship

$$A = Z + N \tag{2.1}$$

Another word for atom that is widely used is "nuclide." Now that we have defined A, Z, and N, we can specify the composition of any nuclide by means of a shorthand notation consisting of the chemical symbol of the element, the mass number written as a superscript, and the atomic number written as a subscript. For example, $^{14}_{6}C$ identifies the nuclide as an atom of carbon having 6 protons (therefore 6 electrons in a neutral atom) and a total of 14 nucleons. Using Equation 2.1, we calculate that the nucleus of this nuclide contains $14 - 6 = 8$ neutrons. Similarly, $^{23}_{11}Na$ is a sodium atom having 11 protons and $23 - 11 = 12$ neutrons. Actually, it is redundant to specify Z when the chemical symbol is used. For this reason, the subscript is sometimes omitted in informal usage. A great deal of information about nuclides can be shown on a diagram in which each nuclide is represented by a square in coordinates of Z and N. Figure 2.1 is a part of such a chart of the nuclides.

We are now in a position to define several additional terms. Referring to the chart of the nuclides, we see that each element having a particular atomic number Z is represented by several atoms arranged in a

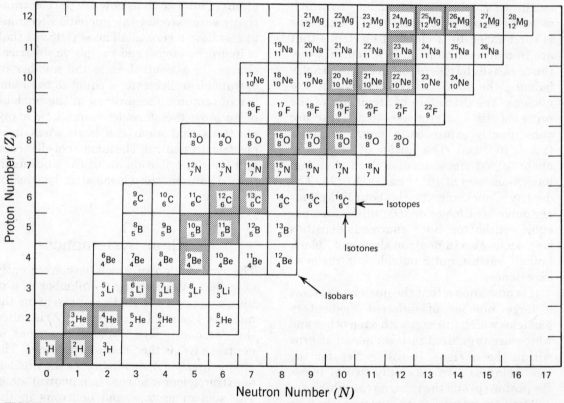

FIGURE 2.1   Partial chart of the nuclides. Each square represents a particular nuclide which is defined in terms of the number of protons (Z) and neutrons (N) that make up its nucleus. The shaded squares represent stable atoms, while the white squares are the unstable or radioactive nuclides. Isotopes are atoms having the same Z but different values of N. Isotones have the same N but different values of Z. Isobars have the same A but different values of Z and N. Only isotopes are atoms of the same element and therefore have nearly identical chemical properties.

horizontal row having different neutron numbers. Such atoms, which have the same Z but different values of N, are called "isotopes." Because they have the same Z, isotopes are atoms of the same chemical element. They have very similar chemical properties and differ only in their masses. Nuclides, which occupy vertical columns on the chart of the nuclides, are called "isotones." They have the same value of N but different values of Z. Isotones are therefore atoms of different elements. We also recognize nuclides that occupy diagonal rows on the chart of the nuclides. These have the same value of A and are called "isobars." Isobars have different values of Z and N and are therefore atoms of different elements. However, because they contain the same number of nucleons, they have similar masses.

2
Atomic Weights

The masses of atoms are too small to be conveniently expressed in grams. For this reason we define the "atomic mass unit" (amu), which is one-twelfth of the mass of $^{12}_{6}C$. In other words, the mass of $^{12}_{6}C$ is arbitrarily fixed at 12.000 . . . amu and the masses of all other nuclides and subatomic particles are obtained by comparison to that of $^{12}_{6}C$. The masses of the isotopes of the elements have been measured by mass spectrometry and are known with great precision.

The total number of different nuclides is close to 1700, but only about 260 of these are stable. These stable nuclides, along with a small number of naturally occurring unstable nuclides, make up the elements in the periodic table. Many elements have two or more naturally occurring isotopes, some have only one, and two elements (technetium and promethium) have none. These two ele-

ments therefore do not occur naturally on the Earth. However, they have been identified in the optical spectra of stars where they may be synthesized by nuclear reactions. At the moment we are concerned only with elements that have one or more naturally occurring isotopes.

The relative proportions of the naturally occurring isotopes of an element are expressed in terms of percentages by number. For example, when we say that the isotopic abundance of $^{85}_{37}Rb$ is 72.15 percent, it means that in a sample of 10,000 rubidium atoms, 7215 are the isotope $^{85}_{37}Rb$. When the masses of the naturally occurring isotopes of an element and their abundances are known, the atomic weight of that element can be calculated. The atomic weight of an element is the sum of the masses of its naturally occurring isotopes, weighted in accordance with the abundance of each isotope, expressed as a decimal fraction. For example, the atomic weight of chlorine is calculated from the masses and abundances of its two naturally occurring isotopes, as shown below:

ISOTOPE	MASS × ABUNDANCE
$^{35}_{17}Cl$	$34.96885 \times 0.7577 = 26.4958$
$^{37}_{17}Cl$	$36.96590 \times 0.2423 = 8.9568$
	Atomic weight 35.4526 amu

The abundances of the naturally occurring isotopes of the elements and their measured masses are listed in tables such as those of Holden and Walker (1972) and Lederer et al. (1967).

The atomic weights of the elements are expressed in atomic mass units. However, in chemistry it is convenient to define the "gram atomic weight" or "mole" which is the atomic weight of an element in grams.

One mole of an atom, or a compound, contains a fixed number of atoms or molecules, respectively. The number of atoms or molecules in a mole is given by Avogadro's number, which is equal to 6.02252×10^{23}.

Now that we have a unit for measuring the small masses of atoms and subatomic particles, it is possible to attempt to calculate the mass of a particular nuclide by adding together the masses of protons + electrons ($M_H = 1.007825$ amu) and of the neutrons ($M_n = 1.008665$ amu) of which it is composed. When this is done, one finds that the observed masses are consistently *less* than the masses calculated from the constituent particles. It appears, therefore, that the mass of an atom is less than the sum of its parts. This phenomenon is a very important clue to an understanding of the nature of the atomic nucleus. The explanation of the observed "mass defect" is that some of the mass of the nuclear particles is converted into "binding energy" that holds the nucleus together. The binding energy can be calculated by means of Einstein's equation:

$$E_B = \Delta M c^2 \qquad (2.2)$$

where E_B is the binding energy, ΔM is the mass defect, and c is the velocity of light equal to 2.998×10^{10} cm/sec.

Before we can make the calculation, we need to review briefly the relationship between units of mass and energy. The basic unit of energy in the cgs system (centimeter, gram, second) is the erg. However, the amount of energy released by a nuclear reaction involving a single atom is only a small fraction of one erg. For this reason we use the "electron volt" (eV) which is defined as the energy acquired by any charged particle carrying a unit electronic charge when it is acted upon by a potential difference of 1 volt. One electron volt is equivalent to 1.602×10^{-12} ergs. It is convenient to define

two additional units, the kiloelectron volt (keV) and the million electron volt (MeV), where $1\,\text{keV} = 10^3$ eV, and $1\,\text{MeV} = 10^6$ eV. We now calculate the energy in million electron volts equivalent to 1 amu, using Equation 2.2. It follows from the definition of the atomic mass unit that

$$1 \text{ amu} = \frac{1}{12} \times \frac{12.000}{N} = \frac{1}{N} \text{grams} \quad (2.3)$$

where N is Avogadro's number. The amount of energy in million electron volts equivalent to 1 amu is obtained from Equation 2.2, by substituting $(1/N)$ for the mass of 1 amu in grams and by converting ergs to million electron volts by means of the appropriate conversion factor given above

$$E = \frac{(2.998 \times 10^{10})^2}{6.02252 \times 10^{23} \times 1.602 \times 10^{-12} \times 10^6}$$

$$= 931.58 \text{ MeV/amu}$$

We have obtained the useful result that 1 amu of mass is equivalent to about 931.6 MeV of energy. The binding energy of a nucleus (E_B) can now be calculated from the mass defect (ΔM) by means of the equation

$$E_B = 931.6\,\Delta M \text{ MeV} \qquad (2.4)$$

As an example, we calculate the binding energy per nucleon for $^{27}_{13}$Al. The theoretical mass of this nuclide is $13 \times 1.007825 + 14 \times 1.008665 = 27.223035$ amu. Its measured mass is only 26.981541 amu. The mass defect and binding energies therefore are

Mass defect $\Delta M = 0.241494$ amu

Binding energy $E_B = \dfrac{0.241494 \times 931.6}{27}$

$$= 8.332 \text{ MeV per nucleon}$$

The binding energies per nucleon of the atoms of most elements have values ranging

THE INTERNAL STRUCTURE
OF ATOMS 2

from about 7.5 to 8.8 MeV. The atoms of hydrogen, helium, lithium, and beryllium have lower binding energies. In the other elements the binding energy per nucleon rises slightly with increasing mass number and reaches a maximum value for $^{56}_{26}$Fe. Thereafter, the binding energies decline slowly with increasing mass number.

3
Nuclear Stability and Abundance

It is reasonable to expect that a relationship exists between the stability of the nucleus of an atom and the abundance of that atom in nature. Conversely, we can use the observed abundances of the elements in the solar system and the abundances of their naturally occurring isotopes to derive information about the apparent stabilities of different kinds of atoms.

First let us recall a statement made in the previous section of this chapter that only about 260 of nearly 1700 known nuclides are stable. Evidently, nuclear stability is the exception rather than the rule. To appreciate this in more detail we examine Figure 2.2 which is a schematic plot of the nuclides in coordinates of N and Z. It can be seen that the stable nuclides (shown in black) form a band flanked on both sides by unstable nuclides (shown in white). The existence of such a region of stability indicates that only those nuclei are stable in which Z and N are nearly equal. Actually, the ratio of N to Z increases from 1 to about 3 with increasing values of A. Only $^1_1$H and $^3_2$He have fewer neutrons than protons.

Another interesting observation is that most of the stable nuclides have even numbers of protons and neutrons. Stable nuclides with even Z and odd N or vice versa are much less common, while nuclides having odd Z and odd N are rare. These facts are shown in Table 2.1. They suggest that the nucleons are arranged into some sort of regular pattern within the nucleus consisting of even numbers of protons and neutrons. A careful study of this phenomenon combined with theoretical considerations has led to the concept of "magic numbers" for Z and N. Nuclides having magic proton numbers or magic neutron numbers, or both, are unusually stable as indicated by their greater abundance or slower decay rates in the case of unstable nuclides. The magic numbers for Z and N are 2, 8, 10, 20, 28, 50, 82, and 126. Calcium is a good example of the effect of the magic number. It has a magic proton number (20) and has five stable isotopes whose mass numbers are 40, 42, 44, 46, and 48 (Fig. 2.2). Note that all of these stable isotopes have *even* neutron numbers. The nucleus of $^{40}_{20}$Ca is doubly magic because it also has a magic number of neutrons ($N = 20$). Its isotopic abundance is 96.94 percent. $^{48}_{20}$Ca is also doubly magic ($N = 28$), but its abundance is only 0.19 percent, which is nevertheless remarkably high considering the fact that it is some distance from the region of stability.

So far we have been concerned primarily with the stable nuclides and have considered some regularities of the composition of their nuclei. Actually, most known nuclides are not stable but decompose spontaneously until they achieve a stable nuclear configuration. These are the so-called radioactive nuclides, or radionuclides. The spontaneous transformations that occur in their nuclei give rise to the phenomenon of "radioactivity." Most of the known radionuclides do not occur naturally because their decay rates are rapid compared to the age of the solar system. However, they can be produced artificially by means of nuclear reactions in the laboratory. In isotope

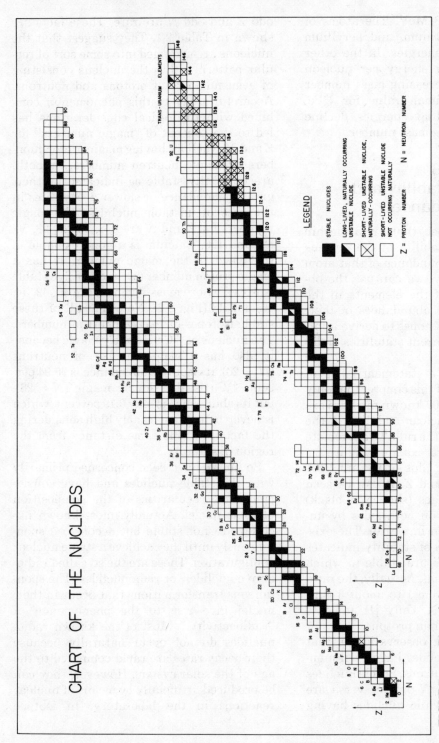

FIGURE 2.2 Plot of the nuclides in coordinates of Z and N. The diagram does not show all of the short-lived radioactive nuclides. (After Holden and Walker, 1972.)

Table 2.1 **Distribution of Stable Nuclides Depending on "Evenness" or "Oddness" of A, Z, and N (Holden and Walker, 1972)**

A	Z	N	NUMBER OF STABLE NUCLIDES
Even	Even	Even	157
Odd	Even	Odd	53
Odd	Odd	Even	50
Even	Odd	Odd	4
Total number of stable nuclides			264

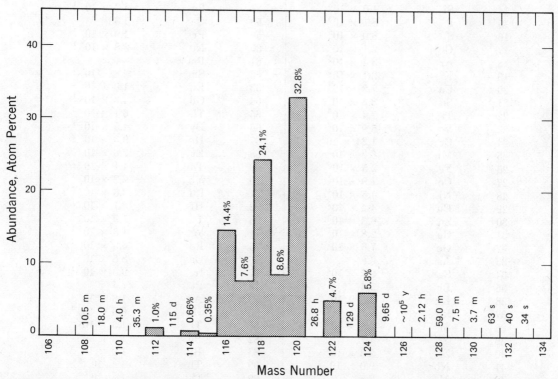

FIGURE 2.3 The isotopes of tin ($Z = 50$). The shaded areas represent the abundances of the stable isotopes, expressed in units of atom percent. The unstable isotopes are indicated by their half-lives expressed in units of years (y), days (d), minutes (m), or seconds (s). The diagram illustrates the effect of even and odd neutron numbers on the abundances of the stable isotopes and on the half-lives of the unstable isotopes of tin. Note that the half-life of an unstable nuclide is the time required for one-half of a given number of atoms to decay. (Data from Holden and Walker, 1972.)

Table 2.2 **Abundances of the Elements in the Solar System in Cosmic Abundance Units, i.e., Atoms per 10^6 Si Atoms (Goles, 1969)**

Z	ELEMENT	ABUNDANCE	Z	ELEMENT	ABUNDANCE
1	H	4.8×10^{10}	44	Ru	1.83
2	He	3.9×10^9	45	Rh	3.3×10^{-1}
3	Li	16	46	Rd	1.33
4	Be	0.81	47	Ag	3.3×10^{-1}
5	B	6.2	48	Cd	1.2
6	C	1.7×10^7	49	Tn	1.0×10^{-1}
7	N	4.6×10^6	50	Sn	1.7
8	O	4.4×10^7	51	Sb	2.0×10^{-1}
9	F	2.5×10^3	52	Te	3.1
10	Ne	4.4×10^6	53	I	4.1×10^{-1}
11	Na	3.5×10^4	54	Xe	3.0
12	Mg	1.04×10^6	55	Cs	2.1×10^{-1}
13	Al	8.4×10^4	56	Ba	5.0
14	Si	1.0×10^6	57	La	4.7×10^{-1}
15	P	8.1×10^3	58	Ce	1.38
16	S	8.0×10^5	59	Pr	1.9×10^{-1}
17	Cl	2.1×10^3	60	Nd	8.8×10^{-1}
18	Ar	3.4×10^5	61	Pm	—
19	K	2.1×10^3	62	Sm	2.8×10^{-1}
20	Ca	7.2×10^4	63	Eu	1.0×10^{-1}
21	Sc	3.5×10^1	64	Gd	4.3×10^{-1}
22	Ti	2.4×10^3	65	Tb	6.1×10^{-2}
23	V	5.9×10^2	66	Dy	4.5×10^{-1}
24	Cr	1.24×10^4	67	Ho	9.3×10^{-2}
25	Mn	6.2×10^3	68	Er	2.8×10^{-1}
26	Fe	2.5×10^5	69	Tm	4.1×10^{-2}
27	Co	1.9×10^3	70	Yb	2.2×10^{-1}
28	Ni	4.5×10^4	71	Lu	3.6×10^{-2}
29	Cu	4.2×10^2	72	Hf	3.1×10^{-1}
30	Zn	6.3×10^2	73	Ta	1.9×10^{-2}
31	Ga	2.8×10^1	74	W	1.6×10^{-1}
32	Ge	7.6×10^1	75	Re	5.9×10^{-2}
33	As	3.8	76	Os	8.6×10^{-1}
34	Se	2.7×10^2	77	Ir	9.6×10^{-1}
35	Br	5.4	78	Pt	1.4
36	Kr	2.5×10^1	79	Au	1.8×10^{-1}
37	Rb	4.1	80	Hg	6.0×10^{-1}
38	Sr	2.5×10^1	81	Tl	1.3×10^{-1}
39	Y	4.7	82	Pb	1.3
40	Zr	2.3×10^1	83	Bi	1.9×10^{-1}
41	Nb	9.0×10^{-1}	90	Th	4×10^{-2}
42	Mo	2.5	92	U	1×10^{-2}
43	Tc	—			

geology, we are interested in the relatively small group of radionuclides that do occur naturally. ~~These nuclides occur in nature for group of radionuclides that do occur naturally.~~ These nuclides occur in nature for several reasons: (1) they have not yet completely decayed since the time of synthesis of the elements because their decay rates are very slow ($^{238}_{92}$U, $^{235}_{92}$U, $^{232}_{90}$Th, $^{87}_{37}$Rb, $^{40}_{19}$K, and others); (2) they are produced by decay of long-lived naturally occurring, radioactive parents ($^{234}_{92}$U, $^{230}_{90}$Th, $^{226}_{88}$Ra, and others); and (3) they are produced by nuclear reactions occurring in nature ($^{14}_{6}$C, $^{10}_{4}$Be, $^{32}_{14}$Si, and others). There is also a fourth group of radionuclides that can now be found in nature because they have been produced artificially, mainly as a result of the operation of nuclear fission reactors and the testing of explosive fission and fusion devices. The dispersal of these radionuclides into the atmosphere results in "fallout" and may cause contamination of food crops and drinking water. The safe disposal of radioactive waste products is an increasingly serious problem which will become more acute in the future as more and more energy is produced by nuclear power reactors and the quantity of radioactive "waste" increases correspondingly.

Let us return once more to the even-odd criterion of nuclear stability and abundance to examine the isotopes of tin ($Z = 50$). Tin has 10 stable and 16 radioactive isotopes which are shown in Figure 2.3. Notice that there is a region of stability extending from $A = 114$ to $A = 120$. Within that range we find 7 stable isotopes whose abundances show the effect of the even or odd mass numbers. Those having even A (even Z, even N) are consistently more abundant than their neighbors which have odd values of A (even Z, odd N). Outside this stable region in the direction of greater or smaller values of A, the even A isotopes are either stable or have longer half-lives than their neighbors. We also notice that the half-lives become shorter as we move away from the region of stability. This suggests that the isotopes become increasingly unstable as the number of neutrons increases or decreases outside the zone of stability. In other words, we can regard the band of stability on the chart of the nuclides (Fig. 2.2) as an "energy valley" into which the unstable nuclides on either side of the line of stability have a tendency to fall. We will elaborate on these ideas in the next chapter.

Finally, let us examine the data in Table 2.2 and in Figure 2.4 which contain estimates of the abundances of the elements in the solar system ("cosmic abundances") relative to Si = 10^6 atoms. Notice that the elements are plotted in order of increasing atomic number along the horizontal axis. Several observations can be made from this diagram: (1) hydrogen and helium are by far the most abundant elements; (2) the abundances of Li, Be, and B are anomalously low; (3) the abundances of the elements decrease with increasing atomic number in a somewhat exponential fashion; (4) the abundances of elements of even atomic number are higher than those of their neighbors (Oddo-Harkins Rule); and (5) the abundance of iron ($Z = 26$) is greater than expected from an extrapolation of the abundance curve and may be caused by the maximum in the binding energy curve at $A = 56$.

The abundances of the elements in the solar system are primarily the results of the nuclear reactions by which the elements were synthesized and reflect indirectly the relative stabilities of their isotopes. Most of the unstable nuclides that were initially produced have since decayed, except for a very small number of nuclides that have not yet become extinct because of their slow rates of decay. In the next chapter we describe the different modes by which radioactive nuclides decay and ultimately become stable isotopes of other elements.

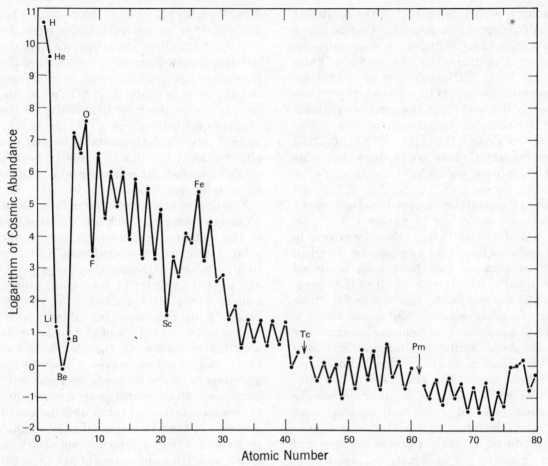

FIGURE 2.4 Plot of the abundances of the elements in the solar system versus their atomic number. The abundances are expressed ás the logarithm of the number of atoms of each element relative to 10^6 atoms of silicon. (Data are listed in Table 2.2 after Goles, 1969.)

PROBLEMS

1. (a) Specify the values of Z, N, and A for the following nuclides: $^{10}_{4}Be$, $^{51}_{23}V$, $^{130}_{54}Xe$, $^{162}_{66}Dy$, $^{222}_{86}Rn$, and $^{238}_{92}U$.

    (b) By reference to Figure 2.2, determine which of these nuclides is stable or unstable.

    (c) Calculate the ratio N/Z and plot it against A for the nuclides listed in Problem 1a.

    (d) Using the conventional symbolism, specify the following nuclides: arsenic-75, yttrium-89, antimony-123, neodymium-146, polonium-216, berkelium-250.

2. Calculate the atomic weight of silicon: ^{28}Si (92.2 percent, 27.976929), ^{29}Si (4.7 percent, 28.976497), ^{30}Si (3.1 percent, 29.973772). (**ANSWER:** 28.085809 amu.)

3. (a) Calculate the binding energies per nucleon for $^{4}_{2}He$ (4.0026033 amu),

$^{56}_{26}$Fe (55.934934 amu) and $^{232}_{90}$Th (232.03807 amu) in units of million electron volts.

(b) Plot the binding energies calculated above versus the mass number and draw a smooth curve through the points.

4. (a) Calculate the number of moles of strontium in 2.50 grams of strontium nitrate (Atomic weights: Sr = 87.62, N = 14.00, O = 15.99). (**ANSWER:** 1.18×10^{-2}.)

(b) How many atoms of strontium are present in 2.50 grams of strontium nitrate? (**ANSWER:** 7.11×10^{21} atoms.)

(c) How many atoms of ^{87}Sr are present in 2.50 grams of strontium nitrate? (Abundance of ^{87}Sr = 7.0 percent.) (**ANSWER:** 4.98×10^{20} atoms.)

(d) What is the ratio of the number of ^{87}Sr atoms to ^{86}Sr atoms in 2.50 grams of strontium nitrate? (Abundance of ^{86}Sr = 9.9 percent.) (**ANSWER:** ^{87}Sr/^{86}Sr = 0.707.)

5. The concentration of K_2O in a sample of orthoclase is found to be 8.75 percent by weight. How many atoms of ^{40}K are present in 1 gram of this material? (Atomic weight of K = 39.09, abundance of ^{40}K = 0.012 percent.) (**ANSWER:** 1.34×10^{17} atoms of ^{40}K per gram.)

REFERENCES

Friedlander, G., J. W. Kennedy, and J. M. Miller (1964) Nuclear and radiochemistry. (2nd ed.) John Wiley, New York, 585 p.

Goles, G. G. (1969) Cosmic abundances. In Handbook of Geochemistry, K. H. Wedepohl, ed. Springer-Verlag, Berlin, Heidelberg, New York, vol. 1, 116–133.

Holden, N. E., and F. W. Walker (1972) Chart of the nuclides (11th ed.) Educational Relations, General Electric Co., Schenectady, N.Y. 12345.

Kaplan, I. (1955) Nuclear Physics. Addison-Wesley, Reading, Mass., 609 p.

Lederer, C. M., J. M. Hollander, and I. Perlman (1967) Table of isotopes. (6th ed.) John Wiley, New York, 594 p.

Lefort, M (1968) Nuclear chemistry. D. Van Nostrand, London, 531 p.

McKay, H. A. S. (1971) Principles of radiochemistry. Butterworth, London, 550 p.

3 DECAY MECHANISMS OF RADIOACTIVE ATOMS

The nuclei of unstable atoms undergo spontaneous transformations that involve the emission of particles and of radiant energy. These processes give rise to the phenomenon that we call radioactivity. As we shall see in this chapter, there are several different ways in which unstable atoms can decay. Some atoms decay in two or three different ways, but most do so in only one particular way. In either case, radioactive decay results in changes of Z and N of the parent and thus leads to the transformation of an atom of one element into that of another element. The daughter may itself be radioactive and will in turn decay to form an isotope of yet another element. This process continues until at last a stable atom is produced.

Soon after radioactivity was discovered, Rutherford and others demonstrated that it involved the emission of three different types of rays which were named alpha, beta, and gamma. The beta rays were subsequently shown to be streams of particles identical to electrons. It is customary to restrict the term "beta particle" to electrons emitted by the nucleus of an atom. The beta particle is represented by the Greek letter β^- or β^+, depending on its charge. The existence of positively charged beta particles (positrons)

was first predicted by P. A. M. Dirac and was subsequently confirmed by Anderson in 1932 on the basis of cosmic-ray tracks in a Wilson cloud chamber. The emission of a positron is an important decay mechanism for a large group of unstable atoms. In fact, electrons are involved in yet a third mode of decay called "orbital electron capture." We will deal with them and with alpha decay in subsequent sections of this chapter.

1
Beta (Negatron) Decay

A large group of unstable atoms decays by emitting a negatively charged beta particle (negatron) from the nucleus, often accompanied by the emission of radiant energy in the form of gamma rays. According to a theory by E. Fermi that was first published in 1934, beta decay can be regarded as a transformation of a neutron into a proton and an electron. The electron is then expelled from the nucleus as a negative beta particle. As a result of such a beta decay, the atomic number of the atom is increased by one while its neutron number is reduced by one:

	ATOMIC NUMBER	NEUTRON NUMBER	MASS NUMBER
Parent	Z	N	$Z + N = A$
Daughter	$Z + 1$	$N - 1$	$Z + 1 + N - 1 = A$

Consequently, the daughter of a radioactive parent that decays by negatron emission is an isobar and can be easily located on a chart of the nuclides, as shown in Figure 3.1.

In 1914, Chadwick observed that the beta particles emitted by a particular radioactive nuclide have a continuous energy distribution. This means that the beta particles have kinetic energies that vary continuously from nearly zero all the way to some maximum value. Careful studies of this phenomenon have shown that most of the beta particles emitted by a particular nuclide have kinetic energies equal to about one-third of the maximum energy and that only a relatively small fraction of the beta particles are emitted with the maximum kinetic energy. The ques-

tion is, when a beta particle is emitted with less than the maximum energy, what happens to the rest of the energy? Beta decay seemed to violate the principle of conservation of mass and energy, which is one of the cornerstones of science. It also appeared to be violating other kinds of conservation laws, all of which resulted in a serious embarrassment for nuclear physicists.

This crisis was resolved in 1931 by W. Pauli who suggested the existence of a hypothetical particle that is produced during beta decay and that has a series of postulated properties such that the conservation laws are satisfied. This mysterious particle was named "neutrino" by Fermi who incorporated it into his theory of beta decay. The neutrino has no charge and a very small (or possibly zero) rest mass, but it can have varying amounts of kinetic energy. It interacts "sparingly" with matter and consequently carries its energy into space. According to Fermi's theory, each beta decay includes the emission of a beta particle and a neutrino having complementary kinetic energies, such that their sum is equal to the maximum decay energy for a particular nuclide. Neutrinos emitted during β^- decay are called antineutrinos which differ from neutrinos formed during β^+ decay. The existence of the elusive neutrinos was confirmed in 1954 by F. Reines and C. L. Cowan. The Earth is thought to be exposed to a continuous flux of neutrinos of the order of 10^{10} to 10^{11} neutrinos/cm^2/sec, most of which originate in the sun. However, recent efforts (Bahcall, 1969) to measure the flux density of neutrinos by means of an elaborate detection system inside the Homestake Gold Mine at Lead, South Dakota, suggest a somewhat lower value.

The apparent deficiency of solar neutrinos has posed a serious problem to astrophysicists because the neutrino flux emitted by the sun is a direct monitor of the nuclear

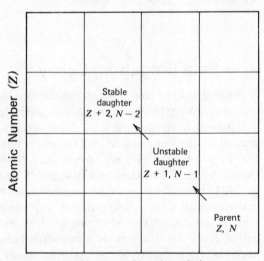

FIGURE 3.1 Schematic representation of the location, on the chart of the nuclides, of the position of daughter atoms relative to their parent which is subject to beta (negatron) decay. The atomic number is increased by one while the neutron number is reduced by one. Consequently, parent and daughter have the same mass number, that is, they are isobars. If the daughter is itself radioactive and decays by beta emission, a second isobaric daughter is formed and so on, until at last a stable daughter is produced.

reactions responsible for energy production in the sun. The lack of solar neutrinos implies a lower rate of energy production than is consistent with the sun's luminosity. Therefore, the lower neutrino flux indicates either that our understanding of the nuclear reactions in the interior of the sun is faulty, or that the sun is currently producing less energy in its interior than is compatible with the amount of energy being radiated into space. This problem was discussed by Fowler (1972) in a paper entitled "What Cooks with Solar Neutrinos?" Fowler suggested that the apparent deficiency of solar neutrinos was caused by a slight decrease in the core temperature of the sun which resulted from convective mixing of solar matter from the limbs into the core. Lowering of the temperature is almost immediately reflected by a decrease in the neutrino flux, while the reduced rate of energy production does not affect the energy radiated into space for several million years. If Fowler's suggestion is correct, we are living at a time of reduced energy production in the solar interior which may have already manifested itself in a climatic cooling trend at the end of the Tertiary period, resulting ultimately in continental glaciation. The anomalously low flux of solar neutrinos may thus be an important clue to the cause of the Pleistocene ice ages. The astrophysical and geological consequences of Fowler's explanation of the neutrino problem have also been discussed by Rood (1972), Ezer and Cameron (1972), Cameron (1973), and Ulrich (1975).

The nuclear transformations that occur during beta decay can be shown in the form of an equation. For example, the beta decay of naturally occurring $^{40}_{19}\text{K}$ to stable $^{40}_{20}\text{Ca}$ is represented by

$$^{40}_{19}\text{K} \rightarrow {}^{40}_{20}\text{Ca} + \beta^- + \bar{v} + Q \qquad (3.1)$$

where β^- is the beta particle, $\bar{v}$ is the antineutrino, and Q stands for the maximum de-

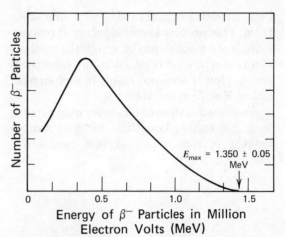

FIGURE 3.2 Energy spectrum of beta particles emitted by $^{40}_{19}\text{K}$. Note that most of the beta particles are emitted with energies that are about one-third of the maximum or end point energy. During each beta decay, a neutrino is emitted with a kinetic energy equal to the difference between the maximum energy and the kinetic energy of the associated beta particle. (Data from Dzelepow et al., 1946.)

cay energy, which in this case is 1.312 MeV. The energy spectrum of the beta particles emitted by $^{40}_{19}\text{K}$ is shown in Figure 3.2. The kinetic energies of the antineutrinos emitted by $^{40}_{19}\text{K}$ are also variable and range from near zero up to 1.312 MeV in such a way that the sum of the kinetic energies of the antineutrino and the beta particle emitted in a particular decay event is equal to the total decay energy.

In some cases the product nucleus formed by beta decay is left in an excited state. This excess energy is emitted either as a gamma ray having a discrete energy, or the excess energy is transferred from the nucleus to an extranuclear electron which is then ejected from its orbital with a kinetic energy equal to the difference between the excitation energy of the nucleus and the binding energy of the electron. These electrons are superimposed on the continuous spectrum of the beta particles and give rise to a line spectrum. This process is called internal conversion. It is a mechanism by which the nucleus of an atom in an excited state can lose excess energy, but it does not result in a change of either Z or N of the atom.

Some nuclei in excited states may remain in a metastable condition for measurable lengths of time. Such excited nuclei are called isomers. They decay to their ground state by emission of gamma rays or by internal conversion with half-lives ranging from 10^{-11} seconds up to 241 years, as in the case of ^{192m}Ir. The superscript "m" identifies the nuclide as an isomer of ^{192}Ir which is itself radioactive and decays by β^- emission to stable ^{192}Pt. The identification of isomers has become somewhat confused by techno-

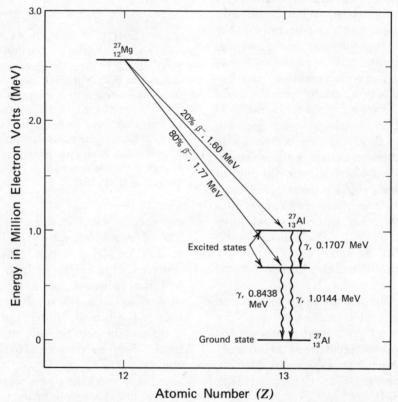

FIGURE 3.3   Decay scheme diagram for the beta (negatron) decay of $^{27}_{12}$Mg to stable $^{27}_{13}$Al. In this decay, two sets of beta particles are emitted having end point energies of 1.77 and 1.60 MeV. In addition, three gamma rays with energies of 1.0144, 0.8438, and 0.1707 MeV are observed. These are emitted because the product nuclei following beta decay are left in excited states. The excess energy is lost by emission of gamma rays to give a total decay energy of 2.61 MeV for each of the three possible decay paths. (Data from Holden and Walker, 1972.)

logical improvements in our ability to measure short intervals of time, and the number of known short-lived isomers has been increasing correspondingly.

Many beta emitters have complex spectra, meaning that they emit two or more suites of beta particles having different maximum or endpoint energies. For example, $^{27}_{12}Mg$ emits two beta particles: 80 percent of the decays have an end point energy of 1.77 MeV, while 20 percent have an end point energy of 1.60 MeV. In addition, three discrete gamma rays are observed with energies of 1.0144, 0.8438, and 0.1707 MeV. The total decay energy Q is 2.610 MeV. Reference to Figure 3.1 indicates that the nucleus produced by beta decay of $^{27}_{12}Mg$ must have 13 protons and 14 neutrons, which indicates that the daughter of $^{27}_{12}Mg$ is stable $^{27}_{13}Al$. The gamma rays serve to de-excite the product nuclei of $^{27}_{13}Al$ to the ground state. Consequently, we must combine the end point energies of the two groups of beta particles with the energies of the gamma rays in such a way that the total amount of energy emitted by a $^{27}_{12}Mg$ atom adds up to 2.61 MeV. When we attempt to calculate such an energy balance, we must remember that whenever a beta particle is emitted with less than the end point energy, the difference is carried away by an antineutrino.

The conventional method of showing the overall energy balance of a particular decay process is by means of a plot of energy, in million electron volts (ordinate), versus the atomic number (abscissa) called a decay-scheme diagram. Figure 3.3 is such a decay scheme diagram for $^{27}_{12}Mg$. We can demonstrate the energy equivalence of the three pathways for the decay of $^{27}_{12}Mg$:

(1) $\dfrac{\begin{array}{l} 1.60\ \text{MeV}\ \beta^- \\ +\,1.0144\ \text{MeV}\ \gamma \end{array}}{2.6144\ \text{MeV}}$

(2) $\dfrac{\begin{array}{l} 1.60\ \text{MeV}\ \beta^- \\ 0.1707\ \text{MeV}\ \gamma \\ +\,0.8438\ \text{MeV}\ \gamma \end{array}}{2.6145\ \text{MeV}}$

(3) $\dfrac{\begin{array}{l} 1.77\ \text{MeV}\ \beta^- \\ +\,0.8438\ \text{MeV}\ \gamma \end{array}}{2.6138\ \text{MeV}}$

A somewhat more complicated example is the beta decay of $^{38}_{17}Cl$ to stable $^{38}_{18}Ar$ during which three sets of beta particles are emitted whose end point energies are 4.91 MeV (53.4 percent), 1.1 MeV (30.8 percent), and 2.8 MeV (15.8 percent). In addition, two gamma rays are observed having energies of 2.167 and 1.642 MeV. The total decay energy is 4.914 MeV (Holden and Walker, 1972). Figure 3.4 shows the different pathways open to this decay. We see that about 53 percent of the decay is directly to the ground state of $^{38}_{18}Ar$. The other two beta emissions leave the product nuclei in excited states and are followed by gamma rays. Again we can demonstrate that all of the possible pathways result in the same decay energy for $^{38}_{17}Cl$:

(1) $\dfrac{4.91\ \text{MeV}\ \beta^-}{4.91\ \text{MeV}}$

(2) $\dfrac{\begin{array}{l} 2.8\ \text{MeV}\ \beta^- \\ +\,2.167\ \text{MeV}\ \gamma \end{array}}{4.967\ \text{MeV}}$

(3) $\dfrac{\begin{array}{l} 1.1\ \text{MeV}\ \beta^- \\ 1.642\ \text{MeV}\ \gamma \\ +\,2.167\ \text{MeV}\ \gamma \end{array}}{4.909\ \text{MeV}}$

The amount of mass (ΔM) equivalent to 4.91 MeV is

$$\Delta M = \frac{4.91}{931.6} = 0.00527\ \text{amu}$$

Since the mass of stable $^{38}_{18}Ar$ is 37.96273 amu, the mass of $^{38}_{17}Cl$ is the sum of the following terms: $37.96273 + 0.00527 + 0.00054 = 37.96854$ amu, where 0.00054 amu is the rest mass of the beta particle. Beta decay is an energy-liberating process and requires that the parent must have more mass (or energy) than its daughter.

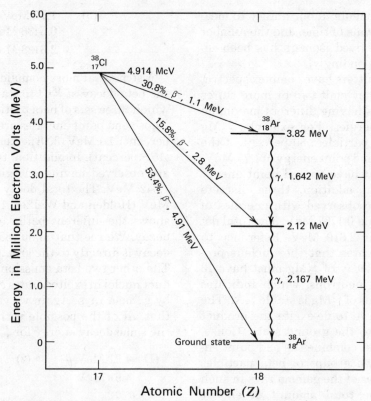

FIGURE 3.4 Decay scheme diagram for the beta (negatron) decay of $^{38}_{17}Cl$ to stable $^{38}_{18}Ar$. The decay occurs by emission of three sets of beta particles, each set having a different end point energy. Two gamma rays are emitted which allow the product nuclei to de-excite to the ground state. (Data from Holden and Walker, 1972.)

2
Positron Decay

A large group of radionuclides decays by emission of a positively charged electron (positron) from the nucleus. The positrons have energy spectra similar to those of the negatrons, that is, most of the positrons emitted by a particular radionuclide have kinetic energies that are less than the maximum or end point energy. As before, each positron decay is accompanied by the emission of a neutrino whose kinetic energy is the difference between the end point energy and the energy of the positron. According to Fermi's theory of beta decay, positron emission is regarded as the transformation of a proton in the nucleus into a neutron, a positron, and a neutrino. The configuration of the product nucleus relative to that of its parent is illustrated in Figure 3.5 and can be determined from the following statements:

DECAY MECHANISMS OF
RADIOACTIVE ATOMS 3

	ATOMIC NUMBER	NEUTRON NUMBER	MASS NUMBER
Parent	Z	N	$Z + N = A$
Daughter	$Z - 1$	$N + 1$	$Z - 1 + N + 1 = A$

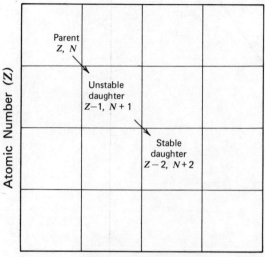

FIGURE 3.5 Schematic representation of the position, in the chart of the nuclides, of daughters produced by positron emission of the parent. Note that the atomic number (Z) decreases by one while the neutron number (N) increases. The daughters are isobars of their parent and are isotopes of different elements.

Evidently the daughter produced by positron decay is an isobar and has one less proton than its parent.

The positron that is emitted in this process is slowed by collisions with atoms. When it is nearly at rest, it interacts with a normal negative electron as a result of which both are annihilated. Their rest masses are converted into two gamma rays of 0.511 MeV each, which are emitted in opposite directions. The total decay energy of positron decay is the end point energy of the positron

plus 1.02 MeV arising from the annihilation. Positron decay is possible only when the mass of the parent is greater than that of the daughter by at least two electron masses.

Positron decay can be represented by an equation similar to one we used to illustrate negatron decay:

$$^{18}_{9}\text{F} \rightarrow {}^{18}_{8}\text{O} + \beta^{+} + v + Q \qquad (3.2)$$

where β^{+} is the positron, v is the neutrino, and Q is the total energy given off by this decay. The value of Q in this case is 1.655 MeV and the end point energy of the positrons is $1.655 - 1.02 = 0.635$ MeV. Positron decay may leave the product nucleus in an excited state. The excess energy is lost by emission of gamma rays. This is illustrated by the decay scheme diagram in Figure 3.6 for the positron decay of $^{14}_{8}\text{O}$ to stable $^{14}_{7}\text{N}$. During this decay two groups of positrons are emitted having end point energies of 1.809 and 4.1 MeV, respectively. In addition, one gamma ray is observed with an energy of 2.313 MeV. The total decay energy is 5.145 MeV (Holden and Walker, 1972). We again demonstrate that the two decay paths are equivalent in energy:

(1)   1.809 MeV β^{+}
        2.313 MeV γ
      +1.02 MeV annihilation
      $\overline{\text{5.142 MeV}}$

(2)   4.1 MeV β^{+}
      +1.02 MeV annihilation
      $\overline{\text{5.12 MeV}}$

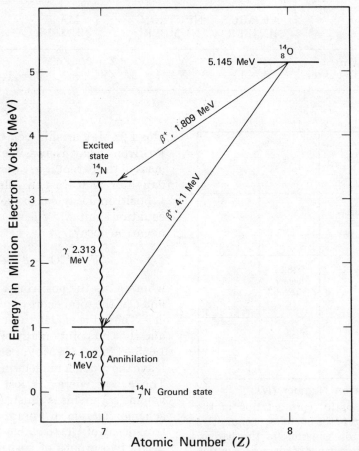

FIGURE 3.6   Positron decay of $^{14}_{8}$O to stable $^{14}_{7}$N. The parent emits two sets of positrons having end point energies of 1.809 and 4.1 MeV, respectively. The emission of the former leaves the product nucleus in an excited state and is followed by a gamma ray of 2.313 MeV. The positrons are slowed by a series of collisions and are eventually annihilated by interaction with an electron resulting in the formation of two gamma rays with a total energy of 1.02 MeV. (Data from Holden and Walker, 1972.)

We can now point out a significant fact about negatron and positron decay. Atoms that have an excess of neutrons, and that therefore lie to the right of the zone of nuclear stability shown in Figure 2.2, are subject to negatron decay because during this process the neutron number is reduced. Similarly, atoms lying to the left of the zone of stability can be thought of as being deficient in neutrons. They are subject to positron decay because during this process the neutron number is increased. Positron and negatron decay tend to change Z and N in such a way that the resulting daughters lie within the zone of stability in the chart of the nuclides.

3
Electron Capture Decay

An alternative mechanism whereby a nucleus can decrease its proton number and increase its neutron number is by capturing one of its extranuclear electrons. The probability of capture is greatest for the electrons in the K shell (quantum number one) because they are closest to the nucleus. However, electrons from other shells (L, M, etc.) can also be captured. Electron capture requires the emission of a neutrino from the nucleus. However, the neutrinos are thought to be mono-energetic because the electrons are captured from a definite energy state.

Electron capture can be visualized as the reaction between an extranuclear electron and a proton in the nucleus to form a neutron and a neutrino. The product nucleus has the following configuration compared to its parent:

the next higher L shell, the resulting x-ray may interact with another electron which is then ejected with a kinetic energy equal to the difference between the energy of the K-x-ray and its own binding energy. Such electrons are called Auger electrons.

The emission of characteristic x-rays following electron capture decay has been used to construct highly portable sources of x-rays. For example, $^{125}_{53}I$ decays by electron capture to an excited state of $^{125}_{52}Te$, which decays to the ground state by internal conversion and electron emission. During this process, a series of x-rays are emitted with energies ranging from 27.2 to 31.5 keV. Meyers (1962) used this decay pair to construct a point source of x-rays in a thin plastic tube weighing about 1 gram with which he made medical radiographs of good quality. Such portable isotopic x-ray sources may also be useful in geology for radiography of rocks and minerals.

	PROTON NUMBER	NEUTRON NUMBER	MASS NUMBER
Parent	Z	N	$Z + N = A$
Daughter	$Z - 1$	$N + 1$	$Z - 1 + N + 1 = A$

The daughter is isobaric with its parent and occupies the *same* position on the chart of the nuclides relative to its parent as the daughter of a positron decay, as illustrated in Figure 3.5.

Electron capture may leave the product nucleus in an excited state and is then followed by the emission of a gamma ray. Removal of an extranuclear electron from the K shell or from higher-energy shells leaves a vacancy that is subsequently filled by other electrons that fall into the vacant position. In the process, these electrons emit a series of x-rays that can be detected. If a vacancy in the K shell is filled by an electron from

4
Branched Decay and
Beta Decay of Isobars

According to a rule formulated by the Austrian physicist J. Mattauch in 1934, the difference in the atomic number of two stable isobars is greater than one. In other words, two adjacent isobars cannot both be stable. The reason for this is that adjacent isobars have different masses and binding energies which makes possible a spontaneous reaction whereby one isobar is converted into the other by a suitable beta decay that liberates energy.

Mattauch's isobar rule implies that two stable isobars must be separated by a radioactive isobar that can undergo branched decay and thus forms two stable isobaric daughters. The easiest way to appreciate this is to consider some specific examples. Let us turn therefore to the three isobaric nuclides: $^{40}_{18}Ar$, $^{40}_{19}K$, and $^{40}_{20}Ca$. In accordance with Mattauch's rule, $^{40}_{19}K$ is found to be radioactive and undergoes branched decay to both $^{40}_{18}Ar$ and $^{40}_{20}Ca$. The decay to $^{40}_{18}Ar$ is by positron emission and electron capture, while

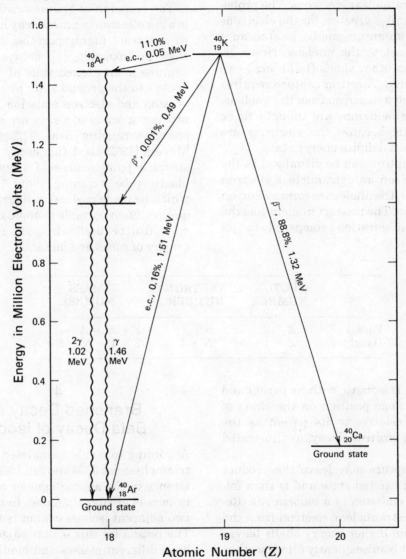

FIGURE 3.7 Decay scheme diagram for the branched decay of $^{40}_{19}K$ to $^{40}_{18}Ar$ by electron capture and by positron emission and to $^{40}_{20}Ca$ by emission of negative beta particles. A detailed explanation is given in the text. (After Dalrymple and Lanphere, 1969.)

$^{40}_{20}$Ca is produced by emission of a negative beta particle (Equation 3.1). The decay scheme diagram, shown in Figure 3.7, is fairly complicated. It shows that 11.0 percent of the $^{40}_{19}$K atoms decay by electron capture to an excited state of $^{40}_{18}$Ar, followed by emission of a gamma ray of 1.46 MeV. Another path taken by 0.16 percent of the $^{40}_{19}$K atoms is by electron capture directly to the ground state of $^{40}_{18}$Ar resulting in the release of 1.51 MeV. A third path, taken by only 0.001 percent of $^{40}_{19}$K atoms, is by emission of a positron having an end point energy of 0.49 MeV, followed by annihilation of the positron and the release of an additional 1.02 MeV of energy. The other branch leads to the formation of $^{40}_{20}$Ca in the ground state by emission of a set of negative beta particles with an end point energy of 1.32 MeV. The reader can easily verify that each of the three pathways leading to the formation of $^{40}_{18}$Ar involves the same change in the energy. Note also that the ground state of $^{40}_{20}$Ca lies above that of $^{40}_{18}$Ar.

Several additional examples of branched beta decay can be found in the chart of the nuclides (Fig. 2.2). Among them we mention only the decay of $^{176}_{71}$Lu to $^{176}_{72}$Hf by negative beta decay and to $^{176}_{70}$Yb by positron emission or electron capture. This decay scheme has possible useful geological applications as does the decay of $^{40}_{19}$K which was presented above.

We now return to a consideration of the question of nuclear stability alluded to in Chapter 2. Figure 2.2 clearly shows that there is a region of stability on the chart of the nuclides that is flanked on either side by radionuclides. Many of these radionuclides are subject to beta decay which leads to the formation of isobaric daughters that are either closer to or lie within the stable region. The preceding discussion of beta decay enables us to draw isobaric energy profiles across the band of nuclides. Figure 3.8

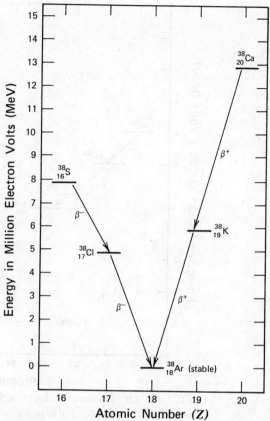

FIGURE 3.8 Schematic decay scheme diagram for isobars having $A = 38$. It can be seen that $^{38}_{20}$Ca and $^{38}_{19}$K decay by positron emission to stable $^{38}_{18}$Ar, while $^{38}_{16}$S and $^{38}_{17}$Cl decay by negatron emission. $^{38}_{18}$Ar is the only stable nuclide in this set of isobars which is consistent with the fact that it lies at the bottom of this energy profile. (Data from Holden and Walker, 1972.)

is such an energy profile for isobars having mass number $A = 38$. It can be seen that $^{38}_{18}$Ar lies at the bottom of the energy profile and is therefore the only stable nuclide in this group of isobars. The other isobars decay in such a way that their nuclear configurations are changed in steps until $^{38}_{18}$Ar is formed.

A second energy profile is shown in Figure 3.9 for branched decay in the group

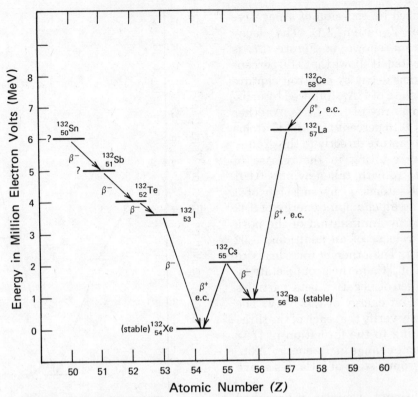

FIGURE 3.9 Schematic decay scheme diagram illustrating branched beta decay in the group of nine isobars having $A = 132$. The diagram shows that $^{132}_{54}Xe$ and $^{132}_{56}Ba$ are the only two stable nuclides in this group of isobars and that they lie at the bottom of the energy profile. $^{132}_{55}Cs$ undergoes branched beta decay consistent with Mattauch's isobar rule. Emission of negative beta particles by $^{132}_{50}Sn$, $^{132}_{51}Sb$, $^{132}_{52}Te$, and $^{132}_{53}I$ increases their atomic numbers in steps until stable $^{132}_{54}Xe$ is formed. Positron decay and electron capture reduces Z of $^{132}_{58}Ce$ and $^{132}_{57}La$ to form stable $^{132}_{56}Ba$. (Data from Holden and Walker, 1972.)

of isobars having $A = 132$. In this group, there are two stable nuclides: $^{132}_{56}Ba$ and $^{132}_{54}Xe$. They are separated by unstable $^{132}_{55}Cs$ in accordance with Mattauch's rule and are flanked by several additional radionuclides belonging to this group of nine isobars. Again the decay modes are such that the unstable nuclides in this group decay to form the appropriate stable nuclides in this group.

5
Alpha Decay

A large group of radionuclides decays by the spontaneous emission of alpha particles from their nuclei. This mode of decay is available to nuclides having atomic numbers of 58 (cerium) or greater and to a few nuclides of low atomic number, including $^{5}_{2}He$, $^{5}_{3}Li$, and $^{6}_{4}Be$. Alpha particles are composed of two protons and two neutrons and consequently have a charge of $+2$. The values of Z and N of the product nucleus relative to those of its parent can be ascertained by the following simple considerations:

and v_p are the mass and velocity of the product nucleus, respectively. The total alpha decay energy E_α is given by

$$E_\alpha = \tfrac{1}{2}M_\alpha v_\alpha{}^2 + \tfrac{1}{2}M_p v_p{}^2 \qquad (3.4)$$

	ATOMIC NUMBER	NEUTRON NUMBER	MASS NUMBER
Parent	Z	N	$Z + N = A$
Daughter	$Z - 2$	$N - 2$	$Z + N - 4 = A - 4$

The emission of an alpha particle reduces both the atomic number and the neutron number by two and the mass number by four. The daughter is an isotope of a different element and is *not* an isobar of its parent as in the case of beta decay or electron capture. The position on the chart of the nuclides of the daughter resulting from the alpha decay of its parent is illustrated in Figure 3.10.

A radionuclide subject to alpha decay may emit alpha particles having a discrete energy resulting in the formation of a daughter that is in the ground state. Others emit alpha particles at slightly differing energies such that the product nucleus remains in an excited state. In such cases, the excess energy is emitted by the product nucleus as gamma rays. Because the alpha particle has an appreciable mass, its ejection will impart a certain amount of recoil energy to the nucleus. The total alpha decay energy (E_α) is the sum of the kinetic energy of the alpha particle, the recoil energy it imparts to the product nucleus, and the energies of any gamma rays that are emitted. Leaving aside relativistic effects, the principle of conservation of momentum requires that

$$M_\alpha v_\alpha = M_p v_p \qquad (3.3)$$

where M_α and v_α are the mass and velocity of the alpha particle, respectively, and M_p

It follows from Equation 3.3 that

$$v_p = \frac{M_\alpha v_\alpha}{M_p} \qquad (3.5)$$

which we substitute into Equation 3.4 and

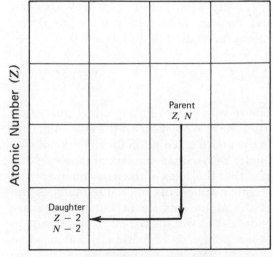

FIGURE 3.10 Location of the daughter nuclide produced by alpha decay of its parent in coordinates of Z and N. Both the atomic number and the neutron number of the daughter are reduced by two, thus reducing its mass number by four. The daughter may itself be subject to further decay by either alpha emission, or beta emission, or both.

obtain

$$E_\alpha = \tfrac{1}{2}M_\alpha v_\alpha{}^2 + \tfrac{1}{2}M_p \left(\frac{M_\alpha v_\alpha}{M_p}\right)^2$$

$$E_\alpha = \tfrac{1}{2}M_\alpha v_\alpha{}^2 \left(1 + \frac{M_\alpha}{M_p}\right) \qquad (3.6)$$

Equation 3.6 indicates that the recoil energy is obtained by multiplying the kinetic energy of the alpha particle by the ratio M_α/M_p. For heavy alpha emitters, the recoil energy is found to be about 0.1 MeV.

Alpha decay can be represented by an equation illustrated here by the decay of $^{238}_{92}\text{U}$ to $^{234}_{90}\text{Th}$:

$$^{238}_{92}\text{U} \rightarrow {}^{234}_{90}\text{Th} + {}^{4}_{2}\text{He} + Q \qquad (3.7)$$

where $^{4}_{2}\text{He}$ is the alpha particle and Q is the total alpha decay energy. The kinetic energy of the alpha particle is 4.20 MeV (Holden and Walker, 1972). The total alpha decay energy, according to Equation 3.6, is

$$E_\alpha = 4.20 + \frac{4.20 \times 4}{234} = 4.271 \text{ MeV}$$

The recoil energy of the product nucleus is 0.071 MeV. We have used the mass numbers in the above calculation for the sake of simplicity rather than the actual masses. Note also that the sums of the mass numbers and atomic numbers of the product nucleus and of the alpha particle in Equation 3.7 are equal to those of the parent, respectively. In other words, the equation is balanced in terms of the number of neutrons and protons of the parent, the daughter, and of the alpha particle.

Alpha decay can also be represented by a decay scheme diagram constructed in the same way as those we used previously to illustrate beta decay schemes. We consider first the decay of $^{222}_{86}\text{Rn}$ which emits a single set of alpha particles having a kinetic energy of 5.4897 MeV and forms $^{218}_{84}\text{Po}$. The alpha decay energy of $^{222}_{86}\text{Rn}$ is

$$E_\alpha = 5.4897 + \frac{5.4897 \times 4}{218} = 5.5904 \text{ MeV}$$

In addition to the alpha particle, a gamma ray of 0.51 MeV is observed which indicates that some of the alphas are emitted with lower kinetic energies and leave the product nucleus in an excited state (Fig. 3.11).

A far more complicated case is the decay of $^{228}_{90}\text{Th}$ which emits four alpha particles in its decay to $^{224}_{88}\text{Ra}$. The kinetic energies of the alpha particles are: 5.421, 5.338, 5.208, and 5.173 MeV (Friedlander et al. 1964). Only the most energetic of these decays is to the ground state of $^{224}_{88}\text{Ra}$. The others leave the product nucleus in different excited states and are therefore followed by gamma rays that either take the nucleus to its ground state or to an intermediate excited state. Four different gamma rays with energies of 0.217, 0.169, 0.133, and 0.084 MeV are observed. These have all been fitted into the decay scheme diagram shown in Figure 3.12. The total alpha decay energy of ^{228}Th is 5.519 MeV.

Several daughters of the naturally occurring isotopes of uranium and thorium can decay either by emission of a negative beta particle or by alpha decay. If the first decay is by negatron emission, the daughter thus produced decays by alpha emission and vice versa. In other words, these alternate decay modes form closed decay cycles and always lead to the same end product regardless of the sequence in which the alpha and beta particles are emitted. Let us consider the following decay series as an example, shown also in Figure 3.13.

We see that $^{227}_{89}\text{Ac}$, which is the third daughter of $^{235}_{92}\text{U}$, can decay by beta emission to $^{227}_{90}\text{Th}$, which then emits an alpha particle to form $^{223}_{88}\text{Ra}$, or $^{227}_{89}\text{Ac}$ undergoes alpha decay

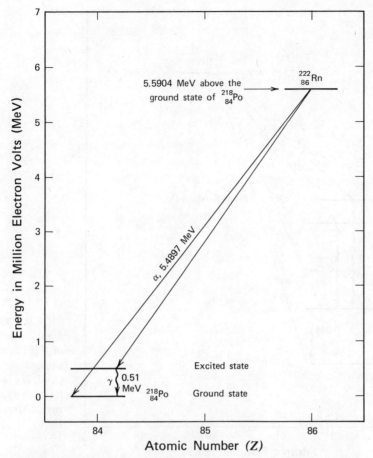

FIGURE 3.11 Decay scheme diagram for the simple alpha decay of $^{222}_{86}$Rn to $^{218}_{84}$Po. The alpha disintegration energy E_α is 5.5904 MeV, which is the sum of the kinetic energies of the alpha particles and the recoil energy of the product nucleus. Some of the alpha particles are emitted with lower kinetic energies and leave the product nucleus in an excited state. The excess energy is then emitted as a gamma ray with an energy of 0.51 MeV. (Data from Holden and Walker, 1972.)

to $^{223}_{87}$Fr, which in turn emits a negative beta particle to form $^{223}_{88}$Ra. Since this is a closed cycle, the net change in energy along either route must be the same. We can demonstrate that this is true by adding up the decay energies shown in Figure 3.13. For the $^{227}_{90}$Th branch, the sum of the decay energies is 0.044 + 6.145 = 6.189 MeV; for the $^{223}_{87}$Fr branch, the decay energies are 5.039 + 1.15 = 6.189 MeV. The two sums are identical as expected. Actually the cycles are more extensive than shown here, because $^{223}_{87}$Fr is also subject to alpha decay leading to $^{219}_{85}$At, which in turn can decay to $^{215}_{83}$Bi by alpha

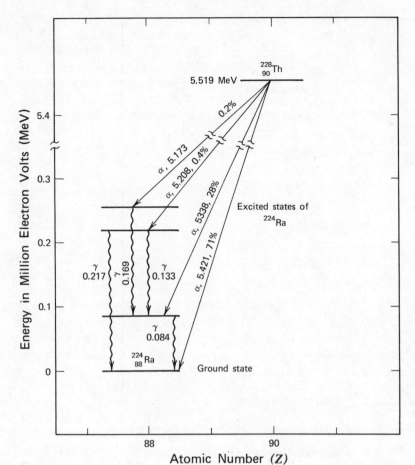

FIGURE 3.12 Decay scheme diagram for the alpha decay of $^{228}_{90}$Th to $^{224}_{88}$Ra. In this case, four different alpha particles are emitted with four complementary gamma rays. (After Friedlander et al. 1964.)

FIGURE 3.13 Decay cycle of $^{227}_{89}$Ac which can decay both by alpha and negative beta decay. Beta decay of $^{227}_{89}$Ac leads to $^{227}_{90}$Th which then decays by alpha emission to $^{223}_{88}$Ra. Alpha decay of $^{227}_{89}$Ac produces $^{223}_{87}$Fr which then forms $^{223}_{88}$Ra by beta decay. The decay energies are in units of million electron volts and add up to 6.189 MeV in both branches of the cycle. (Data from Holden and Walker, 1972.)

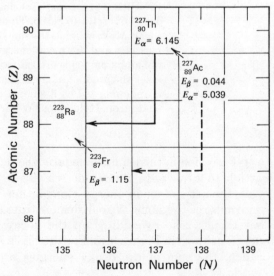

emission or to $^{219}_{86}$Rn by beta decay, and so on. However, all of these alternate routes lead by a succession of alpha and beta decay steps to the same end product which is stable $^{207}_{82}$Pb.

6
Nuclear Fission

In 1938, Otto Hahn and Fritz Strassmann, two of Germany's foremost radiochemists, bombarded uranium with neutrons and found that barium was produced in this experiment. This discovery was communicated to Niels Bohr by Lise Meitner and O. R. Frisch who were then working in Copenhagen because they had been forced to leave Germany by the Hitler regime. Niels Bohr discussed Hahn's data with Albert Einstein and J. A. Wheeler during a visit to the United States in January 1939. The formation of barium from uranium indicated that the uranium atoms had been split into two or more fragments. Fermi suggested that the fission of uranium might also result in the release of neutrons which might be used to initiate additional fission events leading to a chain reaction. Experimental results soon confirmed that each fission event is indeed accompanied by the release of 2.5 neutrons on the average. Thus began the era of atomic bombs and the production of energy by controlled fission of uranium in nuclear reactors.

It is now known that fission of the isotopes of uranium and thorium ($^{235}_{92}$U, $^{238}_{92}$U and $^{232}_{90}$Th) can be induced by bombarding their nuclei with neutrons, protons, deuterons ($^{2}_{1}$H), alpha particles, gamma rays, and even x-rays. Many other heavy elements undergo fission when they are bombarded with nuclear particles of sufficiently high energy (50 to 450 MeV). As a result of the fission of a heavy nuclide such as $^{235}_{92}$U, two product nuclei are produced having atomic numbers ranging from 30 (zinc) to 65 (terbium). In addition, each fission event leads to the emission of alpha particles, neutrons, and possibly other light fragments along with a large amount of energy of about 200 MeV. The atomic numbers of the fission products are usually unequal, that is, the parent nucleus does not break into equal halves. The fission products generally have an excess of neutrons and lie to the right of the region of stability on the chart of the nuclides. They are therefore radioactive and decay by successive beta (negatron) emission to stable isobars. One of the longest series of such beta decays starts with $^{144}_{54}$Xe which is a fission product of $^{235}_{92}$U and decays as follows: $^{144}_{54}$Xe $\rightarrow$ $^{144}_{55}$Cs $\rightarrow$ $^{144}_{56}$Ba $\rightarrow$ $^{144}_{57}$La $\rightarrow$ $^{144}_{58}$Ce $\rightarrow$ $^{144}_{59}$Pr $\rightarrow$ $^{144}_{60}$Nd. The radioactivity of such fission products is a potentially hazardous consequence of nuclear fission unless great care is taken to contain and to sequester them.

Bohr and Wheeler explained fission reactions by postulating that the nuclei of heavy atoms behave like drops of liquid which assume a spherical shape because of surface tension. They suggested that nuclei may reach a critical size, depending on a parameter Z^2/A, when the force of electrostatic repulsion will be greater than the surface forces holding the nucleus together. When that critical size is reached, the nucleus may be susceptible to *spontaneous* fission. The critical size is attained by atoms having atomic number of about 100.

Such spontaneous nuclear fission of the isotopes of uranium was first reported by Petrzhak and Flerov in 1940. Since then, spontaneous fission has been confirmed in more than 20 heavy nuclides for which it is an alternate mode of decay. In fact, it is the principal mode of decay of some of the trans-uranium elements that have been synthesized in the laboratory.

PROBLEMS

1. Without consulting the chart of the nuclides, specify Z, N, and A of the daughters of the following radionuclides which decay by negatron decay: $^{14}_{6}C$, $^{87}_{37}Rb$, $^{176}_{71}Lu$, and $^{206}_{81}Tl$.

2. Without consulting the chart of the nuclides, specify Z, N, and A for the daughters of the following radionuclides which decay by positron or electron capture decay: $^{22}_{11}Na$, $^{88}_{39}Y$, $^{52}_{25}Mn$, and $^{207}_{83}Bi$.

3. Without consulting the chart of the nuclides, specify Z, N, and A for the daughters of the following nuclides which decay by alpha emission: $^{147}_{62}Sm$, $^{234}_{92}U$, $^{242}_{94}Pu$, and $^{230}_{90}Th$.

4. Naturally occurring $^{232}_{90}Th$ decays to stable $^{208}_{82}Pb$ through a series of radioactive daughters. In this decay series, four negatrons are emitted. How many alpha particles must be emitted to produce $^{208}_{82}Pb$ from $^{232}_{90}Th$?

5. Draw a decay scheme diagram for $^{21}_{9}F$ which emits two suites of β^- particles having end point energies of 5.4 and 4.0 MeV and gamma rays of 0.345 and 1.38 MeV. The total decay energy $Q = 5.69$ MeV (Lederer et al. 1967).

6. Draw a decay scheme diagram for $^{14}_{8}O$ which emits two sets of positrons having end point energies of 1.809 and 4.1 MeV and a gamma ray of 2.313 MeV. The total decay energy is 5.145 MeV (Holden and Walker, 1972).

7. $^{227}_{92}U$ emits an alpha particle having kinetic energy of 6.87 MeV. What is the recoil energy of the product nucleus? What is the total alpha decay energy (Holden and Walker, 1972)? (**ANSWER:** Recoil energy: 0.1232 MeV; Total alpha decay energy: 6.9932 MeV.)

8. The most energetic alpha particle emitted by $^{239}_{94}Pu$ has a kinetic energy of 5.155 MeV. Estimate the difference in mass between $^{239}_{94}Pu$ and its daughter in amu. ($M_{He} \simeq 4.0026$ amu; Holden and Walker, 1972). (**ANSWER:** 4.0082 amu.)

REFERENCES

Anderson, C. D. (1932) The positive electron. Science, *76*, 238.

Bahcall, J. B. (1969) Neutrinos from the Sun. Sci. Amer., *221*, 29–37.

Bohr, N., and J. A. Wheeler (1939) The mechanism of nuclear fission. Phys. Rev., *56*, 426–450.

Cameron, A. G. W. (1973) Major variations in solar luminosity? Rev. Geophys. Space Phys., *11*, 505–510.

Dalrymple, G. B., and M. A. Lanphere (1969) Potassium-argon dating. W. H. Freeman, San Francisco, 258 p.

Dzelepow, B., M. Kopjova, and E. Vorobjov (1946) β-spectrum of K^{40}. Phys. Rev., *69*, 538–539.

Ezer, D., and A. G. W. Cameron (1972) Effects of sudden mixing in the solar core on solar neutrinos and ice ages. Nature, *240*, 180–182.

Fermi, E. (1934) Versuch einer Theorie der β-Strahlen. Z. Physik, *88*, 161–177.

Fowler, W. A. (1972) What cooks with solar neutrinos? Nature, *238*, 24–26.

Friedlander, G., J. W. Kennedy, and J. M. Miller (1964) Nuclear and radiochemistry. John Wiley, New York, 585 p.

Hahn, O., and F. Strassmann (1939) Über den Nachweis und das Verhalten der bei der Bestrahlung des Urans mittels Neutronen entstehenden Erdalkalimetalle. Naturwiss., *27*, 11–15.

Holden, N. E., and F. W. Walker (1972) Chart of the nuclides. (11th ed.) General Electric Co., Schenectady, N.Y. 12305.

Lederer, C. M., J. M. Hollander, and I. Perlman (1967) Table of isotopes. (6th ed.) John Wiley, New York, 594 p.

Mattauch, J. (1934) Zur Systematik der Isotopen. Z. Physik, *91*, 361–371.

Meyers, W. G. (1962) On a new source of x-rays. Ohio State Medical J., *58*, 772–773.

Reines, F. (1960) Neutrino interactions. Ann. Rev. Nucl. Phys., *10*, 1–26.

Rood, R. T. (1972) A mixed-up Sun and solar neutrinos. Nature, *240*, 178–180.

Ulrich, R. K. (1975) Solar neutrinos and variations in the solar luminosity. Science, *190*, 619–623.

4 RADIOACTIVE DECAY AND GROWTH

In 1902, Rutherford and Soddy, working together at McGill University in Montreal, dissolved thorium nitrate in water and then reprecipitated the thorium as the hydroxide. They found that the radioactivity of the thorium was greatly diminished but that the remaining solution contained a highly radioactive substance which they called ThX, later identified as $^{224}_{88}Ra$. Crooks had done a similar experiment with uranium in 1900 and discovered UX which was far more radioactive than purified uranium. Becquerel showed subsequently that the activity of UX decreases with time while that of uranium increases. Rutherford and Soddy made careful measurements of the activities of Th and ThX over a period of about one month and found that the activity of ThX decreased exponentially to zero in that period of time, while the activity of Th recovered its previous intensity. From the results of these experiments, Rutherford and Soddy concluded that radioactivity involved the spontaneous decomposition of the atoms of one element to form atoms of another element. They suggested that radioactivity is a property of certain atoms and that the rate of disintegration is proportional to the number of atoms present. This explanation of the process of radioactive decay was immediately accepted and has never been seriously challenged or modified.

1
Decay of a Radioactive Parent to a Stable Daughter

According to the theory of Rutherford and Soddy, the rate of decay of an unstable parent nuclide is proportional to the number of atoms (N) remaining at any time t. Translating this statement into mathematical language results in

$$-\frac{dN}{dt} \propto N \qquad (4.1)$$

where dN/dt is the rate of change of the number of parent atoms and the minus sign is required because the rate decreases as a function of time. The proportionality expressed above is transformed into an equality by the introduction of the proportionality constant λ which is called the decay constant. The numerical value of λ is characteristic of the particular radionuclide under consideration and is expressed in units of reciprocal time. The decay constant represents the probability that an atom will decay within a stated unit of time. The equation describing the rate of decay of a radionuclide therefore is

$$-\frac{dN}{dt} = \lambda N \qquad (4.2)$$

It will be helpful later to remember that λN means "the rate of decay at any time t."

The next step is to rearrange the terms of Equation 4.2 and to integrate:

$$-\int \frac{dN}{N} = \lambda \int dt \qquad (4.3)$$

This leads to

$$-\ln N = \lambda t + C \qquad (4.4)$$

where $\ln N$ is the logarithm to the base e of N, and C is the constant of integration. ($e = 2.718....$) The constant of integration can be evaluated from the condition that

$N = N_0$ when $t = 0$. Therefore,

$$C = -\ln N_0 \tag{4.5}$$

Substituting into Equation 4.4, we obtain

$$-\ln N = \lambda t - \ln N_0 \tag{4.6}$$

$$\ln N - \ln N_0 = -\lambda t$$

$$\ln \frac{N}{N_0} = -\lambda t$$

$$\frac{N}{N_0} = e^{-\lambda t}$$

$$N = N_0 e^{-\lambda t} \tag{4.7}$$

Equation 4.7 gives the number of radioactive parent atoms (N) that remain at any time t of an original number of atoms (N_0) that were present when $t = 0$. It is the basic equation describing all radioactive decay processes.

Next, let us assume that the decay of a radioactive parent produces a stable radiogenic daughter and that the number of daughter atoms is zero at $t = 0$. The number of daughter atoms (D^*) produced by the decay of its parent at any time t is given by

$$D^* = N_0 - N \tag{4.8}$$

provided that no daughter atoms are added to or lost from the system and that the change in the number of parent atoms is due only to radioactive decay. By substituting Equation 4.7 into Equation 4.8, we obtain

$$D^* = N_0 - N_0 e^{-\lambda t}$$
$$D^* = N_0(1 - e^{-\lambda t}) \tag{4.9}$$

Equation 4.9 gives the number of stable *radiogenic* daughter atoms (D^*) at any time t that formed by decay of a radioactive parent whose initial number at $t = 0$ was N_0, provided that no daughter atoms were present initially and provided that no parent or daughter atoms were either added to or lost from the system since t was equal to zero.

Before turning to graphical illustrations of radioactive decay and growth, it is helpful to define the half-life of a radioactive atom. The half-life ($T_{1/2}$) is the time required for one-half of a given number of a radionuclide to decay. It follows therefore that when $t = T_{1/2}, N = \frac{1}{2}N_0$. Substituting these values into Equation 4.7, we have

$$\tfrac{1}{2}N_0 = N_0 e^{-\lambda T_{1/2}}$$

$$\ln(\tfrac{1}{2}) = -\lambda T_{1/2}$$

$$\ln 2 = \lambda T_{1/2}$$

$$T_{1/2} = \frac{\ln 2}{\lambda} = \frac{0.693}{\lambda} \tag{4.10}$$

Equation 4.10 provides a convenient relationship between the half-life of a radionuclide and its decay constant.

Another parameter that is sometimes used to describe the decay of a radioactive species is the mean life, defined as the average life expectancy of a radioactive atom. The mean life (τ) is defined as

$$\tau = -\frac{1}{N_0} \int_{t=0}^{t=\infty} t \, dN \tag{4.11}$$

From Equation 4.2,

$$-dN = \lambda N \, dt$$

Therefore,

$$\tau = \frac{1}{N_0} \int_0^\infty \lambda N t \, dt$$

Since

$$N = N_0 e^{-\lambda t},$$

$$\tau = \lambda \int_0^\infty t e^{-\lambda t} \, dt = -\left[\frac{\lambda t + 1}{\lambda} e^{-\lambda t} \right]_0^\infty$$

$$\tau = \frac{1}{\lambda} \tag{4.12}$$

Thus the mean life τ is equal to the reciprocal of the decay constant. It is longer than the half-life by the factor $1/0.693$. The activity of a radioactive nuclide will be reduced by a factor equal to $1/e$ of its initial value during each mean life. One can describe radioactive decay either in terms of the half-life or in terms of the mean life of the decaying nuclide. We will use the half-life here because it is easier to visualize than the mean life and because most isotope geologists have used it in the development of equations.

Figure 4.1 illustrates the exponential decay of a hypothetical radionuclide (N) to a stable daughter (D^*). The graph shows that initially ($t = 0$) there are 120 atoms ($N_0 = 120$) of the radionuclide but no atoms of the daughter nuclide ($D_0 = 0$). After the first half-life has elapsed, only 60 atoms of N are left on the average, 30 after the second half-life, and so on. In general, the number of parent atoms is reduced by a factor 2^{-1} after each half-life. If decay has occurred for n half-lives, the initial number of radionuclides is reduced by a factor of 2^{-n}. The number of radiogenic daughter atoms in the meantime increases with each successive

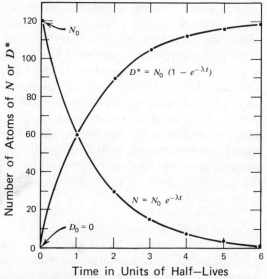

FIGURE 4.1 Decay of a hypothetical radionuclide (N) to a stable radiogenic daughter (D^*) as a function of time measured in units of half-lives. It can be seen that $N \to 0$ as $t \to \infty$, while $D^* \to N_0$ as $t \to \infty$.

half-life from zero to 60 to 90 to 105, and so forth, until eventually D^* approaches N_0 asymptotically as $t \to \infty$. For most practical purposes, this occurs when $t = 10T_{1/2}$, that is, after decay has occurred for 10 half-lives and N_0 has been reduced by a factor of 2^{-10} (1/1024). It is obvious from this discussion that the rate of decay of a radionuclide is not constant but decreases exponentially as a function of time. The only aspect of this process that remains constant is the proportion of radionuclides that decay during one half-life.

In many practical applications of radioactivity to the study of rocks and minerals, it is more convenient to relate the number of radiogenic daughters (D^*) to the number of parent atoms remaining (N), rather than to N_0. We therefore return to Equation 4.8

$$D^* = N_0 - N$$

and replace N_0 by $Ne^{\lambda t}$ which we obtain from Equation 4.7. Substituting, we find

$$D^* = Ne^{\lambda t} - N = N(e^{\lambda t} - 1) \quad (4.13)$$

In the general case, the total number of daughter atoms (D) present in a system in which decay is occurring is

$$D = D_0 + D^* \quad (4.14)$$

where D_0 is the number of daughter atoms present initially ($t = 0$) and D^* is the number of daughter atoms produced within the system by decay of the parent. Since

$$D^* = N(e^{\lambda t} - 1),$$
$$D = D_0 + N(e^{\lambda t} - 1) \quad (4.15)$$

This is the basic equation that is used to make age determinations of rocks and minerals based on the decay of a radioactive parent to a stable daughter. Both D and N are measurable quantities, while D_0 is a constant whose value can be either assumed or calculated from the data. When D and N have been measured and an appropriate value is substituted for D_0, Equation 4.15 can be solved for t, which is the "age" of the system, provided the following conditions are satisfied.

1. The rock or mineral system has neither gained nor lost either parent or daughter atoms so that the ratio of D^*/N has changed only as a result of radioactive decay. This condition is often expressed by the statement that the rock or mineral sample must be a "closed system" with respect to the parent and daughter.

2. It must be possible to assign a realistic value to D_0. This can usually be done reliably, especially when D^* is much greater than D_0.

3. The value of the decay constant (λ) must be known accurately.

4. The measurements of D and N must be accurate and representative of the rock or mineral to be dated.

When these four basic assumptions are satisfied, the solution of Equation 4.15 yields a *date* which may represent the *age* of the rock or mineral. We will make a distinction between the words "date" and "age." Any solution of Equation 4.15 produces a "date." This date represents the "age" of the rock or mineral only when all four basic assumptions are satisfied and when the date can be associated with a significant geological event in the history of that rock or mineral.

The use of radioactive decay to measure the ages of rocks and minerals also implies the assumption that the decay constants have not changed during the past 4.6 billion years. This assumption is justified because radioactive decay is a property of the nucleus which is shielded from outside influences by the electrons that surround it. Moreover, there is no evidence that decay constants have changed as a function of time during the history of the solar system. The only decay modes that may be affected by changes in the electron density in the vicinity of the nucleus are electron capture and internal conversion because both involve extranuclear electrons. Several studies have shown that the decay constants of ^{7}Be, ^{99m}Tc, and ^{131}Ba increase very slightly when these atoms are subjected to very high pressures of the order of 100 kbars or more. A case in point is the work of Hensley et al. (1973) with ^{7}Be which decays by electron capture to ^{7}Li. These authors observed an increase of 0.59 percent in the decay constant of ^{7}Be when this nuclide (in the form of BeO) was subjected to a pressure of 270 ± 10 kbars in a diamond anvil press. The only naturally occurring radioactive nuclide used for dating that decays by electron capture is ^{40}K which forms ^{40}Ar. There is no evidence that potassium now residing in the crust of the Earth has been subjected to pressures of the order of several hundred kbars for a sufficient length of time to affect the amount of radiogenic ^{40}Ar produced. Consequently, there is no reason to doubt that the decay constants of the naturally occurring long-lived radioactive isotopes used for dating are invariant and independent of the physical and chemical conditions to which they may have been subjected since nucleosynthesis. Obviously, this statement does not apply to atoms in the sun or in other stars where electron capture and internal conversion decay modes may be inhibited by the removal of all or most electrons from the atoms.

Figure 4.2 is a plot of D^*/N versus t in units of half-lives for a hypothetical radionuclide (N) decaying to stable radiogenic daughter (D^*). The data points for this graph were taken from Figure 4.1. It is evident that the ratio D^*/N increases as a function of time. This relationship is the basic concept that underlies all methods of age determination based on the decay of naturally occur-

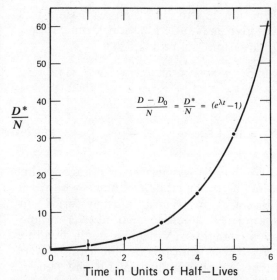

$$\frac{D - D_0}{N} = \frac{D^*}{N} = (e^{\lambda t} - 1)$$

FIGURE 4.2 Graph showing the relationship between D^*/N and t for a hypothetical radionuclide (N) decaying to form a stable radiogenic daughter D^*. It is evident that the ratio D^*/N increases as a function of t and approaches infinity as t approaches infinity. Time is plotted in units of half-lives for convenience. The data points for this graph were taken from Figure 4.1.

ring radionuclides in rocks and minerals. Equation 4.15 can be solved explicitly for t to give

$$\frac{D - D_0}{N} = e^{\lambda t} - 1$$

$$t = \frac{1}{\lambda} \ln \left[\frac{D - D_0}{N} + 1 \right] \quad (4.16)$$

2
Decay Series

Let us consider the decay of a parent (N_1) to a radioactive daughter (N_2) which decays to a second daughter (N_3). The rate of decay of N_1 is

$$-\frac{dN_1}{dt} = \lambda_1 N_1 \quad (4.17)$$

The rate of decay of N_2 is the difference between the rate at which it is produced by decay of N_1 and its own decay rate. Therefore

$$\frac{dN_2}{dt} = \lambda_1 N_1 - \lambda_2 N_2 \quad (4.18)$$

where λ_1 and λ_2 are the decay constants of radionuclides 1 and 2, respectively, and N_1 and N_2 are the numbers of atoms remaining at any time t. The number of atoms of N_1 remaining at any time t is given by Equation 4.7:

$$N_1 = N_1^0 e^{-\lambda_1 t}$$

where N_1^0 indicates the number of atoms of N_1 at $t = 0$. This equation is substituted into 4.18 and, by rearrangement of terms, we obtain

$$\frac{dN_2}{dt} + \lambda_2 N_2 - \lambda_1 N_1^0 e^{-\lambda_1 t} = 0 \quad (4.19)$$

This is a linear differential equation of the first order, first solved by Bateman (1910). A detailed solution of this equation was also given by Friedlander and Kennedy (1955, 129–130):

$$N_2 = \frac{\lambda_1}{\lambda_2 - \lambda_1} N_1^0 (e^{-\lambda_1 t} - e^{-\lambda_2 t})$$
$$+ N_2^0 e^{-\lambda_2 t} \quad (4.20)$$

The first term of this equation gives the number of atoms of N_2 that have formed by decay of N_1, but have not yet decayed. The second term represents the number of atoms of N_2 that remain from an initial number N_2^0. In most applications of interest to us N_2^0 will be zero, that is, no atoms of N_2 are present initially. Therefore, Equation 4.20 reduces to

$$N_2 = \frac{\lambda_1}{\lambda_2 - \lambda_1} N_1^0 (e^{-\lambda_1 t} - e^{-\lambda_2 t}) \quad (4.21)$$

The general case of a radioactive decay series $N_1 \rightarrow N_2 \rightarrow N_3 \cdots N_n$ was solved by

Bateman (1910). The solution, giving the number of atoms of any member of the decay series as a function of time for the condition that $N_2^0 = N_3^0 = \cdots N_n^0 = 0$, has the form

$$N_n = C_1 e^{-\lambda_1 t} + C_2 e^{-\lambda_2 t} + \cdots C_n e^{-\lambda_n t} \quad (4.22)$$

where

$$C_1 = \frac{\lambda_1 \lambda_2 \cdots \lambda_{n-1} N_1^0}{(\lambda_2 - \lambda_1)(\lambda_3 - \lambda_1) \cdots (\lambda_n - \lambda_1)}$$

$$C_2 = \frac{\lambda_1 \lambda_2 \cdots \lambda_{n-1} N_1^0}{(\lambda_1 - \lambda_2)(\lambda_3 - \lambda_2) \cdots (\lambda_n - \lambda_2)}$$

We can use Equation 4.22 to determine the number of atoms of any member of a decay series at any time, assuming that initially only the parent was present. For example, let us assume that the second daughter N_3 is also radioactive. From Equation 4.22, we obtain

$$N_3 = C_1 e^{-\lambda_1 t} + C_2 e^{-\lambda_2 t} + C_3 e^{-\lambda_3 t} \quad (4.23)$$

$$C_1 = \frac{\lambda_1 \lambda_2 N_1^0}{(\lambda_2 - \lambda_1)(\lambda_3 - \lambda_1)}$$

$$C_2 = \frac{\lambda_1 \lambda_2 N_1^0}{(\lambda_1 - \lambda_2)(\lambda_3 - \lambda_2)}$$

$$C_3 = \frac{\lambda_1 \lambda_2 N_1^0}{(\lambda_1 - \lambda_3)(\lambda_2 - \lambda_3)}$$

The complete equation is obtained by substituting C_1, C_2, and C_3 into Equation 4.23:

$$N_3 = \frac{\lambda_1 \lambda_2 N_1^0 e^{-\lambda_1 t}}{(\lambda_2 - \lambda_1)(\lambda_3 - \lambda_1)} + \frac{\lambda_1 \lambda_2 N_1^0 e^{-\lambda_2 t}}{(\lambda_1 - \lambda_2)(\lambda_3 - \lambda_2)}$$

$$+ \frac{\lambda_1 \lambda_2 N_1^0 e^{-\lambda_3 t}}{(\lambda_1 - \lambda_3)(\lambda_2 - \lambda_3)} \quad (4.24)$$

Now let us assume that N_3 is stable and that the decay series consists of the parent N_1 decaying to N_2 which decays to stable N_3. If N_3 is stable, $\lambda_3 = 0$. We can now use Equation 4.24 to obtain an expression that will give us N_3 as a function of time. If $\lambda_3 = 0$, Equation 4.24 reduces to

$$N_3 = -\frac{\lambda_2 N_1^0 e^{-\lambda_1 t}}{(\lambda_2 - \lambda_1)} - \frac{\lambda_1 N_1^0 e^{-\lambda_2 t}}{(\lambda_1 - \lambda_2)} + N_1^0 \,.$$

$$N_3 = N_1^0 \left(1 + \frac{\lambda_1 e^{-\lambda_2 t}}{\lambda_2 - \lambda_1} + \frac{\lambda_2 e^{-\lambda_1 t}}{\lambda_2 - \lambda_1} \right) \quad (4.25)$$

We have now derived expressions that allow us to calculate the number of atoms present at any time in a decay series consisting of a radioactive parent (N_1), a radioactive daughter (N_2), and a stable daughter (N_3). These solutions establish the pattern involving the application of the general solution (Equation 4.23) to even longer decay series. A numerical example will further illustrate the behavior of a three-component decay series. Let us assume that the half-life of the parent (N_1) is one hour ($\lambda_1 = 0.693 \text{ hr}^{-1}$) and that of its first daughter (N_2) is five hours ($\lambda_2 = 0.1486 \text{ hr}^{-1}$). The third daughter ($N_3$) is stable. We further assume that $N_1^0 = 100$, $N_2^0 = 0$, and $N_3^0 = 0$. Figure 4.3 illustrates the results calculated by means of Equations 4.7, 4.21, and 4.25. It can be seen that N_1 decreases exponentially while N_2 increases and reaches a maximum value in about three hours. Thereafter, N_2 declines, while N_3 continues to increase until it approaches the value of N_1^0 asymptotically as t approaches infinity. The time when N_2 attains a maximum can be calculated by differentiating Equation 4.21 with respect to t. If $N_2^0 = 0$,

$$\frac{dN_2}{dt} = -\frac{\lambda_1^2 N_1^0 e^{-\lambda_1 t}}{\lambda_2 - \lambda_1} + \frac{\lambda_1 \lambda_2 N_1^0 e^{-\lambda_2 t}}{\lambda_2 - \lambda_1}.$$

For a minimum or maximum

$$\frac{dN_2}{dt} = 0$$

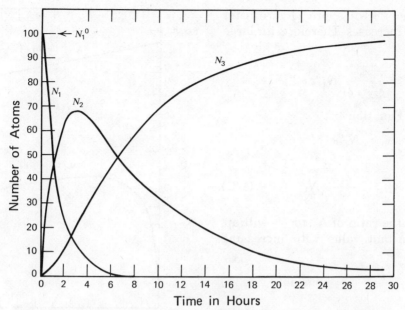

FIGURE 4.3 Radioactive decay and growth in a three-component series. The parent (N_1) has a half-life of one hour; the first daughter (N_2) has a half-life of five hours. The third daughter (N_3) is stable. Initially only atoms of the parent are present, i.e., $N_1^0 = 100$, $N_2^0 = 0$, $N_3^0 = 0$. Note that the number of atoms of N_2 initially increases, achieves a maximum about three hours after decay began in the system and then decreases. The number of stable daughters (N_3) increases smoothly and will eventually approach N_1^0 as t goes to infinity. (Adapted from Kaplan, 1955, p. 200.)

Therefore,

$$\frac{\lambda_1^2 N_1^0 e^{-\lambda_1 t_m}}{\lambda_2 - \lambda_1} = \frac{\lambda_1 \lambda_2 N_1^0 e^{-\lambda_2 t_m}}{\lambda_2 - \lambda_1}$$

$$\lambda_1 e^{-\lambda_1 t_m} = \lambda_2 e^{-\lambda_2 t_m}$$

$$\frac{\lambda_1}{\lambda_2} = e^{(\lambda_1 - \lambda_2) t_m}$$

$$\ln\left(\frac{\lambda_1}{\lambda_2}\right) = (\lambda_1 - \lambda_2) t_m$$

$$t_m = \frac{1}{\lambda_1 - \lambda_2} \ln\left(\frac{\lambda_1}{\lambda_2}\right)$$

$$t_m = \frac{2.3026}{\lambda_1 - \lambda_2} \log_{10}\left(\frac{\lambda_1}{\lambda_2}\right) \qquad (4.26)$$

where t_m = time when N_2 is a maximum. In the example considered above where $\lambda_1 = 0.693$ hr^{-1} and $\lambda_2 = 0.1486$ hr^{-1}, $t_m = 2.83$ hours. We can see by inspection of Equation 4.26 that N_2 will have a maximum only when $\lambda_1 > \lambda_2$. This raises the question of how a radioactive decay series will behave when $\lambda_1 < \lambda_2$, that is, when the parent has a longer half-life than its daughter.

Let us consider a decay series consisting of the parent N_1 decaying to a radioactive daughter N_2 which decays to daughter N_3, such that $\lambda_1 < \lambda_2$. Equation 4.21, for the case that $N_2^0 = 0$, states that

$$N_2 = \frac{\lambda_1}{\lambda_2 - \lambda_1} N_1^0 (e^{-\lambda_1 t} - e^{-\lambda_2 t})$$

If $\lambda_1 < \lambda_2$, then $e^{-\lambda_2 t}$ will approach zero faster than $e^{-\lambda_1 t}$ as t increases. Therefore, for large values of t,

$$N_2 \simeq \frac{\lambda_1}{\lambda_2 - \lambda_1} N_1{}^0 e^{-\lambda_1 t}$$

According to Equation 4.7,

$$N_1 = N_1{}^0 e^{-\lambda_1 t}$$

Therefore,

$$N_2 \simeq \frac{\lambda_1}{\lambda_2 - \lambda_1} N_1 \qquad (4.27)$$

Accordingly, the ratio of N_1 to N_2 will approach a constant value with increasing time:

$$\frac{N_1}{N_2} = \frac{\lambda_2 - \lambda_1}{\lambda_1} = \text{constant} \qquad (4.28)$$

and the series will achieve a condition of radioactive equilibrium. Figure 4.4 is a plot of N_1 and N_2 for the case that the half-life of the parent is 10 hours while that of the daughter is 1 hour.

The naturally occurring radioactive decay series arising from $^{238}_{92}\mathrm{U}$, $^{235}_{92}\mathrm{U}$, and $^{232}_{90}\mathrm{Th}$ are examples in which the half-lives of the parent nuclides are very much longer than those of their respective daughters and where therefore $\lambda_1 \ll \lambda_2$. Therefore, the number of parent atoms remains essentially constant for several half-lives of the daughters. In these decay series, Equation 4.28 can be further modified by introducing the approximation that $\lambda_2 - \lambda_1 \simeq \lambda_2$. Thus,

$$\frac{N_1}{N_2} = \frac{\lambda_2}{\lambda_1}$$

or

$$\lambda_1 N_1 = \lambda_2 N_2 \qquad (4.29)$$

This condition is known as *secular equilibrium* in which the rate of decay of the daughter is equal to that of its parent. In a radioactive decay series consisting of a very long-lived parent and a series of short-lived

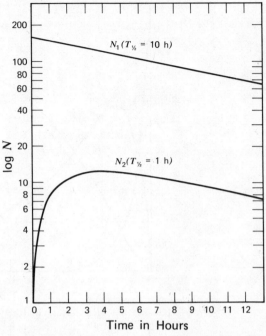

FIGURE 4.4   Decay of a long-lived parent (N_1) to a short-lived daughter (N_2). The half-life of the parent is 10 hours, while that of the daughter is 1 hour. The number of daughter atoms increases as a function of time from an initial value of zero. Eventually in this system the ratio of parent to daughter assumes a constant value equal to $\lambda_2 - \lambda_1/\lambda_1$, and a radiochemical equilibrium is thereby established. Note that the ordinate is in units of $\log N$ which is related to $\ln N$ by $\ln N = 2.303 \log N$.

intermediate daughters, the condition of secular equilibrium is propagated through the entire series such that

$$\lambda_1 N_1 = \lambda_2 N_2 = \lambda_3 N_3 = \cdots \lambda_n N_n$$

When secular equilibrium has been established, the number of atoms present of any of the unstable daughters will be

$$N_n = \frac{\lambda_1 N_1}{\lambda_n} \qquad (4.30)$$

The ratios of the disintegration rates of the

daughters will be

$$\frac{\lambda_1 N_1}{\lambda_2 N_2} = \frac{\lambda_2 N_2}{\lambda_3 N_3} = \cdots = 1.000 \quad (4.31)$$

The rate of growth of the stable radiogenic daughter at the end of a decay series in secular equilibrium is equal to the rate of decay of the parent that starts the series. Therefore, the number of radiogenic daughters that form by decay in a series in secular equilibrium is

$$D^* = N_1{}^0 - N_1 - N_2 - \cdots - N_n \quad (4.32)$$

From Equation 4.30

$$N_2 = \frac{\lambda_1}{\lambda_2} N_1, \ N_3 = \frac{\lambda_1}{\lambda_3} N_1, \ \text{etc.}$$

Therefore

$$D^* = N_1{}^0 - N_1 \left(1 - \frac{\lambda_1}{\lambda_2} - \frac{\lambda_1}{\lambda_3} - \cdots - \frac{\lambda_1}{\lambda_n}\right) \quad (4.33)$$

The ratios λ_1/λ_2, λ_1/λ_3, and so forth are all very much less than unity. Therefore,

$$D^* = N_1{}^0 - N_1 = N_1 e^{\lambda_1 t} - N_1$$
$$D^* = N_1(e^{\lambda_1 t} - 1) \quad (4.34)$$

In other words, the number of radiogenic daughters that have accumulated by decay in a series under conditions of secular equilibrium can be treated as though the initial parent decays directly to the stable daughter.

3
Units of Radioactivity and Dosage

The radioactivity of a sample is measured by electrical or photographic methods and is expressed as the number of disintegrations observed in unit time. It is usually not practical to describe radioactivity by the number of atoms remaining (N) which we have used in deriving the equations of the preceding sections. The observed activity (A) of a radio-active sample is defined as

$$A = c\lambda N \quad (4.35)$$

where A is the observed disintegration rate, λN is the actual rate of decay, and c is the detection coefficient. The value of c depends on the nature of the detection instruments, their efficiency in detecting radiation of a particular type and energy, the geometry of the counter and sample, and other experimental factors. The numerical value of c must be determined by a suitable calibration of the detectors.

We can now modify Equation 4.7 in accordance with Equation 4.35 as follows:

$$c\lambda N = c\lambda N_0 e^{-\lambda t}$$
$$A = A_0 e^{-\lambda t} \quad (4.36)$$

where A is the observed rate of disintegration at any time t and A_0 is the initial disintegration rate when $t = 0$. By taking logarithms of Equation 4.36, we obtain

$$\ln A = \ln A_0 - \lambda t \quad (4.37)$$

This is the equation of a straight line in coordinates of $\ln A$ and t. When the activity of a single radioactive nuclide is measured at intervals of time and the results are plotted in coordinates of $\ln A$ and t, the data points will lie on a straight line whose slope is $-\lambda$.

The basic unit for the measurement of radioactivity is the curie (Ci) which is defined as the quantity of any radioactive nuclide in which the number of disintegrations is 3.700×10^{10} per second. For many practical applications it is more convenient to express radioactivity in millicuries (mCi) or microcuries (μCi) which are defined as follows:

$$1 \ \text{mCi} = 10^{-3} \ \text{Ci}$$
$$1 \ \mu\text{Ci} = 10^{-6} \ \text{Ci}$$

The passage of ionizing radiation through biological tissue causes injury, but the symptoms of injury may not become evident until later. In order to avoid or minimize the harmful effects of radioactivity on humans, exposure to different kinds of ionizing radiations must be carefully monitored. The nature and severity of the threat posed by different kinds of particles and radiations depends on the type of particle or radiation, their energies, the duration of the exposure, the parts of the body that are exposed, and whether the exposure is internal or external. Alpha and beta particles have a low penetrating power. A beta particle having an energy of 1 MeV will penetrate only 4 mm of human skin. The penetrating power of alpha particles of low energy is even less. Such particles lose their energy by producing secondary ions and by breaking chemical bonds, but the damage is confined primarily to the skin. However, when alpha or beta emitters enter the body by ingestion or inhalation, they may do serious internal damage that will be localized in those parts of the body where the radionuclides are concentrated. Therefore, charged particles are far more dangerous when the exposure is internal than when it is external. X-rays and gamma rays have high penetrating powers. They interact with matter in several ways, all of which result in the formation of positively charged ions and electrons. The formation of such ions causes damage to biological tissue. Because of their penetration, x-rays and gamma rays are external health hazards. Except for lethal or near-lethal exposures to radiation, the symptoms of biological injury may be delayed for several days or even years. Lapp and Andrews (1954, 496) listed the following effects of overexposure to ionizing radiation:

1. Leucopenia (decrease in number of white blood cells)
2. Epilation (loss of hair)
3. Sterility
4. Mutations (altered heredity of offspring)
5. Cancer
6. Bone necrosis (destruction and death of the bones)
7. Cataracts (primarily from neutrons)

The radiation dose delivered to part or all of the body is measured in a variety of units that generally express the amount of energy transmitted by the radiation to air or body tissue. The following units are in use:

Roentgen (R) That quantity of x or gamma radiation for which the associated corpuscular emission per 0.001293 grams of air produces, in air, ions carrying 1 electrostatic unit (esu) of electricity of either sign. **Rad** One rad is the dose corresponding to the absorption of 100 ergs per gram of tissue. **Rem** The rem is a measure of the dose of any ionizing radiation to body tissues in terms of its estimated biological effect relative to a dose of one roentgen (R) of x-rays.

The most widely used of these units is the roentgen which produces 1.61×10^{12} ion pairs per gram of air corresponding to the absorption of 84 ergs of energy per gram of air. It is also equivalent to the absorption of 93 ergs per gram of water due to irradiation by x-rays or gamma rays having energies higher than about 50 keV. Survey meters used to monitor the dose rate are commonly calibrated in units of milliroentgen per hour (mR/hr) or microroentgen (μR/hr), where $1 \text{ mR} = 10^{-3} \text{ R}$ and $1 \mu\text{R} = 10^{-6} \text{ R}$.

4
Neutron Activation

Nuclear reactions can be induced by bombarding the nucleus of any atom with a variety of nuclear particles, such as protons, deuterons (nuclei of deuterium), alpha particles, neutrons, and so forth. In the case of bombardment with positively charged particles, nuclear reactions can occur only when the particles have sufficient energy to overcome the electrostatic repulsion of the protons in the target nucleus. Therefore the projectiles must have a certain minimum energy that depends on their electric charge and the atomic number of the target nucleus. These limitations do not apply to neutrons that have no charge and can interact with the nucleus of any atom regardless of their energy distribution.

When a nuclear particle enters the nucleus, a reaction will occur which manifests itself by the emission of particles or of radiant energy from the product nucleus. Depending on the nature of the reaction, the product may be an isotope of the same element as the target or it may be an isotope of a different element. The kind of reaction that takes place depends on the nature of the projectile, its energy, and the nature of the target nucleus. Such nuclear reactions make possible elemental transformations of the atoms of one element into atoms of another element. Thus the fondest dream of the alchemists has at last come true!

Nuclear reactions can be described by equations similar to those used in Chapter 3 to represent radioactive decay. Let us consider some schematic examples to illustrate the usefulness of such equations:

$$_Z^A X + {}_1^1 H \rightarrow {}_{z+1}^A Y + {}_0^1 n \qquad (4.38)$$

Equation 4.38 shows that a target atom $_Z^A X$ reacts with a proton ($_1^1 H$) and forms a product atom ($_{z+1}^A Y$) and a neutron ($_0^1 n$) which is emitted. It is clear that in this case the product atom is an isobar of the target atom, that is, it is an isotope of another element. The same reaction can also be represented by a shorthand notation which in this case is

$$_Z^A X(p, n)_{z+1}^A Y \qquad (4.39)$$

The nuclide in front of the brackets is the target, while the nuclide following the brackets is the product. Inside the brackets are shown the proton (p) which enters the target nucleus and the neutron (n) which comes out of the product nucleus. The two particles are separated by a comma to indicate which particle goes in and which comes out. The notation

$$_Z^A X(\alpha, 2n)_{Z+2}^{A+2} Y \qquad (4.40)$$

means that one alpha particle (α) goes in and two neutrons come out. As a result, the target atom ($_Z^A X$) is transformed into the product atom ($_{Z+2}^{A+2} Y$) which in this case is an atom of another element and is not an isobar of the target. A variety of artificial nuclear reactions which are possible and the relative positions of target and product nuclides are summarized in Figure 4.5.

Nuclear bombardment reactions caused by neutrons are of great interest in isotope geology and geochemistry because such reactions are used for analytical purposes to measure the concentrations of trace elements in geological materials. Neutrons are produced in large numbers by the controlled fission of ^{235}U in nuclear reactors and are therefore readily available for irradiation purposes. The neutrons produced during fission of a uranium atom are initially emitted with high velocities and are called "fast" neutrons. In order to sustain the chain reaction, the fast neutrons must be slowed down because the fission reaction is strongly favored when it is induced by "slow" rather than "fast" neutrons. The slowing down of the neutrons is achieved by a "moderator" with which the fast neutrons can collide without being absorbed by the nuclei of the moderator. The first experimental reactors used graphite or heavy water (D_2O) as moderators, although ordinary water (H_2O) also serves this purpose in the so-called "swimming pool" reactors.

Slow neutrons, having velocities corresponding to the ambient temperature, are readily absorbed by the nuclei of most of the

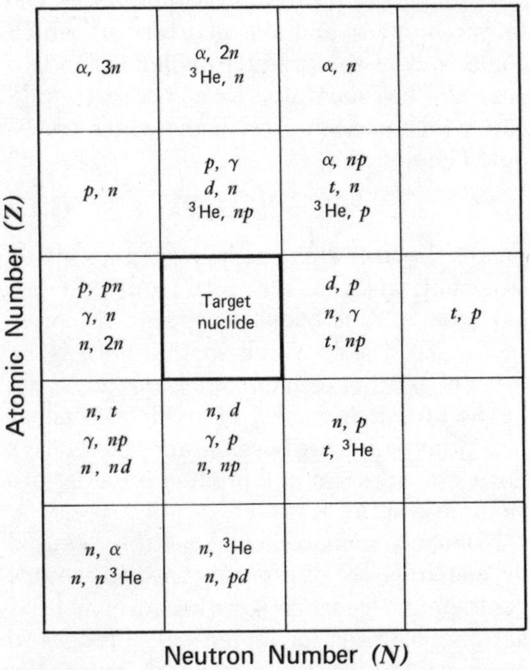

α, 3n	α, 2n ³He, n	α, n	
p, n	p, γ d, n ³He, np	α, np t, n ³He, p	
p, pn γ, n n, 2n	Target nuclide	d, p n, γ t, np	t, p
n, t γ, np n, nd	n, d γ, p n, np	n, p t, ³He	
n, α n, n³He	n, ³He n, pd		

Atomic Number (Z) — vertical axis

Neutron Number (N) — horizontal axis

FIGURE 4.5 Displacement of product nuclides relative to the target nuclide caused by nuclear bombardment reactions. The symbol in front of the comma indicates the particle or radiation that goes in, the symbol after the comma indicates the particles or radiation that come out of the product nucleus. The symbols have the following meaning: p = proton (^1_1H), n = neutron, γ = gamma ray, α = alpha particle (^4_2He), d = deuteron (^2_1H), and t = triton (^1_3H). (After Holden and Walker, 1972.)

stable isotopes of the elements. The neutron number of the product nucleus is increased by one compared to the target nucleus, but Z remains unchanged so that the product is an isotope of the same element as the target. The product nucleus is left in an excited state and de-excites by emission of a gamma ray. The absorption of a slow neutron by the nucleus of an atom can be represented by the following equation:

$$^A_Z X + ^1_0 n \rightarrow ^{A+1}_Z Y + \gamma$$

or

$$^A_Z X(n, \gamma)^{A+1}_Z Y \qquad (4.41)$$

The product of an "(n, γ)" reaction may be either stable or unstable. When the product nuclides produced by neutron irradiation are unstable, the sample becomes radioactive as the product nuclides decay with their characteristic half-lives. Thus a slow-neutron irradiation of a sample composed of the stable atoms of a variety of elements leads to the formation of radioactive isotopes of these elements and the irradiated sample then becomes radioactive, hence the term "neutron activation." The induced activity of a specific radioactive isotope of an element in the irradiated sample depends on several factors, including the concentration of the target element in the sample. This fact is the basis for using neutron activation as an analytical tool (Mapper, 1960; Brunfelt and Steinnes, 1971; DeSoete et al., 1972).

Let us now consider the typical case in which a radioactive nuclide (P) is produced at a constant rate (R) by a nuclear reaction which is maintained by a constant source of particles such as a nuclear reactor, a cyclotron, or a linear particle accelerator. The product of the reaction is radioactive and decays with its characteristic half-life. The situation, therefore, is analogous to that of a long-lived parent (N_1) decaying to a short-lived daughter (N_2). Since a particle accelerator or a nuclear reactor can be operated at a constant level, the production rate (R) does not change with time. This means that the irradiation of a target by a constant flux of particles acts like a radioactive parent with an infinite half-life. We can therefore adapt Equation 4.21 to obtain an expression that will give the number of product atoms (P) as a function of the irradiation time, assuming that no product atoms are present at the start of the irradiation ($P_0 = 0$ at $t = 0$).

Equation 4.21 states that

$$N_2 = \frac{\lambda_1}{\lambda_2 - \lambda_1} N_1{}^0(e^{-\lambda_1 t} - e^{-\lambda_2 t})$$

We now substitute P for N_2 and assume that $\lambda_1 = 0$. In this case $e^{-\lambda_1 t} = 1$ and $\lambda_2 - \lambda_1 = \lambda_2$. Therefore,

$$P = \frac{R}{\lambda}(1 - e^{-\lambda t}) \qquad (4.42)$$

where R is the rate at which P is produced by the nuclear reaction, λ is the decay constant of the product nuclide (P), and t is the time elapsed since the start of the irradiation. As the irradiation time increases and approaches infinity,

$$\underset{t \to \infty}{\text{Lim}} (1 - e^{-\lambda t}) = 1$$

and

$$\lambda P = R \qquad (4.43)$$

The rate of decay of the product nuclide eventually approaches its production rate which is the maximum or saturation disintegration rate that can be achieved. If one must pay for an irradiation based on the irradiation time, it is obviously important to limit the irradiation time to something less than the time required to achieve saturation. In order to estimate the most advantageous irradiation time, we write Equation 4.42 as

$$\frac{\lambda P}{R} = (1 - e^{-\lambda t}) \qquad (4.44)$$

where $\lambda P/R$ is the disintegration rate of the product expressed as a fraction of the maximum attainable rate. Let us calculate the value of this fraction when the irradiation t is equal to one half-life of the product nuclide. If $t = T_{1/2}$, then

$$\lambda t = \ln 2 = 0.693$$

and

$$\frac{\lambda P}{R} = 1 - e^{-0.693} = 1 - 0.5$$

$$\frac{\lambda P}{R} = 0.5$$

Figure 4.6 is a plot of the function $\lambda P/R$ versus the irradiation time expressed in units of half-lives of the product nuclide. It can be seen that little is to be gained from

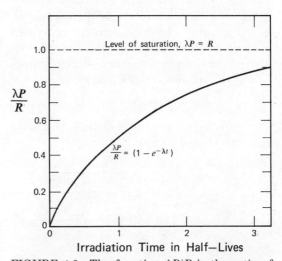

FIGURE 4.6   The function $\lambda P/R$ is the ratio of the disintegration rate of a radionuclide (P) produced by the irradiation of a suitable target at a constant production rate (R) to the saturation activity. The maximum disintegration rate is equal to the production rate. Therefore the ratio $\lambda P/R$ approaches unity with increasing irradiation times, here expressed in terms of half-lives of the product nuclide. The graph shows that after an irradiation time equal to one half-life, the activity will be one-half of the maximum. After two half-lives, the activity is three-quarters of the maximum activity. It is apparent that an irradiation time between two and three half-lives of the desired product nuclide is sufficient to produce an appreciable fraction of the maximum activity and that little is to be gained by extending it further.

irradiation times much greater than two half-lives. When the irradiation is terminated, the product nuclide continues to decay and the number of atoms remaining at any time will be

$$P = \frac{R}{\lambda}(1 - e^{-\lambda t_i})e^{-\lambda t_d} \qquad (4.45)$$

where t_i is the length of time the sample was irradiated and t_d is the decay time measured from the time the irradiation was terminated. The rate of disintegration is measured by means of a suitable Geiger-Müller or scintillation counter whose efficiency is indicated by the detection coefficient c. Thus the activity of the sample some time after termination of the irradiation is given by

$$A = P\lambda c = Rc(1 - e^{-\lambda t_i})e^{-\lambda t_d} \quad (4.46)$$

The activity of a sample during and after irradiation is shown in Figure 4.7.

The production rate R for an irradiation with slow neutrons is defined as

$$R = Na\sigma F \qquad (4.47)$$

where N is the number of target atoms of a particular element in the irradiated sample, a is the isotopic abundance of the target isotope, σ is the neutron capture cross section in units of barns (where 1 barn = 10^{-24} cm^2) of the target nuclide, and F is the neutron flux in units of neutrons per square centimeter per second. The cross section of a nuclear reaction is a measure of the probability that a particular reaction will occur. This probability is related to the cross-sectional area of the nucleus. Since the radius of an atomic nucleus is of the order of 10^{-12} cm, its cross-sectional area is of the order of 10^{-24} cm^2. Hence the reaction probability is measured in multiples of 10^{-24} cm^2. The flux F is expressed as the number of neutrons that

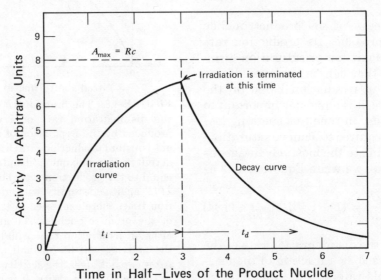

FIGURE 4.7 This diagram shows the increase in the observed disintegration rate (A) of a product nuclide produced by a nuclear irradiation reaction. Time is plotted in units of half-lives of the product nuclide. The graph shows that the irradiation was terminated after three half-lives and that the activity then decreased at a rate determined by the half-life of the radionuclide.

cross an area of 1 cm² in one second. The slow-neutron flux available for irradiations at many reactors is of the order of 10^{12} to 10^{13} neutrons/cm²/sec. Even higher flux densities are available in some of the largest reactors.

In order to determine the concentration of an element in a sample of matter, a known weight of the sample is irradiated with slow neutrons along with a standard containing a known amount of the same element. When the irradiation is completed, the activity due to the desired product nuclide is counted at intervals both in the sample and in the standard. Counting of the activity of a particular radionuclide produced by the irradiation is often complicated by the presence of other radionuclides. If the desired radionuclide is a gamma emitter, it is usually possible to screen out beta or alpha particles by means of suitable absorbers. Gamma ray detectors can be tuned to respond only to energy quanta in a small energy range. In this way gamma rays emitted by other radionuclides can be discriminated against. However, sometimes it is necessary to perform chemical separations on the irradiated sample to remove interfering radioactivities. Such separations may require the use of "carriers" which are small amounts of the unactivated element that facilitate precipitation reactions but do not interfere directly with measurements of the induced radioactivity. The chemical processing and counting of irradiated samples is the domain of the radiochemist and requires skill and ingenuity. Let us assume that in our case such technical problems have been handled satisfactorily and that we are able to measure the activity of the desired product radionuclide without interference. The observed activities are then plotted in coordinates of ln A and t_d as shown in Figure 4.8. If the measured activities are due only to the decay of a single species, the data points will lie on a straight

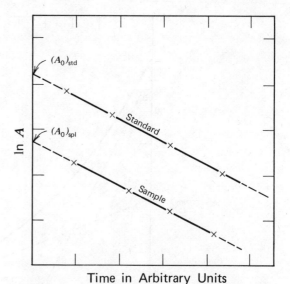

FIGURE 4.8 Plot of the observed activity of a single radionuclide after neutron irradiation of a sample and a standard in coordinates of ln A and t. The resulting straight lines are extrapolated back to A_0 which is the activity of the radionuclide in the sample and the standard at the end of the irradiation. The slopes of the two lines are identical and equal to $-\lambda$ of the radionuclide.

line. The line is extrapolated back to $t_d = 0$ to obtain a measurement of A_0, which is the activity at the time the irradiation was terminated. The data for the sample and the standard will, of course, form two separate, but parallel lines. The amount of the desired element in the irradiated sample can be obtained from the following relationship:

$$\frac{(A_0)_{spl}}{(A_0)_{std}} = \frac{(\text{amount of element } X)_{spl}}{(\text{amount of element } X)_{std}} \quad (4.48)$$

When the amount of element X in the sample is known, its concentration can be calculated by dividing the amount by the known sample weight. If the measured activities are due to the presence of two radionuclides having different half-lives, a plot of the logarithm of A versus time will be a curve rather than a

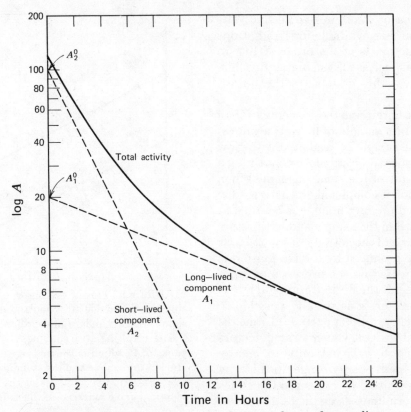

FIGURE 4.9   Decay curve due to simultaneous decay of two radionu-clides having different half-lives. The activity observed initially is primarily due to the short-lived component (A_2). After it has decayed, the activity is mainly due to the long-lived component (A_1), and the plot of log A versus t then becomes a straight line. The straight tail of such a composite decay curve can be extrapolated back to $t = 0$ to determine the initial activity of the long-lived component. The decay curve of the short-lived component (A_2) is then obtained by subtracting the activity of A_1 from the total observed activity. Note that the activity is plotted as log A as in Figure 4.4.

straight line. Such a curve can be resolved into its two components as shown in Figure 4.9.

We now calculate the disintegration rate due to sodium in a 1 gram sample of rock that is irradiated with slow neutrons for one hour. Sodium has only one stable isotope which is $^{23}_{11}\text{Na}$. The interaction of this isotope with thermal neutrons results in the formation of

$^{24}_{11}\text{Na}$ by the following reaction:

$$^{23}_{11}\text{Na}(n, \gamma)^{24}_{11}\text{Na}$$

Sodium-24 is radioactive and decays to $^{24}_{12}\text{Mg}$ by emitting beta particles and several gamma rays. The half-life of $^{24}_{11}\text{Na}$ is 15.0 hours. The concentration of sodium in the sample is 5.0 percent, the neutron flux is 1×10^{12} n/cm^2/ sec, the thermal neutron capture cross section of $^{23}_{11}$Na is 0.54 barns, and the atomic weight of sodium is 22.989 amu. The production rate R of $^{24}_{11}$Na is

$$R = Na\sigma F$$

$$R = \frac{0.05 \times 6.0225 \times 10^{23} \times 0.54 \times 10^{-24} \times 10^{12}}{22.989} \text{ atoms/sec}$$

$$R = 7.07 \times 10^8 \text{ atoms/sec}$$

The disintegration rate at the end of a one hour irradiation is calculated from

$$\lambda P = R(1 - e^{-\lambda t})$$
$$\lambda P = 7.07 \times 10^8(1 - e^{-0.693 \times 1/15.0})$$
$$= 3.19 \times 10^7 \text{ dis/sec}$$

Note that the irradiation time and the half-life must be in seconds. In our example the conversion factors cancel. The disintegration rate after a one hour irradiation can also be expressed as a fraction of the saturation activity:

$$\frac{\lambda P}{R} = \frac{3.19 \times 10^7}{7.07 \times 10^8} = 4.51 \times 10^{-2}, \text{ or } 4.51\%$$

The irradiation time was obviously insufficient to achieve a disintegration rate approaching saturation. Nevertheless, the radioactivity of the sample due to the presence of $^{24}_{11}$Na is appreciable. Expressed in millicuries, the activity of this isotope in the sample is

$$\lambda P = \frac{3.19 \times 10^7}{3.70 \times 10^7} = 0.86 \text{ mCi}$$

If the detection coefficient of the counter is $c = 0.20$, the observed activity would be

$$A = c\lambda P = 0.2 \times 3.19 \times 10^7$$
$$= 6.38 \times 10^6 \text{ counts/sec}$$

When one irradiates a sample of rock with slow neutrons "(n, γ)" reactions will occur with all of the nuclides that are present and many of the resulting product nuclides will be radioactive. In order to estimate the total disintegration rate to be expected from the sample when it is withdrawn from the reactor, it is necessary to calculate the activities due to all of the major elements present, as well as some of the trace elements such as the rare earth elements, which have large neutron-capture cross sections.

PROBLEMS

1. Calculate the fraction of atoms remaining of $^{24}_{11}$Na ($T_{1/2} = 15.0$ hr) after a decay interval of 5.0 hr. (**ANSWER:** 0.7937.)

2. Calculate the fraction of $^{24}_{11}$Na remaining after a decay interval of three half-lifes. (**ANSWER:** 0.125.)

3. Plot a decay curve for $^{24}_{11}$Na ($T_{1/2} = 15.0$ hr) in terms of N and $T_{1/2}$, assuming $N_0 = 128$ at $t = 0$.

4. Take logarithms to the base e of Equation 4.7 and plot the decay curve for $^{24}_{11}$Na ($T_{1/2} = 15.0$ hr) in terms of ln N and t, assuming $N_0 = 128$ at $t = 0$. What is the significance of the slope of the line?

5. Take logarithms to the base 10 of Equation 4.7 and plot the decay curve of $^{24}_{11}$Na ($T_{1/2} = 15.0$ hr) in terms of log N and t, assuming that $N_0 = 128$ atoms at $t = 0$. (Note that ln $x = 2.30258$ log x). What is the significance of the slope of the line?

6. Plot the number of radiogenic daughters (D^*) produced by decay of parent N

during a decay interval equal to five half-lives assuming that $N_0 = 128$ atoms and that $D_0 = 0$.

7. Calculate the age of a chemical system containing a radioactive nuclide N whose half-life is 5.00×10^6 years given that the ratio D^*/N of that system at the present time is equal to 1.75. (**ANSWER**: 7.30×10^6 years.)

8. Calculate the number of atoms of a radioactive daughter (N_2) that would be present as a result of decay of its parent (N_1) after a decay interval of 15 hours, given the $N_1{}^0 = 10^4$ atoms, $\lambda_1 = 0.0693$ hr^{-1} and $\lambda_2 = 0.1386$ hr^{-1}. (**ANSWER**: 2.28×10^3 atoms.)

9. Calculate the abundance of ^{234}U in *secular equilibrium* with its parent ^{238}U, given that the half-lives of ^{238}U and ^{234}U are 4.467×10^9 y and 2.44×10^5 y, respectively, and that the abundance of ^{238}U is 99.28 percent. (**ANSWER**: 0.00542 percent.)

10. Given that the observed counting rate (activity) of a radioactive sample is 5.0×10^8 counts/sec, what is the disintegration rate in microcuries, assuming a counting efficiency of 15 percent? (**ANSWER**: 9.0×10^4 μCi.)

11. Calculate the total activity of 1 gram of purified metallic uranium in units of microcuries given the following information: ^{238}U: 99.28 percent, 4.467×10^9 y; ^{235}U: 0.720 percent, 7.036×10^8 y;

^{234}U: 0.0055 percent, 2.44×10^5 y. Compare the activity of metallic uranium to that of uranium minerals in secular radioactive equilibrium. (**ANSWER**: 6.87×10^{-1} μCi.)

12. What is the production rate of $^{24}_{11}Na$ as a result of irradiation of 1 gram of rock containing 4.3 percent Na_2O by weight with slow neutrons having a flux density of 2×10^{12} n/cm^2/sec, given that the abundance of $^{23}_{11}Na$ is 100 percent and that its neutron capture cross section is 0.536 barns? (**ANSWER**: 8.96×10^8 atoms/sec.)

13. What will be the disintegration rate due to $^{24}_{11}Na$ produced by a one hour irradiation of a rock sample weighing 0.5 gram with a thermal neutron flux of 8×10^{11} n/cm^2/sec, assuming that the rock contains 2.4 percent of sodium? The thermal neutron capture cross section of $^{23}_{11}Na$ is 0.536 barns, its abundance is 100 percent and the half-life of $^{24}_{11}Na$ is 15.0 hr. [**ANSWER**: 6.08×10^6 (dis/sec).]

14. Calculate the activity due to $^{56}_{25}Mn$ one hour after a two-hour thermal neutron irradiation of a 0.6 gram sample containing 0.25 percent MnO_2, given that σ (^{55}Mn) = 13.3 barns, $F = 2 \times 10^{12}$ n/cm^2/sec, the abundance of $^{55}_{25}Mn$ is 100 percent and the half-life of $^{56}_{25}Mn$ is 2.582 hr. Express the activity in units of microcuries. (**ANSWER**: 23.7×10^2 μCi.)

REFERENCES

Bateman, H. (1910) Solution of a system of differential equations occurring in the theory of radio active transformations. Proc. Cambridge Phil. Soc., *15*, 423.

Brunfelt, A. O., and E. Steinnes (1971) Activation analysis in geochemistry and cosmochemistry. Proc. Nato Adv. Study Inst., Oslo Universitetsforlaget, 468 p.

DeSoete, D., R. Gijbels, and J. Hoste (1972) Neutron activation analysis. John Wiley, New York, 836 p.

Friedlander, G., and J. W. Kennedy (1955) Nuclear and radiochemistry. John Wiley, New York, 468 p.

Friedlander, G., J. W. Kennedy, and J. M. Miller (1964) Nuclear and radiochemistry. (2nd ed.) John Wiley, New York, 585 p.

Hensley, W. K., W. A. Bassett, and J. R. Huizenga (1973) Pressure dependence of the radioactive decay constant of beryllium-7. Science, *181*, 1164–1165.

Holden, N. E., and F. W. Walker (1972) Chart of the nuclides. (11th ed.) Educational Relations, General Electric Co., Schenectady, N.Y. 12345.

Kaplan, I. (1955) Nuclear physics. Addison-Wesley, Reading, Mass., 609 p.

Lapp, R. E., and H. L. Andrews (1954) Nuclear radiation physics. (2nd ed.) Prentice-Hall, Englewood Cliffs, N.J., 532 p.

Mapper, D. (1960) Radioactivation analysis. In Methods of Geochemistry, A. A. Smales and L. R. Wager, eds., pp. 297–357. Interscience, New York and London, 464 p.

5 MASS SPECTROMETRY

In the course of experiments with cathode-ray tubes in 1886, Goldstein observed luminous rays which he called "Kanalstrahlen" because they passed through perforations in the cathode. In 1898, Wien demonstrated that these canal rays consisted of positively charged particles. After J. J. Thomson at the Cavandish Laboratory had discovered the electron and had identified cathode rays as streams of electrons, he investigated these positive rays and eventually used a "positive ray apparatus" to demonstrate that neon has two isotopes having atomic weights of 20 and 22. Thomson's work was followed up by F. W. Aston in England and by A. J. Dempster at the University of Chicago. Aston (1919) and Dempster (1918) designed mass spectrographs which they used in subsequent years to discover the naturally occurring isotopes of most of the elements in the periodic table and to measure their masses and their abundances. The design of mass spectrographs was further improved in the 1930s by K. T. Bainbridge, J. Mattauch, and R. Herzog. At the end of that decade, the work of discovering the naturally occurring isotopes of the elements and of measuring their masses and abundances was virtually completed. Since then, mass spectrometers have evolved into tools employed in a wide range of research problems in physics, chemistry, and biology. In addition, mass spectrometers based on a design by A. O. Nier (1940) have made possible the measurement and interpretation of variations in the isotopic compositions of certain elements in natural materials and thus have permitted the spectacular growth of isotope geology. The design of modern mass spectrometers has

been treated by Inghram and Hayden (1954), Duckworth (1958), McDowell (1963), White (1968), Milne (1971), and others.

1
The Principles of
Mass Spectrometry

A mass spectrometer is an instrument designed to separate charged atoms and molecules on the basis of their masses based on their motions in electrical and/or magnetic fields. Mass spectrometers employ electronic methods of detection of the separated ions, while in mass spectrographs they are detected photographically or by other non-electrical methods. Most of the mass spectrometers currently in use in isotope geology have evolved from the work of Dempster and Bainbridge and follow the design of Nier (1940) whose mass spectrometers achieved a level of accuracy and reliability of operation that set the standard for mass spectrometry.

The modern Nier-type mass spectrometer consists of three essential parts: (1) a source of a positively charged mono-energetic beam of ions; (2) a magnetic analyzer; and (3) an ion collector. All three parts of the mass spectrometer are evacuated to pressures of the order of 10^{-6} to 10^{-9} mm Hg. Both gaseous as well as solid samples can be analyzed, depending on the design of the ion source. For a mass analysis of a gaseous sample, such as argon or carbon dioxide, the sample gas is allowed to leak into the source through a small orifice. The molecules are then ionized by bombardment with electrons. The resulting positively charged ions are accelerated by an adjustable voltage and are collimated into a beam by means of suitably spaced slit plates. For the mass analysis of solid samples, a salt of the element is deposited on a filament in the source. The filament (composed of Ta, Re, or W) is heated electrically to a temperature sufficient to ionize the element to be analyzed. The resulting ions are then accelerated and collimated into a beam as before.

The unresolved ion beam enters a magnetic field generated by an electromagnet whose pole pieces are carefully shaped and positioned in such a way that the magnetic field lines are perpendicular to the direction of travel of the ions. The magnetic field deflects the ions into circular paths whose radii are proportional to the masses of the isotopes, that is, the heavier ions are deflected less than the lighter ones. The pole pieces of the magnet are shaped like prisms so that the ions continue on straight paths as they leave the magnetic field. The separated ion beams continue through the analyzer tube to the collector.

The ion collector consists of a metallic cup positioned behind a slit plate. The accelerating voltage in the source and the magnetic field are adjusted in such a way that one of the resolved ion beams is focused through the collector slit and enters the cup while the other isotopic ion beams are neutralized on the collector plate. The beam that enters the collector cup is neutralized by electrons that flow from ground to the collector through a resistor (10^{10} to 10^{12} ohms). The voltage difference generated across the terminals of this resistor is magnified and is measured by a very sensitive voltmeter, such as a vibrating reed electrometer. The output from the voltmeter is fed to a strip chart recorder or is converted into a digital output.

A mass analysis of an element (or compound) consisting of several isotopes (or

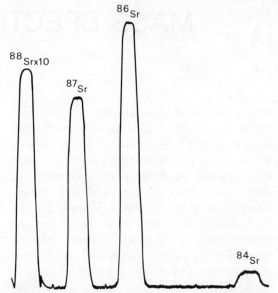

FIGURE 5.1 Mass spectrum of strontium obtained on a 60° sector, 15.24 cm radius Nier-type mass spectrometer equipped with a single-filament source for analysis of solid samples (Nuclide Corp., Model 6-60-S). The spectrum was scanned by continuous variation of the magnetic field at a constant rate and was traced on a linear strip chart recorder. Note that the height of the ^{88}Sr peak has been reduced by one-tenth. The heights of the peaks are proportional to the abundances of the respective isotopes.

isotopic masses) is obtained by varying either the magnetic field or the accelerating voltage in such a way that the separated ion beams are focused into the collector in succession. The resulting signal, traced out by a strip chart recorder, consists of a series of peaks and valleys that form the mass spectrum of the element. Each peak represents a discrete isotope whose abundance is proportional to the relative height of the peak in the mass spectrum. Figure 5.1 is such a mass spectrum of strontium showing the stable isotopes ^{88}Sr, ^{87}Sr, ^{86}Sr, and ^{84}Sr.

2
The Equations of Motion of Ions in a Mass Spectrometer

When an ion of mass m and charge e is acted upon by a potential difference of V volts it acquires energy E equal to

$$E = eV = \tfrac{1}{2}mv^2 \qquad (5.1)$$

where v is its velocity. All ions having the same charge emerge from the exit slit of the source with the same kinetic energy because they were accelerated by the same potential difference V. However, ions of different masses have differing velocities:

$$v = \sqrt[2]{\frac{2eV}{m}} \qquad (5.2)$$

When the ions enter the magnetic field, they are deflected into circular paths subject to the condition:

$$Bev = m\frac{v^2}{r} \qquad (5.3)$$

where B is the strength of the magnetic field and r is the radius of their paths. By eliminating v from Equations 5.1 and 5.3, we obtain

$$\frac{2eV}{m} = \frac{B^2 e^2 r^2}{m^2}$$

$$\frac{m}{e} = \frac{B^2 r^2}{2V} \qquad (5.4)$$

If B is measured in gauss, r in centimeters, m in atomic mass units, V in volts, and e in units of electronic charge, the equation becomes

$$\frac{m}{e} = 4.825 \times 10^{-5}\frac{B^2 r^2}{V} \qquad (5.5)$$

Equation 5.5 may be solved for either r or B for convenience:

$$r = \frac{143.95}{B}\sqrt[2]{\frac{mV}{e}} \qquad (5.6)$$

$$B = \frac{143.95}{r}\sqrt[2]{\frac{mV}{e}} \qquad (5.7)$$

We can use Equation 5.7 to calculate the magnetic field strength required to focus ions of $^{40}\text{Ca}^+$ into the collector of a mass spectrometer whose analyzer tube has a radius of 30.48 cm (12 inches), assuming $V = 1000$ volts. Since the charge of the ^{40}Ca ion is $e = 1$ electronic charge unit and its mass $m = 39.9626$ amu, $B = 944$ gauss. Equation 5.6 indicates that either the accelerating voltage or the magnetic field can be adjusted to force any ion of mass m and charge e into a path of radius r. The equation also shows that if B and V are constants, the radius of the path of ions having unit charge is proportional to the square root of the mass, that is,

$$r \propto \sqrt[2]{m} \qquad (5.8)$$

Consequently, ions of large mass are deflected into paths having larger radii than lighter ions. In other words, heavy isotopes are deflected from a straight-line path less than light isotopes. An important feature of Nier-type mass spectrometers is the wedge-shaped magnetic field that is achieved by the design of the pole pieces of the electromagnet. In order to focus the ion beam into the collector, the exit slit, the apex of the pole pieces, and the collector slit must lie on a straight line as shown in Figure 5.2.

Nier used his mass spectrometers to determine the isotopic compositions of many elements, including Ca, Ti, S, Ar, Sr, Ba, Bi, Tl, Hg, Ne, Kr, Rb, Xe, C, N, O, and K (Nier, 1938a, 1938b, 1950a, 1950b). Moreover, he measured the isotopic compositions of lead from various ore deposits and made a lasting contribution to the study of the isotopic evolution of lead in the solar system and to the determination of the age of the Earth (Nier, 1938c, 1939a, 1939b; Nier et al.,

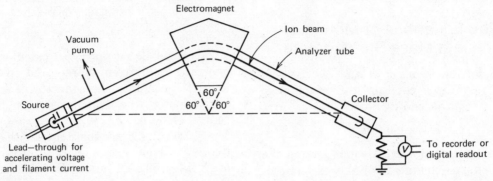

FIGURE 5.2 Schematic diagram of a 60° sector mass spectrometer showing arrangement of ion source, electromagnet, and collector.

1941). Nier and Gulbranson (1939) first reported the variation of the isotopic composition of carbon in nature, and subsequently Nier designed mass spectrometers for the simultaneous detection of two ion beams which permit isotope ratios to be measured directly (Nier et al., 1947; Nier, 1947). Nier's isotope ratio mass spectrometer was further improved by McKinney et al. (1950) and has made possible the precise determination of isotope ratios of D/H, $^{13}C/^{12}C$, $^{18}O/^{16}O$ and $^{34}S/^{32}S$ in geological and biological materials. In recent years, mass spectrometers have been further improved primarily by incorporation of digital data-collection systems using computers (Hagan and deLaeter, 1966; Weichert et al., 1967; Arriens and Compston, 1968; Wasserburg et al., 1969; Stacey et al., 1971, 1972).

3
Isotope Dilution Analysis

Mass spectrometers can be used to determine the concentration of an element in a sample by means of an analytical technique called isotope dilution (Webster, 1960; Moore et al., 1973). This method is based on the determination of the isotopic composition of an element in a mixture of a known quantity of a "spike" with an unknown quantity of the normal element. The spike is a solution (liquid or gaseous) containing a known concentration of a particular element whose isotopic composition has been changed by enrichment of one of its naturally occurring isotopes. The sample to be analyzed contains an unknown concentration of the element whose isotopic composition is known. Therefore, when a known amount of a sample solution is mixed with a known amount of spike, the isotopic composition of the mixture can be used to calculate the amount of the element in the sample solution. Isotope dilution analysis applies to all elements that have two or more naturally occurring isotopes, provided that a spike enriched in one of the isotopes of that element is available. The isotope enrichment is done by means of a large electromagnetic separator called a calutron which is capable of separating the isotopes of an element quantitatively.

When the element to be analyzed is nonvolatile, the spike is prepared in the form of a solution whose concentration and isotopic composition are verified separately. A known weight (or volume) of this spike solution is then added to a known weight (or volume) of the sample solution, and the mixture is stirred to achieve complete isotopic homogenization. The mixture of the normal element and the spike is then analyzed on a mass spectrometer to determine its isotopic composition. The result, expressed in terms of the ratio of the abundances of two isotopes, is used to calculate the concentration of the element in question in the sample solution.

Let R be the ratio of the abundances of two isotopes A and B of an element and let N and S be the numbers of atoms of the normal element and of the spike, respectively, in the mixture. Then

$$R_m = \frac{Ab_N^A N + Ab_S^A S}{Ab_N^B N + Ab_S^B S} \qquad (5.9)$$

where Ab_N^A is the abundance of isotope A in the normal element, and so forth and R_m is the ratio of isotopes A to B in the mixture. R_m is measured, leaving N as the only unknown variable in Equation 5.9. Solving for N yields

$$N = S\left[\frac{Ab_S^A - R_m Ab_S^B}{R_m Ab_N^B - Ab_N^A}\right] \qquad (5.10)$$

where N and S are in terms of numbers of atoms. Next, we convert to weights of N and S by dividing by Avogadro's number and by multiplying by the respective atomic weights (W_N and W_S). (Note that W_N and W_S are not equal because of the difference in the isotopic compositions of the normal element in the sample and its spike). Since $N_W = N \times W_N/A$ and $S_W = S \times W_S/A$, we find that $N = N_W \times A/W_N$ and $S = S_W \times A/W_S$ where A is Avogadro's number, N and S are in terms of numbers of atoms, and N_W and S_W are the weights of the normal element and the spike, respectively. Substituting into Equation 5.10 yields

$$N_W = \frac{S_W \times W_N}{W_S}\left[\frac{Ab_S^A - R_m Ab_S^B}{R_m Ab_N^B - Ab_N^A}\right] \qquad (5.11)$$

where N_W and S_W are the weights of the normal element and the spike in the mixture in units of grams, milligrams, or micrograms, depending on the most convenient way of expressing the concentration of the spike solution. Finally, we obtain the concentration of the normal element in the solution by dividing N_W by the weight or volume of the sample solution.

Let us illustrate these calculations by means of an example. Suppose we want to determine the concentration of rubidium in a solution prepared by dissolving a known weight of powdered rock sample. Rubidium has two naturally occurring isotopes, ^{87}Rb and ^{85}Rb, whose natural abundances are 27.83 percent and 72.17 percent, respectively. The abundances of ^{87}Rb and ^{85}Rb in the spike are 99.40 percent and 0.60 percent, respectively. Suppose we add 3.50 grams of a spike solution containing 7.50 μg/gram of rubidium enriched in ^{87}Rb to 50 grams of the sample solution prepared by dissolving 0.25 grams of the rock sample to be analyzed. After the solutions have thoroughly mixed, the rubidium is separated by cation exchange chromatography and is then analyzed on a "solid source" mass spectrometer (Faure and Powell, 1972). The results indicate that the ^{87}Rb/^{85}Rb ratio of the mixture is $R_m = 1.55$. According to the statement of the problem: $S_w = 26.26$ μg, $Ab_N^A = 0.2783$ (^{87}Rb), $Ab_N^B = 0.7217$ (^{85}Rb), $Ab_S^A = 0.9940$, $Ab_S^B = 0.0060$, $W_N = 85.4677$, and $W_S = 86.8971$. The atomic weight of the spike rubidium must be calculated from its isotopic composition and the atomic masses of its naturally occurring isotopes (Chapter 2.2). Substituting into Equation 5.11, we find that

$$N_w = \frac{26.25 \times 85.4677}{86.8971}$$

$$\times \left[\frac{0.9940 - 1.55 \times 0.0060}{1.55 \times 0.7217 - 0.2783}\right]$$

$$N_w = 30.25 \ \mu g$$

Therefore, the concentration of rubidium in the sample solution is 30.25/50 = 0.6050 μg/ gram, and the rubidium concentration of the rock is 30.25/0.25 = 121.0 μg/gram or parts per million by weight (ppm).

Isotope dilution has several important advantages over other analytical methods: (1) it is free of interference from other elements because isobaric isotopes of other elements can be removed chemically prior to the mass spectrometric analysis; (2) it has great sensitivity because the amount of sample can be increased as the concentration of the desired element decreases; (3) it is potentially a very accurate analytical method depending only on the calibration of the spike solution; (4) quantitative recovery of the mixture of spike and the normal element is not required, provided isotopic homogenization has first taken place. On the other hand, the method has its problems which include the following: (1) The concentration and isotopic composition of the spike solution must be known accurately. The calibration of spike solutions is subject to errors that may arise from the possible nonstoichiometric composition of the compound containing the spike, from isotope fractionation during the isotope analysis of the spike on a mass spectrometer, and from errors in weighing and diluting the spike. (2) The measured value of the isotope ratio of the mixture may be affected by fractionation in the mass spectrometer. If the element has only two naturally occurring isotopes, this error cannot be corrected. (3) The normal element and the spike must mix completely. This may be difficult to achieve in some geological samples in which the element to be determined resides in refractory minerals or when the element is incorporated into an insoluble precipitate during solution of the sample. (4) The concentration of the spike solution may change as a function of time because of evaporation of water (Hamilton, 1962) or by adsorption of the element onto the walls of the container used to store the spike. (5) The isotopic composition of the spike or of the mixture of the normal element and the spike may be changed during processing of the sample as a result of isotope exchange reactions with the isotopes of that element in the walls of the container. (6) The ratio of N/S must be optimized so as to avoid magnification of errors or loss of sensitivity of the functional relationship between R and N/S (Equation 5.9). (7) The method is time-consuming which discourages replication of analyses and thereby results in inadequate documentation of analytical errors. Nevertheless, isotope dilution is indispensable in age determinations of rocks and minerals based on radioactive decay.

Isotope dilution has another important advantage over other analytical methods. When elements are determined which have more than two naturally occurring isotopes, one can measure two or more isotopic ratios from which not only the concentration but also the isotopic composition of the normal element can be calculated. Such treatment is especially useful in studies of the isotopic compositions of strontium and lead, both of which have four naturally occurring isotopes. The procedures for making such calculations have been discussed by Long (1966), Boelrijk (1968), Krogh and Hurley (1968), Dodson (1970), Gale (1970), and Russell (1971).

PROBLEMS

1. What must be the accelerating potential in order to focus $^{206}Pb^+$ ions into the collector of a Nier-type mass spectrometer under the following conditions: $B = 2000$ gauss, $r = 30.48$ cm, $m = 205.9744$ amu? (**ANSWER:** 870.5 volts.)

2. What is the mass of an ion that is focused into the collector of a mass spectrometer under the following conditions: $B = 5000$ gauss, $r = 15.24$ cm, $V = 3187$ volts? (**ANSWER:** 87.90 amu.)

3. Calculate the radius of the path of $^{87}Sr^+$ ions when they are acted upon by an accelerating potential of 4995 volts and 3112 gauss, given that the mass of ^{87}Sr is 86.9089 amu. (**ANSWER:** 30.48 cm.)

4. Calculate the Rb concentration of a rock sample from the following information pertaining to an isotope dilution analysis: $R_m = {}^{87}Rb/{}^{85}Rb = 1.12$, $S_w = 29.45$ μg, $Ab_S^{87} = 95.4$ percent, $Ab_S^{85} = 4.6$ percent, weight of sample $= 0.35$ gram, masses of Rb isotopes: $^{87}Rb = 86.9092$ amu, $^{85}Rb = 84.9117$ amu. (**ANSWER:** 141.0 ppm.)

5. Calculate the total Sr concentration of a sample of muscovite having the following isotopic composition: $^{87}Sr/{}^{86}Sr = 5.30$, $^{86}Sr/{}^{88}Sr = 0.1194$, $^{84}Sr/{}^{88}Sr = 0.0068$. The isotopic composition of the spike is $^{88}Sr = 10.00$ percent, $^{87}Sr = 2.50$ percent, $^{86}Sr = 87.49$ percent, $^{84}Sr = 0.01$ percent, and the concentration of the spike solution is 3.55 $\mu g/gram$. The weight of muscovite was 1.25 gram, the amount of spike solution added was 5.05 gram, and $R_m = {}^{86}Sr/{}^{88}Sr = 2.05$. The masses of the Sr isotopes are: $^{88}Sr = 87.9056$ amu, $^{87}Sr = 86.9089$ amu, $^{86}Sr = 85.9092$ amu, $^{84}Sr = 83.9134$ amu. (**ANSWER:** 8.88 ppm.)

6. A sample of biotite is reported to contain 15.5 ppm of Sr, 265.4 ppm of Rb, and has an $^{87}Sr/{}^{86}Sr$ ratio of 2.25. What is its age, assuming that its initial $^{87}Sr/{}^{86}Sr$ ratio is 0.70 and $\lambda({}^{87}Rb) = 1.39 \times 10^{-11}$ y^{-1}? Assume that the other isotopic ratios of Sr in the biotite are: $^{86}Sr/{}^{88}Sr = 0.1194$, $^{84}Sr/{}^{88}Sr = 0.0068$. The masses of the Sr isotopes are given in Problem 5 and the abundances of the Rb isotopes are: $^{87}Rb = 27.83$ percent, $^{85}Rb = 72.17$ percent, atomic weight: 85.4677. (**ANSWER:** 1.92×10^9 y.)

REFERENCES

Arriens, P. A., and W. Compston (1968) A method for isotopic ratio measurement by voltage peak switching, and its application with digital input. Int. J. Mass Spectrom. Ion Phys., *1*, 471–481.

Aston, W. F. (1919) A positive ray spectrograph. Phil. Mag., VI, *38*, 707–714.

Bainbridge, K. T., and E. B. Jordan (1936) Mass spectrum analysis. Harvard Univ. Jefferson Phys. Lab. Contrib., Ser. 2, *3*, No. 2.

Boelrijk, N. A. I. M. (1968) A general formula for "double" isotope dilution analysis. Chem. Geol., *3*, 323–325.

Dempster, A. J. (1918) A new method of positive ray analysis. Phys. Rev., *11*, 316–325.

Dodson, M. H. (1970) Simplified equations for double-spiked isotopic analyses. Geochim. Cosmochim. Acta, *34*, 1241–1244.

Duckworth, H. E. (1958) Mass spectroscopy. Cambridge University Press, 206 p.

Faure, G., and J. L. Powell (1972) Strontium isotope geology. Springer-Verlag, Berlin, Heidelberg, and New York, 188 p.

Gale, N. H. (1970) A solution in closed form for lead isotopic analysis using a double spike. Chem. Geol., *6*, 305–310.

Hagan, P. J., and J. R. deLaeter (1966) Mass spectrometric data processing using a time-shared computer. J. Sci. Instrum., *43*, 662–664.

Hamilton, E. I. (1962) Storage of standard solutions in polyethylene bottles. Nature, *193*, No. 4811, 200.

Herzog, R. (1934) Ionen-und elektronenoptische Zylinderlinsen und Prismen. Z. Phys., *89*, 447–473.

Herzog, R., and J. Mattauch (1934) Theoretische Untersuchungen zum Massenspektrometer ohne Magnetfeld. Ann. Phys., Ser. 5, *19*, 345–386.

Inghram, M. G., and R. J. Hayden (1954) Mass spectroscopy. Nucl. Sci. Ser., Rept. No. 14, Nat. Aca. Sci.-Nat. Res. Council, 51 p.

Krogh, T. E., and P. M. Hurley (1968) Strontium isotope variation and whole-rock isochron studies, Grenville Province, Ontario. J. Geophys. Res., *73*, 7107–7125.

Long, L. E. (1966) Isotope dilution analysis of common and radiogenic strontium using ^{84}Sr-enriched spike. Earth Planet. Sci. Letters, *1*, 289–292.

McDowell, C. A. (1963) Mass spectrometry. McGraw-Hill, New York.

McKinney, C. R., J. M. McCrea, S. Epstein, H. A. Allen, and H. C. Urey (1950) Improvements in mass spectrometers for the measurement of small differences in isotope abundance ratios. Rev. Sci. Instrum., *21*, 724–730.

Milne, G. W., ed. (1971) Mass spectrometry: Techniques and applications. Wiley Interscience, New York, 521 p.

Moore, L. J., J. R. Moody, I. L. Barnes, J. W. Gramlich, T. J. Murphy, P. J. Paulsen, and W. R. Shields (1973) Trace determination of rubidium and strontium in silicate glass standard reference materials. Anal. Chem., *45*, 2384–2387.

Nier, A. O. (1938a) The isotopic constitution of calcium, titanium, sulphur and argon. Phys. Rev., *53*, 282–286.

Nier, A. O. (1938b) Isotopic constitution of strontium, barium, bismuth, thallium and mercury. Phys. Rev., *54*, 275–278.

Nier, A. O. (1938c) Variations in the relative abundances of the isotopes of common lead from various sources. J. Amer. Chem. Soc., *60*, 1571–1576.

Nier, A. O., and E. A. Gulbransen (1939) Variations in the relative abundance of the carbon isotopes. J. Amer. Chem. Soc., *61*, 697–698.

Nier, A. O. (1939a) The isotopic contitution of uranium and the half-lives of the uranium isotopes I. Phys. Rev., *55*, 150–153.

Nier, A. O. (1939b) The isotopic constitution of radiogenic leads and the measurement of geologic time II. Phys. Rev., *55*, 153–163.

Nier, A. O. (1940) A mass spectrometer for routine isotope abundance measurements. Rev. Sci. Instrum., *11*, 212–216.

Nier, A. O., R. W. Thompson, and B. F. Murphey (1941) The isotopic constitution of lead and the measurement of geologic time III. Phys. Rev., *60*, 112–116.

Nier, A. O., E. P. Ney, and M. G. Inghram (1947) A null method for the comparison of two ion currents in a mass spectrometer. Rev. Sci. Instrum., *18*, 294–297.

Nier, A. O. (1947) A mass spectrometer for isotope and gas analysis. Rev. Sci. Instrum., *18*, 398–411.

Nier, A. O. (1950a) A redetermination of the relative abundances of the isotopes of carbon, nitrogen, oxygen, argon and potassium. Phys. Rev., *77*, 789–793.

Nier, A. O. (1950b) A redetermination of the relative abundances of the isotopes of neon, krypton, rubidium, xenon, and mercury. Phys. Rev., *79*, 450–454.

Russell, R. D. (1971) The systematics of double spiking. J. Geophys. Res., *76*, 4949–4955.

Stacey, J. S., E. E. Wilson, Z. E. Peterman, and R. Terrazas (1971) Digital recording of mass spectra in geologic studies. I. Can. J. Earth Sci., *8*, 371–377.

Stacey, J. S., E. E. Wilson, and R. Terrazas (1972) Digital recording of mass spectra in geologic studies. Can. J. Earth Sci., *9*, 824–834.

Wasserburg, G. J., D. A. Papanastassiou, E. V. Nenow, and C. A. Baum (1969) A programmable magnetic field mass spectrometer with on-line data processing. Rev. Sci. Instrum., *40*, 288–295.

Webster, R. K. (1960) Mass spectrometric isotope dilution analysis. In Methods in Geochemistry, A. A. Smales and L. R. Wager, eds., pp. 202–246. Interscience, New York and London, 464 p.

Weichert, D. H., R. D. Russell, and J. Blenkinsop (1967) A method for digital recording for mass spectra. Can. J. Phys., *45*, 2609–2619.

White, F. A. (1968) Mass spectrometry in science and technology. John Wiley, New York, 352 p.

6 THE Rb-Sr METHOD OF DATING

Although the natural radioactivity of rubidium was demonstrated in 1906 by Campbell and Wood, over 30 years elapsed before ^{87}Rb was identified as the naturally occurring radioactive isotope (Hahn et al. 1937; Mattauch, 1937). The feasibility of dating Rb-bearing minerals by the decay of ^{87}Rb to ^{87}Sr was discussed by Hahn and Walling (1938), and the first age determination by this method followed a few years later (Hahn et al., 1943). However, the Rb-Sr method of dating did not come into wide use until the 1950s when mass spectrometers based on Nier's design became available for the isotopic analysis of solids, and the concentrations of rubidium and strontium could be measured by isotope dilution combined with the separation of these elements by cation exchange chromatography. The entire subject of dating by the Rb-Sr method, including its history, theoretical basis, and applicability, have been presented in detail by Faure and Powell (1972).

1 Geochemistry of Rubidium and Strontium

Rubidium is an alkali metal belonging to Group IA which consists of lithium, sodium, potassium, rubidium, cesium, and francium. Its ionic radius (1.48 Å) is sufficiently similar to that of potassium (1.33 Å) to allow rubidium to substitute for potassium in all K-bearing minerals. Consequently, rubidium is a dispersed element that does not form any minerals of its own, but it occurs in easily detectable amounts in common K-bearing

minerals such as the micas (muscovite, biotite, phlogopite, and lepidolite), K-feldspar (orthoclase and microcline), certain clay minerals, and in evaporite minerals such as sylvite and carnallite.

Rubidium has two naturally occurring isotopes $^{85}_{37}$Rb and $^{87}_{37}$Rb whose isotopic abundances are 72.1654 percent and 27.8346 percent, respectively. Its atomic weight is 85.46776 amu (Catanzaro et al., 1969). ^{87}Rb is radioactive and decays to stable ^{87}Sr by emission of a negative beta particle as shown by Equation 6.1:

$$^{87}_{37}\text{Rb} \rightarrow {}^{87}_{38}\text{Sr} + \beta^- + \bar{\nu} + Q \qquad (6.1)$$

where β^- is the beta particle, $\bar{\nu}$ is an antineutrino, and Q is the decay energy (see Chapter 3). The decay energy is only 0.275 MeV which has caused problems in the determination of the specific decay rate of this isotope.

Strontium is a member of the alkaline earths of Group IIA which consists of beryllium, magnesium, calcium, strontium, barium, and radium. Its ionic radius (1.13 Å) is slightly larger than that of calcium (0.99 Å) which it can replace in many minerals. Thus strontium is also a dispersed element and occurs in Ca-bearing minerals such a plagioclase, apatite, and calcium carbonate, especially aragonite. The ability of strontium to replace calcium is somewhat restricted by the fact that strontium ions (Sr^{+2}) favor eightfold coordinated sites, whereas calcium ions (Ca^{+2}) can be accommodated in both six and eightfold coordinated lattice sites because of their smaller size. Moreover, Sr^{+2} ions can be captured in place of K^{+1} ions by K-feldspar, but the replacement of K^{+1} by Sr^{+2} must be coupled by the replacement of Si^{+4} by Al^{+3} in order to preserve electrical neutrality. Strontium is the major cation in strontianite ($SrCO_3$) and celestite ($SrSO_4$), both of which occur in certain hydrothermal deposits and in carbonate rocks.

Strontium has four naturally occurring isotopes ($^{88}_{38}Sr$, $^{87}_{38}Sr$, $^{86}_{38}Sr$, and $^{84}_{38}Sr$), all of which are stable. Their isotopic abundances are 82.53 percent, 7.04 percent, 9.87 percent, and 0.56 percent, respectively. Actually, the isotopic abundances of strontium isotopes are variable because of the formation of radiogenic ^{87}Sr by the decay of naturally occurring ^{87}Rb. For this reason, the precise isotopic composition of strontium in a rock or mineral that contains rubidium depends on the age and Rb/Sr ratio of that rock or mineral.

The average concentrations of rubidium, potassium, strontium, and calcium in different kinds of igneous and sedimentary rocks are shown in Table 6.1. These data illustrate the general geochemical coherence of rubidium and potassium and of strontium and calcium. The rubidium concentrations of common igneous and sedimentary rocks range from less than 1 part per million (ultramafic rocks and carbonates) to more than 170 ppm in low-calcium granitic rocks. The concentrations of strontium range from a few parts per million (ultramafic rocks) to about 465 ppm in basaltic rocks and reach very high values in carbonate rocks (up to 2000 ppm or more). Evidently, most common rocks contain appreciable concentrations of rubidium and strontium of the order of tens to several hundred parts per million. The Rb/Sr ratios of common igneous rocks range between wide limits from 0.06 (basaltic rocks) to 1.7, or more, in highly differentiated granitic rocks having low calcium concentrations.

During the fractional crystallization of magma, strontium tends to be concentrated in early formed calcic plagioclase, while rubidium remains in the liquid phase. Consequently, the Rb/Sr ratio of the residual magma may increase gradually in the course of progressive crystallization. Suites of differentiated igneous rocks therefore tend to have increasing Rb/Sr ratios with increasing degree of differentiation. The highest Rb/Sr ratios, amounting to 10 or higher,

Table 6.1 **Average Concentrations of Rubidium, Potassium, Strontium, and Calcium in Igneous and Sedimentary Rocks (Turekian and Wedepohl, 1961)**

	ROCK TYPE	Rb ppm	K ppm	Sr ppm	Ca ppm
1.	Ultrabasic	0.2	40	1	25,000
2.	Basaltic	30	8,300	465	76,000
3.	High Ca granitic	110	25,200	440	25,300
4.	Low Ca granitic	170	42,000	100	5,100
5.	Syenite	110	48,000	200	18,000
6.	Shale	140	26,600	300	22,100
7.	Sandstone	60	10,700	20	39,100
8.	Carbonate	3	2,700	610	302,300
9.	Deep sea carbonate	10	2,900	2000	312,400
10.	Deep sea clay	110	25,000	180	29,000

occur in late-stage differentiates including pegmatites.

2
Dating of Rb-Bearing Minerals in Igneous Rocks

The growth of radiogenic ^{87}Sr in a Rb-rich mineral can be described by an equation derivable from the law of radioactivity (Chapter 4). The total number of atoms of ^{87}Sr in a mineral whose age is t years is obtained from Equation 4.15:

$$^{87}Sr = {}^{87}Sr_0 + {}^{87}Rb(e^{\lambda t} - 1) \quad (6.2)$$

where ^{87}Sr is the total number of atoms of this isotope in a unit weight of the mineral at the present time; $^{87}Sr_0$ is the number of atoms of this isotope that was incorporated into the same unit weight of this mineral at the time of its formation; ^{87}Rb is the number of atoms of this isotope in a unit weight of the mineral at the present time; λ is the decay constant of ^{87}Rb in units of reciprocal years; and t is the time elapsed in years since the time of formation of the minerals, that is, t is the "age" of the mineral. We can modify Equation 6.2 by dividing each term by the number of ^{86}Sr atoms which is constant because this isotope is stable and is not produced by decay of a naturally occurring isotope of another element. Thus we obtain

$$\frac{^{87}Sr}{^{86}Sr} = \left(\frac{^{87}Sr}{^{86}Sr}\right)_0 + \frac{^{87}Rb}{^{86}Sr}\,(e^{\lambda t} - 1) \quad (6.3)$$

This equation is the basis for age determinations by the Rb-Sr method. Equation 6.3 is valid only when the numbers of ^{87}Sr and ^{87}Rb atoms in the mineral have changed as a result of radioactive decay. In other words, the mineral must be a "closed system" with respect to rubidium and strontium. If this condition is not satisfied and rubidium and strontium have been added to or lost from the mineral during its lifetime, then the value of t calculated by solution of Equation 6.3 may have no real meaning.

The half-life of ^{87}Rb has been measured many times with somewhat contradictory results. Most geochronologists have used a value of $(5.0 \pm 0.2) \times 10^{10}$ years that was obtained by Aldrich et al. (1956) from a comparison of Rb-Sr dates of micas with concordant U-Pb dates in coexisting uraninites and monazites from six pegmatites. However, several more recent direct determinations have indicated somewhat lower values ranging from 4.7 to 4.8×10^{10} years (Flynn and Glendenin, 1959; Kovach, 1964; McMullen et al., 1966). The most recent determination is by Neumann and Huster (1974) who reported a half-life of $(4.88 \pm {}^{0.06}_{0.10}) \times 10^{10}$ years. This is probably the best direct measurement of the half-life and may be adopted by geochronologists in the future. We shall continue to use a half-life of 5.0×10^{10} years in this book because it is firmly embedded in the literature and because it generally brings Rb-Sr dates into agreement with K-Ar and U-Pb dates of coexisting minerals in situations in which such agreement is to be expected for geological reasons. The value of the decay constant that corresponds to a half-life of 5.0×10^{10} years is

$$\lambda = \frac{0.693}{5.0 \times 10^{10}} = 1.39 \times 10^{-11}\ y^{-1}$$

A complete listing of all half-life values reported prior to 1964 was compiled by Heier and Adams (1965) and by Hamilton (1965).

In order to solve Equation 6.3 for t and thus to date a Rb-bearing mineral, the concentrations of rubidium and strontium and its $^{87}Sr/^{86}Sr$ ratio must be measured. The concentrations of rubidium and strontium are usually determined either by x-ray fluorescence or by isotope dilution (Chapter 5.3). The $^{87}Sr/^{86}Sr$ ratio is measured on a suitable mass spectrometer using a pure strontium salt obtained from the mineral by dissolving it in acid followed by the separation of strontium using cation exchange chromatography. The ratio of the concentrations of rubidium to strontium is converted into the $^{87}Rb/^{86}Sr$ ratio by the following equation:

$$\frac{^{87}Rb}{^{86}Sr} = \left(\frac{Rb}{Sr}\right)_c \times \frac{Ab^{87}Rb \times WSr}{Ab^{86}Sr \times WRb} \quad (6.4)$$

where $^{87}Rb/^{86}Sr$ is the ratio of these isotopes in terms of numbers of atoms present in a unit weight of the mineral at the present time, $(Rb/Sr)_c$ is the ratio of the concentrations of these elements, $Ab^{87}Rb$ and $Ab^{86}Sr$ are the isotopic abundances of ^{87}Rb and ^{86}Sr, respectively, and WRb and WSr are the respective atomic weights. Note that the abundance of ^{86}Sr and the atomic weight of strontium depend on the abundance of ^{87}Sr and therefore appropriate values must be calculated for each sample.

The following example will illustrate how this calculation is made. Let us assume that the isotopic composition of a sample of strontium is expressed in terms of the following ratios: $^{87}Sr/^{86}Sr = 0.7090$, $^{86}Sr/^{88}Sr = 0.1194$, $^{84}Sr/^{88}Sr = 0.0068$. The abundances of the isotopes can be obtained in the following way:

ISOTOPIC RATIOS	MASS	ABUNDANCE
87/88 = 0.0846	87	0.0698
86/88 = 0.1194	86	0.0986
84/88 = 0.0068	84	0.0056
88/88 = 1.0000	88	0.8259
Sum = 1.2108		Sum = 0.9999

The abundance of ^{86}Sr in this sample of strontium is 0.0986 or 9.86 atom percent. The atomic weight can be calculated from the masses of the isotopes: $^{88}Sr = 87.9056$ amu, $^{87}Sr = 86.9088$ amu, $^{86}Sr = 85.9092$ amu,

$^{84}Sr = 83.9134$ amu (Chapter 2.3). The atomic weight of this sample of strontium is 87.6079 amu. Although the abundance of ^{87}Rb and the atomic weight of rubidium have also changed continuously since nucleosynthesis, all samples of terrestrial, meteoritic, and lunar rubidium have the same isotopic composition and atomic weight at the present time.

In order to solve Equation 6.3 for t, we must also substitute an appropriate value for $(^{87}Sr/^{86}Sr)_0$ which is the $^{87}Sr/^{86}Sr$ ratio of the strontium that was incorporated into the mineral at the time of its formation. When the mineral to be dated is strongly enriched in radiogenic ^{87}Sr, the date calculated from Equation 6.3 is insensitive to the value of the initial $^{87}Sr/^{86}Sr$ ratio. For example, one may select an initial ratio of 0.704 which is representative of basic volcanic rocks of Recent age which contain strontium derived primarily from the upper mantle of the Earth.

After the $^{87}Sr/^{86}Sr$ ratio and the concentrations of rubidium and strontium of a Rb-bearing mineral have been measured and an appropriate value has been chosen for the initial $^{87}Sr/^{86}Sr$ ratio, Equation 6.3 can be solved for t:

$$t = \frac{2.303}{\lambda} \log\left[\frac{\frac{^{87}Sr}{^{86}Sr} - \left(\frac{^{87}Sr}{^{86}Sr}\right)_0}{\frac{^{87}Rb}{^{86}Sr}} + 1\right] \quad (6.5)$$

The numerical value of t is a "date" in the geologic past. This date is the "age" of the mineral only when that mineral has remained a closed system with respect to rubidium and strontium, when the assumed value of the initial $^{87}Sr/^{86}Sr$ ratio is appropriate, and when the analytical results are accurate and representative of the material to be dated. The first two conditions define a model for the geologic history of the mineral which is why the date calculated from Equation 6.5 is sometimes called a "model date." It is a valid age determination only to the extent that the assumed model resembles the actual history of the mineral.

Dates calculated from the abundances of radiogenic isotopes are often referred to as "radiometric" dates. This practice is misleading and should be discouraged because it implies that the date was measured with a radiometer which is an instrument for measuring the intensity of radiant energy. A more appropriate adjective for such dates might be "isotopic" which at least suggests that they are based on the abundances of *isotopes* produced by radioactive decay.

Igneous rocks of granitic composition may contain both mica minerals and K-feldspar, all of which can be dated by the Rb-Sr method. Ideally, all minerals of an igneous rock should indicate the same date which can then be regarded as the age of the rock. When mineral dates obtained from one rock specimen or from a suite of cogenetic igneous rocks are in agreement, they are said to be "concordant." Unfortunately, "discordance" of mineral dates is more common than "concordance." The reason is that the constituent minerals of a rock may gain or lose radiogenic ^{87}Sr as a result of reheating during regional or contact metamorphism after crystallization from a magma. In such cases, the mineral dates generally are not reliable indicators of the age of the rock. We must therefore turn to the rocks themselves if we want to determine their ages.

3
Dating of Igneous Rocks

We know that fractional crystallization of magma and separation of crystals from the remaining liquid result in the formation of suites of comagmatic igneous rocks of differing chemical composition. If the strontium in such a magma remained isotopically homogeneous throughout the cooling period, we may *assume* that all the diverse rocks that formed from the magma had the same initial ^{87}Sr/^{86}Sr ratio. Moreover, we may assume that the time required for the crystallization of the magma was relatively short and that all rocks produced by this process have very nearly the same age. Under these conditions, Equation 6.3 is the equation of a family of straight lines in the slope-intercept form:

$$y = b + mx \qquad (6.6)$$

All rock specimens belonging to a comagmatic suite will plot as points on a straight line in coordinates of ^{87}Sr/^{86}Sr(y) and ^{87}Rb/^{86}Sr(x). This line is called an "isochron" because all points on that line represent systems having the same age (t) and the same initial ^{87}Sr/^{86}Sr ratio. The slope m of the isochron is related to the age of the comagmatic rocks by

$$m = e^{\lambda t} - 1 \qquad (6.7)$$

The value of the initial ^{87}Sr/^{86}Sr ratio is given by the y-intercept

$$b = \left(\frac{^{87}\mathrm{Sr}}{^{86}\mathrm{Sr}}\right)_0 \qquad (6.8)$$

A suite of comagmatic rocks of age t will define an isochron *only* when each member of that suite had the same initial ^{87}Sr/^{86}Sr ratio and when the rocks have remained closed to rubidium and strontium since crystallization.

In order to date comagmatic igneous rocks by the whole-rock isochron method, a suite of rocks must be collected which span as wide a range of Rb/Sr ratios as possible so that the slope of the isochron will be well defined. After the necessary analytical results have been obtained, the data are plotted in coordinates of ^{87}Sr/^{86}Sr and ^{87}Rb/^{86}Sr. A straight line is then fitted to the resulting points by a suitable regression procedure (to be discussed later) and the slope and intercept of the isochron are determined. The age of the suite of rocks is obtained from the slope by solving Equation 6.7. The resulting date indicates the time elapsed since all rocks in the suite had the same initial $^{87}Sr/^{86}Sr$ ratio which usually refers to the time of their crystallization from a magma. Therefore, whole-rock Rb-Sr isochron dates are generally regarded as reliable indicators of the age of the rocks.

The isotopic evolution of strontium in a suite of three hypothetical igneous rocks that formed from a common magma and have different Rb/Sr ratios is illustrated in Figure 6.1. At the time of crystallization, all three rocks plot as points on a straight line whose slope is zero because they all have the same $^{87}Sr/^{86}Sr$ ratio. After cooling to a temperature at which they become closed systems, their $^{87}Sr/^{86}Sr$ ratios begin to increase as a result of decay of ^{87}Rb to ^{87}Sr. Each decay of ^{87}Rb reduces the $^{87}Rb/^{86}Sr$ ratio and increases the $^{87}Sr/^{86}Sr$ ratio by the same

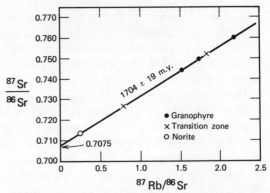

FIGURE 6.2 Whole-rock isochron for a suite of granophyres, transitional rocks and norite from Sudbury, Ontario, with which are associated important deposits of sulfides of nickel, copper, and other metals. The data points define an isochron whose slope indicates a date of 1704 ± 19 m.y. which was interpreted as the time elapsed since crystallization of these rocks. (Data from Fairbairn et al., 1968.)

amount. Consequently, these ratios will change along straight lines with a slope of −1 in such a way that the rock samples remain on the isochron as its slope increases as a function of time. The value of the y-intercept, however, remains constant and indicates the initial $^{87}Sr/^{86}Sr$ ratio of the suite of rocks.

The dating of comagmatic igneous rocks by the whole-rock isochron method is illustrated in Figure 6.2 by an isochron for a suite of rocks from the norite-granophyre ("micropegmatite") assemblage from Sudbury, Ontario. (Fairbairn et al., 1968). The assumption was made that the norite, granophyre, and transitional rocks initially had the same $^{87}Sr/^{86}Sr$ ratio and that they formed in an interval of time which is short compared to their common age. The satisfactory fit of the data points to the isochron indicates that these assumptions are justified, at least to a first approximation. The slope of the isochron indicates a date of

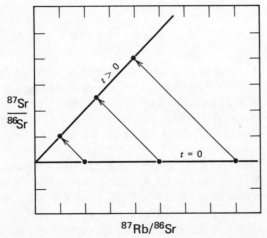

FIGURE 6.1 Rb-Sr isochron diagram showing the time-dependent isotopic evolution of rock systems after their crystallization from a homogeneous magma.

1704 ± 19 m.y. which was interpreted as the crystallization age of these rocks.

In general, it is apparent that the isochron method of dating suites of cogenetic igneous rocks is superior to calculated model dates based on assumed values of the initial ratio. The isochron yields not only the age of the suite but also indicates their initial ratio. Moreover, the goodness of fit of data points serves as a check of the assumption that all of the specimens have remained closed to rubidium and strontium.

The minerals of igneous rocks that have not been metamorphosed lie on the same isochron as the whole rocks from which they were separated. However, in case the rocks have been heated during regional or contact metamorphism, their minerals commonly depart from the whole-rock isochron and may form separate mineral isochrons to be discussed below.

4
Dating of Metamorphic Rocks

Rocks may be subjected to physical and chemical processes as a result of which their mineralogical and chemical compositions and even their textures are changed. These processes constitute metamorphism in the broadest sense of the word. Metamorphism almost invariably involves an increase in the ambient temperature which promotes recrystallization of existing minerals or may cause formation of new minerals at the expense of existing ones. These mineralogical changes imply considerable mobility of the chemical constituents of rocks either by virtue of the presence of an aqueous phase or by diffusion of ions, or both. Metamorphism may also be accompanied by metasomatism as a result of which both the bulk chemical composition as well as the trace-element composition of the rocks may be changed.

It is to be expected that metamorphism exerts a profound effect on the parent-daughter relationships of all of the naturally occurring radioactive isotopes which may be present in a rock. In fact, even a modest increase in temperature of 100 to 200°C or so may have drastic effects on the parent-daughter relationships of natural decay schemes without necessarily being reflected in the usual mineralogical or textural criteria for metamorphism. The apparent sensitivity of isotopic systems in rocks to increases in temperature is probably related to the fact that rates of diffusion of ions through a crystal lattice and across grain boundaries are sensitive functions of the temperature. Moreover, the daughter atoms produced by decay in a mineral are isotopes of different elements and have different ionic charges and radii compared to their parents. The energy released during the decay may produce dislocations or even destroy the crystal lattice locally, thus making it all the more easy for the radiogenic daughters to escape.

For these reasons it is not surprising that Rb-Sr decay schemes in minerals are profoundly affected by even modest increases in temperature during metamorphism. The observed behavior of minerals can generally be treated as though it had been caused solely by the migration of radiogenic ^{87}Sr among the constituent minerals of a rock. However, this is undoubtedly an oversimplification, and it is likely that the concentrations of rubidium and strontium in the minerals are also affected. Nevertheless, in the presentation that follows we shall assume that radiogenic ^{87}Sr is the mobile component and that the concentrations of rubidium and strontium of the minerals remain essentially constant during regional or contact metamorphism.

To help us describe the changes that occur in the $^{87}Sr/^{86}Sr$ ratios of the minerals of a rock during metamorphism, we now introduce a useful approximation to Equation 6.3. We expand $e^{\lambda t}$ as a power series:

$$e^{\lambda t} = 1 + \lambda t + \frac{(\lambda t)^2}{2!} + \frac{(\lambda t)^3}{3!} + \cdots \quad (6.9)$$

Since the decay constant of ^{87}Rb is very small ($\lambda = 1.39 \times 10^{-11}$ y^{-1}),

$$1 + \lambda t \gg \frac{(\lambda t)^2}{2!} + \frac{(\lambda t)^3}{3!} + \cdots \quad (6.10)$$

even for values of t of the order of 10^9 years. Thus we can rewrite Equation 6.3 to a good approximation as

$$\frac{^{87}\text{Sr}}{^{86}\text{Sr}} \simeq \left(\frac{^{87}\text{Sr}}{^{86}\text{Sr}}\right)_0 + \frac{^{87}\text{Rb}}{^{86}\text{Sr}}\lambda t \quad (6.11)$$

This is the equation of a straight line in the familiar slope-intercept form in coordinates of ^{87}Sr/^{86}Sr and t. It is very useful in describing the time-dependent increase of the ^{87}Sr/^{86}Sr ratios of Rb-bearing systems because the strontium evolution lines are straight lines whose slopes are $m = (^{87}\text{Rb}/^{86}\text{Sr})\lambda$. Figure 6.3 is such a strontium development or evolution diagram for four systems having different ^{87}Rb/^{86}Sr ratios.

Based on experience derived from the study of metamorphosed igneous rocks, it has been found that the following model is an adequate description of the response of the Rb-Sr decay schemes in minerals during thermal metamorphism (Fairbairn et al., 1961). We consider an igneous rock containing two Rb-rich minerals such as biotite and microcline (or orthoclase) and one Rb-poor, but Sr-rich mineral such as apatite. This rock initially formed t_i years ago and at that time the rock and all of its constituent minerals contained strontium of identical isotopic composition represented by $(^{87}\text{Sr}/^{86}\text{Sr})_i$. Some time after crystallization, the rock was subjected to an increase in temperature for

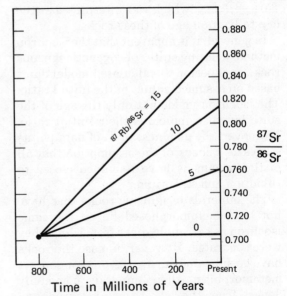

FIGURE 6.3 Strontium development diagram for a suite of four rock systems having different ^{87}Rb/^{86}Sr ratios. The four systems came into existence 800 million years ago when each had the same ^{87}Sr/^{86}Sr ratio of 0.704. Thereafter, the ^{87}Sr/^{86}Sr ratios of the systems evolved along a set of diverging straight lines whose slopes depend on the ^{87}Rb/^{86}Sr ratio of each system. Note that the rock system whose ^{87}Rb/^{86}Sr ratio is zero has a constant ^{87}Sr/^{86}Sr ratio equal to the initial value of all the systems.

a short interval of time Δt. Subsequently, the rock cooled to the ambient temperature t_m years ago and remained undisturbed to the present time. The effects of such a history on the ^{87}Sr/^{86}Sr ratios of the rock and its constituent minerals are shown in Figure 6.4.

Starting t_i years ago when the rock had cooled sufficiently for its minerals to retain radiogenic ^{87}Sr (Dodson, 1973), their ^{87}Sr/^{86}Sr ratios evolved along straight-line paths at rates controlled by their ^{87}Rb/^{86}Sr ratios. When the temperature was increased, the Rb-rich phases biotite and K-feldspar lost radiogenic ^{87}Sr, and their ^{87}Sr/^{86}Sr ratios decreased until they became identical

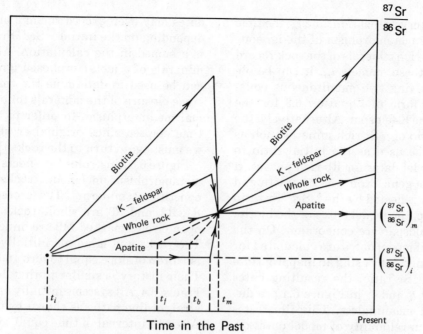

FIGURE 6.4 Strontium development diagram showing the isotopic homogenization of the minerals of a rock as a result of thermal metamorphism. t_i = time elapsed since initial crystallization and cooling when the isotopic composition of strontium in all minerals was $(^{87}Sr/^{86}Sr)_i$; t_m = time elapsed since closure of the minerals following isotopic reequilibration of strontium to $(^{87}Sr/^{86}Sr)_m$ by thermal metamorphism; t_b and t_f are fictitious model dates of biotite and feldspar calculated relative to an arbitrary and inappropriate choice of the initial $^{87}Sr/^{86}Sr$ ratio.

with that of the rock as a whole. (Sometimes K-feldspar actually gains ^{87}Sr.) The radiogenic ^{87}Sr lost by the Rb-rich phases entered the apatite and its $^{87}Sr/^{86}Sr$ ratio increased until it, too, was equal to that of the whole rock. As a result of these changes brought about by the increase in temperature, strontium was isotopically homogenized so that all minerals once again had the same $^{87}Sr/^{86}Sr$ ratio equal to $(^{87}Sr/^{86}Sr)_m$. Note especially that $(^{87}Sr/^{86}Sr)_m$ is greater than $(^{87}Sr/^{86}Sr)_i$ and that the whole rock remained a closed system while its minerals exchanged radiogenic ^{87}Sr until the $^{87}Sr/^{86}Sr$ ratios of all the minerals were equalized. It is entirely possible that the concentrations

of rubidium and strontium of the minerals were also affected by this process, although Figure 6.4 was constructed with the assumption that they were not. Changes in the concentrations of rubidium and strontium of the minerals would not affect the interpretation, provided that the rock as a whole remained a closed system. Following isotopic homogenization, the temperature eventually decreased sufficiently for the minerals to become closed systems again at t_m. Their $^{87}Sr/^{86}Sr$ ratios then increased to the present at rates consistent with their $^{87}Rb/^{86}Sr$ ratios.

There are two meaningful dates in the history outlined above. These are t_i, the time elapsed since crystallization, and t_m, the time elapsed since metamorphism of the igneous rock. The Rb-rich minerals of the rock record only one of these, namely t_m. It can be obtained by solving two simultaneous equations of the form of Equation 6.3 for the biotite and the K-feldspar. Alternatively, the $^{87}Sr/^{86}Sr$ ratio of a Sr-rich mineral such as apatite can be used as the initial ratio to calculate model dates for Rb-rich phases. If isotopic homogenization was complete and is accurately reflected by the $^{87}Sr/^{86}Sr$ ratio of the apatite, the model dates of the biotite and the K-feldspar are concordant. On the other hand, if model dates are calculated for these two minerals by using an inappropriate initial $^{87}Sr/^{86}Sr$ ratio, the resulting dates indicated by t_b and t_f in Figure 6.4 are discordant and meaningless. This illustrates the general unreliability of model dates derived by analysis of separated minerals. Such

dates may even exceed the age of the rock, depending on the initial $^{87}Sr/^{86}Sr$ ratio that is assumed in the calculation. At best, the minerals of a metamorphosed igneous rock can be used to determine the time elapsed since closure of the minerals following thermal metamorphism. In order to obtain the time elapsed since original crystallization, we must again turn to the rocks themselves.

Figure 6.5 illustrates the effects of thermal metamorphism on igneous rocks and their constituent minerals. The crosses marked R1, R2, and R3 are whole rocks, while the open circles marked M2 are minerals of R2. The rocks and minerals initially lay on an isochron of slope equal to zero and thus $t = 0$. Their history is similar to that described in Figure 6.4. All systems initially moved along straight-line paths as shown by the arrows. After an interval of time equal to $t_i - t_m$ the rocks were heated for a short period of time.

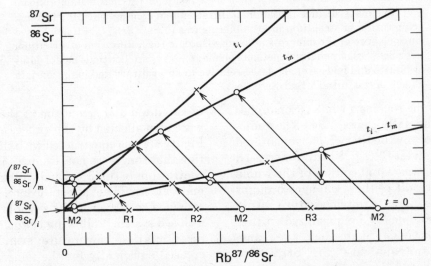

FIGURE 6.5 Evolution of strontium in three whole rocks (R1, R2, and R3) and in the minerals of R2 (M2). The strontium in the minerals was isotopically homogenized by an episode of thermal metamorphism of short duration. The slope of the whole-rock isochron corresponds to t_i which is the time elapsed since crystallization of these rocks. The slope of the mineral isochron indicates t_m which is the time elapsed since the end of the thermal metamorphism.

The whole-rock samples remained closed, but the minerals of R2 exchanged radiogenic ^{87}Sr until they all had the same $^{87}Sr/^{86}Sr$ ratio as R2. Consequently, at a time $t = t_m$ years ago, the minerals of R2 had realigned themselves on a new isochron having a slope equal to zero. The whole-rock systems continued their evolution without interruption and formed an isochron whose slope corresponds to t_i, which is the time elapsed since initial crystallization. The minerals in the meantime evolved on their own isochron which still includes TR2, but whose slope represents t_m, the time elapsed since the minerals became closed systems after being re-equilibrated by thermal metamorphism. The y-intercepts of the whole-rock and the mineral isochrons correspond to $(^{87}Sr/^{86}Sr)_i$ and $(^{87}Sr/^{86}Sr)_m$, respectively. Note that we have considered only the minerals of one of the three rocks. The minerals of the other rocks form similar isochrons. Their slopes are identical, but they have different values of $(^{87}Sr/^{86}Sr)_m$, appropriate to the whole rocks whose constituents they are.

A granite complex at Carn Chuinneag (pronounced Carn Coon-e-ag) in the northern Scottish Highlands provides a good example of an igneous rock that was later regionally metamorphosed (Long, 1964). Figure 6.6a is an isochron formed by four whole rock specimens from this and a related intrusive. The age of these rocks is 560 ± 10 million years (m.y.) (recalculated to λ $^{87}Rb = 1.39 \times 10^{-11}$ y^{-1}) and their initial ratio is 0.710 ± 0.002. Figure 6.6b shows muscovite, biotite, and feldspar (plagioclase and orthoclase partially converted to microcline) of one of the rock specimens. It can be seen that the parent rock is colinear with its minerals and that they together define another isochron equivalent to a date of 412 ± 5 m.y. Long interpreted this date as being indicative of the late Caledonian metamorphism that caused complete homogenization of the strontium in the minerals of this intrusive. Note that the initial $^{87}Sr/^{86}Sr$ ratio of the mineral isochron is 0.782, which is very much higher than that of the whole rocks. The initial $^{87}Sr/^{86}Sr$ ratio of the minerals is determined by the $^{87}Sr/^{86}Sr$ of the whole rock at the end of the metamorphic episode, as suggested previously in Figures 6.4 and 6.5.

All of the foregoing discussion is predicated on the assumption that the *minerals* were isotopically homogenized during thermal metamorphism and that *whole-rock* samples of the size of conventional hand specimens remained closed. If homogenization of the minerals was incomplete, the time of metamorphism cannot be determined from them. In this case the whole rocks may still tell the time of initial crystallization. However, if the *rocks* were chemically altered so that rubidium and/or strontium were either added or lost at any time after their formation, they cannot be dated by the Rb-Sr method.

An example of dating granitic gneisses was presented by Moorbath and his colleagues from Oxford University (Moorbath et al., 1972) for samples of the Amitsoq gneiss from the Godthaab district in southwestern Greenland. The Amitsoq gneiss was formed by deformation, metamorphism, and migmatization of a complex of igneous rocks of granitic composition. Several suites of these rocks from the Godthaab district were dated by the whole-rock Rb-Sr method. Figure 6.7 shows the isochron of Amitsoq gneisses from the Qilangarssuit area. The slope of this isochron yields a date of 3740 ± 100 m.y. and an initial $^{87}Sr/^{86}Sr$ ratio of 0.7009 ± 0.0011. The Amitsoq gneiss is one of the oldest terrestrial rocks known. The date indicated by the isochrons may be the time of crystallization of the igneous rocks or it may reflect the metamorphic event that produced the Amitsoq gneiss. The latter is preferred in this case.

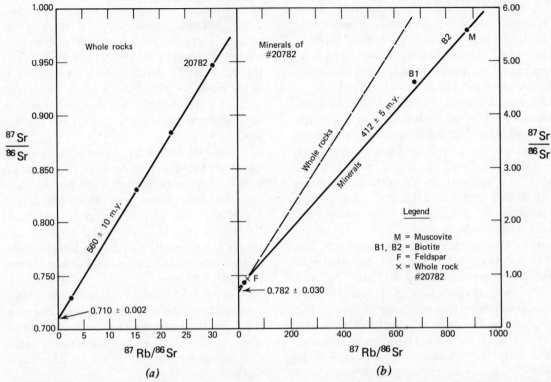

FIGURE 6.6 (a) Whole rock isochron of granitic rocks from the Carn Chuinneag complex in the northern Scottish Highlands. (b) Mineral isochron formed by feldspar, muscovite, and biotite of one of the whole-rock samples. The interpretation of these dates is that the Carn Chuinneag intrusive crystallized 560 m.y. ago and was later affected by the Caledonian orogeny 412 m.y. ago. The strontium in the minerals was isotopically homogenized as a result of regional metamorphism, while the whole-rock samples remained closed systems. Note the difference in the initial ratios of the two isochrons. The history of these rocks is similar to those depicted in Figures 6.4 and 6.5. (Data from Long, 1964. See also Pidgeon and Johnson, 1974.)

After publication of these results, other examples of very old dates were reported for the Onverwacht Group, South Africa (Hurley et al., 1972; Allsopp et al., 1973; Jahn and Shih, 1974) and for the Mushandike granite (Hickman, 1974) in Rhodesia. In addition, dates in excess of 3.5 billion years have been obtained for granitic gneisses in Labrador (Hurst et al., 1975; Barton, 1975) and for the Montevideo Gneiss of Minnesota (Goldich and Hedge, 1974). However, the interpretation of the Montevideo data was subse-

quently questioned (Farhat and Wetherill, 1975; Goldich and Hedge, 1975).

Metamorphic rocks also form from sequences of *sedimentary rocks* containing minerals that are unstable at elevated temperatures and that therefore recrystallize to form new mineral assemblages when the temperature is elevated. The Rb-bearing minerals of such metamorphic rocks (primarily micas and K-feldspar) can be dated either by the isochron method, or by calculating model dates. Such dates generally reflect the time

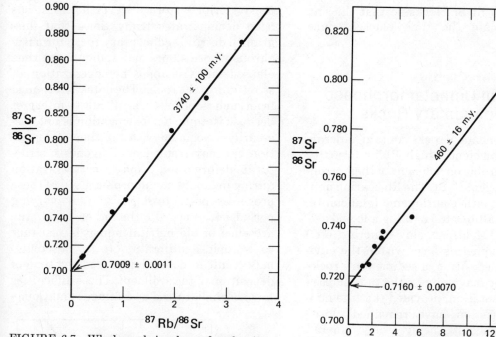

FIGURE 6.7 Whole-rock isochron for the Amitsoq gneiss from the Quilangarssuit area of the
Godthaab district of southwestern Greenland.
These are among the oldest terrestrial rocks
known. The date probably represents the time
elapsed since metamorphism of igneous rocks of
granitic composition to form the Amitsoq gneiss.
(Data from Moorbath et al., 1972.)

elapsed since the end of the last metamorphic
episode to which the rocks were subjected.
Whole-rock isochrons may *likewise* indicate
the age of the metamorphic event during
which the sediment was recrystallized.

This point is illustrated in Figure 6.8
by an isochron for samples of phyllite and
slate from the La Gorce Formation of the
Wisconsin Range in the Transantarctic
Mountains (Montigny and Faure, 1969). The
La Gorce Formation consists of fine-grained
detrital sedimentary rocks that were iso-
clinally folded and metamorphosed to the
greenschist facies during the Ross Orogeny
in late Cambrian to early Ordovician time.
However, several lines of field evidence leave

FIGURE 6.8 Whole-rock isochron of slates and
phyllites from the La Gorce Formation of probable late Precambrian age from the Wisconsin
Range of the Transantarctic Mountains. The date
indicated by the slope of the isochron represents
the time of metamorphism and extensive homogenization of the isotopic composition of strontium in these rocks. The scatter of points about
the isochron is greater than expected from analytical errors alone and suggests that homogenization was not complete or approached
different values in different parts of the formation
(Data from Montigny and Faure, 1969.)

little room for doubt that these rocks were
deposited in late Precambrian time. Nevertheless, the isochron defined by whole-rock
samples of the La Gorce Formation indicates
a date of only 460 ± 16 m.y., which apparently reflects the time of metamorphism of
these rocks rather than their depositional
age. Evidently the strontium in these samples was isotopically reequilibrated during
metamorphism on a large scale, thereby

resetting the Rb-Sr clocks so that they no longer indicate the time elapsed since deposition.

5
Dating of Unmetamorphosed Sedimentary Rocks

Certain sedimentary rocks contain minerals that have sufficiently high Rb/Sr ratios to cause measurable enrichment of their strontium in radiogenic ^{87}Sr. The Rb-bearing minerals may be either authigenic (glauconite, sylvite, carnallite, etc.) or allogenic and detrital (mica, K-feldspar, clay minerals, etc.) Authigenic minerals form within the environment of deposition of sedimentary rocks and therefore may indicate the time elapsed since sedimentation, provided that the minerals in question have remained closed systems. Glauconite, being a clay mineral, is subject to recrystallization by deep burial, tectonic deformation, or metamorphism of the rocks in which it occurs. Sylvite and carnallite, being evaporite minerals, are likewise easily altered by recrystallization. Nevertheless, unaltered glauconite from rocks that have not been deeply buried or deformed may be suitable for dating and indicate the time of original formation of the glauconite which comes close to being the time of deposition of sedimentary rocks. However, Rb-bearing evaporite minerals are very fragile systems and may undergo more or less continuous recrystallization in equilibrium with connate brines. These minerals therefore do not appear to be reliable geochronometers.

Of special interest is the possibility that fine-grained, unmetamorphosed, detrital sedimentary rocks, such as shales, can be dated by the whole-rock isochron method. Such rocks consist primarily of mineral particles that formed prior to deposition of the sediment and which therefore may have ex-

tensive prior histories. Nevertheless, it has been demonstrated many times that fine-grained detrital sedimentary rocks form isochrons whose slopes may indicate the time elapsed since isotopic homogenization of strontium in the rocks. The dates determined from such isochrons are difficult to interpret because isotopic homogenization is not necessarily associated with the time of deposition but may result from diagenesis, structural deformation, and recrystallization during incipient metamorphism. All of these processes postdate deposition by varying periods of time. On the other hand, the presence of old detrital minerals resistant to isotopic equilibration (e.g., muscovite) may result in dates that *exceed* the time of deposition of the sediment. These difficulties have been discussed by Bofinger and Comp-

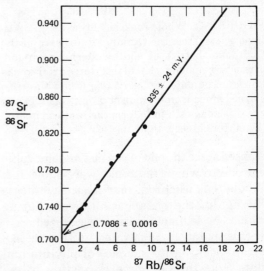

FIGURE 6.9 Whole-rock isochron for 10 samples of unmetamorphosed siltstone from the Lower Torridonian of Stoer in northwest Scottland. The good fit of these samples on the isochron indicates that they had the same initial ^{87}Sr/^{86}Sr ratio 935 ± 24 m.y. ago which Moorbath (1969) interpreted as the time of diagenesis soon after deposition and compaction of these sedimentary rocks. (Data from Moorbath, 1969.)

ston (1967), Dasch (1969), Clauer (1973), Perry and Turekian (1974), Gebauer and Grünenfelder (1974), and others.

Moorbath (1969) used this method to date suites of shale and siltstone from the Torridonian sediments of northwest Scotland. He obtained a date of 935 ± 24 m.y. for the Lower Torridonian of Stoer and a date of 761 ± 17 m.y. for the Upper Torridonian Applecross Formation which lies unconformably on the Lower Torridonian formations. He interpreted the dates as being representative of the time of diagenesis which followed closely after deposition and compaction of these rocks. One of the two isochrons is shown in Figure 6.9.

6
The Fitting of Isochrons

The dating of rocks or minerals by the Rb-Sr method begins with the selection of a suite of samples which, on the basis of prior geological evaluation, are likely to satisfy the assumption that all specimens chosen for analysis formed at the same time, had the same initial $^{87}Sr/^{86}Sr$ ratio, and are likely to have remained closed systems. Samples that show evidence of chemical weathering or of other forms of postdepositional alteration must be excluded at the outset. After the $^{87}Sr/^{86}Sr$ and $^{87}Rb/^{86}Sr$ ratios of the samples have been determined and have been plotted on the isochron diagram, the problem arises of fitting the "best" straight line to the data points. The fit of the data points to a straight line is never "perfect" because of errors arising from the analyses of the samples. These analytical errors give rise to a corresponding uncertainty in the estimate of the slope and hence the age of samples to be dated. The uncertainty of the initial $^{87}Sr/^{86}Sr$ ratio will likewise be affected by the analytical errors associated with the data points.

The term "analytical error" generally means the deviation of a measured value from its true value. Such errors may be random or systematic. Random errors have a "normal" or Gaussian distribution about the arithmetic mean of the measurements which approaches the true value as the number of measurements increases. Systematic errors, on the other hand, are consistent differences between the true value and a set of measurements such that their arithmetic mean is displaced from the true value. In other words, random errors determine the *precision* of a set of measurements, while systematic errors limit their *accuracy*. Systematic errors in the measurements of $^{87}Sr/^{86}Sr$ and $^{87}Rb/^{86}Sr$ ratios lead to similar systematic errors in the date and initial $^{87}Sr/^{86}Sr$ ratio derived from an isochron. Such errors must obviously be eliminated as nearly as possible to assure the accuracy of the results. The important considerations in fitting isochrons are (1) how to use the analytical errors of the coordinates in determining the best slope and intercept; and (2) how to decide whether a particular point fits the isochron within the analytical error. If it can be shown that a point does not fit the isochron within random analytical errors, then the coordinates of that data point either have systematic analytical error or the sample does not satisfy the prerequisite assumptions of dating by the isochron method. It may have a different age or a different initial $^{87}Sr/^{86}Sr$ ratio compared to other samples in the suite or it may not have remained a closed system.

The simplest method of fitting an isochron to data points is to draw the best straight line by eye on a piece of graph paper. If the scale is sufficiently large, the slope and intercept can be determined from the graph with sufficient accuracy to serve at least as a good first approximation. This method is useful for obtaining preliminary estimates of the slope and intercept, but should be replaced by a more objective approach in any serious evaluation of isochrons.

A somewhat better method is the "least squares" regression procedure which consists of minimizing the deviations of either the x- or the y-coordinate from the best line in the slope-intercept form. The equations for calculating the slope m and the intercept b of the best straight line are

$$m = \frac{\sum XY - \frac{(\sum X)(\sum Y)}{N}}{\sum X^2 - \frac{(\sum X)^2}{N}} \qquad (6.12)$$

$$b = \frac{(\sum X)(\sum XY) - \sum Y(\sum X^2)}{(\sum X)^2 - N(\sum X^2)} \qquad (6.13)$$

where y represents the $^{87}Sr/^{86}Sr$ ratios, x the $^{87}Rb/^{86}Sr$ ratios, and N is the number of data points. The calculation of m and b is facilitated by the construction of a table with the following headings:

$$x \qquad y \qquad xy \qquad x^2$$

The columns are then summed to obtain $\sum x$, $\sum y$, $\sum xy$, and $\sum x^2$ which are needed to calculate m and b from Equations 6.12 and 6.13. Such a regression is based on the assumption that the deviations of the data points from the best straight line are due only to errors in the y-coordinates and that the x-coordinates are free of error. This is clearly not a good assumption for fitting isochrons because both the x- and y-coordinates have analytical errors.

The preferred method of calculating the slope and intercept of the isochron must take into account the known analytical errors of the coordinates of the data points on the isochron diagram. This is accomplished by the use of weighting factors which are calculated from the reciprocals of the variance of each coordinate. The problem of fitting isochrons and of determining the best slope

and intercept from analytical data has been treated by York (1966, 1967, 1969), McIntyre et al. (1966), and by Brooks et al. (1968, 1972). Appendix I contains a computer program based on the method of York (1969) with certain modifications by M. J. McSaveney. The program computes dates and initial $^{87}Sr/^{86}Sr$ ratios using three alternate weighting procedures and rejects data points whose coordinates differ significantly from the calculated "best-fit" line.

7 SUMMARY

Rubidium and strontium are dispersed elements whose concentrations in igneous, sedimentary, and metamorphic rocks range from a few parts per million or less to several hundred parts per million or more. Rubidium is concentrated primarily in mica, K-feldspar, and clay minerals, whereas strontium occurs in plagioclase feldspar, apatite, and carbonate minerals. The decay of naturally occurring ^{87}Rb to stable ^{87}Sr in Rb-bearing minerals can be used to calculate dates for such minerals from measurements of the concentrations of rubidium and strontium and of the $^{87}Sr/^{86}Sr$ ratio. In addition, an assumption must be made regarding the value of the initial $^{87}Sr/^{86}Sr$ ratio. Such calculated dates may represent the age of the mineral provided it remained closed to rubidium and strontium after crystallization and provided an appropriate value was chosen for the initial $^{87}Sr/^{86}Sr$ ratio. The *dates* calculated from the minerals of an igneous rock indicate the *age* of the rock only when the minerals were not disturbed after crystallization by subsequent thermal metamorphism.

Suites of comagmatic igneous rocks may have sufficiently variable Rb/Sr ratios to yield isochrons whose slopes in most cases record the time of original crystallization. Subsequent thermal metamorphism may lead to isotopic homogenization of the minerals within each whole-rock specimen without affecting the Rb-Sr decay schemes in the whole rocks. Consequently, the whole-rock specimens may form an isochron whose slope records the time of original crystallization, while the minerals of each specimen in the suite form a series of parallel isochrons whose slopes represent the time elapsed since closure after the internal isotopic homogenization. Dating is not possible in cases where the minerals were incompletely homogenized or where whole-rock samples did not remain closed to rubidium and strontium. Metamorphosed sedimentary rocks may form whole-rock isochrons indicating the time of isotopic homogenization during metamorphism.

Unmetamorphosed, fine-grained, clastic sedimentary rocks in many cases form linear arrays on the isochron diagram. However, the interpretation of dates derived from such isochrons is uncertain because isotopic homogenization probably results from post-depositional processes of diagenesis and incipient thermal metamorphism and recrystallization. The presence of old, unreactive, Rb-bearing phases may lead to dates that exceed the time of deposition of sedimentary rocks.

The fitting of isochrons to analytical data is best done by regression methods which include the use of weighting factors based on analytical errors. The objectives of such procedures include the determination of the best slope and intercept of the line and the identification of data points that deviate from the line by more than their analytical errors at a stated level of confidence.

PROBLEMS

1. Calculate the abundances of the isotopes and the atomic weight of strontium which has the following isotope ratios:

$\frac{87}{86} = 1.0000$, $\frac{86}{88} = 0.1194$, $\frac{84}{88} = 0.0068$. The isotopic masses are given in section 2 of this chapter. (**ANSWER:** $^{88}Sr = 80.28$ percent, $^{87}Sr = 9.58$ percent, $^{86}Sr = 9.58$ percent; $^{84}Sr = 0.54$ percent; atomic weight: 87.5796.)

2. Calculate the $^{87}Rb/^{86}Sr$ ratio (atomic) of a specimen of biotite which has the following concentrations: Rb = 465 ppm, Sr = 30 ppm, and whose $^{87}Sr/^{86}Sr$ ratio is 1.0000. Use the results of Problem 1 above. (**ANSWER:** 46.147.)

3. If the initial $^{87}Sr/^{86}Sr$ ratio of this biotite (Problem 2) was 0.7035, what is the "age" of this mineral, assuming $\lambda(^{87}Rb) = 1.39 \times 10^{-11} y^{-1}$? Calculate dates both by means of the approximate Equation 6.11 and the exact Equation 6.5. (**ANSWER:** Approximate equation: 462.2×10^6 y; exact equation: 458.9×10^6 y.)

4. What was the $^{87}Sr/^{86}Sr$ ratio of this biotite 250 million years ago, using the approximate equation? (**ANSWER:** 0.8397.)

5. The following data were obtained for three minerals from a pegmatite

	Rb ppm	Sr ppm	$^{87}Sr/^{86}Sr$
Muscovite	238.4	1.80	1.4125
Biotite	1080.9	12.8	1.2587
K-feldspar	121.9	75.5	0.7502

Calculate model dates for these minerals by assuming an initial $^{87}Sr/^{86}Sr$ ratio of 0.704. Are the resulting dates concordant or discordant? (**ANSWER:** Muscovite: 123.7 m.y.; biotite: 154.1 m.y.; K-feldspar: 704.6 m.y.)

6. Plot the three minerals of Problem 5 on an isochron diagram using graph paper. Draw an isochron representing the best fit in your judgment and determine the best slope and intercept. Calculate a date from the slope of the isochron. (**ANSWER:** Intercept: 0.743; date: 131 m.y.)

7. Which of the dates calculated in Problems 5 and 6 most nearly represents the age of the pegmatite?

8. Given that the $^{87}Sr/^{86}Sr$ ratio of a mineral is 0.955 ± 0.001 and its $^{87}Rb/^{86}Sr$ ratio is 62.5 ± 1.9, what is the model date of this mineral and what is the uncertainty of this date corresponding to the analytical errors? Assume $(^{87}Sr/^{86}Sr)_0 = 0.704$. (**ANSWER:** $t = 287.4 + 10.2/-9.6$ m.y.)

9. The following data apply to whole rocks and separated minerals of the Baltimore Gneiss, Maryland. (Wetherill et al., 1968)

	$^{87}Rb/^{86}Sr$	$^{87}Sr/^{86}Sr$
Rock 1	2.244	0.7380
Rock 2	3.642	0.7612
Rock 3	6.59	0.7992
Biotite (3)	289.7	1.969
K-spar (3)	5.60	0.8010
Plagioclase (3)	0.528	0.7767
Rock 4	0.2313	0.7074
Rock 5	3.628	0.7573
Biotite (5)	116.4	1.2146
K-spar (5)	3.794	0.7633
Plagioclase (5)	0.2965	0.7461

Interpret these data by means of suitable isochron diagrams. Determine dates and initial $^{87}Sr/^{86}Sr$ ratios and use them to reconstruct the geological history of these rocks and minerals.

REFERENCES

Aldrich, L. T., G. W. Wetherill, G. R. Tilton, and G. L. Davis (1956) The half-life of [87]Rb. Phys. Rev., *104*, 1045–1047.

Allsopp, H. L., M. J. Viljoen, and R. P. Viljoen (1973) Strontium isotopic studies of the mafic and felsic rocks of the Onverwacht Group of the Swaziland Sequence, Geol. Rundschau, *62*, 902–916.

Barton, J. M., Jr. (1975) Rb-Sr isotopic characteristics and chemistry of the 3.6 b.y. Hebron gneiss, Labrador. Earth Planet. Sci. Letters, *27*, 427–435.

Bofinger, V. M., and W. Compston (1967) A reassessment of the age of the Hamilton Group, New York and Pennsylvania, and the role of inherited radiogenic Sr^{87}. Geochim. Cosmochim. Acta, *31*, 2353–2359.

Campbell, N. R., and A. Wood (1906) The radioactivity of the alkali metals. Proc. Cambridge Phil. Soc., *14*S, 15–21.

Brooks, C., I. Wendt, and W. Harre (1968) A two-error regression treatment and its application to Rb-Sr and initial Sr^{87}/Sr^{86} ratios of younger Variscan granitic rocks from the Schwarzwald Massif, southwest Germany. J. Geophys. Res., *73*, 6071–6084.

Brooks, C., S. R. Hart, and I. Wendt (1972) Realistic use of two-error regression treatments as applied to rubidium-strontium data. Rev. Geophys. Space Phys., *10*, 551–577.

Catanzaro, E. J., T. J. Murphy, E. L. Garner, and W. R. Shields (1969) Absolute isotopic abundance ratio and atomic weight of terrestrial rubidium. J. Res. Nat. Bur. Std. Phys. and Chem., *73*A, 511–516.

Clauer, N. (1973) Utilization de la methode rubidium-strontium pour la datation de niveaux sedimentaires du Precambrian superieur de l' Adrar mauritanien (Sahara occidental) et la mise en evidence de transformations precoces des mineraux argileux. Geochim. Cosmochim. Acta, *37*, 2243–2255.

Dasch, E. J. (1969) Strontium isotopes in weathering profiles, deep sea sediments, and sedimentary rocks. Geochim. Cosmochim. Acta, *33*, 1521–1552.

Dodson, M. H. (1973) Closure temperature in cooling geochronological and petrological systems. Contrib. Mineral. Petrol., *40*, 259–274.

Fairbairn, H. W., P. M. Hurley, and W. H. Pinson, Jr. (1961) The relation of discordant Rb-Sr mineral and rock ages in an igneous rock to its time of subsequent Sr^{87}/Sr^{86} metamorphism. Geochim. Cosmochim. Acta, *23*, 135–144.

Fairbairn, H. W., G. Faure, W. H. Pinson, Jr., and P. M. Hurley (1968) Rb-Sr whole-rock age of the Sudbury lopolith and basin sediments. Can. J. Earth Sci., *5*, 707–714.

Farhat, J. S., and G. W. Wetherill (1975) Interpretation of apparent ages in Minnesota. Nature, *257*, 721.

Faure, G., and J. L. Powell (1972) Strontium isotope geology. Springer-Verlag, Berlin, Heidelberg and New York, 188 pp.

Flynn, K. F., and L. E. Glendenin (1959) Half-life and beta spectrum of Rb^{87}. Phys. Rev., *116*, 744–748.

Gebauer, D., and M. Grünenfelder (1974) Rb-Sr whole-rock dating of late diagenetic to anchimetamorphic, Paleozoic sediments in southern France (Montagne Noire). Contrib. Mineral. Petrol., *47*, 113–130.

Goldich, S. S., and C. E. Hedge (1974) 3,800-myr granitic gneiss in south-western Minnesota. Nature, *252*, 467–468.

Goldich, S. S., and C. E. Hedge (1975) Reply (to Farhat and Wetherill, 1975). Nature, *257*, 722.

Hahn, O., F. Strassman, and E. Walling (1937) Herstellung wägbarer Mengen des Strontiumisotops 87 als Umwandlungsprodukt des Rubidiums aus einem kanadischen Glimmer. Naturwissenschaft., *25*, 189.

Hahn, O., and E. Walling (1938) Über die Möglichkeit geologischer Altersbestimmungen rubidiumhaltiger Mineralen und Gesteine. Z. Anorg. Allgem. Chem., *236*, 78–82.

Hahn, O., F. Strassman, J. Mattauch, and H. Ewald (1943) Geologische Altersbestimmungen mit der Strontiummethode. Chem. Zeitung, *67*, 55–56.

Hamilton, E. I. (1965) Applied geochronology. Academic Press, New York and London, 267 pp.

Heier, K. S., and J. A. S. Adams (1965) The geochemistry of the alkali metals. Phys. Chem. Earth, *5*, 253–281.

Hickman, M. H. (1974) 3,500-Myr-old granite in southern Africa. Nature, *251*, 295–296.

Hurley, P. M., W. H. Pinson, Jr., B. Nagy, and T. M. Teska (1972) Ancient age of the Middle Marker Horizon: Onverwacht Group; Swaziland Sequence, South Africa. Earth Planet. Sci. Letters, *14*, 360–366.

Hurst, R. W., D. Bridgwater, K. D. Collerson, and G. W. Wetherill (1975) 3600-m.y. Rb-Sr ages from very early Archaean gneisses from Saglek Bay, Labrador. Earth Planet. Sci. Letters, *27*, 393–403.

Jahn, B.-M., and C.-Y. Shih (1974) On the age of the Onverwacht group, Swaziland sequence, South Africa. Geochim. Cosmochim. Acta, *38*, 873–885.

Kovach, A. (1966) A redetermination of the half-life of rubidium-87. Acta Phys. Acad. Sci. Hung., *17*, 341–351.

Long, L. E. (1964) Rb-Sr chronology of the Carn Chuinneag Intrusion, Rossshire, Scotland, J. Geophys. Res., *69*, 1589–1597.

Mattauch, J. (1937) Das Paar Rb^{87}-Sr^{87} und die Isobarenregel. Naturwissenschaft., *25*, 189–191.

McIntyre, G. A., C. Brooks, W. Compston, and A. Turek (1966) The statistical assessment of Rb-Sr isochrons. J. Geophys. Res., *71*, 5459–5468.

McMullen, C. C., K. Fritze, and R. H. Tomlinson (1966) The half-life of rubidium-87. Can. J. Phys., *44*, 3033–3038.

Montigny, R., and G. Faure (1969) Contribution au probleme de l'homogeneisation isotopique du strontium des roches totales au course de metamorphisme: Cas du Wisconsin Range, Antarctique. C.R. Acad. Sci. Paris, *268*, 1012–1015.

Moorbath, S. (1969) Evidence for the age of deposition of the Torridonian sediments of northwest Scotland. Scott. J. Geol., *5*, Part 2, 154–170.

Moorbath, S., R. K. O'Nions, R. J. Pankhurst, N. H. Gale, and V. R. McGregor (1972) Further rubidium strontium age determinations on the very early Precambrian rocks of the Godthaab district, West Greenland. Nature, Phys. Sci., *240*, 78–82.

Neumann, W., and H. Huster (1974) The half-life of [87]Rb measured as a difference between the isotopes [87]Rb and [85]Rb. Z. Physik, *270*, 121–127.

Pankhurst, R. J., S. Moorbath, and V. R. McGregor (1973) Late event in the geological evolution of the Godthaab district, West Greenland. Nature, Phys. Sci., *243*, 24–26.

Perry, E. A., and K. K. Turekian (1974) The effect of diagenesis on the redistribution of strontium isotopes in shales. Geochim. Cosmochim. Acta, *38*, 929–935.

Pidgeon, R. T., and M. R. W. Johnson (1974) A comparison of zircon U-Pb and whole-rock Rb-Sr systems in three phases of the Carn Chuinneag granite, northern Scotland. Earth Planet. Sci. Letters, *24*, 105–112.

Turekian, K. K., and K. H. Wedepohl (1961) Distribution of the elements in some major units of the Earth's crust. Geol. Soc. Amer. Bull., *72*, 175–182.

Wetherill, G. W., G. L. Davis, and C. Lee-Hu (1968) Rb-Sr measurements on whole rocks and separated minerals from the Baltimore Gneiss, Maryland. Geol. Soc. Amer. Bull., *79*, 757–762.

York, D. (1966) Least-squares fitting of a straight line. Can. J. Phys., *44*, 1079–1086.

York, D. (1967) The best isochron. Earth Planet. Sci. Letters, *2*, 479–482.

York, D. (1969) Least-squares fitting of a straight line with correlated errors. Earth Planet. Sci. Letters, *5*, 320–324.

7 STRONTIUM IN TWO-COMPONENT MIXTURES

Several common geological processes result in the mixing of materials having different chemical and isotopic compositions of elements such as strontium or lead. Examples of such mixing processes are the mingling of the water of a tributary stream with that of its master stream, the discharge of river water into a lake or into the oceans, the mixing of two types of sediment in a depositional basin, and the contamination of a mantle-derived magma as a result of interactions with rocks of the Earth's crust. In all such cases the chemical and isotopic compositions of the resulting mixtures can be related by means of simple mixing models. We shall confine our attention to two-component models and assume that the compositions of the resulting mixtures are not modified by reactions or processes after mixing has taken place. These restrictions simplify the mathematical treatment of the problem without necessarily invalidating the geological significance of the results.

1
Chemical Compositions of Two-Component Mixtures

Let us assume that we mix two components A and B in differing proportions specified by a parameter f defined as

$$f = \frac{A}{A + B} \qquad (7.1)$$

where A and B are the weights of the two components in a given mixture. The concentration of any element X in such a mixture is

$$X_M = X_A f + X_B(1 - f) \qquad (7.2)$$

where X_A and X_B are the concentrations of element X in components A and B, respectively, expressed in weight units. Since X_A and X_B are constants for any suite of samples formed by mixing A and B in differing proportions, X_M is a linear function of f:

$$X_M = f(X_A - X_B) + X_B \qquad (7.3)$$

Therefore, the value of the mixing parameter f can be calculated from the concentration of any element X in a mixture of two components, provided that the concentrations of X in the end members (X_A and X_B) are known.

Next we shall consider the concentrations of two elements X and Y in mixtures of components A and B whose concentrations in X and Y are X_A, Y_A, and X_B, Y_B, respectively. The equations relating the concentrations of X and Y in the mixtures (X_M, Y_M) to those of their end members are similar to Equation 7.2. They can be combined into a single equation by eliminating the variable f:

$$Y_M = X_M \frac{(Y_A - Y_B)}{(X_A - X_B)} + \frac{Y_B X_A - Y_A X_B}{X_A - X_B} \qquad (7.4)$$

This is the equation of a straight line in coordinates of X_M and Y_M. It is the locus of all points representing mixtures of components A and B in differing proportions, including the pure components.

The value to geology of these mixing relations lies in the fact that they produce linear correlations between the concentrations of pairs of elements. Straight lines can therefore be fitted to arrays of data points representing mixtures of two components in differing proportions. If the concentration of one of the two elements in the end members is known, the mixing equation can be used to calculate the concentrations of the other element. Moreover, the value of the mixing parameter f can be determined from Equation 7.3 for any given sample in a suite of two-component mixtures from the observed concentration of any element in that mixture. However, these relationships are valid only when the chemical compositions of the samples are determined exclusively by mixing of two components and are not modified by other processes.

2
Two-Component Mixtures Having Different ^{87}Sr/^{86}Sr Ratios

Let us now consider mixtures of two components A and B having not only different concentrations of strontium but also different ^{87}Sr/^{86}Sr ratios. The total number of ^{87}Sr atoms in a unit weight of such a mixture is

$$^{87}\text{Sr}_M = \frac{\text{Sr}_A Ab_A^{87} Nf}{W_A} + \frac{\text{Sr}_B Ab_B^{87} N(1-f)}{W_B} \quad (7.5)$$

where Sr_A and Sr_B are the concentrations of Sr, Ab_A^{87} and Ab_B^{87} are the isotopic abundances of ^{87}Sr, and W_A and W_B are the atomic weights of Sr in components A and B, respectively. N is Avogadro's number and f is the mixing parameter defined by Equation 7.1. We can write a similar equation for the total number of ^{86}Sr atoms and then form the ^{87}Sr/^{86}Sr ratio which eliminates Avogadro's number:

$$\left(\frac{^{87}\text{Sr}}{^{86}\text{Sr}}\right)_M$$

$$= \frac{\text{Sr}_A Ab_A^{87} f W_B + \text{Sr}_B Ab_B^{87}(1-f)W_A}{\text{Sr}_A Ab_A^{86} f W_B + \text{Sr}_B Ab_B^{86}(1-f)W_A} \quad (7.6)$$

Next, we introduce the approximations that the atomic weights of strontium and the abundances of ^{86}Sr in components A and B are identical:

$$W_A = W_B \quad \text{and} \quad Ab_A^{86} = Ab_B^{86} \quad (7.7)$$

These approximations are valid, provided that the ^{87}Sr/^{86}Sr ratios of components A and B are not too different. For example, if the ^{87}Sr/^{86}Sr ratios are 0.700 and 0.800, the abundances of ^{86}Sr differ by less than 1 percent and the corresponding atomic weights of strontium differ by only 0.08 percent. Equation 7.6 therefore reduces to

$$\left(\frac{^{87}\text{Sr}}{^{86}\text{Sr}}\right)_M = \frac{\text{Sr}_A Ab_A^{87} f + \text{Sr}_B Ab_B^{87}(1-f)}{Ab^{86}[\text{Sr}_A f + \text{Sr}_B(1-f)]} \quad (7.8)$$

Setting $Ab_A^{87}/Ab^{86} = (^{87}\text{Sr}/^{86}\text{Sr})_A$, $Ab_B^{87}/Ab^{86} = (^{87}\text{Sr}/^{86}\text{Sr})_B$ and by making use of Equation 7.2, this equation becomes

$$\left(\frac{^{87}\text{Sr}}{^{86}\text{Sr}}\right)_M = \left(\frac{^{87}\text{Sr}}{^{86}\text{Sr}}\right)_A \left(\frac{\text{Sr}_A f}{\text{Sr}_M}\right)$$

$$+ \left(\frac{^{87}\text{Sr}}{^{86}\text{Sr}}\right)_B \left(\frac{\text{Sr}_B(1-f)}{\text{Sr}_M}\right) \quad (7.9)$$

By eliminating f from Equations 7.8 and 7.2, and after rearranging the resulting equation, we obtain the useful relationship

$$\left(\frac{^{87}\text{Sr}}{^{86}\text{Sr}}\right)_M = \frac{\text{Sr}_A \text{Sr}_B \left[\left(\dfrac{^{87}\text{Sr}}{^{86}\text{Sr}}\right)_B - \left(\dfrac{^{87}\text{Sr}}{^{86}\text{Sr}}\right)_A\right]}{\text{Sr}_M(\text{Sr}_A - \text{Sr}_B)}$$

$$+ \frac{\text{Sr}_A \left(\dfrac{^{87}\text{Sr}}{^{86}\text{Sr}}\right)_A - \text{Sr}_B \left(\dfrac{^{87}\text{Sr}}{^{86}\text{Sr}}\right)_B}{\text{Sr}_A - \text{Sr}_B}$$

$$(7.10)$$

This is the equation of a hyperbola in coordinates of $(^{87}\text{Sr}/^{86}\text{Sr})_M$ and Sr_M of the form:

$$\left(\frac{^{87}\text{Sr}}{^{86}\text{Sr}}\right)_M = \frac{a}{\text{Sr}_M} + b \quad (7.11)$$

where a and b are constants specified by the concentrations and ^{87}Sr/^{86}Sr ratios of strontium in components A and B as shown in Equation 7.10.

The mixing hyperbola (7.10) can be transformed into a straight line by plotting $(^{87}\text{Sr}/^{86}\text{Sr})_M$ versus $1/\text{Sr}_M$. This is a very useful property because it enables us to fit a straight line to data points in coordinates of $^{87}Sr/^{86}Sr$ and $1/Sr$ and thus to derive the mixing equation from measurements of these parameters in a suite of geological samples formed by mixing of two components. The goodness of fit of the data points to a straight line is a test for the validity of the mixing hypothesis and of the assumption that neither the strontium concentrations nor the $^{87}Sr/^{86}Sr$ ratios were modified after mixing had occurred. The $^{87}Sr/^{86}Sr$ ratios of components A and B can usually be estimated quite reliably from a knowledge of the isotope geology of strontium to be discussed later in Chapter 8. Therefore, compatible strontium concentrations of components A and B can be calculated by means of the mixing equation using estimates of their $^{87}Sr/^{86}Sr$ ratios.

We are now able to calculate the concentrations and $^{87}Sr/^{86}Sr$ ratios of mixtures of two components for different values of f by using Equations 7.2 and 7.9. The resulting values of Sr_M and $(^{87}Sr/^{86}Sr)_M$ are the coordinates of points that form a segment of a hyperbola in the right upper quadrant in coordinates of the $^{87}Sr/^{86}Sr$ ratio and the concentration of strontium. Figure 7.1a is such a hyperbola drawn for the case that

$$\left(\frac{^{87}Sr}{^{86}Sr}\right)_A = 0.725, \qquad Sr_A = 200 \text{ ppm}$$

$$\left(\frac{^{87}Sr}{^{86}Sr}\right)_B = 0.704, \qquad Sr_B = 450 \text{ ppm}$$

The equation of this hyperbola is obtained by calculating the values of the constants a and b according to Equation 7.10. For the case above, we obtain $a = 7.56$, $b = 0.687$.

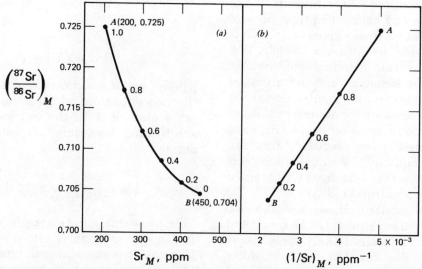

FIGURE 7.1 (a) Mixing hyperbola formed by components A and B. The coordinates of points on the hyperbola were calculated for selected values of the mixing parameter f from Equations 7.2 (Sr_M) and 7.9 ($^{87}Sr/^{86}Sr)_M$. Only a segment of the complete hyperbola in the upper right quadrant is shown where both $(^{87}Sr/^{86}Sr)_M$ and $(1/Sr)_M$ are positive. (b) Transformation of the mixing hyperbola into a straight line by plotting the reciprocals of the strontium concentrations. The line has a slope of 7.56 and an intercept on the y-axis of 0.687.

Thus the mixing equation is

$$\left(\frac{^{87}Sr}{^{86}Sr}\right)_M = \frac{7.56}{Sr_M} + 0.687 \qquad (7.12)$$

The calculated data points used to construct the mixing hyperbola (Figure 7.1a) have been replotted in part (b) of that figure in coordinates of $(^{87}Sr/^{86}Sr)_M$ and $(1/Sr)_M$ to illustrate its transformation into a straight line.

In the foregoing example we have used equations representing two-component mixtures to calculate the concentrations and $^{87}Sr/^{86}Sr$ ratios of any desired mixture when the strontium concentration and $^{87}Sr/^{86}Sr$ ratios of the components are known. However, we are usually faced with the opposite problem in geology. We can measure the strontium concentrations and $^{87}Sr/^{86}Sr$ ratios of a series of samples that are the products of mixing of two components in varying proportions. The problem is to determine the strontium concentrations of the end members and hence the proportions of these components in any given sample. This is accomplished by fitting a straight line to the data points in coordinates of $^{87}Sr/^{86}Sr$ and 1/Sr. The strontium concentrations of the end members are then obtained by solving this equation for assumed values of their $^{87}Sr/^{86}Sr$ ratios. After Sr_A and Sr_B have been estimated by this procedure, they can be used in Equation 7.2 to calculate f for any sample in the suite from its observed strontium concentration (Sr_M).

Equations similar to those derived above have been used by Pushkar (1968), Ewart and Stipp (1968), Bell and Powell (1969), Scott et al. (1971), Pushkar et al. (1972), Pushkar and Condie (1973), Bowman et al. (1973), and Faure et al. (1974) in the interpretation of chemical and isotopic compositions of volcanic rocks. Boger and Faure (1974, 1976) and Shaffer and Faure (1976) used this approach to interpret systematic variations of $^{87}Sr/^{86}Sr$ ratios in the detrital, noncar-bonate fractions of sediment in the Red Sea and the Ross Sea, respectively.

3
Isotope Ratios and Chemical Compositions

We have shown that the $^{87}Sr/^{86}Sr$ ratios and strontium concentrations of two-component mixtures are related by a hyperbolic mixing equation that transforms into a straight line by inverting the strontium concentrations. Let us now consider the relationship between $^{87}Sr/^{86}Sr$ ratios and the concentrations of other elements in such mixtures.

According to Equation 7.4, the concentrations of strontium (Sr_M) and any other element (X_M) in a two-component mixture are linearly related:

$$Sr_M = X_M \frac{(Sr_A - Sr_B)}{(X_A - X_B)} + \frac{Sr_B X_A - Sr_A X_B}{X_A - X_B}$$

$$(7.13)$$

or

$$Sr_M = c X_M + d \qquad (7.14)$$

where c and d are constants determined by the concentrations of strontium and any other element X in the end members. By substituting Equation 7.14 into 7.11, we obtain

$$\left(\frac{^{87}Sr}{^{86}Sr}\right)_M = \frac{a}{c X_M + d} + b \qquad (7.15)$$

This is the equation of a hyperbola. However, it is not transformable into a straight line because it has an additional term containing the product of the x- and y-coordinates. Therefore, the mixing equations linking the $^{87}Sr/^{86}Sr$ ratios of two-component mixtures to the concentrations of other elements cannot be derived by fitting straight lines to data points in coordinates of $^{87}Sr/^{86}Sr$ and 1/X. Instead, we must fit straight lines

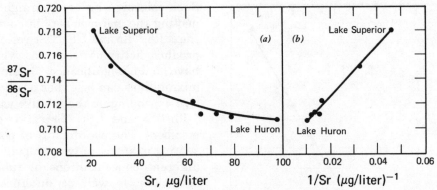

FIGURE 7.2 (a) Hyperbolic mixing curve defined by $^{87}Sr/^{86}Sr$ ratios and strontium concentrations of water samples from the North Channel of Lake Huron, Canada. (b) Plot of $^{87}Sr/^{86}Sr$ ratios and the reciprocals of the strontium concentration which fit a straight line. The equation of this line is given in the text and was used to plot the hyperbola in part (a) of this diagram. (Data from Faure et al., 1967.)

to data points in coordinates of Sr_M and X_M and derive equations of the form of 7.14. These equations can then be solved to obtain compatible values of X_A and X_B for Sr_A and Sr_B derived previously from the linear equation (7.11) linking $(^{87}Sr/^{86}Sr)_M$ and Sr_M.

We shall illustrate the use of these mixing equations by means of an example from the North Channel of Lake Huron in Canada reported by Faure et al. (1967). The North Channel of Lake Huron is a body of water restricted along its southern border by Manitoulin, Cockburn, and Drummond Island. Its northern shore is the Precambrian shield of Canada. The chemical composition and $^{87}Sr/^{86}Sr$ ratio of the water in the North Channel are the result of mixing of water derived from the crystalline rocks of the Precambrian shield with the water of Lake Huron. Most of the water derived from the Precambrian shield enters the North Channel at its western end by the discharge of the St. Mary's River which drains Lake Superior. Water draining the Precambrian shield as represented by Lake Superior has $^{87}Sr/^{86}Sr = 0.718$ (Hart and Tilton, 1966), $Sr = 21.8\,\mu g/liter$ and $Ca = 14.4\,\mu g/ml$, where-

as water in the main body of Lake Huron has $^{87}Sr/^{86}Sr = 0.7107$, $Sr = 98.2\,\mu g/liter$ and $Ca = 26.9\,\mu g/ml$. The water within the North Channel is intermediate in composition, presumably because it represents mixtures of water from Lake Superior and Lake Huron in differing proportions. The relationship between the concentrations of strontium and the $^{87}Sr/^{86}Sr$ ratios of water from the North Channel is shown in Figures 7.2a and b. A straight line has been fitted to a plot of $^{87}Sr/^{86}Sr$ ratios and $1/Sr$ shown in (b). The equation of this line is

$$\left(\frac{^{87}Sr}{^{86}Sr}\right)_M = \frac{0.1996}{Sr_M} + 0.7086 \quad (7.16)$$

This equation was used to plot the hyperbola shown in Figure 7.2a. Figure 7.3 is a plot of the concentrations of calcium and strontium in water from the North Channel. It shows the predicted linear correlation characteristic of two-component mixtures. Evidently, the $^{87}Sr/^{86}Sr$ ratio and the concentration of strontium, or of any other conservative element, of a water sample from the North Channel of Lake Huron are directly determined by the mixing parameter f and by the

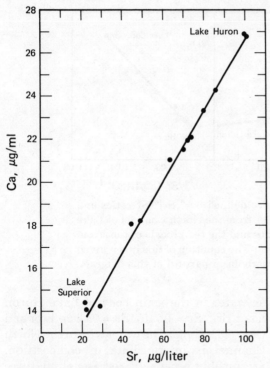

FIGURE 7.3 Plot of the concentrations of calcium and strontium in water samples from the North Channel of Lake Huron, Canada. The linear relationship displayed by the data points is characteristic of two-component mixtures. (Data from Faure et al., 1967.)

Such mixtures lie on a straight line connecting the end members in coordinates of the $^{87}Rb/^{86}Sr$ and $^{87}Sr/^{86}Sr$ ratios and thus produce fictitious isochrons whose slopes have no time significance. The slopes of such mixing lines can be either positive or negative depending on the relative values of the $^{87}Rb/^{86}Sr$ and $^{87}Sr/^{86}Sr$ ratios of the end members. The coordinates of points representing mixtures of two components having different concentrations of rubidium and strontium as well as different $^{87}Sr/^{86}Sr$ ratios can be calculated from Equations 7.2 and 7.9. The equation representing two-component mixtures in coordinates of $^{87}Sr/^{86}Sr$ and $^{87}Rb/^{86}Sr$ is

$$\left(\frac{^{87}Sr}{^{86}Sr}\right)_M = g\left(\frac{^{87}Rb}{^{86}Sr}\right)_M + e \quad (7.17)$$

where

$$e = \frac{Rb_B Sr_A \left(\frac{^{87}Sr}{^{86}Sr}\right)_A - Rb_A Sr_B \left(\frac{^{87}Sr}{^{86}Sr}\right)_B}{Rb_B Sr_A - Rb_A Sr_B} \quad (7.18)$$

$$g = \frac{Sr_A Sr_B \left[\left(\frac{^{87}Sr}{^{86}Sr}\right)_B - \left(\frac{^{87}Sr}{^{86}Sr}\right)_A\right]}{k(Rb_B Sr_A - Rb_A Sr_B)} \quad (7.19)$$

and k is a constant whose value depends on the abundance of ^{86}Sr and the atomic weight of strontium, that is,

$$k = \frac{(Ab^{87}Rb)WSr}{(Ab^{86}Sr)WRb} \simeq 2.90 \pm 0.007 \quad (7.20)$$

for strontium having $^{87}Sr/^{86}Sr$ ratios between limits of 0.700 to 0.750. The slope and intercept of straight lines representing two-component mixtures on isochron diagrams can be calculated directly from Equations 7.19 and 7.18, respectively. For mixtures of two components A and B ($Sr_A = 200$ ppm, $Rb_A = 400$ ppm, $(^{87}Sr/^{86}Sr)_A = 0.725$; $Sr_B = 450$ ppm, $Rb_B = 40$ ppm, $(^{87}Sr/^{86}Sr)_B = 0.704$),

known $^{87}Sr/^{86}Sr$ ratios and strontium concentrations of the two water masses being mixed. Therefore, the two-component mixing model is applicable in this instance and can be used to give a quantitative interpretation of regional variations of $^{87}Sr/^{86}Sr$ ratios and concentrations of conservative elements in the North Channel of Lake Huron.

4
Fictitious Isochrons

Let us assume that a suite of samples has formed by mixing of two components which have different Rb/Sr and $^{87}Sr/^{86}Sr$ ratios.

the slope is 0.0038 which yields a fictitious date of 272 million years.

An example of correlated $^{87}Sr/^{86}Sr$ and $^{87}Rb/^{86}Sr$ ratios was reported by Bell and Powell (1969) for potassic lava flows from the Birunga and Toro-Ankole fields along the borders of Uganda, Zaire, and Rwanda of East Africa. These two volcanic centers are about 100 miles from each other and are located along the western rift north of Lake Tanganyika and west of Lake Victoria. The lava flows are known to be Pliocene to Recent in age and there is evidence of volcanic activity in historical times. These volcanic rocks are therefore very young and should form an isochron having a slope approaching zero. However, Bell and Powell found a significant positive correlation between the average $^{87}Sr/^{86}Sr$ and Rb/Sr ratios of different rock types. Their results have been plotted in Figure 7.4 in coordinates of $^{87}Sr/^{86}Sr$ and $^{87}Rb/^{86}Sr$. Evidently, the $^{87}Sr/^{86}Sr$ ratios of these rocks, which are all less than one million years old, are positively correlated with their $^{87}Rb/^{86}Sr$ ratios. The line which has been arbitrarily drawn through the cluster of points is *not* an isochron, but may represent a mixing line. The date indicated by the slope of the line as drawn in Figure 7.4 is 773 million years. This date is fictitious because in this case the $^{87}Sr/^{86}Sr$ ratios of these rocks are probably the result of a mixing process and are not due to decay of ^{87}Rb in the rocks after their formation. Other examples of correlated $^{87}Sr/^{86}Sr$ and Rb/Sr ratios of Tertiary lava flows were reported by Dickinson et al. (1969) and by Leeman and Manton (1971).

The positive correlation of initial $^{87}Sr/^{86}Sr$ ratios of igneous rocks and their Rb/Sr ratios is also interpretable as evidence of long-term heterogeneity of the upper mantle. Brooks et al. (1976) listed thirty occurrences of volcanic and plutonic igneous rocks that

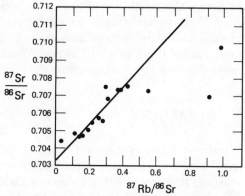

Figure 7.4 Fictitious isochron drawn through the average $^{87}Sr/^{86}Sr$ and $^{87}Rb/^{86}Sr$ ratios different varieties of potassic lavas from the Birunga and Toro-Ankole regions of the East African Rift Valleys. These rocks are known to be Pliocene or younger in age and some may have been erupted in historical time. The correlation of the $^{87}Sr/^{86}Sr$ and $^{87}Rb/^{86}Sr$ ratios of these rocks may be due to mixing of two components containing strontium of differing isotopic compositions. (Data from Bell and Powell, 1969.)

yield "pseudoisochrons" when the *initial* $^{87}Sr/^{86}Sr$ ratios are plotted versus their $^{87}Rb/^{86}Sr$ ratios. They rejected contamination as the cause for this phenomenon and proposed instead that such pseudoisochrons reflect the age of the mantle from which magma originated. Their treatment of this problem is helpful because it clarifies our understanding of the geochemical evolution of the mantle of the Earth into the lithosphere, asthenosphere and mesosphere. The lithospheric mantle, which underlies the oceanic and continental crust, may contain regions that acquired different Rb/Sr ratios at different times and subsequently remained closed. Magma generated in these regions could then have different $^{87}Sr/^{87}Sr$ ratios depending on the Rb/Sr ratios and the age of the source materials. Consequently, pseudoisochrons may in fact be "mantle isochrons" although the reliability of the dates derived from them remains open to question. The isotopic evolution of strontium in the mantle and crust of the Earth will be considered in more detail in the next chapter.

5   SUMMARY

Mixtures of two components having differing $^{87}Sr/^{86}Sr$ ratios and strontium concentrations form hyperbolas in coordinates of $^{87}Sr/^{86}Sr$ and the strontium concentrations. The mixing equation for any set of such two-component mixtures can be derived by fitting straight lines to the data points in coordinates of $^{87}Sr/^{86}Sr$ and $1/Sr$. The concentrations of the end members can then be estimated by solving this equation for assumed values of their $^{87}Sr/^{86}Sr$ ratios. Once the strontium concentrations of the end members have been determined, the concentration of either one of the two components in any given mixture can be calculated. Such two-component mixtures are formed during mingling of water masses of differing composition, by deposition of sediment derived from different sources (i.e., old granitic rocks and young volcanic rocks of basaltic composition), and by interaction of basaltic magma with old granitic rocks of the Earth's crust.

The equations relating $^{87}Sr/^{86}Sr$ ratios of mixtures to the concentrations of other elements are also hyperbolas. However, they are not transformable into straight lines. Instead, the concentrations of other elements in the end members can be determined from the linear relationships between conservative elements in two-component mixtures using their strontium concentrations determined as outlined above.

Mixing of two components having different Rb/Sr and $^{87}Sr/^{86}Sr$ ratios yields linear arrays in coordinates of $^{87}Sr/^{86}Sr$ and

$^{87}Rb/^{86}Sr$ whose slopes yield fictitious dates. However, isochrons formed when the initial $^{87}Sr/^{86}Sr$ ratios of igneous rocks are plotted versus their $^{87}Rb/^{86}Sr$ ratios have been interpreted as "mantle isochrons" and may reflect long-term heterogeneity of the lithospheric mantle.

PROBLEMS

1. Calculate the concentration of strontium in a sample composed of A (calcium carbonate) and B (silicate detritus) containing 40 percent of A. Assume that $Sr_A = 600$ ppm and $Sr_B = 100$ ppm. (**ANSWER:** $Sr = 300$ ppm.)

2. Calculate the concentration of component A in a mixture of A and B containing 265 ppm of strontium. Assume $Sr_A = 85$ ppm, $Sr_B = 450$ ppm. (**ANSWER:** $A = 50.6$ percent.)

3. Calculate Sr_M and $(^{87}Sr/^{86}Sr)_M$ for a series of mixtures of A and B for values of $f = 0.2$, 0.4, 0.6, and 0.8. Assume that $Sr_A = 600$ ppm, $(^{87}Sr/^{86}Sr)_A = 0.709$, $Sr_B = 100$ ppm, $(^{87}Sr/^{86}Sr)_B = 0.725$. Plot the results in coordinates of $^{87}Sr/^{86}Sr$ and Sr and draw a smooth curve through the points.

4. Calculate concentrations of SiO_2 and $^{87}Sr/^{86}Sr$ ratios for a series of mixtures of A and B for values of $f = 0.2$, 0.4, 0.6, and 0.8. Assume that $(SiO_2)_A = 72$ percent, $Sr_A = 150$ ppm, $(^{87}Sr/^{86}Sr)_A = 0.730$, $(SiO_2)_B = 48$ percent, $Sr_B = 475$ ppm, $(^{87}Sr/^{86}Sr)_B = 0.704$. Plot the points in coordinates of $^{87}Sr/^{86}Sr$ and SiO_2 and draw a smooth curve through the points.

5. The strontium concentrations and $^{87}Sr/^{86}Sr$ ratios of detrital, noncarbonate fractions of a suite of deep sea sediment samples are:

NO.	Sr, ppm	$^{87}Sr/^{86}Sr$
1	111	0.7136
2	143	0.7109
3	200	0.7083
4	500	0.7043

NO.	Sr μg/liter	Ca μg/ml	$^{87}Sr/^{86}Sr$
1	28.8	14.3	0.7151
2	48.8	18.2	0.7130
3	62.3	21.1	0.7123
4	66.6	22.0	0.7112
5	72.5	22.1	0.7113
6	78.6	23.4	0.7111

Fit a mixing hyperbola to these data and derive its equation. Assume that the end members of these mixtures have $^{87}Sr/^{86}Sr$ ratios of 0.7150 (component A) and 0.7030 (component B). Calculate Sr_A and Sr_B and determine the concentration of component A in all four samples.

6. The following analytical data pertain to water samples from the North Channel area of Lake Huron:

Assume that the $^{87}Sr/^{86}Sr$ ratio of component A (Lake Superior) is 0.7180 and that of component B (Lake Huron) is 0.7107. Estimate the concentrations of Sr and Ca of water in Lake Superior and Lake Huron from these data and calculate the volume percent concentration of Lake Superior water in each of the samples.

REFERENCES

Bell, K., and J. L. Powell (1969) Strontium isotopic studies of alkalic rocks: The potassium-rich lavas of the Birunga and Toro-Ankole regions, east and central equatorial Africa. J. Petrol., *10*, 536–572.

Boger, P. D., and G. Faure (1974) Strontium-isotope stratigraphy of a Red Sea core. Geol., *2*, 181–183.

Boger P. D., and G. Faure (1976) Systematic variations of sialic and volcanic detritus in piston cores from the Red Sea. Geochim. Cosmochim. Acta, *40*, 731–742.

Bowman, H. R., A. Asaro, and I. Perlman (1973) On the uniformity of composition in obsidians and evidence for magmatic mixing. J. Geol., *81*, 312–327.

Brooks, C., D. E. James, and S. R. Hart (1976) Ancient lithosphere: Its role in young continental volcanism. Science, *193*, 1086–1094.

Dickinson, D. R., M. H. Dodson, I. G. Gass, and D. C. Rex (1969) Correlation of initial Sr^{87}/Sr^{86} with Rb/Sr in some late Tertiary volcanic rocks of South Arabia. Earth Planet. Sci. Letters, *6*, 84–90.

Ewart, A., and J. J. Stipp (1968) Petrogenesis of the volcanic rocks of the central North Island, New Zealand, as indicated by a study of $^{87}Sr/^{86}Sr$ ratios, and Sr, Rb, K, and Th abundances, Geochim. Cosmochim. Acta, *32*, 699-736.

Faure, G., J. R. Bowman, D. H. Elliot, and L. M. Jones (1974) Strontium isotope composition and petrogenesis of the Kirkpatrick Basalt, Queen Alexandra Range, Antarctica. Contrib. Mineral. Petrol., *48*, 153–169.

Faure, G., L. M. Jones, R. Eastin, and M. Christner (1967) Strontium isotope composition and trace element concentrations in Lake Huron and its principal tributaries. Rept. No. 2, Laboratory for Isotope Geology and Geochemistry, The Ohio State University, Columbus, Ohio, 109 pp.

Hart, S. R., and G. R. Tilton (1966) The isotope geochemistry of strontium and lead in Lake Superior sediments and water. The Earth beneath the Continents. J. S. Steinhart and T. L. Smith, eds., Geophys. Monogr. No. 10, Amer. Geophys. Union, 127–137.

Leeman, W. P., and W. I. Manton (1971) Strontium isotopic composition of basaltic lavas from the Snake River Plain, southern Idaho, Earth Planet. Sci. Letters, *11*, 420–434.

Pushkar, P. (1968) Strontium isotope ratios in volcanic rocks of three island arc areas. J. Geophys. Res., *73*, 2701–2714.

Pushkar, P., A. R. McBirney, and A. M. Kudo (1972) The isotopic composition of strontium in Central American ignimbrites. Bull. Volcan., *35*, 265–294.

Pushkar, P., and K. C. Condie (1973) Origin of the Quaternary basalts from the Black Rock Desert region, Utah: Strontium isotopic evidence. Geol. Soc. Amer. Bull., *84*, 1053–1058.

Scott, R. B., R. W. Nesbitt, E. J. Dasch, and R. J. Armstrong (1971) A strontium isotope evolution model for Cenozoic magma genesis, eastern Great Basin, U.S.A. Bull. Volcan., *35*, 1–26.

Shaffer, N. R., and G. Faure (1976) Regional variation of $^{87}Sr/^{86}Sr$ ratios and mineral composition of sediment in the Ross Sea. Geol. Soc. Amer. Bull., *87*, 1491–1500.

8 ISOTOPE GEOLOGY OF STRONTIUM

The isotopic composition of strontium has been changing continuously since nucleosynthesis because of the decay of ^{87}Rb to ^{87}Sr. We showed in Chapter 6 (Equation 6.11) that the rate of change of the ^{87}Sr/^{86}Sr ratio is closely approximated by (^{87}Rb/^{86}Sr) λ, that is, it is proportional to the Rb/Sr ratio of the system in which the strontium resides. Therefore, the rocks and minerals of the crust and mantle of the Earth have acquired widely differing ^{87}Sr/^{86}Sr ratios consistent with their Rb/Sr ratios and their ages. It follows that strontium has become increasingly heterogeneous in its isotopic composition throughout geologic time and will, of course, continue its evolution in the future. On the other hand, these considerations imply a greater *homogeneity* of the isotopic composition of strontium in the Earth at the beginning of geologic time. In fact, it is generally agreed that the matter that accreted from the solar nebula to form the proto-planet Earth had very similar ^{87}Sr/^{86}Sr ratios clustered closely about the so-called primordial value. The subsequent diversification of the isotopic composition of strontium resulted from the geochemical differentiation of the Earth which led to the formation of rocks and minerals having widely differing Rb/Sr ratios.

The isotopic evolution of strontium in the Earth throughout geologic history is complicated by the fact that geologic processes have continuously destroyed rock systems formed earlier and recombined the material to form new rocks. The strontium in such newly formed rocks is therefore derived from older rocks which themselves may have formed from still earlier generations of rocks. The evidence for this phenomenon is preserved in the "initial ^{87}Sr/^{86}Sr ratios" of igneous rocks which contain a record of the previous history of the strontium before it was incorporated into the rock in which it now resides. Therefore, the study of the Rb-Sr decay scheme in igneous rocks is useful not only for the purpose of discovering their ages but also provides information about the geochemical history of the strontium they contain.

1
Stony Meteorites and Primordial Strontium

It has not been possible to measure the isotopic composition of strontium that was incorporated into the Earth at the time of its formation. The reason is that the rocks formed at that time have been destroyed by geological processes, or have been altered by subsequent metamorphism, or have been buried deep in the mantle and are not accessible to us. For this reason, we must rely on the study of meteorites to determine the primordial ^{87}Sr/^{86}Sr ratio of terrestrial strontium.

Meteorites are fragments of larger parent bodies that are believed to have formed in the solar system at about the same time as the planets and their satellites (Mason, 1962). These parent bodies underwent chemical differentiation for short periods of time lasting only tens to perhaps a few hundred million years. During this process, melting occurred in the larger ones, resulting in the segregation of metallic iron-nickel and silicate liquids. The silicate liquids subsequently cooled and formed a variety of solids having differing Rb/Sr ratios. Subsequent collisions among the parent bodies produced a variety of breccias containing fragments of diverse origins and compositions. It is generally believed, but difficult to prove (Wetherill, 1976), that the meteorites that have hit the Earth and the other inner planets of the solar system originate from the belt of asteroids between the orbits of Mars and Jupiter. In any case, it seems very probable that meteorites represent matter from our solar system and that they preserve a record of the isotopic composition of strontium (and lead) of their parent bodies at the time they became inactive due to rapid loss of heat.

Age determinations of stony meteorites by the Rb-Sr method generally indicate dates in the time interval 4.6 ± 0.1 billion years

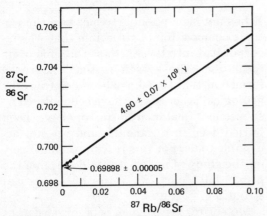

FIGURE 8.1 Rb-Sr isochron for the basaltic achondrite Juvinas. The data points represent glass(?), tridymite and quartz, pyroxene, total rock, and plagioclase in order of decreasing $^{87}Sr/^{86}Sr$ ratios. The date indicated by the isochron is the time elapsed since crystallization of a silicate liquid in the parent body of this meteorite. (Data from Allegre et al., 1975. The decay constant of ^{87}Rb is $1.39 \times 10^{-11} \, y^{-1}$.)

(Fig. 8.1, Table 8.1). However, at least three meteorites (Kodaikanal, Nakhla, and Kapoete) yield lower dates suggesting that some of the parent bodies remained active for longer periods of time than the others (Burnett and Wasserburg, 1967; Papanastassiou and Wasserburg, 1974; Papanastassiou et al., 1974). The Rb-Sr isochrons of meteorites are generally compatible with the hypothesis that most of them crystallized within a short time interval and that they had similar initial $^{87}Sr/^{86}Sr$ ratios. However, careful measurements have disclosed significant differences among the initial $^{87}Sr/^{86}Sr$ ratios of several stony meteorites. These differences provide information about events that occurred very early in the history of the solar system, based on the following line of reasoning.

Measurements of the rubidium and strontium concentrations of the photosphere of the sun suggest a Rb/Sr ratio of about 0.65. If this value is representative of the solar nebula, its $^{87}Sr/^{86}Sr$ ratio increased by 2.6×10^{-5} per million years. Therefore, the isotopic composition of strontium of a Rb-free phase formed from the solar nebula depends on the time of its formation (Fig. 8.2). It follows, that the $^{87}Sr/^{86}Sr$ ratios of the parent bodies of the meteorites depended on the time and duration of their formation. Subsequently, strontium continued to evolve within each parent body at rates determined by its Rb/Sr ratio. Some of the larger parent bodies may have melted completely and were therefore isotopically homogenized. Others may have remained heterogeneous overall, with only localized homogenization. These considerations make clear that stony meteorites representing fragments of different parent bodies do not, strictly speaking, have identical initial $^{87}Sr/^{86}Sr$ ratios. The fact that so many of them do fit isochrons reasonably well indicates only that most of them formed in a relatively short interval of time,

Table 8.1   **Summary of Rb-Sr Dates and Initial $^{87}Sr/^{86}Sr$ Ratios of Meteorites**

MATERIAL	METHOD	DATE 10^9 YEARS	$\left(\dfrac{^{87}Sr}{^{86}Sr}\right)_0$	REFERENCE
Juvinas (achondrite)	Mineral isochron	4.60 ±0.07	0.69898 ±0.00005	1
Allende (carb. chondrite)	Mixed isochron	4.5–4.7	0.6988	2
Colomera (silicate inclusion, iron meteorite)	Mineral isochron	4.61 ±0.04	0.6994 ±0.0001	3
Enstatite chondrites	Whole-rock isochron	4.54 ±0.13	0.6993 ±0.0015	4
Enstatite chondrites	Mineral isochron	4.56 ±0.15	0.7005 ±0.0030	4
Carbonaceous chondrites	Whole-rock isochron	4.69[a] ±0.14	0.6983[a] ±0.0024	5
Amphoterite chondrites	Whole-rock isochron	4.56 ±0.15	0.7005 ±0.0015	6
Bronzite chondrites	Whole-rock isochron	4.69 ±0.14	0.6983 ±0.0024	7
Hypersthene chondrites	Whole-rock isochron	4.48 ±0.14	0.7008 ±0.001	8
Krähenberg (amphoterite)	Mineral isochron	4.70 ±0.01	0.6989 ±0.0005	9
Norton County (achondrite)	Mineral isochron	4.7 ±0.1	0.700 ±0.002	10

Note: All dates are based on a value of 1.39×10^{-11} y^{-1} for the decay constant of ^{87}Rb.

[a] Five of 11 specimens fit the bronzite isochron within experimental error. The others scatter widely possibly because of rubidium and strontium contamination.

1. Allegre et al. (1975)
2. Wetherill et al. (1973)
3. Sanz et al. (1970)
4. Gopalan and Wetherill (1970)
5. Kaushal and Wetherill (1970)
6. Gopalan and Wetherill (1969)
7. Kaushal and Wetherill (1969)
8. Gopalan and Wetherill (1968)
9. Kempe and Müller (1968)
10. Bogard et al. (1967)

that their initial $^{87}Sr/^{86}Sr$ ratios were similar, and that they have not been altered since final cooling and crystallization of their respective parent bodies.

The question regarding the primordial $^{87}Sr/^{86}Sr$ ratio of the Earth therefore does not have an unambiguous answer. The best we can hope to do is to select a value that is representative of the $^{87}Sr/^{86}Sr$ ratio in the solar nebula at the time of formation of planetary objects. The basaltic achon- drites are probably most suitable for this purpose because they show evidence of having crystallized from silicate liquids and therefore resemble terrestrial igneous rocks. Moreover, their Rb/Sr ratios are very low (about 0.002), which means that their $^{87}Sr/^{86}Sr$ ratios have changed little since crystallization. For these reasons Papa- nastassiou and Wasserburg (1969) analyzed seven basaltic achondrites and determined an initial ratio of 0.698990 ± 0.000047. This

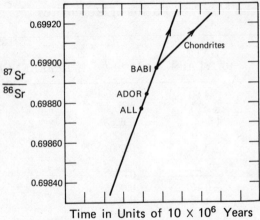

FIGURE 8.2 Isotopic evolution of strontium in the solar nebula, assuming it had a Rb/Sr ratio of 0.65. ALL is the initial $^{87}Sr/^{86}Sr$ ratio of Ca-Al rich chondrules from the carbonaceous chondrite Pueblito de Allende. ADOR is the initial ratio of the augite achondrite Angra dos Reis, and BABI is the best initial $^{87}Sr/^{86}Sr$ ratio of basaltic achondrites. The diagram illustrates the point that the initial $^{87}Sr/^{86}Sr$ ratios of meteorites depend on the sequence in which they became separated from the solar nebula and can be used to date events of short duration in the early history of the solar system. The diagram also shows a hypothetical Sr-evolution line for chondritic meteorites having an average Rb/Sr ratio of 0.25 and an initial $^{87}Sr/^{86}Sr$ ratio equal to BABI.

value is known as BABI (basaltic achondrite best initial). A very similar value of 0.698976 ± 0.000055 (JUSI) was obtained for the achondrites Juvinas and Sioux County alone. The initial ratio of Juvinas was subsequently confirmed by Allegre et al. (1975). BABI has since been adjusted very slightly with reference to an interlaboratory standard (N.B.S., No. SRM-987, $^{87}Sr/^{86}Sr$ = 0.71014) to 0.69897 ± 0.00003. BABI appears to be the $^{87}Sr/^{86}Sr$ ratio of a certain class of parent bodies, at the time of their final cooling and crystallization, within which the basaltic achondrite meteorites were formed. Because of the low Rb/Sr ratios of these parent bodies, the extent of isotopic evolution during their period of internal differentiation is probably very small. Therefore, BABI can be taken as representing the $^{87}Sr/^{86}Sr$ ratio of the solar nebula at a very early stage during the formation of planetary bodies. It is a reference point to which one can compare the initial $^{87}Sr/^{86}Sr$ ratios of other planetary objects in order to determine whether they formed earlier or later than the basaltic achondrites.

There is no reason to believe that basaltic achondrites are the earliest objects to form in the solar system. In fact, several objects have been found that appear to be more primitive (Fig. 8.2). One of these is the augite achondrite Angra dos Reis, which has an initial $^{87}Sr/^{86}Sr$ ratio of 0.69884 ± 0.0003 (ADOR). Wetherill et al. (1973) reported a similar initial ratio of 0.69880 for Na-poor, Ca-rich inclusions in the carbonaceous chondrite Pueblito de Allende. Gray et al. (1973) found an even lower value of 0.69877 ± 0.00002 (ALL) in Ca-Al rich and alkali-poor chondrules extracted from Allende. Therefore, chondrules, small silicate spheres, may be the earliest objects to form in the solar system.

The initial $^{87}Sr/^{86}Sr$ ratios of lunar samples are generally greater than BABI which indicates that the rocks exposed on the surface of the moon were derived from material in its interior containing strontium that had evolved beyond BABI. Age determinations by the Rb-Sr method of lunar rocks suggest that large-scale melting and subsequent crystallization of magma peaked at about 3.8 to 4.0 billion years ago and ceased largely by about 3.0 billion years in the past (Birck et al., 1975; Jessberger et al., 1974; Mark et al., 1974; Papanastassiou and Wasserburg, 1973, 1971, 1970; Cliff et al., 1972; Murthy et al., 1971). Many additional papers dealing with the isotopic evolution of strontium and with dating of lunar rocks by the Rb-Sr method can be found in the Proceedings of Lunar Science Conferences published as supplements to Geochimica et Cosmochimica Acta. Reviews based on the study of material collected during the landings of Apollo 11 and 12 have been published by Wetherill (1971) and by Faure and Powell (1972).

2
The Isotopic Evolution of Strontium in the Earth

The foregoing review of the early history of the solar system revealed by the study of meteorites supports the assumption that the Earth formed from the solar nebula in the time interval 4.6 ± 0.1 billion years ago and that its $^{87}Sr/^{86}Sr$ ratio was close to 0.699. We do not know whether the Earth acquired its internal structure as a result of the sequential accumulation of particles of differing composition or whether it was initially heterogeneous. It is becoming clear, however, that rocks enriched in alkali metals, silica, and alumina existed on the surface of the Earth only a few hundred million years after its formation. For example, Moorbath et al. (1975) have shown that gneisses and associated "supracrustal rocks" in the Isua area of West Greenland formed within about 200 million years prior to about 3700 million years ago. The supracrustal rocks consist of metavolcanic and metasedimentary rocks including banded ironstones and conglomerate. The ironstones appear to be chemical precipitates and therefore signal the presence of water on the surface of the Earth. The conglomerate contains boulders of clastic sediments derived from an acid igneous parent. The boulders and the associated matrix yield a whole-rock Rb-Sr isochron date of 3710 m.y., which is indistinguishable from dates derived from the associated Amitsoq gneiss (see Chapter 6, Fig. 6.7). The similarity of dates obtained for the metasediments and the gneisses suggests that they reflect the time of isotopic homogenization of strontium during regional metamorphism. Barton (1975) and Hurst et al. (1975) have reported similar dates for gneisses from the east coast of Labrador which they, too, regard as indicative of a metamorphic event.

The initial $^{87}Sr/^{86}Sr$ ratios of these and other ancient granitic rocks are generally quite low (0.700–0.702) but all are greater than BABI. However, Barton (1975) reported a remarkably high value of 0.7044 ± 0.0010 for the Hebron gneiss (east-coast of Labrador) dated at 3618 ± 106 million years. This implies that the *parent* of the Hebron gneiss must have had a Rb/Sr ratio of at least 0.12 in order to increase its $^{87}Sr/^{86}Sr$ from BABI to 0.7044 in less than about one billion years. On the other hand, if the Hebron gneiss has the same Rb/Sr ratio (0.25) as its ancestor, the latter could have separated from a source represented by BABI about 4150 million years ago.

Evidently, rocks of "granitic" composition, enriched in rubidium, began to form very early in the history of the Earth and have played a dominant role in the isotopic evolution of strontium ever since. The resulting enrichment of the continental crust in radiogenic ^{87}Sr contrasts sharply with the much lesser increase of the $^{87}Sr/^{86}Sr$ ratio in the upper mantle. This difference has been very useful because it permits us to identify strontium that has resided in the continental crust by its elevated $^{87}Sr/^{86}Sr$ ratio and to distinguish it from strontium that originated in the upper mantle (Faure and Hurley, 1963). Such a distinction is useful because it enables us to make statements about the sources from which igneous rocks are derived and about the consequent geochemical evolution of the continental crust. In order to apply this criterion, we must not only know the range of $^{87}Sr/^{86}Sr$ ratios that characterize crustal rocks but we must also have information regarding the isotopic evolution of strontium in the upper mantle throughout geologic time.

The isotopic composition of strontium in the mantle of the Earth has been studied by analysis of basalt and of large gabbroic intrusives that are believed to have originated in the upper mantle without significant contamination with crustal strontium. The results suggest that the present-day $^{87}Sr/^{86}Sr$ ratios of the upper mantle lie within a narrow interval of 0.704 ± 0.002. However, even though strontium in the upper mantle is more homogeneous isotopically and has lower $^{87}Sr/^{86}Sr$ ratios than strontium that has resided in the continental crust for long periods of time, significant differences in the $^{87}Sr/^{86}Sr$ ratios of Recent basaltic rocks have been observed. These will be discussed later in this chapter. Of more immediate concern is the question how the $^{87}Sr/^{86}Sr$ ratio of the upper mantle has evolved throughout geologic time.

The isotopic evolution of strontium in the upper mantle can be studied by plotting the initial $^{87}Sr/^{86}Sr$ ratios of basaltic and gabbroic rocks versus their ages determined from Rb-Sr isochrons. The resulting data points might be expected to define a development line for strontium evolution in the upper mantle. However, the range of $^{87}Sr/^{86}Sr$ ratios of young basaltic rocks suggests that the Rb/Sr ratio of the upper mantle is variable. This means that the isotopic evolution of strontium in the upper mantle may have to be represented by diverging lines in a Sr development diagram rather than by a single line. Moreover, the Rb/Sr ratio of the mantle may have decreased in the course of geologic time, if rubidium was transported preferentially into the crust. In this case, the slopes of the

development lines would decrease as a function of time. Faure and Powell (1972) plotted the initial $^{87}Sr/^{86}Sr$ of several gabbroic intrusives on a Sr-development diagram and concluded that the evidence favors nonlinear isotopic evolution of strontium in the upper mantle (Fig. 8.3). This view has been supported more recently by Pidgeon and Hopgood (1975) and is consistent with models discussed by Hart (1969). The time-integrated, average Rb/Sr ratio of the upper mantle, indicated by its present $^{87}Sr/^{86}Sr$ ratio of 0.704 ± 0.002, relative to BABI 4.6 billion years ago, is 0.027 ± 0.011.

The isotopic evolution of strontium in the continental crust cannot be described by means of simple models because of its heterogeneity in terms of the ages and Rb/Sr ratios of crustal rocks (Chapter 6, Table 6.1). Analyses of strontium in mollusk shells from lakes and rivers of the Canadian shield, which contain strontium derived from the Precambrian gneisses, indicate $^{87}Sr/^{86}Sr$ ratios ranging from 0.712 to 0.726 (Faure et al., 1963). However, the evolutionary history of any given sample of strontium in the continental crust is difficult to reconstruct. After initial separation from the upper mantle early in the history of the Earth, the strontium may have passed through an unknown number of systems having different Rb/Sr ratios for varying periods of time. Its $^{87}Sr/^{86}Sr$ ratio may have been changed not only by decay of ^{87}Rb but also by a variety of geological processes related to igneous activity, metamorphism, weathering, transport, and deposition of sediment followed by diagenesis, folding, and renewed metamorphism. However, in spite of the complex history of strontium in the continental crust, its $^{87}Sr/^{86}Sr$ ratio should, in most cases, be significantly greater than that of strontium in the upper mantle. This property is the basis for the interpretation of the initial $^{87}Sr/^{86}Sr$ ratios

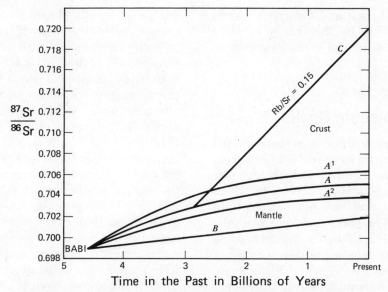

FIGURE 8.3   Isotopic evolution of terrestrial strontium. The three curved lines represent possible evolutionary paths of strontium in the upper mantle under the continents and are based on the initial $^{87}Sr/^{86}Sr$ ratios of gabbroic intrusives (Faure and Powell, 1972). The curvature of the lines implies a time-dependent decrease in the Rb/Sr ratio of the upper mantle. The straight line (B) connecting BABI to a present value of 0.702 represents strontium evolution in mantle regions depleted in rubidium. The diagram also shows a development line (C) for strontium that was withdrawn from the upper mantle about 2.9 billion years ago and subsequently resided in a closed system having a Rb/Sr ratio of 0.15. The present $^{87}Sr/^{86}Sr$ ratio of this strontium is fairly representative of the isotopic composition of strontium in the continental crust.

of igneous and sedimentary rocks. An elevated initial $^{87}Sr/^{86}Sr$ ratio indicates the presence of crustal strontium and implies that the rock was formed, at least partly, from crustal material. In other words, a study of the initial $^{87}Sr/^{86}Sr$ ratios provides information about the *sources* of granitic rocks that make up the continental crust.

The evidence regarding the evolution of the continental crust favors the view that it formed early in the history of the Earth (Moorbath, 1975) and that many granitic rocks have crustal ancestors (Faure and Powell, 1972). This is particularly true of late Precambrian and Phanerozoic granitic gneisses and associated plutonic rocks which may represent crystallization products of magmas derived locally. However, the upper mantle has undeniably acted as a continuous source for large volumes of basaltic and ultramafic rocks, including syenites, carbonatites, and kimberlites. Even some plutonic rocks of granitic composition have sufficiently low initial $^{87}Sr/^{86}Sr$ to suggest a source in the upper mantle (Moorbath, 1975). The interaction between the continental crust and the upper mantle has been clarified by the theory of plate tectonics to be discussed in the next section of this chapter.

3
Volcanic Rocks

The isotopic composition of strontium (and lead) of volcanic rocks provides information about the sources from which magmas originate and about the processes by which their chemical and isotopic compositions are modified. The theory of plate tectonics has been very helpful in this regard because it provides a framework within which volcanic activity can be viewed. This is especially true of volcanism within the ocean basins and along continental margins but does not necessarily apply to volcanic activity in the interiors of continents.

It is now known that the surface of the Earth consists of a number of rigid lithospheric plates that are about 100 km thick and encompass the upper mantle as well as the overlying oceanic and continental crust. The oceanic crust is formed by intrusion of dikes and by the outpouring of basaltic lava along midocean ridges, exemplified by the Mid-Atlantic Ridge. The plates are carried away from the spreading ridges by convective motions in the asthenosphere of the underlying mantle. The continuous enlargement of lithospheric plates at midocean ridges requires a compensating destruction of the opposite plate margins. This takes place along the subduction zones where the plates plunge back into the mantle. Subduction zones are marked by the presence of deep-sea trenches bordered by volcanic island arcs such as the Aleutian trench and the Aleutian Islands in the North Pacific ocean. Subduction zones, as well as midocean ridges are centers of frequent earthquake activity. The focal points of earthquakes along island arcs lie along inclined planes called the Benioff zone which is believed to coincide with the downgoing lithospheric slab (Fig. 8.4). The volcanic activity of island arcs is due to the generation of magma within the upper part of the plunging plate and/or within the mantle above it (Peterman et al., 1967; Church, 1973).

Evidently, plate tectonics provides a rational explanation for volcanic activity within the ocean basins (midocean ridges) and along their margins (volcanic island arcs). In addition, volcanic activity is also observed within the ocean basins unrelated to active spreading centers and may be manifested by the presence of volcanic islands or submerged sea mounts. The Hawaiian Islands are an example of this phenomenon which may be caused by the eruption of magma through a lithospheric plate as it moves over a "hot spot" in the mantle (Jackson et al., 1972). The chemical compositions and $^{87}Sr/^{86}Sr$ ratios of volcanic rocks generated at these three sites vary considerably, but tend to exhibit certain characteristic patterns that reflect differences in the sources of magma and in the processes by which magmas are formed and subsequently evolve.

Figure 8.5 displays the $^{87}Sr/^{86}Sr$ ratios of over 900 young volcanic rocks of basaltic to intermediate composition taken from the literature. It corroborates the statement made earlier that most young volcanic rocks believed to have originated in the upper mantle have $^{87}Sr/^{86}Sr$ ratios that lie in the interval 0.704 ± 0.002. However, the data in Figure 8.5 suggest that there are systematic differences in the average $^{87}Sr/^{86}Sr$ ratios of volcanic rocks formed in different environments and that real variations exist within volcanic rocks in each of the four categories shown. The lowest $^{87}Sr/^{86}Sr$ ratios (0.70280) occur in the oceanic tholeiites

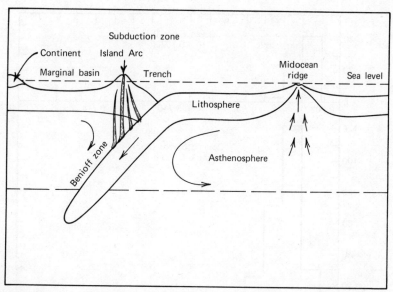

FIGURE 8.4 Schematic view of the cross section of the Earth's crust and upper mantle across a midocean ridge and a volcanic island arc. The lithospheric plate, consisting of rigid mantle and the overlying oceanic crust composed of basalt and pelagic sediment, is subducted into the mantle along a deep-sea trench. Magma is generated by melting of oceanic crust along the upper part of the plunging plate and/or in the overlying mantle as a result of fluxing with water derived by dehydration of hydrous minerals of the oceanic crust. The resulting volcanic and plutonic igneous rocks may display a certain amount of chemical zonation with respect to the trench axis or with regard to depth to the Benioff zone. The centers of igneous activity may migrate away from the trench axis and the initial $^{87}Sr/^{86}Sr$ ratios tend to increase in the same direction. The diagram is not drawn to scale.

(Engel et al., 1965) extruded along midocean ridges. Those from oceanic islands (0.70386), island arcs (0.70437), and continental areas (0.70577) are successively higher. All of the histograms are skewed toward higher values of the $^{87}Sr/^{86}Sr$ which, in some cases, may result from contamination of the magma with radiogenic ^{87}Sr derived from old sialic rocks or their derivatives. This is most likely for volcanic rocks extruded on the continents whose $^{87}Sr/^{86}Sr$ ratios range beyond 0.710, but may also apply to certain volcanic rocks in island arcs and oceanic islands where magma may be contaminated with terrigenous sediment derived from nearby continents or by interaction with sea water.

The isotopic composition of strontium in volcanic and plutonic igneous rocks associated with subduction zones is of particular interest because of the possibility that ore deposits may be formed in this context (Sillitoe, 1972; Mitchell and Bell, 1973; Lowell, 1974). Here we must distinguish between subduction of oceanic crust under oceanic crust, which gives rise to volcanic island arcs, and subduction of oceanic crust under continental crust, which produces

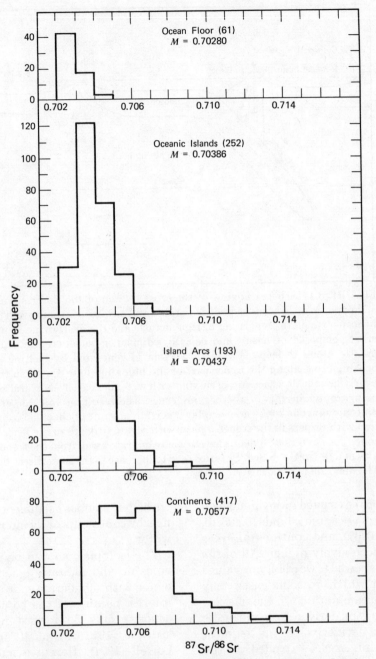

FIGURE 8.5 Histograms of the $^{87}Sr/^{86}Sr$ ratios of volcanic rocks in different geological environments. The data were taken from the literature and were corrected to a value of 0.7080 for the $^{87}Sr/^{86}Sr$ ratio of the Eimer and Amend strontium isotope standard. The number in brackets indicates the number of samples and M is the arithmetic mean.

igneous activity exemplified by the Andes of South America. In both situations the chemical compositions and ages of the resulting volcanic and plutonic rocks tend to vary systematically with distance from the trench and therefore with depth to the Benioff zone, that is, basalt near the trench grading into andesite and dacite farther away. Dickinson and Hatherton (1967) observed that the concentration of K_2O (adjusted to 55 or 50 percent SiO_2) of andesites in the circumpacific area increases as a function of depth to the Benioff zone. Hart et al. (1970) reported a similar effect for strontium concentrations in island arc systems of the Pacific Ocean. However, Nielson and Stoiber (1973) found that the absolute values of K_2O and the rate of change of this parameter as a function of depth to the Benioff zone vary widely at different localities and concluded that the adjusted K_2O content of andesites is not a reliable indicator of the depth to the Benioff zone underlying an andesite volcano. Apparently, the concentrations of K_2O and SiO_2 of basaltic and andesitic volcanic rocks extruded along island arcs are controlled by several factors, among which the depth to the Benioff zone (or distance from the trench) is not necessarily dominant. Nevertheless, the bulk chemical compositions of lavas in island arcs do display a certain degree of zoning in relation to the subduction zone.

The isotopic compositions of strontium in volcanic rocks of island arcs also appear to vary systematically, but the relationship is not simple and may be absent in some areas. Local differences in the generation and subsequent evolution of magma appear to overshadow any broad similarities that may exist among different island arcs. For example, Whitford (1975) found that the $^{87}Sr/^{86}Sr$ ratios of volcanic rocks in the Sunda arc of Indonesia (Sumatra, Java, Bali,

Lombok, and Sumbawa) increase somewhat with inferred depths to the Benioff zone, but the correlation of these parameters is weak. One of several reasons for the poor correlation is that the $^{87}Sr/^{86}Sr$ ratios of the calc-alkaline rocks lying 170 to 200 km above the Benioff zone *increase along the strike* of the arc from east to west. The increase may reflect the westward thickening of the crust overlying the subduction zone. Thus, Sumatra (at the western end) may be underlain by continental crust, whereas Java (to the east) is underlain by a thinner crust which appears to be intermediate between oceanic and continental crust in terms of observed seismic velocities. Nevertheless, the evidence clearly shows that presumably cogenetic volcanic rocks in the Sunda arc have significantly *different* $^{87}Sr/^{86}Sr$ ratios that reflect in a complex and not yet fully understood way the tectonic setting and the petrogenetic processes by which they formed. The variability of $^{87}Sr/^{86}Sr$ ratios of volcanic (and plutonic) rocks formed in island arcs is at variance with the assumption that cogenetic igneous rocks have identical initial $^{87}Sr/^{86}Sr$ ratios. This insight is obviously relevant to the dating of such rocks by the Rb-Sr isochron method and may affect not only the fit of data points on isochrons but also their slopes and hence the dates derived from them.

In a study of volcanic and plutonic rocks in the central Andes (between latitudes of 26° and 29° south), McNutt et al. (1975) found that the initial $^{87}Sr/^{86}Sr$ ratios of these rocks increase fairly systematically from west to east across this mountain chain. The rocks range in age from Early Jurassic (along the coast) to Quaternary (in the interior), thus implying a general eastward shift of the igneous activity. Starting in Mid-Cretaceous time, the initial $^{87}Sr/^{86}Sr$ ratios *increase* with decreasing age and increasing distance from the coast from 0.7022

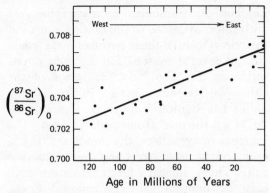

FIGURE 8.6 Systematic variation of initial $^{87}Sr/^{86}Sr$ ratios of volcanic and plutonic igneous rocks of the central Andes as a function of age, indicated by K-Ar dates. The geographic distribution of these samples implies an east-ward migration of magma sources as a function of geologic time. The variation of initial $^{87}Sr/^{86}Sr$ and the eastward shift of magma sources are thought to be related to the continuous subduction of oceanic crust under South America. (Data from McNutt et al., 1975.)

to 0.7077 (Fig. 8.6). The initial ratios of the Jurassic plutons range from 0.7043 to 0.7059, but do not correlate with age. Volcanic rocks from other island arc systems have been studied by Eward and Stipp (1968), Page and Johnson (1974), Hedge and Knight (1969), Pushkar et al. (1973), and others.

The formation of magma in subduction zones is probably localized within the upper portion of the downgoing lithospheric plate and in the mantle above it (Fig. 8.4). The oceanic crust at the top of the plate consists of about 5 km of basalt overlain by a thin layer of pelagic sediment that accumulated while the crust was moving from the ridge to the subduction zone. The temperature and pressure distribution in the downgoing plate (Oxburgh and Turcotte, 1970) results in the conversion of basalt through amphibolite to eclogite which then melts partially to form magma of calc-alkaline composition (Green, 1973). On the other hand, water evolved from

the pelagic sediment and hydrated basalt in the oceanic crust of the downgoing plate may act as a flux and cause melting also in the overlying mantle. In either case, the chemical composition of the magmas depends not only on the composition of the parent material but also on the degree of melting and on subsequent interactions of the magmas with the country rock through which they migrate on their way toward the surface. While it may be difficult to reconstruct in detail the exact sequence of events in any given case, the circumstances are favorable to account for the observed zonation of the chemical compositions and isotopic compositions of strontium in island arcs or at continental margins. At shallow depths, initial melting may lead to the formation of basaltic liquids which rise to the surface without extensive modification on the way. At greater depths, the degree of melting decreases because of the progressive loss of water. The resulting melts are more alkali-rich and may evolve more extensively as they move upward through a thicker section of mantle and/or crustal material. The $^{87}Sr/^{86}Sr$ ratios of volcanic rocks in island arcs appear to be dominated by strontium characteristic of the upper mantle, but range upward depending on the amount of strontium that was derived from the pelagic sediment of the oceanic crust and/or was contributed subsequently by interaction of the magma with rocks higher up in the section. In addition, the $^{87}Sr/^{86}Sr$ ratios of magma generated in subduction zones may be elevated because the strontium in the basalt of the oceanic crust was increased by reaction with sea water (Hart et al., 1974; Dasch et al., 1973). Moreover, it is possible that strontium is not isotopically equilibrated before melting occurs and that Rb-rich phases (e.g., phlogopite) may melt preferentially, thus forming a magma whose $^{87}Sr/^{86}Sr$ ratio is higher than that of the parent material as a whole (Pushkar and Stoeser, 1975).

It is evident that the formation of magma and its subsequent evolution are complex processes and that the chemical compositions and $^{87}Sr/^{86}Sr$ ratios of the resulting igneous rocks in island arcs and along continental margins are controlled by many factors. Neither the $^{87}Sr/^{86}Sr$ nor the concentration of any particular element are likely to be uniquely determined by the composition of the melting oceanic crust or mantle, or by the depth to the Benioff zone. Nevertheless, the magmatic activity occurring along subduction zones is one of the principal mechanisms by means of which the mantle of the Earth interacts with the continental crust. Some of the material rising into the crust or being deposited on its surface in these regions was derived from the mantle and thereby augments the crust. The question arises, therefore, whether the continental crust has been growing in volume throughout geologic time by such additions to the continental margins. This problem was reviewed by Faure and Powell (1972) on the basis of contributions by Hurley and Rand (1969), Patterson and Tatsumoto (1964), and by Armstrong (1968). Continuous crustal growth has been advocated by Hurley and his associates largely on the basis of the distribution of orogenic belts on the continents (Fig. 8.7) and because granitic igneous rocks tend to have low initial $^{87}Sr/^{86}Sr$ ratios indicating that they were either derived from the upper mantle at the time of formation or were derived from rocks having short crustal residence times. This is especially true for ancient gneisses as pointed out by Moorbath (1975). On the other hand, Armstrong (1968) suggested that magmas arising from subduction zones are at least partly formed by melting of sediment derived from the continental crust and that their initial $^{87}Sr/^{86}Sr$ ratios

may be lowered by isotopic equilibration with strontium in the mantle. Hence, a given magma may consist largely of recycled crustal material even though it has a low $^{87}Sr/^{86}Sr$ ratio. However, even if magmas in subduction zones are produced partly by melting of mantle material, the contribution they make to the volume of the crust is at least partly compensated by loss of the resiual material of the lithospheric plate. The controversy is by no means over as indicated by recent contributions, such as Yanagi (1975), Armstrong and Hein (1973), and Hurley (1972).

Finally, we must consider the possibility that the upper mantle is both chemically and isotopically heterogeneous on a large scale. This possibility can best be evaluated by consideration of volcanic rocks in oceanic islands and along midocean ridges because the probability of contamination with old sialic material is much less at such sites than in subduction zones or on the continents. The data displayed in Figure 8.5, even though incomplete, suggest clearly that the average $^{87}Sr/^{86}Sr$ ratio of oceanic island volcanics (0.70437) is higher than that of seafloor and ridge basalt (0.70280). This difference implies that sea-floor basalts were formed by melting of mantle material having a significantly lower Rb/Sr ratio than that from which volcanic rocks in oceanic islands are derived. Moreover, the difference in Rb/Sr ratio (and bulk chemical composition) must have existed for a long time (i.e., 1 to 2 billion years, or more) to be reflected by the $^{87}Sr/^{86}Sr$ ratio. We must conclude, therefore, that certain regions in the upper mantle were depleted in rubidium and other elements during the early history of the Earth. The oceanic tholeiites being formed along spreading ridges in the oceans may originate by melting of "depleted" mantle, whereas the volcanic rocks of oceanic islands are derived from "virgin" mantle. In fact, the

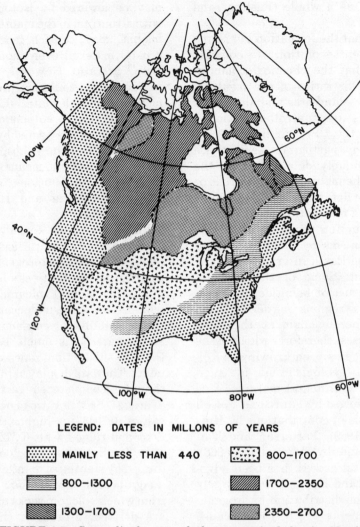

LEGEND: DATES IN MILLONS OF YEARS

▨ MAINLY LESS THAN 440	▨ 800–1700
▨ 800–1300	▨ 1700–2350
▨ 1300–1700	▨ 2350–2700

FIGURE 8.7 Generalized map of the orogenic belts in North America. Note that the oldest rocks occur in the central region of the continent and that younger orogenic belts are located along the continental margin. However, the orogenic belts along both the east and west coast of North America are underlain in part by older Precambrian basement rocks. (Reproduced from Figure X11.4 (p. 136) of Faure, G., and J. L. Powell (1972) Strontium Isotope Geology. Springer-Verlag Berlin, Heidelberg, New York, 188 p. Originally adapted from Figure 2 (p. 1231) of Hurley, P. M., and J. R. Rand (1969) Pre-drift continental nuclei. Science, *164*, No. 3885, 1229–1242. Copyright 1969 by the American Association for the Advancement of Science, with permission from both publishers.)

presumed existence of hot spots in the mantle requires that these regions are undepleted and may contain unusually high concentrations of uranium, thorium, and potassium whose radioactivity is the principal source of heat.

However, even among oceanic islands and along midocean ridges, significant variations of the $^{87}Sr/^{86}Sr$ ratio have been reported (Hawaii: Powell and DeLong, 1966; eastern Pacific: Subbarao, 1972; northern Pacific: Hedge and Peterman, 1970; Indian Ocean: Subbarao and Hedge, 1973; Iceland and Reykjanes Ridge: O'Nions and Pankhurst, 1973, 1974; Flower et al., 1975). The variability of the $^{87}Sr/^{86}Sr$ ratios of basalts formed within the ocean basins may result from one or several of the following causes:

1. Differences in the Rb/Sr ratio of the source regions due to vertical and/or lateral variations of this parameter in the mantle under midocean ridges and under oceanic islands.
2. Preferential melting of certain Rb-rich phases (e.g., phlogopite) in mantle rocks whose minerals are not isotopically homogenized inspite of the elevated temperatures that exist there. In this case the $^{87}Sr/^{86}Sr$ ratio of the melt depends on the mineral composition of the source rocks, the differences in the $^{87}Sr/^{86}Sr$ ratios of the minerals, and on the extent of partial melting.
3. Preferential removal of radiogenic ^{87}Sr into the melt from grain boundaries and microfractures in the source rocks where this isotope may be concentrated due to diffusion out of lattice positions previously occupied by ^{87}Rb.
4. Assimilation or isotope exchange of the basalt magma with rocks in the oceanic crust, including pelagic sediment in the case of oceanic islands.
5. Growth of varying amounts of radio-genic ^{87}Sr within the basaltic magma in the course of fractional crystallization leading to the formation of alkali-rich residual magmas having elevated Rb/Sr ratios.
6. Alteration of ocean-floor basalt by sea water or hot brines (Degens and Ross, 1969) during extrusion and cooling of pillow basalt.

All of the processes listed above have been invoked to explain the observed variation of $^{87}Sr/^{86}Sr$ ratio of basaltic rocks within the ocean basins and each one may contribute to this phenomenon under appropriate circumstances. Nevertheless, the most important cause is undoubtedly the heterogeneity of the upper mantle with respect to the concentrations of alkali metals generally and the Rb/Sr ratio in particular. In a general way, the $^{87}Sr/^{86}Sr$ ratios of oceanic basaltic rocks correlate positively with the $K_2O/K_2O + Na_2O$ ratio of such rocks (Peterman and Hedge, 1971). In other words, basalts having high $^{87}Sr/^{86}Sr$ ratios may have originated from sources having higher potassium concentrations (virgin mantle) than those having low $^{87}Sr/^{86}Sr$ ratios (depleted mantle).

The apparent long-term heterogeneity of the upper mantle with respect to the Rb/Sr ratio and potassium content poses a problem in view of the requirement for convective motions in the asthenosphere. How can the mantle be heterogeneous chemically and isotopically when it is also in continuous convective motion as required by the theory of plate tectonics and sea-floor spreading? The explanation of this paradox has not yet been given. An excellent discussion of this problem was published by Hofmann and Hart (1975) and by Brooks et al. (1976).

Volcanic rocks on the continents have $^{87}Sr/^{86}Sr$ ratios that are far more variable and frequently higher than those in the ocean basins (Fig. 8.5). The interpretation of these values is complicated by the probability that basaltic magmas extruded through the continental crust may interact extensively with old sialic rocks. For this reason, the variation of $^{87}Sr/^{86}Sr$ ratios within and among volcanic provinces cannot be attributed unambiguously to differences in the Rb/Sr ratios of their sources in the upper mantle. In fact, it is entirely possible that some magmas of basaltic to intermediate composition form by melting of granulites and similar high-grade metamorphic rocks at the base of the continental crust having low Rb/Sr ratios. (Leeman, 1974; Pushkar and Condie, 1973; Pushkar et al., 1972; Hedge and Noble, 1971; Scott et al., 1971; Leeman and Manton, 1971; Leeman, 1970; Peterman et al., 1970; Dasch, 1969a; Hoefs and Wedepohl, 1968; Doe, 1968; Hedge, 1966).

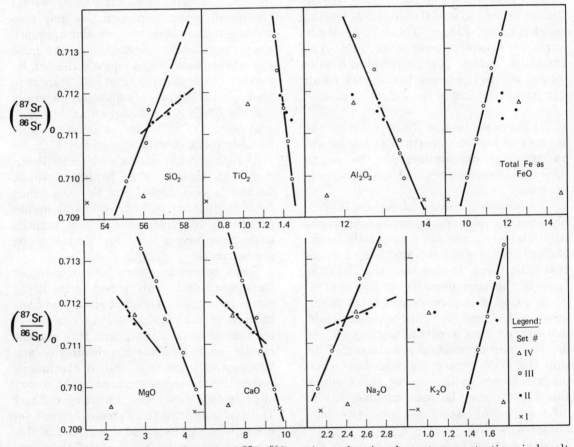

FIGURE 8.8 Correlation between initial $^{87}Sr/^{86}Sr$ ratios and major-element concentrations in basalt flows of the Kirkpatrick Basalt (Jurassic) on Storm Peak, Queen Alexandra Range, Transantarctic Mountains. (Reprinted from Figure 5 (p. 164) of Faure, G., J. R. Bowman, D. H. Elliot, and L. M. Jones (1974) Strontium isotope composition and petrogenesis of the Kirkpatrick Basalt, Queen Alexandra Range, Antarctica. Contributions to Mineralogy and Petrology, 48, 153–169, with permission from Springer-Verlag, Berlin, Heidelberg, New York.)

An extreme case of contamination of basaltic magma was reported by Faure et al. (1974) for Jurassic basalts on Storm Peak, Queen Alexandra Range, in the Transantarctic Mountains. The initial $^{87}Sr/^{86}Sr$ ratios of these rocks range from 0.7094 to 0.7133 and vary systematically in a stratigraphic sense. Moreover, the initial $^{87}Sr/^{86}Sr$ ratios of these flows correlate with the concentrations of the major elements and with the strontium concentrations (Fig. 8.8). Such correlations cannot arise by differentiation of basalt magma in a closed chamber because the isotopes of strontium are not fractionated by the formation of crystals from a magma. It is quite unlikely that such correlations could arise by melting of source rocks of varying chemical and isotopic composition because of the very systematic variations of the chemical composition and initial $^{87}Sr/^{86}Sr$ ratios in sequentially extruded lava flows. Faure et al. (1974) explained these data by means of varying additions of a contaminant, derived from the Precambrian granitic basement, to a basalt magma that originated in the upper mantle. This interpretation, based on a two-component mixing model developed in Chapter 7, permitted estimates of the chemical composition of the parent basalt magma and of the contaminant. The latter appeared to be enriched in SiO_2, total iron, K_2O, and Na_2O, but depleted in Al_2O_3, CaO, and MgO. While the isotopic and chemical data do not indicate directly how this contaminating process works, it is possible that such a contaminant may arise by partial melting of a biotite-bearing granitic rock (Pushkar and Stoeser, 1975).

4
The Origin of Granitic Rocks

We can now state the criterion for distinguishing between granitic rocks derived from the upper mantle and from the continental crust: (1) When the initial $^{87}Sr/^{86}Sr$ ratio of a granitic rock lies *within* the field representing strontium evolution in the upper mantle (Fig. 8.3), then the strontium in that rock could have been derived from the upper mantle, either directly or indirectly, without significant contamination with foreign strontium. (2) When the initial $^{87}Sr/^{86}Sr$ ratio of a granitic rock lies significantly *above* the field representing strontium evolution in the upper mantle, then two alternative explanations are possible, namely: (a) the strontium was derived entirely from crustal rocks of sialic composition which had previously become enriched in radiogenic ^{87}Sr because of their age or elevated Rb/Sr ratio, or both; (b) some of the strontium originated from a source in the upper mantle, but its isotopic composition was significantly modified by subsequent addition of foreign strontium enriched in radiogenic ^{87}Sr. In either case, a high initial $^{87}Sr/^{86}Sr$ ratio reflects significant involvement of strontium derived from preexisting crustal material of sialic composition in the formation of such granitic rocks. This implies that granitic rocks with high initial $^{87}Sr/^{86}Sr$ ratios represent "granitized" crustal rocks or are hybrid rocks containing significant proportions of sialic material assimilated by a magma that originated in the upper mantle. In Case 2a, the initial ratio of the granitic rock may represent the average $^{87}Sr/^{86}Sr$ of the parent crustal material at the time of magma generation. In Case 2b, the initial $^{87}Sr/^{86}Sr$ ratio of the granitic rock is subject to interpretations involving mixing models outlined in Chapter 7.

We shall illustrate the application of this criterion by considering some specific examples. First we turn to a suite of biotite-bearing quartz monzonites from the Martin dome in the Miller Range located at the

confluence of the Nimrod and Marsh glaciers at 83°15'S latitude and 157°0'E longitude in the Transantarctic Mountains. The
Martin dome intrudes quartzo-feldspathic
gneisses and other metamorphosed sedimentary rocks of the Nimrod group of Precambrian age. Seven whole-rock samples from
the Martin dome form an isochron shown
in Figure 8.9. The age of these rocks is 488 ±
17 million years and the initial $^{87}Sr/^{86}Sr$
ratio is 0.734 ± 0.001. Evidently, these granitic rocks formed during the Ordovician
period (Appendix II) from magma that contained strontium that was significantly enriched in radiogenic ^{87}Sr compared to strontium in the upper mantle according to Figure
8.3. We may conclude therefore that the
strontium in this magma was either extensively contaminated with radiogenic ^{87}Sr

derived from old sialic rocks of the continental crust (Case 2b) or that the magma was
produced by partial melting of old sialic
rocks, perhaps even of the Nimrod group
itself, into which the magma was subsequently intruded (Case 2a). Gunner and
Faure (1972) presented evidence that the
Nimrod group may be two billion years old
and that it could have provided strontium
to a granitic melt during the Ordovician
period having a $^{87}Sr/^{86}Sr$ ratio of 0.734. In
any case, it is clear that the Martin dome
magma either assimilated large amounts of
sialic material or formed by partial melting
of granitic basement rocks such as those of
the Nimrod group.

A quite different set of circumstances was
reported by Fullagar et al., (1971) for the
Salisbury pluton of North Carolina. The
Salisbury pluton is composed of white and
pink adamellite and is intrusive into metamorphosed sedimentary and igneous rocks
of the so-called Charlotte Belt in Rowan
County of North Carolina. Seven whole-rock
specimens form the isochron shown in Figure 8.10 from which the authors calculated
an age of 411 ± 6 million years and an initial
$^{87}Sr/^{86}Sr$ ratio of 0.7038 ± 0.0010 (corrected
to 0.7080 for the Eimer and Amend isotope
standard). They interpreted the date as being
the time elapsed since intrusion and crystallization of the Salisbury pluton and concluded that the original magma from which
this pluton was formed was derived from
the mantle and was not contaminated with
crustal rocks.

These two examples illustrate the conclusion that igneous-looking rocks of granitic
composition apparently can form either by
differentiation of magma originating at
great depth or by "granitization" of sialic
rocks of the continental crust. Faure and
Powell (1972) compiled a list of initial $^{87}Sr/$
^{86}Sr ratios of some 130 granitic rocks and

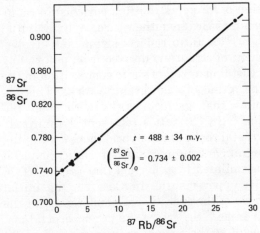

FIGURE 8.9 Whole-rock Rb-Sr isochron of
biotite-bearing quartz monzonites from the
Martin Dome of the Miller Range in the Transantarctic Mountains. The high initial $^{87}Sr/^{86}Sr$
ratio indicates that this is an example of a granitic
intrusive that formed from magma which either
assimilated large amounts of old sialic rocks or
was produced by partial melting of the granitic
basement rocks of this area. (Gunner, 1974.)

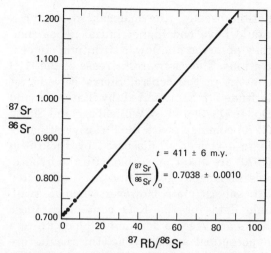

FIGURE 8.10 Whole-rock Rb-Sr isochron of the Salisbury pluton from Rowan County, North Carolina. The initial $^{87}Sr/^{86}Sr$ ratio of this intrusive is well within the range of isotopic compositions of strontium in the upper mantle as outlined in Figure 8.3. Consequently the conclusion is justified that the magma from which the Salisbury pluton crystallized evolved from a parent magma which originated in the upper mantle and was not contaminated with foreign strontium from the continental crust. The initial $^{87}Sr/^{86}Sr$ ratio has been increased by $+0.0006$ to make it compatible with an $^{87}Sr/^{86}Sr$ ratio of 0.7080 for the Eimer and Amend strontium isotope standard. (Fullagar et al., 1971.)

found that about half of them had sialic ancestors. However, an important point to be made here is that the initial $^{87}Sr/^{86}Sr$ ratios of the large granitic batholiths of North America are only slightly higher than those of the upper mantle. Faure and Powell (1972) listed six large batholiths whose average initial $^{87}Sr/^{86}Sr$ ratio is only 0.707. The isotopic composition of strontium therefore does not indicate unequivocally the origin of *these* important rock bodies which make up a significant fraction of the continental crust.

5
The Batholiths of California

The Mesozoic batholiths of California have been studied in sufficient detail to provide an appreciation of the complexity of the processes involved in the formation of rock bodies of granitic compositions and batholithic proportions. Detailed mapping has identified many genetically related discrete plutons which were apparently intruded serially, beginning with mafic varieties and becoming progressively more felsic. Emplacement began during the Triassic period about 210 million years ago and continued episodically for 130 million years, ending during the Cretaceous period about 80 million years ago.

The relatively low initial $^{87}Sr/^{86}Sr$ ratios of these rocks (0.7060 to 0.7080) first reported by Hurley et al. (1965) are compatible with three alternate hypotheses regarding their origin: (1) melting of sedimentary rocks, containing much young volcanic material, which were previously accumulated in a geosyncline along the continental margin; (2) melting in the upper mantle and in rocks of the lower continental crust, which have low Rb/Sr ratios due to depletion in rubidium by granulite-facies metamorphism, with subsequent contamination of the magma with varying amounts of radiogenic ^{87}Sr derived from rocks of the upper continental crust; and (3) magma generation by melting of a mixture of deep-sea sediment, sea-floor basalt, and upper mantle in a downward-plunging lithospheric plate in a subduction zone that was overridden by the North American plate. These alternatives were considered by Kistler and Peterman (1973) in the light of their more extensive data on the regional variation of initial $^{87}Sr/^{86}Sr$ ratios of granitic rocks in the batholiths of California.

The data of Kistler and Peterman (1973) show a general southeasterly increase of the initial $^{87}Sr/^{86}Sr$ ratio in the rocks of the California batholiths north of the Garlock Fault and east of the San Andreas Fault (Fig. 8.11). The range of values is from 0.7031 to 0.7082 with one value of 0.7094. They found that the initial ratios are independent of the age of the rocks and are controlled primarily by geography. This observation is expressed in Figure 8.11 by contours of equal initial $^{87}Sr/^{86}Sr$ ratio with an interval of 0.001. The 0.706 and 0.704 contours apparently mark boundaries between regions having contrasting crustal histories reflected in the distribution and composition of Paleozoic sedimentary rocks, ultramafic rocks, metallic ore deposits, topographic elevations, and gravity anomalies. The 0.706 contour is the boundary between predominantly miogeosynclinal sediments (carbonates, shale, and sandstones) in the south and east and eugeosynclinal rocks (volcanics, chert, and greywacke) to the north and west. The 0.704 contour is the eastern border of Mesozoic granitic rocks whose trace element concentrations are similar to those of oceanic tholeiites and alkali basalts as well as being the eastern limit of the principal exposures of ultramafic rocks. These correlations suggest that the initial $^{87}Sr/^{86}Sr$ ratios of the granitic rocks of the California batholiths reflect compositional differences in the crust of this area that date from late Precambrian and Paleozoic depositional histories.

The correlations of the initial $^{87}Sr/^{86}Sr$ ratios and chemical compositions of the Mesozoic granitic plutons with the compositions of the late Precambrian to Triassic sedimentary rocks which they intrude suggest a strong crustal influence on the granitic magmas. However, Kistler and Peterman (1973) regarded anatexis of sedimentary rocks as an unsatisfactory explanation because the magmas so produced would have had higher initial ratios than they observed and lower strontium concentrations. The eastward increase of the K_2O content in the central Sierra Nevada (at latitude 37°N) reported by Bateman and Dodge (1970) and by Wollenberg and Smith (1970) combined with the "transverse petrologic asymmetry" discussed by Dickinson (1970) are also compatible with the hypothesis of magma generation in a subducted lithospheric plate. However, when all available data for the K_2O content of these rocks are used to determine the depth to a hypothetical seismic zone, the results are disappointing (see also Nielson and Stoiber, 1973). The area having initial $^{87}Sr/^{86}Sr$ ratios of less than 0.7060 appears to be underlain by a nearly horizontal surface at a depth of about 100 km. In areas where the initial $^{87}Sr/^{86}Sr$ ratio is greater than 0.7060, the seismic zone is a gently dipping surface at a depth of about 200 km. These and other considerations discussed by Kistler and Peterman (1973) make the subduction hypothesis unattractive as an explanation for the origin of the California batholiths. They drew particular attention to the high strontium content of these rocks, more than half of which have concentrations in excess of 400 ppm. It is difficult to reconcile this fact with the low strontium concentrations of sea-floor basalt (about 135 ppm). Greywackes of the Franciscan formation that might have been subducted along with the basalt crust likewise have lower strontium concentrations and lower $^{87}Sr/^{86}Sr$ ratios than the granitic rocks at the time of their formation. Kistler and Peterman (1973) therefore concluded that the subduction model does not readily explain the isotopic and trace-element data of the granitic rocks in California.

Instead, they endorsed a model originally proposed by Hamilton and Meyers (1967)

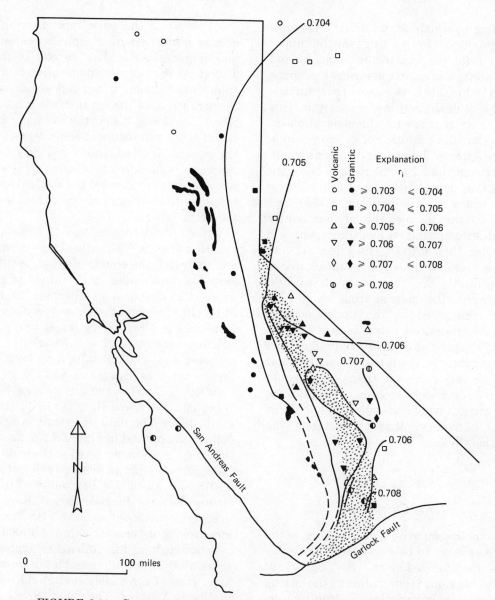

FIGURE 8.11 Contour diagram showing the regional variation of initial
$^{87}Sr/^{86}Sr$ ratios of volcanic and plutonic rocks of Mesozoic age in California.
(Reprinted from Figure 2 (p. 3493) of Kistler, R. W., and Z. E. Peterman (1973)
Geological Society of America Bulletin, *84*, No. 11, 3489–3512 with permission
of G.S.A.)

according to which quartz dioritic magmas were produced by melting in the upper mantle near the continental margin and granodiorite and quartz monzonite magmas formed by melting in the lower crust further inland east of the continental margin. This model is consistent with the general observation that the granitic rocks are predominantly quartz diorite and trondhjemite where the initial $^{87}Sr/^{86}Sr$ ratio is less than 0.7040, that quartz diorite and granodiorite occur where it lies between 0.7040 and 0.7060, and that the rocks are predominantly granodiorite and quartz monzonite where it is greater than 0.7060. The chemical and isotopic compositions of granitic rocks having initial $^{87}Sr/^{86}Sr$ ratio less than 0.7040 indicate that the magmas from which they formed originated in the upper mantle. Those having initial ratios between 0.7040 and 0.7060 may also have originated from the upper mantle. The intrusives with initial $^{87}Sr/^{86}Sr$ ratios greater than 0.7060 could have been derived by melting of rocks in the lower crust having somewhat higher Rb/Sr ratios than those that characterize the upper mantle.

6
The Oceans

The isotopic composition of strontium in the oceans appears to be everywhere the same and is characterized by an $^{87}Sr/^{86}Sr$ ratio of 0.7090. This value is controlled by the mixing of three isotopic varieties of strontium derived from the following sources: (1) young volcanic rocks whose $^{87}Sr/^{86}Sr$ is 0.704 ± 0.002; (2) old sialic rocks of the continental crust with an average $^{87}Sr/^{86}Sr$ ratio of about 0.720 ± 0.005; and (3) marine carbonate rocks of Phanerozoic age having $^{87}Sr/^{86}Sr$ ratios of 0.708 ± 0.001 (Faure et al.,

1965). Most of the strontium entering the oceans is derived by chemical weathering and diagenesis of marine carbonate rocks. Therefore, the isotopic composition of strontium in the oceans is not representative of the strontium of the continental crust as a whole, nor does it reflect the isotopic composition of detrital sediment being deposited in the oceans. Instead, the $^{87}Sr/^{86}Sr$ ratio of sea water is an indirect indicator of the kinds of rocks that are exposed to chemical weathering on the surfaces of continents and in the ocean basins.

These considerations lead to the suggestion that the $^{87}Sr/^{86}Sr$ ratio of the oceans has varied during the course of geologic time in response to changes in the kinds of rocks exposed to chemical weathering. Peterman et al. (1970) reported such systematic variations in the $^{87}Sr/^{86}Sr$ ratios of unreplaced calcium carbonate of fossil shells of Phanerozoic age. Their results have been verified by other investigators and the general validity of these variations is now widely accepted (Dasch and Biscaye, 1971; Tremba et al., 1975). The data of Peterman and his colleagues, plotted in Figure 8.12, show that the $^{87}Sr/^{86}Sr$ ratio declined at the end of the Paleozoic era from an average value of about 0.7078 and reached a low value of 0.70675 during the Late Jurassic period. Beginning in early Cretaceous time, the $^{87}Sr/^{86}Sr$ ratio increased to its present value of 0.7090.

Peterman and his colleagues reported a set of analyses of nine samples of Late Cretaceous age from widely scattered localities in North America. The $^{87}Sr/^{86}Sr$ ratios of these samples are concordant at 0.7074 within narrow limits (0.7073 to 0.7076). These data suggest that the strontium in the Late Cretaceous seas in the area of the North American continent was isotopically homogeneous. This conclusion was further reinforced by measurements of $^{87}Sr/^{86}Sr$ ratios

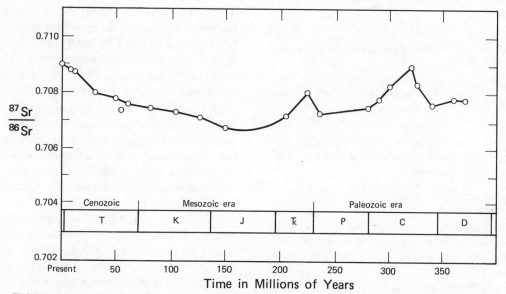

FIGURE 8.12 Systematic variation of the $^{87}Sr/^{86}Sr$ ratio of the oceans during Phanerozoic time as indicated by unreplaced fossil shells. (Peterman et al., 1970.) The time scale is from Harland et al. (1964).

of marine limestones deposited in the Tethys Sea during the Mesozoic era (Tremba et al., 1975). We may conclude, therefore, that strontium in the oceans has been isotopically homogeneous during the Mesozoic era and perhaps also at earlier times.

The apparent isotopic homogeneity of marine strontium enables us to use the $^{87}Sr/^{86}Sr$ ratios of sedimentary carbonate rocks as a criterion for distinguishing between rocks of marine and nonmarine origin. Faure and Barrett (1973) demonstrated that nonmarine carbonate rocks from the Beacon supergroup (Devonian to Late Triassic) of the Transantarctic Mountains are strongly enriched in radiogenic ^{87}Sr compared to rocks of marine origin deposited at equivalent times. They attributed the high $^{87}Sr/^{86}Sr$ ratios of these nonmarine carbonates to the provenance of the strontium which apparently originated as a weathering product of Precambrian basement rocks. This criterion may be applicable elsewhere and may be useful in the study of nonmarine carbonate sequences.

Let us now return to a consideration of the $^{87}Sr/^{86}Sr$ ratio of sea water. We have already stated that the isotopic composition of modern marine strontium can be represented as a mixture of three isotopic varieties derived from young volcanic rocks, old sialic rocks, and marine carbonate rocks of Phanerozoic age. We now present a model based on this premise in which the $^{87}Sr/^{86}Sr$ ratio of sea water is expressed by the equation (Faure et al., 1965)

$$\left(\frac{^{87}Sr}{^{86}Sr}\right)_{sw} = \left(\frac{^{87}Sr}{^{86}Sr}\right)_V v + \left(\frac{^{87}Sr}{^{86}Sr}\right)_S s + \left(\frac{^{87}Sr}{^{86}Sr}\right)_M m \qquad (8.1)$$

where

$$\left(\frac{^{87}\text{Sr}}{^{86}\text{Sr}}\right)_{sw} = \text{value of this ratio in sea water}$$

$$\left(\frac{^{87}\text{Sr}}{^{86}\text{Sr}}\right)_{V,S,M} = \text{value of this ratio in strontium contributed to the oceans by chemical weathering of young volcanic } (V), \text{ old sialic } (S), \text{ and marine carbonate } (M) \text{ rocks.}$$

The coefficients v, s, and m are decimal fractions representing the proportions of strontium in sea water contributed by volcanic, sialic, and marine carbonate rocks, respectively. The numerical values of these coefficients are constrained by the requirement that their sum must be equal to unity and each coefficient is limited to values greater than zero and less than one. Substituting reasonable estimates of the $^{87}\text{Sr}/^{86}\text{Sr}$ ratios into Equation 8.1 we obtain

$$\left(\frac{^{87}\text{Sr}}{^{86}\text{Sr}}\right)_{sw} = 0.704v + 0.720s + 0.708m \quad (8.2)$$

This equation has been plotted in Figure 8.13 in coordinates of $(^{87}\text{Sr}/^{86}\text{Sr})_{sw}$ and v in the form of two sets of parallel straight-line contours representing selected values of s and m.

Figure 8.13 outlines a triangular field giving the resulting values of $(^{87}\text{Sr}/^{86}\text{Sr})_{sw}$ for the limiting cases that v, s, and m are each equal to zero. It is clear that in this model the $^{87}\text{Sr}/^{86}\text{Sr}$ ratio of sea water is limited to values ranging from 0.704 ($v = 1$, $s = 0$, $m = 0$) and 0.720 ($v = 0$, $s = 1$, $m = 0$). Actually, the $^{87}\text{Sr}/^{86}\text{Sr}$ ratio of sea water has varied only from about 0.7065 to 0.7090 during Phanerozoic time and these limits are shown by the dashed lines in Figure 8.13. The line representing strontium in the modern oceans is the locus of compatible sets of values of v, s, and m in this model which would give sea water an $^{87}\text{Sr}/^{86}\text{Sr}$ ratio of

0.7090. Point A on the diagram represents the quite unreasonable case that none of the strontium is derived by weathering and diagenesis of marine carbonate rocks ($m = 0$). In this unlikely case, $v \simeq 0.7$ and $s \simeq 0.3$. In other words, about 70 percent of the strontium would be derived from volcanic sources and 30 percent from sialic rocks. Since m must be greater than zero, we conclude that these percentages are maximum values and they must in reality be less than this.

A somewhat more reasonable situation is represented by point B. In this case, $m = 0.6$, $v \simeq 0.25$, and $s \simeq 0.15$. The high strontium content of marine carbonate rocks (610 ppm, Turekian and Kulp, 1956) and their susceptibility to solution under humid climatic conditions combined with their widespread exposure on the continents make it very likely that these rocks contribute a very large proportion of the strontium entering the oceans. A further contributing factor is the recrystallization of aragonite to calcite during diagenesis of carbonate sediment and the loss of strontium that accompanies dolomitization of limestones after deposition. Therefore 60 percent is not an unreasonable estimate for the proportion of strontium contributed to the oceans from marine carbonate rocks. In this case, about 25 percent would be attributed to volcanic rocks and 15 percent to sialic rocks.

Point C represents the condition that marine carbonates provide 80 percent of the strontium dissolved in the oceans which permits about 10 percent to be derived each from volcanic and sialic sources. In general, the model suggests that the isotopic composition of strontium in the oceans is strongly influenced by recycling of marine carbonate rocks and that its $^{87}\text{Sr}/^{86}\text{Sr}$ ratio is further modulated by inputs from volcanic and sialic sources.

Although this model is relevant primarily to strontium in the *modern* oceans, it can

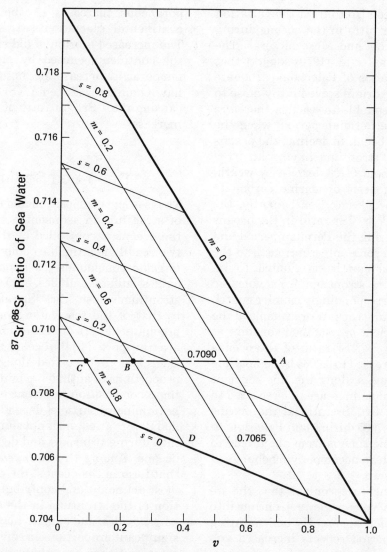

FIGURE 8.13 Model of the $^{87}Sr/^{86}Sr$ ratio in the oceans considered as a mixture of strontium contributed by weathering of young volcanic rocks (0.704 ± 0.002), old sialic rocks (0.720 ± 0.005), and marine carbonate rocks (0.708 ± 0.001). The coefficients v, s, and m are defined as the fractions of strontium contributed to the oceans by volcanic, sialic, and marine carbonate rocks, respectively. The most likely values of m since the Permian period range from 0.6 to 0.8 and limit the values of v to less than 0.40 and those of s to less than 0.15. The systematic variations of the $^{87}Sr/^{86}Sr$ ratios of the oceans since the Permian period can be explained in terms of changes in v, s, and m within these limits, as discussed in the text.

probably be used also to treat the variation of the $^{87}Sr/^{86}Sr$ ratio in the oceans during the late Paleozoic and Mesozoic eras. The data of Peterman et al. (1970) suggest that the $^{87}Sr/^{86}Sr$ ratios of Paleozoic carbonate rocks of marine origin stayed fairly close to 0.708. It is reasonable to assume therefore that during the Permian period, when the $^{87}Sr/^{86}Sr$ ratio began to decline, the oceans were receiving strontium having an $^{87}Sr/^{86}Sr$ ratio of about 0.708 derived by weathering and diagenesis of marine carbonate rocks of Paleozoic age. Consequently, the decline of the $^{87}Sr/^{86}Sr$ ratio in the oceans that started during the Permian period and continued with some interruptions into the Jurassic period can be attributed to increased input of strontium from volcanic sources and correspondingly smaller contributions from sialic rocks. For example, if the value of coefficient m remained constant at 0.6, then the $^{87}Sr/^{86}Sr$ ratio of the oceans could have decreased to 0.7065 by compatible changes of v and s along the line segment BD. The maximum increase in v required to decrease the $^{87}Sr/^{86}Sr$ ratio of the oceans from 0.708 to 0.7065 during late Paleozoic to middle Mesozoic time is from about 0.30 to about 0.40, with an accompanying change in s from 0.10 to zero.

It is tempting to speculate that the increased input of volcanogenic strontium into the oceans which apparently began during the Permian period reflects increased volcanic activity on a worldwide scale associated with the onset of continental drift and the opening of the Atlantic Ocean. However, no convincing demonstration of such a correlation has been given.

The data of Peterman et al. (1970) also suggest the occurrence of episodic increases of the $^{87}Sr/^{86}Sr$ ratios of the oceans during the Carboniferous and Early Triassic periods. These seem to mirror roughly contemporaneous increases in the intensity of continental glaciation (Armstrong, 1971). The increased erosion of old sialic rocks on the continents caused by glaciation may temporarily increase the input of strontium having a high radiogenic ^{87}Sr content, thus causing the $^{87}Sr/^{86}Sr$ ratio of the oceans to increase.

7
Deep-Sea Sediment

The isotopic composition and concentration of strontium in sediment accumulating in the oceans is controlled by the presence of two readily identifiable components: (1) *authigenic* compounds (carbonates, silicates, oxides, sulfates, sulfides, etc.) which contain strontium whose $^{87}Sr/^{86}Sr$ ratio is identical to that of the sea water from which they precipitated; (2) *allogenic* minerals which are transported to the oceans as detrital particles and are deposited along with varying proportions of authigenic minerals. The isotopic composition of strontium of the *allogenic* mineral particles depends on their ages and Rb/Sr ratios and is not significantly modified during transport and deposition in the oceans. Among the *authigenic* compounds that form in the oceans, the carbonates are most abundant and contribute a large fraction of the strontium in the sediment. Silicates, oxides, and sulfides may be present in significant proportions in some types of sediment, but they contribute little strontium. The sulfates are formed only in evaporite basins. The carbonate fraction of the authigenic component includes biogenic skeletal calcium carbonate as well as inorganically precipitated material in the form of calcite and aragonite.

The *allogenic* component originates from two sources that have different $^{87}Sr/^{86}Sr$ ratios and concentrations of strontium. These are (1) weathering products of young volcanic rocks of predominantly basaltic composition and volcanic ash deposited directly in the oceans without much weathering; and (2) weathering products of old sialic rocks derived from continental areas. The $^{87}Sr/^{86}Sr$ ratios of the volcanogenic component is 0.704 ± 0.002; that of the sialic component is greater than this and averages 0.720 ± 0.005. Therefore, the isotopic composition of strontium in the *allogenic* component of deep-sea sediment depends on the relative proportions of the volcanogenic and the sialic fractions that it contains.

We are especially interested in the $^{87}Sr/^{86}Sr$ ratio of the noncarbonate, silicate fraction of deep-sea sediment which consists primarily of mineral particles and amorphous material derived from nearby land masses. Dasch (1969b) made a large number of measurements of the $^{87}Sr/^{86}Sr$ ratios of noncarbonate fractions of sediment samples taken mainly in the Atlantic Ocean. He found that this ratio varies widely and systematically in different parts of the oceans from a low value of 0.7044 to a high of 0.7429. The variability of this ratio is proof that strontium in the detrital silicate fraction of sediment deposited in the oceans does not equilibrate isotopically with sea water. On the contrary, the $^{87}Sr/^{86}Sr$ ratios of the noncarbonate fraction preserve a record of the provenance of the sediment, as is shown in Figure 8.14. It can be seen that the $^{87}Sr/^{86}Sr$ ratios in the Atlantic Ocean exhibit patterns of variation on a regional scale that can be related in a general way to the geology of the land masses from which the sediment was derived. Biscaye and Dasch (1971) examined these regional patterns in more detail in the Argentine basin and confirmed the existence of systematic variations reflecting the provenance of the noncarbonate, detrital silicate fraction of the sediment.

The studies described above deal with the regional variation of the $^{87}Sr/^{86}Sr$ ratios of noncarbonate sediment collected from the bottom of the ocean. The regional patterns that are observed reflected primarily the provenance of sediment that was deposited in the very recent past. This information can be extended backward in time by systematic analyses of sediment samples recovered in piston cores. If the noncarbonate fractions of deep-sea sediment deposited at a specific location in the oceans in the geologic past can be treated as a series of two-component mixtures, then the concentrations of strontium and their $^{87}Sr/^{86}Sr$ ratios are given by Equations 7.2 and 7.10, derived previously in Chapter 7.

Boger and Faure (1974) used this approach to define a "strontium-isotope stratigraphy" of sediment from a piston core taken in the median valley of the Red Sea ($18°09'N$ and $39°53'E$). Figures 8.15a and b show that their data fit a hyperbolic mixing curve whose equation is

$$\frac{^{87}Sr}{^{86}Sr} = 0.70294 + \frac{0.8158}{Sr} \qquad (8.3)$$

They used this equation to calculate the concentration of strontium of the end members by assuming $^{87}Sr/^{86}Sr$ ratios of 0.704 and 0.715 for the basaltic and the sialic components, respectively. The resulting concentrations are 772 ppm for the basalt component and 68 ppm for the sialic component. After thus defining the composition of the end members, they used the concentration of strontium of each sample and Equation 7.2 to determine the corresponding concentration of the basalt component in each sample. Figure 8.16 is a plot of the concentrations of strontium and of the basalt component as a function of depth in the core. It can be seen that the proportion of the basaltic

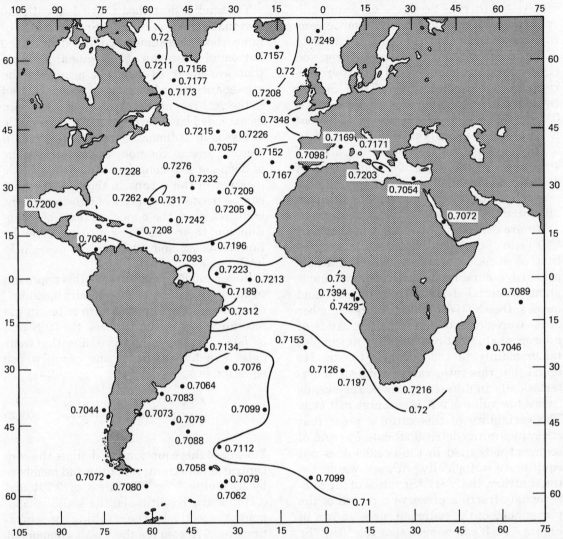

FIGURE 8.14 Regional variation of the $^{87}Sr/^{86}Sr$ ratios of the noncarbonate fractions of deep-sea sediment in the oceans. (Reprinted from Figure 2 (p. 1535) of Dasch, E. J. (1969) Geochimica et Cosmochimica Acta, vol. 33, No. 12, 1521–1522, Pergamon Press Ltd., with permission from the Executive Editor.)

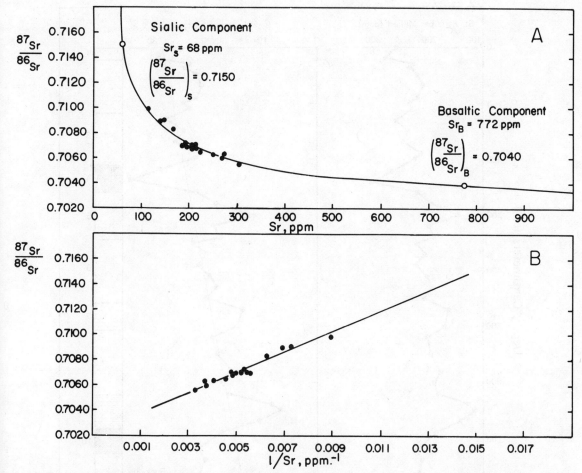

FIGURE 8.15 (a) Hyperbolic mixing curve defined by the $^{87}Sr/^{86}Sr$ ratios and strontium concentrations of noncarbonate fractions of a piston core from the median valley of the Red Sea at latitude 18°09′N and longitude 30°53′E. The concentrations of strontium of the end members are based on assumed values of their $^{87}Sr/^{86}Sr$ ratios. (b) Transformation of the mixing hyperbola into a straight line by inversion of the strontium concentrations. The line was fitted to these points by least-squares regression. The resulting equation was plotted as a hyperbola in (a) above. (Reprinted from Figure 1 (p. 182) of Boger, P. D., and G. Faure (1974) Geology, vol. 2, pp. 181–183, with permission from the Geological Society of America.)

component varies systematically with depth in the core. This variation was used to define sedimentary layers representing three alternate modes of deposition: (1) excess volcanogenic material; (2) excess sialic material; (3) steady-state mode of deposition. The "strontium isotope stratigraphy" thus defined includes two layers rich in volcanic detritus and four layers rich in sialic detritus which are interbedded with layers representing a steady-state mode of deposition. Although much work remains to be done, this type of analysis of deep-sea sediment may prove to be fruitful in reconstructing depositional histories in the ocean basins generally.

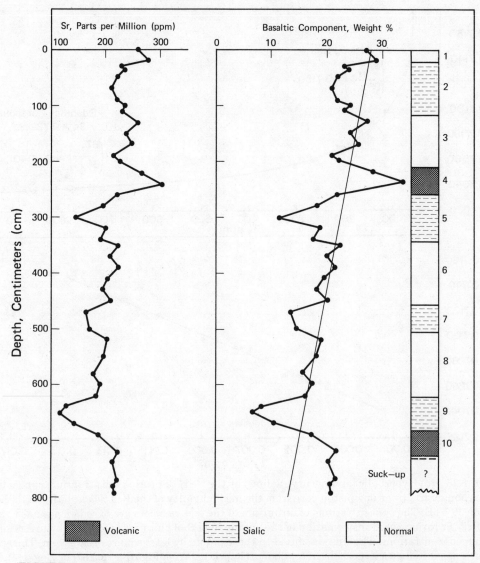

FIGURE 8.16 Depth profiles of the measured strontium concentrations and calculated abundances of the volcanic-basaltic component in the noncarbonate fraction of the core (Fig. 8.15a,b). The systematic variation of the abundance of the volcanic component was used to define the sedimentary layers shown on the right which represent different mixing conditions as indicated in the legend. (Reprinted from Figure 2 (p. 183) of Boger, P. D., and G. Faure (1974) Geology, vol. 2, pp. 181–183, with permission from the Geological Society of America.)

8 SUMMARY

The isotopic evolution of strontium in the Earth began about $4.6 \pm 0.1 \times 10^9$ years ago with a primordial $^{87}Sr/^{86}Sr$ ratio of about 0.699. This information was derived from the study of stony meteorites because no rocks dating from the time of formation of the Earth have been preserved.

The subsequent evolution of strontium in the Earth has taken place at differing rates in the continental crust and in the upper mantle. The formation of the continental crust by internal differentiation of the mantle may have begun around four billion years ago as indicated by recent discoveries of granitic gneisses which crystallized about 3.7×10^9 years ago. The low initial $^{87}Sr/^{86}Sr$ ratios (0.700 to 0.702) of these ancient gneisses support the conclusion that these rocks were derived from the upper mantle only a short time prior to their final crystallization.

The evolution of the $^{87}Sr/^{86}Sr$ ratio of the upper mantle has proceeded more slowly than that of the continental crust because the mantle has a significantly lower average Rb/Sr ratio of about 0.027 ± 0.011. At the present time, strontium in the upper mantle is isotopically heterogeneous within narrow limits, represented by $^{87}Sr/^{86}Sr$ ratios of 0.704 ± 0.002. The heterogeneity of isotopic compositions implies that some parts of the mantle were depleted in rubidium relative to strontium early in its history, while other parts appear to represent virgin mantle.

The theory of plate tectonics provides a rational framework for volcanic activity within the ocean basins and along continental margins and island arcs. Systematic variations of $^{87}Sr/^{86}Sr$ ratios and chemical compositions of volcanic rocks give clues to the sources of magmas and to their subsequent chemical and isotopic evolution. Several alternative explanations have been proposed to account for the chemical and isotopic variations that are observed. However, no unique explanation has emerged, nor is it likely that only one mechanism is dominant in all magmatic-volcanic phenomena in the ocean basins and along continental margins.

The $^{87}Sr/^{86}Sr$ ratios of volcanic rocks extruded through the continental crust are more variable than those in the ocean basins and are higher on the average. This is probably due, in many cases, to interaction of magmas with old granitic rocks or may result from magma generation by melting of crustal rocks. In general, the initial $^{87}Sr/^{86}Sr$ ratios of volcanic and plutonic rocks on the continents provide a criterion by which the involvement of crustal rocks in the formation and evolution of magma may be evaluated. Large granitic batholiths in orogenic belts have complex origins and may involve magma derived from both mantle and crustal sources.

The $^{87}Sr/^{86}Sr$ ratio of the modern oceans is 0.7090 and appears to be constant throughout the world's oceans. Analysis of marine carbonates indicates that the $^{87}Sr/^{86}Sr$ ratio of the oceans has varied systematically throughout Phanerozoic time, but has apparently been constant in the open ocean at any given time. The time-dependent variation of this ratio in the oceans can be explained in terms of changing proportions of strontium contributed to the oceans from different sources.

The $^{87}Sr/^{86}Sr$ ratios of noncarbonate sediment deposited in the oceans vary systematically on a regional scale and reflect the ages and Rb/Sr ratios of their sources. Under certain favorable geologic conditions, the noncarbonate fractions can be treated as two-component mixtures consisting of weathering products of old sialic rocks and of young volcanic rocks of basaltic composition.

PROBLEMS

1. The Uivak Gneiss at Saglek Bay, Labrador, has an age of 3622 million years and an initial $^{87}Sr/^{86}Sr$ ratio of 0.7014. Undifferentiated gneisses in the same area crystallized 3121 million years ago with an initial $^{87}Sr/^{86}Sr$ ratio of 0.7064 (Hurst et al., 1975). If the average $^{87}Rb/^{86}Sr$ ratio of the Uivak Gneiss is 1.0521, calculate its $^{87}Sr/^{86}Sr$ ratio at the time of crystallization of the undifferentiated gneisses and consider whether or not the undifferentiated gneiss could have been derived from the Uivak Gneiss. (**ANSWER:** $(^{87}Sr/^{86}Sr)_0 = 0.7064$, yes.)

2. If the ancestor of the Uivak Gneiss (Problem 1 above) had the same $^{87}Rb/^{86}Sr$ ratio as the present Uivak Gneiss, estimate the length of the time interval between withdrawal of the ancestor from the mantle and the crystallization of the Uivak Gneiss, assuming an $^{87}Sr/^{86}Sr$ ratio of 0.699 for the mantle. (**ANSWER:** 150×10^6 years.)

3. Sample 10071 from the Sea of Tranquillity on the moon crystallized 3.68×10^9 years ago with an initial $^{87}Sr/^{86}Sr$ ratio of 0.69926 (Papanastassiou et al., 1970). If the moon separated from the solar nebula 4.6×10^9 years ago with an initial $^{87}Sr/^{86}Sr$ ratio of 0.69897 (BABI), estimate the Rb/Sr ratio of the lunar interior in which the strontium evolved before being incorporated into sample 10071. (**ANSWER:** Rb/Sr $\simeq 0.0078$.)

4. The Muntsche Tundra mafic pluton crystallized 2.02×10^9 years ago and had an initial $^{87}Sr/^{86}Sr$ ratio of 0.7020 (Birck and Allegre, 1973). Estimate the Rb/Sr ratio of the source rocks in the upper mantle from which the magma was withdrawn, assuming a primordial $^{87}Sr/^{86}Sr$ ratio of 0.69897 and an age of 4.6×10^9 for the Earth. (**ANSWER:** Rb/Sr $\simeq 0.030$.)

5. Four whole-rock samples of the Icarus Pluton, Northern Light Lake, Ontario were analyzed by Hanson et al. (1971) as follows:

SAMPLE	$\dfrac{^{87}Rb}{^{86}Sr}$	$\dfrac{^{87}Sr}{^{86}Sr}$
21	0.1564	0.7068
22	0.0755	0.7037
23	0.216	0.7091
24	0.328	0.7133

Fit an isochron to these data by least squares regression (Chapter 6) and determine the age and initial ratio of this rock. Use the results to answer the question whether this pluton was derived from the upper mantle or formed from remelted crustal rock.

REFERENCES

Allegre, C. J., J. L. Birck, S. Fourcade, and M. P. Semet (1975) Rubidium-87/strontium-87 age of Juvinas basaltic achondrite and early igneous activity in the solar system. Science, *187*, 436–438.

Armstrong, R. L. (1968) A model for the evolution of strontium and lead isotopes in a dynamic earth. Rev. Geophys., *6*, 175–199.

Armstrong, R. L. (1971) Glacial erosion and the variable isotopic composition of strontium in sea water. Nature, Phys. Sci., *230*, 132–133.

Armstrong, R. L., and S. M. Hein (1973) Computer simulation of Pb and Sr isotope evolution of the Earth's crust and upper mantle. Geochim. Cosmochim. Acta, *37*, 1–18.

Bateman, P. C., and F. C. W. Dodge (1970) Variations of major chemical constituents across the central Sierra Nevada batholith. Geol. Soc. Amer. Bull., *81*, 409–420.

Birck, J. L., and C. J. Allegre (1973) [87]Rb-[87]Sr system of Muntsche Tundra mafic pluton, Kola Peninsula, U.S.S.R. Earth Planet. Sci. Letters, *20*, 266–274.

Birck, J. L., S. Fourcade, and C. J. Allegre (1975) [87]Rb-[86]Sr age of rocks from the Apollo 15 landing site and significance of internal isochrons. Earth and Planet. Sci. Letters, *26*, 29–35.

Biscaye, P. E., and E. J. Dasch (1971) The rubidium, strontium, strontium-isotope system in deep sea sediments: Argentine basin. J. Geophys. Res., *76*, 5087–5096.

Bogard, D. D., D. S. Burnett, P. Eberhardt, and G. J. Wasserburg (1967) [87]Rb-[87]Sr isochron and [40]K-[40]Ar ages of the Norton County achondrite. Earth Planet. Sci. Letters, *3*, 179–189.

Boger, P. D., and G. Faure (1974) Strontium isotope stratigraphy of a Red Sea core. Geol., *2*, 181–183.

Brooks, C., D. E. James, and S. R. Hart (1976) Ancient lithosphere: Its role in young continental volcanism. Science, *193*, 1086–1094.

Burnett, D. S., and G. J. Wasserburg (1967) Evidence for the formation of an iron meteorite at 3.8×10^9 years. Earth Planet. Sci. Letters, *2*, 137–147.

Church, S. E. (1973) Limits of sediment involvement in the genesis of orogenic volcanic rocks. Contrib. Mineral. Petrol., *39*, 17–32.

Cliff, R. A., C. Lee-Hu, and G. W. Wetherill (1972) Rubidium-strontium isotope characteristics of lunar soil. J. Geophys. Res., *77*, 2007–2013.

Dasch, E. J. (1969a) Strontium isotope disequilibrium in a porphyritic alkali basalt and its bearing on magmatic processes. J. Geophys. Res., *74*, 560–565.

Dasch, E. J. (1969b) Strontium isotopes in weathering profiles, deep sea sediments, and sedimentary rocks. Geochim. Cosmochim. Acta, *33*, 1521–1552.

Dasch, E. J., and P. E. Biscaye (1971) Isotopic composition of strontium in Cretaceous-to-Recent pelagic foraminifera. Earth Planet. Sci., *11*, 201–204.

Dasch, E. J., C. E. Hedge, and J. Dymond (1973) Effect of sea water interaction on strontium composition of deep-sea basalts. Earth Planet. Sci. Letters, *19*, 177–183.

Degens, E. T., and D. A. Ross, eds. (1969) Hot brines and Recent heavy metal deposits in the Red Sea. Springer-Verlag, New York, 600 p.

Dickinson, W. R., and T. Hatherton (1967) Andesitic volcanism and seismicity around the Pacific. Science, *157*, 801–803.

Dickinson, W. R. (1970) Relation of andesites, granites and derivative sandstones to arc-trench tectonics. Rev. Geophys. Space Phys., *8*, 813–860.

Doe, B. R. (1968) Lead and strontium isotopic studies of Cenozoic volcanic rocks in the Rocky Mountain region—a summary. Quart. Col. School Mines, *63*, 149–174.

Engel, A. E. J., C. G. Engel, and R. G. Havens (1965) Chemical characteristics of oceanic basalts and the upper mantle. Geol. Soc. Amer. Bull., *76*, 719–734.

Ewart, A., and J. J. Stipp (1968) Petrogenesis of the volcanic rocks of the central North Island, New Zealand, as indicated by a study of $^{87}Sr/^{86}Sr$ ratios, and Sr, Rb, K, U and Th abundances. Geochim. Cosmochim. Acta, *32*, 699–736.

Faure, G., and P. M. Hurley (1963) The isotopic composition of strontium in oceanic and continental basalt: Application to the origin of igneous rocks. J. Petrol., *4*, 31–50.

Faure, G., P. M. Hurley, and H. W. Fairbairn (1963) An estimate of the isotopic composition of strontium in rocks of the Precambrian Shield of North America. J. Geophys. Res., *68*, 2323–2329.

Faure, G., P. M. Hurley, and J. L. Powell (1965) The isotopic composition of strontium in surface water from the North Atlantic Ocean. Geochim. Cosmochim. Acta, *29*, 209–220.

Faure, G., and J. L. Powell (1972) Strontium isotope geology. Springer-Verlag, New York, 188 p.

Faure, G., and P. J. Barrett (1973) Strontium isotope compositions of non-marine carbonate rocks from the Beacon Supergroup of the Transantarctic Mountains. J. Sed. Petrol., *43*, 447–457.

Flower, M. F. J., H.-U. Schmincke, and R. N. Thompson (1975) Phlogopite stability and the $^{87}Sr/^{86}Sr$ step in basalts along the Reykjanes Ridge. Nature, *254*, 404–405.

Fullagar, P. D., R. E. Lemmon, and P. C. Ragland (1971) Petrochemical and geochronological studies of plutonic rocks in the southern Appalachians: Part 1. The Salisbury Pluton. Geol. Soc. Amer. Bull., *82*, 409–416.

Gopalan, K., and G. W. Wetherill (1968) Rubidium-strontium age of hypersthene (L) chondrites. J. Geophys. Res., *73*, 7133–7136.

Gopalan, K., and G. W. Wetherill (1969) Rubidium-strontium age of amphoterite (LL) chondrites. J. Geophys. Res., *74*, 4349–4358.

Gopalan, K., and G. W. Wetherill (1970) Rubidium-strontium studies on enstatite chondrites: Whole meteorite and mineral isochrons. J. Geophys. Res., *75*, 3457–3467.

Gray, C. M., D. A. Papanastassiou and G. J. Wasserburg (1973) The identification of early condensates from the solar nebula. Icarus, *20*, 213–239.

Green, D. H. (1973) Contrasted melting relations in a pyrolitic upper mantle under mid-oceanic ridge, stable crust and island arc environments. Tectonophys., *17*, 285.

Gunner, J. D., and G. Faure (1972) Rb-Sr geochronology of the Nimrod Group, central Transantarctic Mountains. Antarctic Geology and Geophysics, R. J. Adie, ed., Universitets forlaget Oslo, 305–311.

Gunner, J. D. (1974) Investigations of lower Paleozoic granites in the Beardmore Glacier region. Ant. J. U.S., *9*, 76–81.

Hamilton, W., and W. B. Meyers (1967) The nature of batholiths. U.S. Geol. Survey Prof. Paper 554-C, 30 p.

Hanson, G. N., S. S. Goldich, J. G. Arth, and D. H. Yardley (1971) Age of the Early Precambrian rocks of the Saganaga Lake—Northern Light Lake area, Ontario-Minnesota. Canadian J. Earth Sci., *8*, 1110–1124.

Harland, W. B., A. Gilbert Smith, and B. Wilcock (1964) The Phanerozoic timescale. Suppl. Quart. J. Geol. Soc. London, *120s*, 458 p.

Hart, S. R. (1969) Isotope geochemistry of crust-mantle processes. In The Earth's Crust and Upper Mantle, P. J. Hart, ed. Amer. Geophys. Union, Monograph 13, 58–62.

Hart, S. R., C. Brooks, T. E. Krogh, G. L. Davis, and D. Nava (1970) Ancient and modern volcanic rocks: A trace element model. Earth Planet. Sci. Letters, *10*, 17–28.

Hart, S. R., A. J. Erlank, and E. J. D. Kable (1974) Sea floor alteration: Some chemical and Sr isotopic effects. Cont. Mineral. Petrol., *44*, 219–230.

Hedge, C. E. (1966) Variations in radiogenic strontium found in volcanic rocks. J. Geophys. Res., *71*, 6119–6126.

Hedge, C. E., and R. J. Knight (1969) Lead and strontium isotopes in volcanic rocks from northern Honshu, Japan. Geochem. J., *3*, 15–24.

Hedge, C. E., and Z. E. Peterman (1970) The strontium isotopic composition of basalts from the Gorda and Juan de Fuca rises, northeastern Pacific Ocean. Contrib. Mineral. Petrol., *27*, 114–120.

Hedge, C. E., and D. C. Noble (1971) Upper Cenozoic basalts with high $^{87}Sr/^{86}Sr$ and Sr/Rb ratios, southern Great Basin, Western United States. Geol. Soc. Amer. Bull., *82*, 3503–3510.

Hoefs, J., and K. H. Wedepohl (1968) Strontium isotope studies on young volcanic rocks from Germany and Italy. Contrib. Mineral. Petrol., *19*, 328–338.

Hofmann, A. W., and S. R. Hart (1975) An assessment of local and regional isotopic equilibrium in a partially molten mantle. Ann. Rept. Director Dept. Terrest. Magnet., Carnegie Institution of Washington, 1974–1975, 195–210.

Hurley, P. M., P. C. Bateman, H. W. Fairbarin, and W. H. Pinson, Jr. (1965) Investigation of initial Sr^{87}/Sr^{86} ratios in the Sierra Nevada plutonic province. Geol. Soc. Amer. Bull., 76, 165–174.

Hurley, P. M., and J. R. Rand (1969) Predrift continental nuclei. Science, *164*, 1229–1242.

Hurley, P. M. (1972) Can the subduction process of mountain building be extended to

Pan-African and similar orogenic belts? Earth Planet. Sci. Letters, *15*, 305–314.

Jackson, E. D., E. A. Silver, and G. B. Dalrymple (1972) Hawaiian-Emperor chain and its relation to Cenozoic circumpacific tectonics. Geol. Soc. Amer. Bull., *83*, 601–618.

Jessberger, E. K., J. C. Huneke, and G. J. Wasserburg (1974) Evidence for a 4.5 aeon age of plagioclase clasts in a lunar highland breccia. Nature, *248*, 199–201.

Kaushal, S. K., and G. W. Wetherill (1969) Rb^{87}-Sr^{87} age of bronzite (H Group) chondrites. J. Geophys. Res., *74*, 2717–2726.

Kaushal, S. K., and G. W. Wetherill (1970) Rubidium-87 strontium-87 age of carbonaceous chondrites. J. Geophys. Res., *75*, 463–468.

Kempe, W., and O. Müller (1969) The stony meteorite Krähenberg, its chemical composition and the Rb-Sr age of the light and dark portions. In Meteorite Research, P. M. Millman ed. Reidel, Dortrecht, Holland, 418–428.

Kistler, R. W., and Z. E. Peterman (1973) Variations in Sr, Rb, K, Na and initial Sr^{87}/Sr^{86} in Mesozoic granitic rocks and intruded wall rocks in central California. Geol. Soc. Amer. Bull., *84*, 3489–3512.

Leeman, W. P. (1970) The isotopic composition of strontium in late-Cenozoic basalts from the Basin-Range province, western United States. Geochim. Cosmochim. Acta, *34*, 857–872.

Leeman, W. P., and W. I. Manton (1971) Strontium isotopic composition of basaltic lavas from the Snake River Plain, southern Idaho. Earth and Planet. Letters, *11*, 420–434.

Leeman, W. P. (1974) Late Cenozoic alkali-rich basalt from the western Grand Canyon area, Utah and Arizona: Isotopic composition of strontium. Geol. Soc. Amer. Bull., *85*, 1691–1696.

Lowell, J. D. (1974) Regional characteristics of porphyry copper deposits of the southwest. Econ. Geol., *69*, 601–617.

Manton, W. I. (1968) The origin of associated basic and acid rocks in the Lebombo-Nuanetsi igneous province, southern Africa, as implied by strontium isotopes. J. Petrol., *9*, 23–39.

Mark, R. K., C. Lee-Hu, and G. W. Wetherill (1974) Rb-Sr age of lunar igneous rocks 62295 and 14310. Geochim. Cosmochim. Acta, *38*, 1643–1648.

Mason, B. (1962) Meteorites. John Wiley, New York.

McNutt, R. H., J. H. Crocket, A. H. Clark, J. C. Caelles, E. Farrar, S. J. Haynes, and M. Zentilli (1975) Initial $^{87}Sr/^{86}Sr$ ratios of plutonic and volcanic rocks of the central Andes between latitudes 26° and 29° south. Earth Planet. Sci. Letters, *27*, 305–313.

Mitchell, A. H., and J. D. Bell (1973) Island-arc evolution and related mineral deposits, J. Geol., *81*, 381–405.

Moorbath, S. (1975) Evolution of Precambrian crust from strontium isotopic evidence. Nature, *254*, 395–397.

Moorbath, S., R. K. O'Nions, and R. J. Pankhurst (1975) The evolution of early Precambrian crustal rocks at Isua, West Greenland—geochemical and isotopic evidence. Earth Planet. Sci. Letters, *27*, 229–239.

Murthy, V. R., N. M. Evensen, B.-M. Jahn, and M. R. Coscio, Jr. (1971) Rb-Sr ages and elemental abundances of K, Rb, Sr, and Ba in samples from the Ocean of Storms. Geochim. Cosmochim. Acta, *35*, 1139–1153.

Nielson, D. R., and R. E. Stoiber (1973) Relationship of potassium content in andesitic lavas and depth to the seismic zone. J. Geophys. Res., *78*, 6887–6892.

O'Nions, R. K., and R. J. Pankhurst (1973) Secular variation in the Sr-isotope composition of Icelandic volcanic rocks. Earth Planet. Sci. Letters, *21*, 13–21.

O'Nions, R. K., and R. J. Pankhurst (1974) Petrogenetic significance of isotope and trace element variations in volcanic rocks from the Mid Atlantic. J. Petrol., *15*, 603–634.

Oxburgh, E. R., and D. L. Turcotte (1970) Thermal structure of island arcs. Geol. Soc. Amer. Bull., *81*, 1665–1688.

Page, R. W., and R. W. Johnson (1974) Strontium isotope ratios of Quaternary volcanic rocks from Papua, New Guinea. Lithos, 7, 91–100.

Papanastassiou, D. A., and G. J. Wasserburg (1969) Initial strontium isotopic abundances and the resolution of small time differences in the formation of planetary objects. Earth Planet. Sci. Letters, *5*, 361–376.

Papanastassiou, D. A., and G. J. Wasserburg (1970) Rb-Sr ages from the Ocean of Storms. Earth Planet. Sci. Letters, *8*, 269–278.

Papanastassiou, D. A., G. J. Wasserburg, and D. S. Burnett (1970) Rb-Sr ages of lunar rocks from the Sea of Tranquillity. Earth Planet. Sci. Letters, *8*, 1–9.

Papanastassiou, D. A., and G. J. Wasserburg (1971) Lunar chronology and evolution from Rb-Sr studies of Apollo 11 and 12 samples. Earth Planet. Sci. Letters, *11*, 37–62.

Papanastassiou, D. A., and G. J. Wasserburg (1973) Rb-Sr ages and initial strontium in basalts from Apollo 15. Earth Planet. Sci. Letters, *17*, 324–337.

Papanastassiou, D. A., and G. J. Wasserburg (1974) Evidence for a late formation and young metamorphism in the achondrite Nakhla. Geophys. Res. Letters, *1*, 23–26.

Papanastassiou, D. A., R. S. Rajan, J. C. Huneke, and G. J. Wasserburg (1974) Rb-Sr ages and lunar analogs in a basaltic achondrite; Implications for early solar system chronologies. Lunar Sci., *5*, 583–585.

Patterson, C., and M. Tatsumoto (1964) The significance of lead isotopes in detrital feldspar with respect to chemical differentiation within the earth's mantle. Geochim. Cosmochim. Acta, *28*, 1–22.

Peterman, Z. E., C. E. Hedge, R. C. Coleman, and P. D. Snavely, Jr. (1967) $^{87}Sr/^{86}Sr$ ratios in some eugeosynclinal sedimentary rocks and their bearing on the origin of granitic magma in orogenic belts. Earth Planet. Sci. Letters, *2*, 433–439.

Peterman, Z. E., B. R. Doe, and H. J. Prostka (1970) Lead and strontium isotopes in rocks of the Absaroka volcanic field, Wyoming. Contrib. Mineral. Petrol., *27*, 121–130.

Peterman, Z. E., C. E. Hedge, and H. A. Tourtelot (1970) Isotopic composition of strontium in sea water throughout Phanerozoic time. Geochim. Cosmochim. Acta, *34*, 105–120.

Peterman, Z. E., and C. E. Hedge (1971) Related strontium isotopic and chemical variations in oceanic basalts. Geol. Soc. Amer. Bull., *82*, 493–500.

Pidgeon, R. T., and A. M. Hopgood (1975) Geochronology of Archaean gneisses and tonalites from north of the Frederikshabs isblink, S. W. Greenland. Geochim. Cosmochim. Acta, *39*, 1333–1346.

Powell, J. L., and S. E. DeLong (1966) Isotopic composition of strontium in volcanic rocks from Oahu. Science, *153*, 1239–1242.

Pushkar, P., A. R. McBirney, and A. M. Kudo (1972) The isotopic composition of strontium in Central American ignimbrites. Bull. Volcanol., *35*, 265–294.

Pushkar, P., A. M. Stueber, J. F. Tomblin, and G. M. Julian (1973) Strontium isotope ratios in volcanic rocks from St. Vincent and St. Lucia, Lesser Antilles. J. Geophys. Res., *78*, 1279–1287.

Pushkar, P., and K. C. Condie (1973) Origin of the Quaternary basalts from the Black Rock Desert region, Utah: Strontium isotopic evidence. Geol. Soc. Amer. Bull., *84*, 1053–1058.

Pushkar, P., and D. B. Stoeser (1975) $^{87}Sr/^{86}Sr$ ratios in some volcanic rocks and some semifused inclusions of the San Francisco volcanic field. Geol., *3*, 669–671.

Sanz, H. G., D. S. Burnett, and G. J. Wasserburg (1970) A precise $^{87}Rb/^{87}Sr$ age and initial $^{87}Sr/^{86}Sr$ ratio for the Colomera iron meteorite. Geochim. Cosmochim. Acta, *34*, 1227–1239.

Scott, R. B., R. W. Nesbitt, E. J. Dasch, and R. L. Armstrong (1971) A strontium isotope evolution model for Cenozoic magma genesis, eastern Great Basin, U.S.A. Bull. Volcanol., *35*, 1–26.

Sillitoe, R. H. (1972) Relation of metal provinces in western America to subduction of oceanic lithosphere. Geol. Soc. Amer. Bull., *83*, 813–818.

Subbarao, K. V. (1972) The strontium isotopic composition of basalts from the East Pacific and Chile Rises and abyssal hills in the eastern Pacific Ocean. Contrib. Mineral. Petrol., *37*, 111–120.

Subbarao, K. V., and C. E. Hedge (1973) K, Rb, Sr and $^{87}Sr/^{86}Sr$ in rocks from the Mid-Indian oceanic ridge. Earth Planet. Sci. Letters, *18*, 223–228.

Tremba, E. L., G. Faure, G. C. Katsikatsos, and C. H. Summerson (1975) Strontium-isotope composition in the Tethys Sea, Euboea, Greece. Chem. Geol., *16* 109–120.

Turekian, K. K., and J. L. Kulp (1956) The geochemistry of strontium. Geochim. Cosmochim. Acta, *10*, 245–296.

Wetherill, G. W. (1971) Of time and the Moon. Science, *173*, 383–392.

Wetherill, G. W., R. K. Mark, and C. Lee-Hu (1973) Chondrites: Initial strontium-^{87}/Strontium-^{86} ratios and the early history of the solar system. Science, *182*, 281–283.

Wetherill, G. W. (1976) Where do the meteorites come from? A re-evaluation of the Earth-crossing Apollo objects as sources of meteorites. Geochim. Cosmochim. Acta, *40*, 1297–1317.

Whitford, D. J. (1975) Strontium isotopic studies of the volcanic rocks of the Sunda arc, Indonesia, and their petrogenetic implications. Geochim. Cosmochim. Acta, *39*, 1287–1302.

Wollenberg, H. A., and A. R. Smith (1970) Radiogenic heat production in pre-batho-lithic rocks of the central Sierra Nevada. J. Geophys. Res., *75*, No. 2, 431–438.

Yanagi, T. (1975) Rubidium-strontium model of formation of the continental crust and the granite at island arcs. Mem. Fac. Sci., Kyushu Un., Ser. D., Geol., *22*, No. 2, 37–98.

9 THE K-Ar METHOD OF DATING

Potassium (Z = 19) is an alkali metal (Group IA), along with lithium, sodium, rubidium, and cesium. It is one of the eight most abundant chemical elements in the crust of the Earth and is a major constituent of many important rock-forming minerals, such as the micas, the feldspars, the feldspathoids, clay minerals, and certain evaporite minerals (Heier and Adams, 1964). The isotopic composition of potassium was first studied by Aston (1921) who discovered ^{39}K and ^{41}K. The radioactivity of potassium salts was suggested by Thomson in 1905 and was subsequently demonstrated by Campbell and Wood (1906) and by Campbell (1908). However, the naturally occurring radioactive isotope of potassium was not identified until 1935 when Nier presented conclusive evidence for the existence of ^{40}K using a much more sensitive mass spectrometer than had been available to Aston. The possible modes of decay open to ^{40}K were discussed by the German physicist Von Weizsäcker (1937). He concluded that ^{40}K undergoes branched decay to ^{40}Ca and to ^{40}Ar, based partly on the fact that the abundance of argon in the atmosphere of the Earth is about 1000 times greater than expected when compared to the "cosmic" abundances of the other noble gases. Von Weizsacker also postulated that excess ^{40}Ar should be present in old K-bearing minerals. Ten years later, Aldrich and Nier (1948) confirmed Von Weizsäcker's prediction by demonstrating that four geologically old minerals (orthoclase, microcline, sylvite, and langbeinite) did, in fact, contain radiogenic ^{40}Ar. The theoretical basis for the K-Ar method of dating was therefore established by about 1950, and since then it has become an important and widely used method of measuring the ages of K-bearing rocks and minerals. Excellent accounts of the principles, techniques, and applications of the K-Ar method of dating have been provided by Dalrymple and Lanphere (1969), Schaeffer and Zähringer (1966), and by McDougall (1966).

The decay of ^{40}K to stable ^{40}Ca is not useful for dating because ^{40}Ca is the most abundant naturally occurring stable isotope of calcium. The formation of radiogenic ^{40}Ca atoms in a rock or mineral therefore increases its abundance only slightly. We shall return to the K-Ca method of dating in a later chapter and turn our attention now to the K-Ar geochronometer.

1 Principles and Methodology

The decay of naturally occurring ^{40}K to stable ^{40}Ar occurs by electron capture and by positron emission with a decay energy of 1.51 MeV (see Fig. 3.7). Eleven percent of ^{40}K atoms decay by electron capture to an excited state of ^{40}Ar which de-excites by emission of a gamma ray having an energy of 1.46 MeV. In addition, 0.16 percent of the decays are by electron capture directly to the ground state of ^{40}Ar. Positron decay occurs only 0.001 percent of the time (E_{max} = 0.49 MeV) and is followed by two annihilation gamma rays with a combined energy of 1.02 MeV. Decay to stable ^{40}Ca is favored by

Table 9.1   Isotopic Abundances of the Naturally Occurring Isotopes of Potassium, Argon, and Calcium

ISOTOPE	ABUNDANCE, %
Potassium	
^{39}K	93.08
^{40}K	0.0119
^{41}K	6.91
Argon (atmospheric)	
^{36}Ar	0.337
^{38}Ar	0.063
^{40}Ar	99.60
Calcium	
^{40}Ca	96.94
^{42}Ca	0.65
^{43}Ca	0.14
^{44}Ca	2.08
^{46}Ca	0.003
^{48}Ca	0.19

The isotopic abundances of potassium and argon are from Nier (1950) and those of calcium from Holden and Walker (1972).

88.8 percent of ^{40}K atoms and occurs by negatron emission directly to the ground state with a decay energy of 1.32 MeV. Thus, only about 11.2 percent of the ^{40}K atoms decay to ^{40}Ar. The isotopic abundances of potassium, argon and calcium are compiled in Table 9.1.

The growth of radiogenic ^{40}Ar and ^{40}Ca in a K-bearing system closed to potassium, argon, and calcium during its lifetime is expressed by Equation 9.1:

$$^{40}\text{Ar*} + {}^{40}\text{Ca*} = {}^{40}\text{K}(e^{\lambda t} - 1) \quad (9.1)$$

where λ is the total decay constant of ^{40}K. Each branch of the decay scheme gives rise to separate decay constants λ_e and λ_β such that the total decay constant is

$$\lambda = \lambda_e + \lambda_\beta$$

where λ_e refers to the decay of ^{40}K to ^{40}Ar

and λ_β represents the decay to ^{40}Ca. The most widely used values of these decay constants are

$$\lambda_e = 0.585 \times 10^{-10} \text{ y}^{-1}$$
$$\lambda_\beta = 4.72 \times 10^{-10} \text{ y}^{-1}$$

However, Beckinsale and Gale (1969) reconsidered the available measurements and arrived at somewhat different values: $\lambda_e = (0.566 \pm 0.0035) \times 10^{-10} \text{ y}^{-1}$, $\lambda'_e = (8.67 \pm 1.74) \times 10^{-13} \text{ y}^{-1}$, $\lambda_\beta = (4.905 \pm 0.009) \times 10^{-10} \text{ y}^{-1}$, and $\lambda = (5.480 \pm 0.010) \times 10^{-10} \text{ y}^{-1}$. (Note that λ'_e is the decay constant for electron capture by ^{40}K directly to the groundstate of ^{40}Ar, shown also in Fig. 3.7). Most geochronologists are still using the old decay constants given above, and we shall continue to use them in this chapter. However, the values of Beckinsale and Gale (1969) may be adopted in the future. The same authors also recommended a value of 0.0118 percent for the isotopic abundance of ^{40}K, but noted that it may vary by up to ± 1 percent due to fractionation of the isotopes of potassium in nature.

If the old values are used, the total decay constant of ^{40}K is

$$\lambda = (0.585 + 4.72) \times 10^{-10}$$
$$= 5.305 \times 10^{-10} \text{ y}^{-1}$$

which corresponds to a half-life of

$$T_{1/2} = \frac{0.693}{5.305 \times 10^{-10}} = 1.31 \times 10^9 \text{ y}$$

The branching ratio R is defined as λ_e/λ_β and has a value of 0.124. The fraction of ^{40}K atoms that decay to ^{40}Ar is given by $(\lambda_e/\lambda)^{40}$K and the growth of radiogenic ^{40}Ar atoms in a K-bearing rock or mineral can therefore be written as

$$^{40}\text{Ar*} = \frac{\lambda_e}{\lambda} {}^{40}\text{K}(e^{\lambda t} - 1) \quad (9.2)$$

The total number of ^{40}Ar atoms is

$$^{40}\text{Ar} = {}^{40}\text{Ar}_0 + {}^{40}\text{Ar*} \qquad (9.3)$$

However, in this case we assume that no ^{40}Ar is present in the mineral at the time of its formation; thus

$$^{40}\text{Ar}_0 = 0$$

Equation 9.2 is the K-Ar age equation. In order to use it to date a K-bearing mineral, we measure the concentration of potassium and the amount of radiogenic ^{40}Ar that has accumulated. Given this information, Equation 9.2 can be solved for t:

$$t = \frac{1}{\lambda} \ln \left[\frac{^{40}\text{Ar*}}{^{40}\text{K}} \left(\frac{\lambda}{\lambda_e} \right) + 1 \right] \qquad (9.4)$$

The value of t so calculated is the *age* of the mineral *only* when the following assumptions are satisfied.

1. No radiogenic ^{40}Ar produced by decay of ^{40}K in the mineral during its life time has escaped.
2. The mineral became closed to ^{40}Ar soon after its formation, which means that it must have cooled rapidly after crystallization.
3. No excess ^{40}Ar was incorporated into the mineral either at the time of its formation or during a later metamorphic event.
4. An appropriate correction is made for the presence of atmospheric ^{40}Ar.
5. The mineral was closed to potassium throughout its lifetime.
6. The isotopic composition of potassium in the mineral is normal and was not changed by fractionation or other processes, except by decay of ^{40}K.
7. The decay constants of ^{40}K are known accurately and have not been affected by the physical or chemical conditions of the environment in which the potassium has existed since it was incorporated into the Earth (see Chapter 4.1).

These assumptions require careful evaluation in each case and place certain restrictions on the geological interpretation of K-Ar dates.

The last two assumptions are quite general in scope and express certain fundamental conditions of dating by any method based on radioactivity. The isotopic composition of potassium in natural samples is believed to be constant, even though fractionation of potassium isotopes has been observed on a small scale across contacts of igneous intrusions (Verbeek and Schreiner, 1967; Morozova and Alferovsky, 1974). The decay constants of radioactive nuclides are likewise believed to be independent of time and not subject to change in response to physical or chemical conditions of natural environments in the crust and mantle of the Earth.

In order to date a K-bearing mineral, the concentration of potassium and the amount of radiogenic ^{40}Ar must be measured. Potassium concentrations can be determined by one of several methods, including flame photometry, atomic absorption spectrometry, isotope dilution, or even neutron activation. These are well-established methods that do not require detailed discussion in this context (Dalrymple and Lanphere, 1969; Müller, 1966; Cooper, 1963). The amount of radiogenic ^{40}Ar is now universally measured by means of isotope dilution. The analytical procedures and calculations required to determine the radiogenic ^{40}Ar content of a mineral are outlined below.

A known weight of the purified mineral concentrate is fused in a molybdenum crucible sealed within a vacuum system. After the fusion, a known quantity of spike argon enriched in ^{38}Ar is mixed with the gas extracted from the mineral. The mixture of gases is purified by removal of all chemically reactive gases, including H_2, CO_2, O_2, and N_2, leaving only a mixture of the noble gases, including argon. The residual gas mixture is then introduced into the source of a mass spectrometer and the $^{40}Ar/^{38}Ar$ and $^{38}Ar/^{36}Ar$ ratios of the mixture are determined. The measured value of the $^{38}Ar/^{36}Ar$ ratio is used to correct for the presence of atmospheric argon whose isotopic composition is known (Table 9.1). The amount of radiogenic ^{40}Ar is calculated from the measured $^{40}Ar/^{38}Ar$ ratio, using the known amount of spike argon that was added and its isotopic composition. The relevant equation (Dalrymple and Lanphere, 1969) is

where

39.102 = atomic weight of potassium,

0.000119 = abundance of ^{40}K expressed as a decimal fraction,

39.9623 = atomic weight of ^{40}Ar,

A = Avogadro's number,

and the factor 10^4 is required to change the percent concentration of K to parts per million. The date is calculated from Equa-

$$^{40}Ar^* = {}^{38}Ar_s \left\{ \left(\frac{^{40}Ar}{^{38}Ar}\right)_m - \left(\frac{^{40}Ar}{^{38}Ar}\right)_s - \left[\frac{1 - \left(\frac{^{38}Ar}{^{36}Ar}\right)_m \left(\frac{^{36}Ar}{^{38}Ar}\right)_s}{\left(\frac{^{38}Ar}{^{36}Ar}\right)_m \left(\frac{^{36}Ar}{^{38}Ar}\right)_A - 1} \right] \times \left[\left(\frac{^{40}Ar}{^{38}Ar}\right)_A - \left(\frac{^{40}Ar}{^{38}Ar}\right)_m \right] \right\} \quad (9.5)$$

where $^{38}Ar_s$ is the number of moles of spike ^{38}Ar added, and the subscripts m, s, and A refer to the isotope ratios of the mixture, the spike, and atmospheric argon, respectively. In the conventional K-Ar method of dating, the isotopic composition of atmospheric argon trapped within K-bearing minerals is assumed to be equal to that of argon in modern air. However, it can also be determined separately by means of K-Ar isochrons to be discussed in Chapter 10.

We shall now illustrate the calculation of K-Ar dates from experimental data by means of an example. A commercial firm in the United States that specializes in measuring K-Ar dates reported the following analytical results for a sample of muscovite from a pegmatite in the Wisconsin Range of the Transantarctic Mountains: K = 8.378 percent and $^{40}Ar^* = 0.3305$ ppm. We first calculate the $^{40}Ar^*/^{40}K$ ratio:

$$\frac{^{40}Ar^*}{^{40}K} = \frac{0.3305 \times 39.102 \times A}{8.378 \times 10^4 \times 0.000119 \times 39.9623 A}$$

$$= 0.03243$$

tion 9.4:

$$t = \frac{2.303}{5.305 \times 10^{-10}} \log \left[\frac{0.03243 \times 5.305}{0.585} + 1 \right]$$

$$t = \frac{2.303 \times 0.11193}{5.305 \times 10^{-10}} = 485.9 \times 10^6 \text{ y}$$

Minerals to be dated by the K-Ar method must have retained all of the radiogenic ^{40}Ar produced within them by decay of ^{40}K (assumption 1) and they must not contain any excess ^{40}Ar (assumption 3). Argon loss from minerals may occur because argon is a noble gas and therefore does not form bonds with other atoms in a crystal lattice. In general, argon loss can be attributed to the following causes.

1. Inability of a mineral lattice to retain argon even at low temperature and atmospheric pressure.

2. Either partial or complete melting of rocks followed by crystallization of new minerals from the resulting melt.

3. Metamorphism at elevated temperatures and pressures resulting in complete or

partial argon loss depending on the temperature and duration of the event.

4. Increase in temperature due to deep burial or contact metamorphism causing argon loss from most minerals without producing any other physical or chemical changes in the rock.

5. Chemical weathering and alteration by aqueous fluids leading not only to argon loss but also to changes in the potassium content of minerals.

6. Solution and redeposition of water-soluble minerals, such as sylvite.

7. Mechanical breakdown of minerals, radiation damage, and shock waves. Even excessive grinding during preparation of samples for dating by the K-Ar method may cause argon loss.

A very good demonstration of argon loss due to increase in temperature in a contact metamorphic zone was presented by Hart (1964). He dated hornblende, biotite, and K-feldspar of rocks from the Idaho Springs Formation (Precambrian) collected at increasing distances from the intrusive contact of the Eldora quartz monzonite stock (Tertiary) in the Front Range of Colorado. The Idaho Springs Formation consists of a quartz-feldspar-biotite gneiss with some biotite-quartz-feldspar schist and amphibolite. Age determinations reviewed by Hart (1964) suggest that these rocks were regionally metamorphosed about 1350 to 1400 million years ago. The Eldora stock was intruded into these rocks about 55 million years ago and produced very minor mineralogical contact metamorphic effects in the Idaho Springs Formation. However, Figure 9.1 shows that the K-Ar dates of hornblende, biotite, and K-feldspar were *profoundly* affected as a result of losses of varying amounts of radiogenic ^{40}Ar. The fraction of argon lost from each of the minerals

* 1 foot = 0.3048 meters

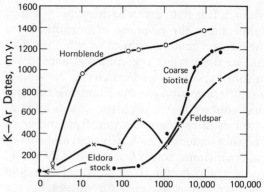

FIGURE 9.1 Variation of K-Ar dates of minerals from the Idaho Springs Formation of the Front Range of Colorado in a contact metamorphic zone produced by intrusion of the Eldora stock. The Idaho Springs Formation was regionally metamorphosed 1350 to 1400 million years ago. The age of the Eldora stock is 55 million years. (Hart, 1964.)

decreases as a function of distance from the contact and reflects the differing retentivities of these minerals for radiogenic argon. The coarse biotite apparently lost all of its radiogenic argon out to a distance of about 100 feet* from the contact, and the effects of argon loss can be traced for more than 7000 feet beyond that distance. The K-Ar dates of the biotite finally stabilize at 1200 million years at a distance of about 14,000 feet from the contact. The K-Ar dates of the K-feldspar increase somewhat erratically away from the contact and show the effects of argon loss even at a distance of 22,500 feet. The hornblende, on the other hand, displays a superior retentivity for argon. Loss of argon is confined primarily to a distance of about 10 feet from the contact and then diminishes rapidly. Approximately 100 feet from the contact the K-Ar dates of the hornblende reach a plateau value of 1200 million years and ultimately rise to 1375 million years about 14,000 feet from the contact. It is not clear whether the 1375 m.y. date reflects the response of hornblende to the contact metamorphism or whether it is due to the earlier regional metamorphism of the Idaho Springs Formation. Nevertheless, these results clearly illustrate the fact that contact metamorphism causes systematic loss of radiogenic argon from minerals and that under identical physical and chemical conditions, hornblende retains its radiogenic argon better than biotite. Potassium feldspar is not suitable for dating by the K-Ar method because it loses argon readily even at surface temperatures. Hart (1964) also dated biotite from the Idaho Springs Formation by the Rb-Sr method and observed a pattern of variation similar to that of the K-Ar dates. However, the Rb-Sr dates are consistently greater than the K-Ar dates, which suggests that, in this case, biotite retained radiogenic ^{87}Sr somewhat better than ^{40}Ar. It should be clear, of course, that K-Ar dates in such a contact metamorphic aureole primarily reflect variable losses of radiogenic argon and that such dates cannot be used to determine the "age" of the rocks. In the case illustrated in Figure 9.1, only the dates in the immediate vicinity of the intrusive and those sufficiently removed from it have geological time significance.

In order to be useful for dating by the K-Ar method, minerals and rocks must retain argon quantitatively (although all minerals begin to lose argon with increasing temperature), they must be resistant to chemical alteration, and they must contain potassium (although not necessarily as a major constituent). In addition, they should be sufficiently common to be useful for geological investigations. Table 9.2 is a compilation of the minerals (and rocks) that meet these criteria and are most useful for dating by the K-Ar method.

The possibility of dating volcanic rocks by analysis of whole-rock specimens (Table 9.2) is of great interest. Several studies summarized by Dalrymple and Lanphere (1969) have demonstrated the reliability of K-Ar dates of whole-rock samples of basalt. Such rocks consist primarily of plagioclase and pyroxene, both of which retain argon satisfactorily. However, it is important to avoid samples that have been altered or that contain devitrified glass, secondary minerals (zeolites, calcite, or clay minerals), and xenoliths or xenocrysts. The argon retentivity of basaltic glass is generally good. However, more siliceous or hydrated glass is questionable, especially when it is devitrified. The presence of devitrified glass and of secondary alteration products results in a *lowering* of K-Ar dates. On the other hand, xenoliths and xenocrysts are likely to contain excess radiogenic ^{40}Ar whose presence in a sample will *increase* the measured K-Ar date. Basalts that are extruded on the ocean floor under high hydrostatic pressure also contain excess ^{40}Ar which is concentrated in the glassy crusts of individual "pillows" apparently because argon in solution in the lava is trapped there by rapid quenching. The amount of excess argon decreases inward from the glassy rind of the pillows because a slower rate of cooling allows the argon to migrate toward the rind. The amount of excess argon in basalt glass increases as a function of water depth which suggests that it is controlled primarily by the hydrostatic pressure. For these reasons only holocrystalline samples taken from the centers of pillows should be used for dating submarine pillow basalts. Recent examples of geological interpretations of K-Ar dates of volcanic rocks include studies by Armstrong et al. (1975), DeLong and McDowell (1975), Dalrymple et al. (1975), Wellman and McDougall (1974a,b) and Wellman (1974).

Excess radiogenic argon has also been found in some K-bearing *minerals*. The effect of excess ^{40}Ar on K-Ar dates is most notice-

Table 9.2   Common Rock-forming Minerals Suitable for Dating by the K-Ar Method

	ROCK TYPE			
	VOLCANIC	PLUTONIC	METAMORPHIC	SEDIMENTARY
Feldspars				
Sanidine	⊗			
Anorthoclase	⊗			
Plagioclase	⊗			
Felsdpathoids				
Leucite	×			
Nepheline	×	×		
Mica				
Biotite	⊗	⊗	⊗	
Phlogopite			⊗	
Muscovite		⊗	⊗	
Lepidolite		×		
Glauconite				×
Amphibole				
Hornblende	⊗	⊗	⊗	
Pyroxene	×	×		
Whole rock	⊗		×	

From Potassium-Argon Dating: Principles, Techniques and Applications by G. Brent Dalrymple and Marvin A. Lanphere. W. H. Freeman and Company. Copyright © 1969.
Key: ⊗ = often useful; × = sometimes useful.

able in minerals having a low potassium content or in minerals that are young. The minerals beryl, cordierite, and tourmaline frequently contain excess ^{40}Ar, while pyroxene, hornblende, feldspar, phlogopite, biotite, and sodalite contain such excess ^{40}Ar only rarely (Hart and Dodd, 1962; McDougall and Green, 1964; York and MacIntyre, 1965; Livingston et al., 1967). Generally, excess ^{40}Ar is observed when minerals have been exposed to a high partial pressure of argon during regional metamorphism, during crystallization of pegmatites (Laughlin, 1969), or during the emplacement of kimberlite pipes. The argon that may either diffuse into the minerals or may be occluded within them is probably derived by outgassing of older K-bearing minerals in the crust and mantle

of the Earth. Another source of excess ^{40}Ar in some minerals is the presence of fluid inclusions (Rama et al., 1965), or of older xenoliths or xenocrysts. The presence of excess ^{40}Ar increases K-Ar dates and may lead to overestimates of the ages of minerals dated by this method.

2
Time Scale for Geomagnetic Polarity Reversals

The first indication that the Earth's magnetic field may have reversed its polarity was presented in 1906 by the French physicist Bernard Brunhes. Twenty years later, M. Matuyama concluded that the Earth's magnetic field had a reversed polarity during the early Quaternary period and that it has been normal since then. However, no systematic study of the history of polarity reversals was possible until the K-Ar method became available for dating of very young volcanic rocks. The year 1963 appears to be the beginning of continuing efforts to study the history of reversals of the Earth's magnetic field as recorded in sequences of volcanic rocks whose ages can be measured by the K-Ar method. In that year Cox, Doell, and Dalrymple (1963a,b) and McDougall and Tarling (1963) proposed time scales for reversals of the Earth's magnetic field and

Vine and Mathews (1963) suggested that the linear magnetic anomalies along midocean ridges could be explained by sea-floor spreading and periodic polarity reversals of the magnetic field. Since then, the K-Ar method of dating has played a key role in the study of the history of the magnetic field and the establishment of a time scale for polarity reversals.

The reversal chronology consists of extended intervals of time called *epochs* during which the field had predominantly one polarity. Such epochs are interrupted by shorter *events* of opposite polarity. The

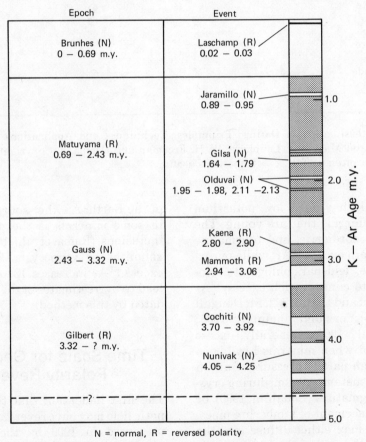

FIGURE 9.2 Time scale for polarity reversals of the Earth's magnetic field, based on K-Ar dates of volcanic rocks. (Cox, 1969.)

distinction between epochs and events is based primarily on the length of time involved. Epochs are named after deceased scientists who made significant contributions to studies of the Earth's magnetic field. Events are named after the location at which a reversal was first recognized. The existing reversal time scale which extends to about 4.5 million years in the past contains four epochs, starting with the present: Brunhes (normal), Matuyama (reversed), Gauss (normal), and Gilbert (reversed). These are shown in Figure 9.2 along with several events of shorter duration (Cox, 1969; Dalrymple, 1972). Research now in progress has already demonstrated that reversals of the Earth's magnetic field can be recognized in older rocks extending all the way into Precambrian time. However, the time scale for such older reversals has not yet been established. Another important extension of the time scale is based on seafloor spreading and the resulting magnetic anomaly patterns associated with midocean ridges (Heirtzler et al., 1968).

3
The Metamorphic Veil

In volcanic rocks and shallow igneous intrusives which cool rapidly, the K-bearing minerals begin to accumulate radiogenic argon almost immediately after crystallization. However, in deep-seated plutonic and in metamorphic rocks which cool slowly, argon retention may be delayed until the temperature drops to some critical value below which diffusion of argon becomes ineffective and accumulation of radiogenic ^{40}Ar begins. In this context it is convenient to introduce the concept of the *blocking temperature* which is defined as the temperature at which the loss of ^{40}Ar by diffusion out of a particular mineral becomes negligible compared to its rate of accumulation.

We can therefore clarify the meaning of K-Ar dates by saying that they represent the time elapsed since the dated mineral cooled through its blocking temperature, assuming that it remained closed to argon and potassium after that event. Studies of argon diffusion in contact metamorphic zones and in the laboratory indicate that the minerals used for dating by the K-Ar method have different blocking temperatures. The importance of diffusion of radiogenic daughters (including ^{40}Ar) to geochronometry has been reviewed by Giletti (1974a,b). A mathematical analysis of this phenomenon was made by Dodson (1973).

Let us consider a deep-seated, slow-cooling plutonic igneous rock containing hornblende and biotite. The blocking temperature of the hornblende with respect to radiogenic argon is higher than that of the biotite and therefore the K-Ar clock of the hornblende will start before that of the biotite. If the temperature decreased smoothly and both minerals remained closed to argon and potassium after their respective blocking temperatures had been passed, the hornblende should register an older date than the biotite. The difference between the K-Ar dates of the two minerals can be used to infer the rate of cooling, provided that their respective blocking temperatures are known. Such *discordance* of K-Ar dates of coexisting minerals has also been observed in metamorphic rocks (Fig. 9.1). In this case, two different temperature histories are possible: (1) all minerals were completely outgassed during metamorphism followed by slow cooling, or (2) outgassing was incomplete, in which case the fraction of argon lost from a given mineral depends on its retentivity for argon. Thus, hornblende in a metamorphic rock may register an older K-Ar date than coexisting biotite because it is more retentive with respect to radiogenic argon than is biotite. In all such cases, both complete or partial argon losses from K-bearing minerals of metamorphic rocks may result in discordance of K-Ar dates which thus reflect the temperature history of the rocks and are not indicative of their ages. Both Hart (1964) and Hanson and Gast (1967) have studied the effects of losses of radiogenic ^{40}Ar and ^{87}Sr on mineral dates in contact metamorphic aureoles where the variation of temperature as a function of time and space can be approximated from heat-flow theory.

Over the years, a large number of biotites from the granitic gneisses and schists of Precambrian shields and orogenic belts have been dated by the K-Ar method. These dates generally do not indicate the ages of rocks in such areas but rather reflect the time elapsed since cooling through the blocking temperature of biotite. When K-Ar dates of biotites from metamorphic terranes are plotted on a map, they form a conceptual surface which Armstrong (1966) termed the "metamorphic veil." In other words, the K-Ar dates obscure the ages of the rocks and reflect instead the regional metamorphism and subsequent cooling history of the rocks. The observed regional homogeneity of K-Ar dates in Precambrian shields and orogenic belts (Fig. 8.7) suggests that large segments of the continental crust were metamorphosed and then cooled at similar rates.

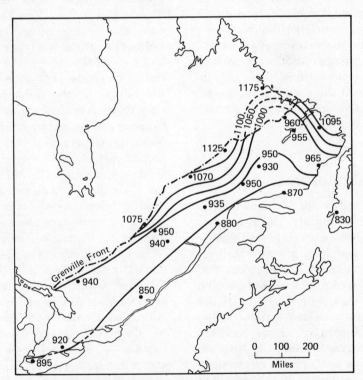

FIGURE 9.3 Metamorphic veil of the Grenville Province of Canada, based on K-Ar dates of biotite from granite gneisses. (Reproduced from Figure 1 (p. 130) of Harper, C. T. (1967), Earth and Planetary Science Letters, 3, No. 2, 128–132, by permission of Elsevier Scientific Publishing Co.)

The cooling of large segments of the continental crust at uniform rates may be due to uplift of large cratonic blocks bounded by major fracture zones that extend to great depths. Accordingly, we may interpret the K-Ar dates of a particular mineral from a cratonic block as the time elapsed since the rocks that are now exposed at the surface were uplift across the isotherm corresponding to the blocking temperature of that mineral. For example, Harper (1967) used K-Ar dates of biotite from granitic gneisses of the Grenville Province of Canada to construct the metamorphic veil shown in Figure 9.3. The contours represent equal times since cooling through the blocking temperature, called "thermochrons" by Harper. In this case the thermochrons are approximately parallel to the Grenville Front and form a surface that slopes to the southeast. Consequently, one may conclude from such an interpretation of the K-Ar dates that the rocks that are now exposed at the surface were uplifted through the blocking-temperature isotherm between 850 and 1125 million years ago and that the western edge of the Grenville craton was uplifted earlier than the eastern side. Alternatively, the lower dates along the eastern exposures of the Grenville Province may also result from partial argon losses due to thermal effects associated with the Appalachian orogeny.

Armstrong (1966) outlined a generalized history for a typical orogen and considered to what extent K-Ar dates provide useful information about the timing of specific evolutionary stages in the development of an orogenic belt. He concluded that K-Ar dates of igneous and metamorphic rocks from the central core of the orogen indicate the time of uplift and cooling and not the age of metamorphism. In fact, he pointed out that the time of initial deposition of sediment, folding, and metamorphism cannot be de-termined by the K-Ar method. Ideally, K-Ar dates of orogenic belts form a metamorphic veil that postdates the occurrence of the principal tectonic and metamorphic events. Only the minerals formed by low-grade metamorphism along the flanks of the orogen record events that lie within the time interval during which metamorphism occurred.

4
The Precambrian Time Scale

K-Ar dates of biotite and muscovite from igneous and metamorphic rocks have played an important role in the development of the Precambrian time scale used by the Geological Survey of Canada (Stockwell, 1968, 1972). Starting in 1959, the Geochronology Laboratories of the Geological Survey of Canada began a program of dating rocks from the Canadian Precambrian shield in order to identify major orogenic events and to assist in the construction of a Precambrian time scale. The initial phase of this program consisted primarily of measurements of K-Ar dates of biotite and muscovite with some work on hornblende and whole-rock samples of mafic dikes. By the end of 1967, R. K. Wanless and his colleagues had dated more than 750 samples from all over the Canadian shield. These dates were used by Stockwell in 1968 to construct a time scale for the Precambrian rocks exposed in Canada.

Stockwell showed that the K-Ar dates tend to cluster around certain values which he identified as being the ages of orogenies. He defined an "orogeny" as a process of folding affecting large segments of the crust commonly associated with virtually contemporaneous regional metamorphism and emplacement of granitic bodies (Stockwell, 1972). He estimated the ages of these orogenies by the arithmetic means of the K-Ar

Eon	Era	Subera	Orogenic Event	Age, m.y. K—Ar, Mica(1)	Age, m.y. Rb—Sr, (2)
Proterozoic	Paleozoic			570	
Proterozoic	Hadrynian				
Proterozoic	Helikian	Neohelikian	Grenvillian	945	1070
Proterozoic	Helikian	Paleohelikian	Elsonian	1370	?
Proterozoic	Aphebian		Hudsonian	1735	1850
Archaean			Kenoran	2490	2690

(1) Stockwell (1968); (2) Stockwell (1972)

FIGURE 9.4   Precambrian time scale used by the Geological Survey of Canada. (Stockwell, 1972.)

dates in each cluster and applied to them the following names:

Grenvillian   945 m.y.
Elsonian   1370 m.y.
Hudsonian   1735 m.y.
Kenoran   2490 m.y.

Stockwell used these orogenies to define intervals of time, shown in Figure 9.4, each of which starts with the deposition of sediment on the erosion surface cut into the rocks subjected to the preceding orogeny and terminates with folding, metamorphism, uplift, and erosion during the next orogeny. For example, rocks of Aphebian age were deposited on a basement that was metamorphosed during the Kenoran orogeny. The Aphebian era ended with the folding and metamorphism of the Hudsonian orogeny. Sedimentary and volcanic rocks of

Helikian age were then deposited on a basement of rocks deformed and metamorphosed in the Hudsonian orogeny, and so on.

An important aspect of this classification is the recognition of eight structural provinces on the Canadian shield. Each province was stabilized at the end of a specific orogeny as indicated by the fact that K-Ar mica dates in each province cluster around a "magic number." For example, the Superior Province, identified in Figure 9.5, is characterized by K-Ar dates that cluster around a mean of 2490 m.y. corresponding to the Kenoran orogeny. Evidently the rocks of the Superior Province were metamorphosed during the Kenoran orogeny and have not been reheated since then. The rocks of the Churchill Province were metamorphosed for the last time during the Hudsonian orogeny and those of the Grenville Province by the Grenvillian orogeny. In this way the

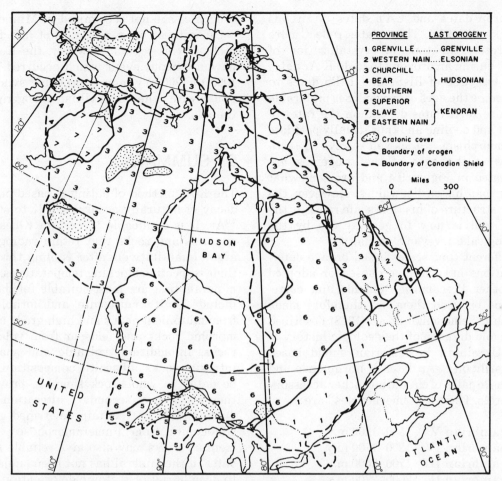

FIGURE 9.5 Orogenies and structural provinces of the Canadian shield. (Reproduced by permission of the National Research Council of Canada from the Canadian Journal of Earth Sciences, Vol. 5, 1968, pp. 693–698.) (Stockwell, 1968.)

entire shield has been subdivided into separate structural provinces which are separated from each other by major unconformities or orogenic fronts.

The K-Ar dates of micas in each province form a metamorphic veil which hides the true ages of the rocks. It is known, for example, that the Grenville Province contains rocks that formed long before the Grenville orogeny (Krogh and Hurley, 1968; Krogh and Davis, 1969; Grant, 1964). However, all of the rocks in this province bear the strong metamorphic and structural imprint of the Grenvillian orogeny and therefore they are properly considered to form a coherent structural province. Similar statements can be made regarding the other provinces although their preorogenic histories have not yet been revealed in as much detail.

The time scale that was originally based on K-Ar dates of micas has now been modified by redefining the ages of the orogenies on the basis of whole-rock Rb-Sr isochron dates and U-Pb dates of zircons (Stockwell, 1972). These dates reflect more accurately the time of crystallization of rocks during an orogeny than do K-Ar dates of micas. Use of Rb-Sr and U-Pb dates to determine the ages of orogenies is preferable because K-Ar dates depend on the rates of uplift and cooling and thus usually postdate metamorphism. Stockwell's revised estimates of the ages of the orogenies are indicated in Figure 9.4 and are somewhat older than his earlier values. However, the basic structure of his time scale remains the same and is now in general use by the Geological Survey of Canada.

Different time scales based on age determinations have been proposed or adopted for other Precambrian shields and correlations between them are therefore made difficult. Semenenko et al. (1968) described a fivefold division of the geologic history of the Ukrainian Precambrian shield based primarily on K-Ar dates of hornblende and U, Th-Pb dates of zircon, and other accessory minerals. Their five "megacycles" are:

Precambrian V	550–1200 m.y.
Precambrian IV	1200–1700 m.y.
Precambrian III	1700–2000 m.y.
Precambrian II	2000–2700 m.y.
Precambrian I	2700–3500 m.y.

The time scale used in Minnesota (Goldich, 1968) and adjacent areas around Lake Superior in the United States is based on a threefold division of Precambrian time (Early, Middle, and Late). Early Precambrian time ended with the Algoman orogeny (2400–2750 m.y.), while the Middle Precambrian was terminated by the Penokean orogeny (1600–1900 m.y.). The Algoman orogeny of Minnesota corresponds to the Kenoran orogeny of Canada and the Penokean orogeny corresponds to the Hudsonian orogeny. The Elsonian and Grenvillian

orogenies are not recognized in Minnesota because these orogenies did not affect the rocks of this region. However, the Keweenawan igneous event which occurred from 1000 to 1200 m.y. ago is probably synchronous with the Grenvillian orogeny of Canada.

5   SUMMARY

The K-Ar method of dating is based on the decay of naturally occurring ^{40}K to stable ^{40}Ar. It is applicable to certain K-bearing minerals and rocks that retain radiogenic argon quantitatively after cooling through their respective blocking temperatures. The minerals that are most suitable for dating include biotite, muscovite, and hornblende from plutonic igneous and high-grade metamorphic rocks and feldspar from volcanic rocks. In addition, volcanic and shallow intrusive rocks of basaltic composition may be dated as whole-rock samples, provided they are free of secondary alteration products and do not contain devitrified glass. Glauconite from unmetamorphosed sedimentary rocks may also give reliable K-Ar dates, if the mineral has not been subjected to deep burial or tectonic deformation.

K-Ar dates derived from young volcanic rocks have been used to construct a time-scale for reversals of the Earth's magnetic field. Dating of basalt from the floor of the oceans and the interpretation of magnetic anomaly patterns in the ocean basins have provided direct evidence for sea-floor spreading and the resulting drift of continents leading to the formulation of the theory of plate tectonics.

The loss of radiogenic ^{40}Ar (and other radiogenic daughters) at elevated temperatures makes K-Ar dates useful indicators of the temperature histories of slow-cooling plutonic and metamorphic rocks. K-Ar dates of biotite (or some other K-bearing mineral) in regionally metamorphosed rocks form a metamorphic veil that obscures the time of original crystallization of such rocks. However, the regional homogeneity of K-Ar dates of a specific mineral from a cratonic block or orogenic belt provides information about cooling rates related to uplift through the blocking-temperature isotherm.

K-Ar dates of biotite have been used to delineate structural provinces on the Canadian Precambrian shield and to construct a Precambrian time scale based on time intervals between periodic episodes of mountain formation, regional metamorphism, uplift, cooling, and subsequent stabilization of cratonic blocks.

PROBLEMS

1. Biotite from the Silver Point quartz monzonite of Idaho contains 8.45 percent K_2O and 6.016×10^{-10} moles/g of radiogenic ^{40}Ar (Miller and Engels, 1975, sample 68). Calculate a K-Ar date for this mineral using $\lambda_e = 0.584 \times 10^{-10}$ y^{-1}, $\lambda_\beta = 4.72 \times 10^{-10}$ y^{-1}, $^{40}K/K = 1.19 \times 10^{-4}$. (**ANSWER:** 48 m.y.)

2. Hornblende from the same rock (Problem 1) contains 0.6078 percent K_2O and 0.4642×10^{-10} moles/g of radiogenic ^{40}Ar. Calculate a K-Ar date using the same constants given above. (**ANSWER:** 51 m.y.)

3. A small pluton east of Chewelah in northeastern Washington contains biotite and hornblende analyzed by Miller and Engels (1975, sample 85):

	$K_2O\%$	RADIOGENIC ^{40}Ar moles/g
Biotite	8.71	12.83×10^{-10}
Hornblende	1.44	4.348×10^{-10}

Calculate dates for both minerals and speculate regarding the *age* of this pluton assuming that it may have been reheated during a later phase of intrusive activity in the area.

4. Another pluton in northeastern Washington contains biotite and muscovite analyzed by Miller and Engels (1975, sample 91):

	$K_2O\%$	RADIOGENIC ^{40}Ar moles/g
Biotite	7.90	6.996×10^{-10}
Muscovite	10.72	13.66×10^{-10}

Calculate dates and derive a conclusion regarding the relative retentivities of these minerals for ^{40}Ar.

5. Sample 87 from the small pluton east of Chewelah in northeastern Washington (Miller and Engels, 1975) contains biotite containing 8.83 percent K_2O and 8.481×10^{-10} moles/g of radiogenic ^{40}Ar yielding a date of 64 m.y. If the age of this pluton is 200 m.y., what is the fraction of radiogenic ^{40}Ar *lost* from the biotite relative to the amount of ^{40}Ar that should be present now? (**ANSWER:** 69.1 percent.)

REFERENCES

Aldrich, L. T., and A. O. Nier (1948) Argon 40 in potassium minerals. Phys. Rev., *74*, 876–877.

Armstrong, R. L. (1966) K-Ar dating of plutonic and volcanic rocks in orogenic belts, 117–131. In Potassium Argon Dating, O. A. Schaeffer and J. Zähringer, eds. Springer-Verlag, Heidelberg.

Armstrong, R. L., W. P. Leeman, and H. E. Malde (1975) K-Ar dating, Quaternary and Neogene volcanic rocks of the Snake River plain, Idaho. Am. J. Sci., *275*, 225–251.

Aston, F. W. (1921) The mass spectra of the alkali metals. Phil. Mag., ser. 6, *42*, 436–441.

Beckinsale, R. D., and N. H. Gale (1969) A reappraisal of the decay constants and branching ratio of ^{40}K. Earth Planet. Sci. Letters, *6*, 289–294.

Campbell, N. R., and A. Wood (1906) The radioactivity of the alkali metals. Proc. Cambridge Phil. Soc., *14*, 15–21.

Campbell, N. R. (1908) The radioactivity of potassium, with special reference to solutions of its salts. Proc. Cambridge Phil. Soc., *14*, 557–567.

Cooper, J. A. (1963) The flame photometric determination of potassium in geological materials used for potassium-argon dating. Geochim. Cosmochim. Acta, *27*, 525–546.

Cox, A., R. R. Doell, and G. B. Dalrymple (1963a) Geomagnetic polarity epochs and Pleistocene geochronometry. Nature, *198*, 1049–1051.

Cox, A., R. R. Doell, and G. B. Dalrymple (1963b) Geomagnetic polarity epochs: Sierra Nevada II. Science, *142*, 382–385.

Cox, A. (1969) Geomagnetic reversals. Science *163*, 237–245.

Dalrymple, G. B., and M. A. Lanphere (1969) Potassium-argon dating. W. H. Freeman, San Francisco, 258 p.

Dalrymple, G. B. (1972) Potassium-argon dating of geomagnetic reversals and North American glaciations, 107–134. In Calibration of Hominoid Evolution, W. W. Bishop and J. A. Miller, eds. Scott. Acad. Press.

Dalrymple, G. B., R. D. Jarrard, and D. A. Clague (1975) K-Ar ages of some volcanic rocks from the Cook and Austral Islands. Geol. Soc. Amer. Bull. *86*, 1463–1467.

DeLong, S. E., and F. W. McDowell (1975) K-Ar ages from the Near Islands, western Aleutian Islands, Alaska. Geol., *3*, No. 12, 691–694.

Dodson, M. A. (1973) Closure temperature in cooling geochronological and petrological systems. Contrib. Mineral. Petrol., *40*, 259–274.

Giletti, B. J. (1974a) Diffusion related to geochronology. In Geochemical Transport and Kinetics, pp. 61–76, A. W. Hofmann, B. J. Giletti, H. S. Yoder, Jr., and R. A. Yund, eds. Carnegie Institution of Washington, Pub. 634, 353 p.

Giletti, B. J. (1974b) Studies in diffusion I: Argon in phlogopite mica. In Geochemical Transport and Kinetics, pp. 107–116, A. W. Hofmann, B. J. Giletti, H. S. Yoder, Jr., and R. A. Yund, eds. Carnegie Institution of Washington, Pub. 634, 353 p.

Goldich, S. S. (1968) Geochronology in the Lake Superior region. Can. J. Earth Sci., *5*, 715–724.

Grant, J. A. (1964) Rubidium-strontium isochron study of the Grenville front near Lake Timagami, Ontario. Science, *146*, 1049–1053.

Hanson, G. N., and P. W. Gast (1967) Kinetic studies in contact metamorphic zones. Geochim. Cosmochim. Acta, *31*, 1119–1153.

Harper, C. T. (1967) On the interpretation of potassium-argon ages from Precambrian shields and Phanerozoic orogens. Earth Planet. Sci. Letters, *3*, 128–132.

Hart, S. R. (1964) The petrology and isotopic-mineral age relations of a contact zone in the Front Range, Colorado. J. Geol., *72*, 493–525.

Hart, S. R., and R. T. Dodd, Jr. (1962) Excess radiogenic argon in pyroxenes. J. Geophys. Res., *67*, 2998–2999.

Heier, K. S., and J. A. S. Adams (1964) The geochemistry of the alkali metals. In Physics and Chemistry of the Earth, vol 5, pp. 253–381, Pergamon Press.

Heirtzler, T. R., G. O. Dickson, E. M. Herron, W. C. Pitman, and X. LePichon (1968) Marine magnetic anomalies, geomagnetic field reversals, and motions of the ocean floor and continents. J. Geophys. Res., *73*, 2119–2136.

Holden, N. E., and F. W. Walker (1972) Chart of the nuclides. Ed. Relat., General Electric Co., Schenectady, N.Y.

Krogh, T. E., and P. M. Hurley (1968) Strontium isotope variation and whole-rock isochron studies, Grenville Province of Ontario. J. Geophys. Res., *73*, 7107–7125.

Krogh, T. E., and G. L. Davis (1969) Old isotopic ages in the northwestern Grenville Province, Ontario. Geol. Assoc. Canada, Spec. Paper No. 5, 189–192.

Laughlin, A. W. (1969) Excess radiogenic argon in pegmatite minerals. J. Geophys. Res., *74*, 6684–6690.

Livingston, D. E., P. E. Damon, R. L. Mauger, R. Bennett, and A. W. Laughlin (1967) Argon 40 in cogenetic feldspar-mica mineral assemblages. J. Geophys. Res., *72*, 1361–1375.

McDougall, I., and D. H. Tarling (1963) Dating of polarity zones in the Hawaiian Islands. Nature, *200*, 54–56.

McDougall, I., and D. H. Green (1964) Excess radiogenic argon in pyroxenes and isotopic ages on minerals from Norwegian eclogites. Norsk. Geol. Tidsskr., *44*, 183–196.

McDougall, I. (1966) Precision methods of potassium-argon isotopic age determination on young rocks. In Methods and Techniques in Geophysics, Vol. 2, pp. 279–304. Interscience, New York.

Miller, F. K., and J. C. Engels (1975) Distribution and trends of discordant ages of the plutonic rocks of northeastern Washington and northern Idaho. Geol. Soc. Amer. Bull., *86*, 517–528.

Morozova, I. M., and A. A. Alferovsky (1974) Fractionation of lithium and potassium isotopes in geological processes. Geokhimia, No. 1, 30–39.

Müller, O. (1966) Potassium analysis. In Potassium Argon Dating, O. A. Schaeffer and J. Zähringer, eds. Springer-Verlag, Berlin and Heidelberg, pp. 40–66.

Nier, A. O. (1935) Evidence for the existence of an isotope of potassium of mass 40. Phys. Rev., *48*, 283–284.

Nier, A. O. (1950) A redetermination of the relative abundances of the isotopes of carbon, nitrogen, oxygen, argon, and potassium. Phys. Rev., *77*, 789–793.

Rama, S. N. I., S. R. Hart, and E. Roedder (1965) Excess radiogenic argon in fluid inclusions. J. Geophys. Res., *70*, 509–511.

Schaeffer, O. A., and J. Zähringer, eds. (1966) Potassium Argon Dating. Springer-Verlag, New York, 234 p.

Semenenko, N. P., A. P. Scherbak, A. P. Vinogradov, A. I. Tugarinov, G. D. Eliseeva, F. I. Cotlovskay, and S. G. Demidenko (1968) Geochronology of the Ukrainian Precambrian. Can. J. Earth Sci., *5*, 661–671.

Stockwell, C. H. (1968) Geochronology of stratified rocks on the Canadian Shield. Can. J. Earth Sci., *5*, 693–698.

Stockwell, C. H. (1972) Revised Precambrian time scale for the Canadian shield. Geol. Surv. Can., Paper 72–52, 4 p.

Thomson, J. J. (1905) On the emission of negative corpuscles by the alkali metals. Phil. Mag., ser. 6, *10*, 584–590.

Verbeek, A. A., and G. D. L. Schreiner (1967) Variations in $^{39}K:^{41}K$ ratio and movement of potassium in a granite-amphibolite contact region. Geochim. Cosmochim. Acta, *31*, 2125–2163.

Vine, F. J., and D. H. Mathews (1963) Magnetic anomalies over oceanic ridges. Nature, *199*, 947–949.

Von Weizsäcker, C. F. (1937) Über die Möglichkeit eines dualen β-Zerfalls von Kalium. Physik. Zeitschrift, *38*, 623–624.

Wellman, P. (1974) Potassium-argon ages on the Cainozoic volcanic rocks of eastern Victoria, Australia. J. Geol. Soc. Aust., *21*, 359–376.

Wellman, P., and I. McDougall (1974a) Potassium-argon ages on the Cainozoic volcanic rocks of New South Wales. J. Geol. Soc. Aust., *21*, 247–272.

Wellman, P., and I. McDougall (1974b) Cainozoic igneous activity in eastern Australia. Tectonophys., *23*, 49–65.

York, D., and R. M. MacIntyre (1965) Excess Ar^{40} in sodalite. Trans. Amer. Geophys. Union, *46*, 177.

10 THE $^{40}Ar/^{39}Ar$ METHOD OF DATING

The conventional K-Ar method of dating depends on the assumption that the sample contained no argon at the time of its formation and that subsequently all radiogenic argon produced within it was quantitatively retained. Because argon may be lost by diffusion even at temperatures well below the melting point, K-Ar dates represent the time elapsed since cooling to temperatures at which diffusion loss of argon is insignificant. Under certain circumstances excess radiogenic ^{40}Ar may be present which causes K-Ar dates to be too old. The $^{40}Ar/^{39}Ar$ method of dating, first described in detail by Merrihue and Turner in 1966, can overcome some of the limitations of the conventional K-Ar method. It has the additional advantages that potassium and argon are determined on the same sample and that only measurements of the isotope ratios of argon are required. The problem of inhomogeneity of samples and the need to measure the absolute concentrations of potassium and argon are thus eliminated. This method is therefore well suited to the dating of very small or valuable samples such as meteorites or lunar rocks and minerals.

1
Principles of the Method

The $^{40}Ar/^{39}Ar$ method of dating is based on the formation of ^{39}Ar by the irradiation of K-bearing samples with thermal and fast neutrons in a nuclear reactor. The desired reaction is

$$^{39}_{19}K(n, p)^{39}_{18}Ar \qquad (10.1)$$

Argon 39 is unstable and decays to ^{39}K by beta emission with a half-life of 269 years. Because of its slow rate of decay, ^{39}Ar can be treated as though it were stable during the short period of time involved in the analyses. While Wänke and König (1956) actually used a counting technique to determine the amount of ^{39}Ar produced by this reaction, Merrihue (1965) proposed that the ratio $^{40}Ar^*/^{39}Ar$ could be measured by mass spectrometry. Subsequently, Merrihue and Turner (1966) described such a procedure and reported dates for several stony meteorites that appeared to be in good agreement with conventional K-Ar dates of the same meteorites. The principles of the $^{40}Ar/^{39}Ar$ method of dating have been presented by Dalrymple and Lanphere (1971, 1974) and by McDougall (1974).

When a K-bearing sample is irradiated with neutrons in a nuclear reactor, isotopes of argon are formed by several reactions involving potassium, calcium, and chlorine. Leaving aside all interfering reactions, we focus attention only on the production of ^{39}Ar by the n, p reaction with ^{39}K. Using the formulation of Mitchell (1968), the number of ^{39}Ar atoms formed in the sample by the neutron irradiation is

$$^{39}Ar = {}^{39}K \, \Delta T \int \varphi(\varepsilon)\sigma(\varepsilon) \, d\varepsilon \qquad (10.2)$$

where ^{39}K is the number of atoms of this isotope in the irradiated sample, ΔT is the length of the irradiation, $\varphi(\varepsilon)$ is the neutron flux density at energy ε, $\sigma(\varepsilon)$ is the capture cross section of ^{39}K for neutrons having energy ε, and the integration is carried out over the entire energy spectrum of the neutrons. The number of radiogenic ^{40}Ar atoms present in the sample due to decay of ^{40}K during its lifetime is

$$^{40}Ar^* = \frac{\lambda_e}{\lambda} \, {}^{40}K(e^{\lambda t} - 1) \qquad (10.3)$$

where λ_e is the decay constants of ^{40}K for electron capture and λ is the total decay constant of ^{40}K. After neutron irradiation of a sample its $^{40}Ar^*/^{39}Ar$ ratio is given by

$$\frac{^{40}Ar^*}{^{39}Ar} = \frac{\lambda_e}{\lambda} \, \frac{^{40}K}{^{39}K} \, \frac{1}{\Delta T} \, \frac{e^{\lambda t} - 1}{\int \varphi(\varepsilon)\sigma(\varepsilon)\,d\varepsilon} \qquad (10.4)$$

We now define a parameter J as

$$J = \frac{\lambda}{\lambda_e} \, \frac{^{39}K}{^{40}K} \, \Delta T \int \varphi(\varepsilon)\sigma(\varepsilon)\,d\varepsilon \qquad (10.5)$$

which is substituted into Equation 10.4 and thus leads to

$$\frac{^{40}Ar^*}{^{39}Ar} = \frac{e^{\lambda t} - 1}{J} \qquad (10.6)$$

The neutron flux density and the capture cross sections are difficult to evaluate from first principles because the energy spectrum of the incident neutrons and the cross sections of ^{39}K for capture of neutrons of varying energies are not known. However, Equation 10.6 suggests that J can be determined by irradiating a sample of known age (the flux monitor) together with samples whose age is unknown. After the $^{40}Ar^*/^{39}Ar$ ratio of the monitor has been measured, J can be calculated from Equation 10.7:

$$J = \frac{e^{\lambda t_m} - 1}{^{40}Ar^*/^{39}Ar} \qquad (10.7)$$

where t_m is the known age of the flux monitor and $^{40}Ar^*/^{39}Ar$ is the measured value of this ratio in the monitor.

The energy spectrum of the neutron flux to which a particular sample is exposed during the irradiation depends on its position in the sample holder. For this reason, several samples of the flux monitor must be inserted into the sample holder at known positions between unknown samples. The entire package is then irradiated for several days in a nuclear reactor to allow ^{39}Ar to be produced. After the irradiation, the argon in the flux monitors is released by fusion in a vacuum system and their $^{40}Ar^*/^{39}Ar$ ratios are measured by mass spectrometry. The J values are then calculated using Equation 10.7 and are plotted as a function of position in the sample holder. The respective J values of the unknown samples are obtained by interpolation of the resulting graph from their known positions in the holder. The $^{40}Ar^*/^{39}Ar$ ratios of the unknown samples are then used to calculate dates from Equation 10.8:

$$t = \frac{1}{\lambda} \ln \left(\frac{^{40}Ar^*}{^{39}Ar} \, J + 1 \right) \qquad (10.8)$$

The estimated analytical error in the calculated date, according to Dalrymple and Lanphere (1971), is

$$\sigma_t \simeq \left[\frac{J^2 F^2 (\sigma_F{}^2 + \sigma_J{}^2)}{t^2 \lambda^2 (1 + FJ)^2} \right]^{1/2} \qquad (10.9)$$

where $F = {}^{40}Ar^*/^{39}Ar$, $\sigma_F{}^2$ and $\sigma_J{}^2$ are the variances of F and J, respectively, expressed in percent, t is the age of the sample, and λ is the total decay constant of ^{40}K. The dates obtained in this manner are referred to as "total argon release dates." They are subject to the same limitations as conventional K-Ar dates because they depend on the assumption that no radiogenic ^{40}Ar has escaped from the sample and that no excess ^{40}Ar is present. However, such dates avoid the problems arising from the inhomogeneous distribution of potassium and argon in a sample and require only the measurement of isotope ratios of argon.

In the ideal case outlined above, it is assumed that all of the ^{40}Ar in the irradiated sample is either radiogenic or atmospheric, that all of the ^{36}Ar is atmospheric, and that ^{39}Ar is produced only by ^{39}K$(n, p)^{39}$Ar. In this case the measured values of the ^{40}Ar/^{39}Ar and ^{36}Ar/^{39}Ar ratios can be used to calculate the desired ratio of radiogenic ^{40}Ar to ^{39}Ar:

$$\frac{^{40}\text{Ar*}}{^{39}\text{Ar}} = \left(\frac{^{40}\text{Ar}}{^{39}\text{Ar}}\right)_m - 295.5 \left(\frac{^{36}\text{Ar}}{^{39}\text{Ar}}\right)_m \quad (10.10)$$

where the subscript m denotes measured values and 295.5 is the ratio ^{40}Ar/^{36}Ar of atmospheric argon. Actually, argon isotopes are also produced by several interfering reactions caused by interactions of neutrons with the isotopes of calcium, potassium, and chlorine in the sample. Therefore, a series of corrections must be made which are especially serious for young samples ($\sim 10^6$ years) and those having K/Ca < 1.0. The interfering reactions are listed in Table 10.1 and can be studied by reference to Figure 10.1. Detailed discussions of the corrections have been given by Mitchell (1968), Brereton (1970), Turner (1971), and Dalrymple and Lanphere (1971).

The most important interfering reactions are those involving the isotopes of calcium. Starting at the top of Table 10.1, we see that

^{36}Ar is produced by ^{40}Ca$(n, n\alpha)^{36}$Ar, but is removed by ^{36}Ar$(n, \gamma)^{37}$Ar. These reactions interfere with the atmospheric argon correction which is based on ^{36}Ar. Calcium is also primarily responsible for the production of ^{37}Ar by means of ^{40}Ca$(n, \alpha)^{37}$Ar. The abundance of ^{37}Ar in an irradiated sample is therefore an indication of the extent of calcium interference. The ^{37}Ar yields from ^{39}K$(n, nd)^{37}$Ar and ^{36}Ar$(n, \gamma)^{37}$Ar are small. ^{37}Ar is radioactive and decays by electron capture to stable ^{37}Cl with a half-life of 35.1 days. For this reason, a correction for decay of ^{37}Ar since irradiation must be made because the abundance of this isotope is used to estimate the contributions of ^{40}Ca$(n, n\alpha)^{36}$Ar, ^{43}Ca$(n, \alpha)^{40}$Ar, and ^{44}Ca$(n, n\alpha)^{40}$Ar, which affect the total amount of ^{40}Ar. The production of ^{38}Ar is of little consequence unless an ^{38}Ar spike is to be used for measuring absolute quantities of argon. Both ^{39}Ar and ^{40}Ar result from reactions with isotopes of calcium and potassium. According to measurements by Dalrymple and Lanphere (1971), the correction factors for Ca- and K- derived argon in the T.R.I.G.A. reactor of the U.S. Geological Survey are: $(^{36}\text{Ar}/^{37}\text{Ar})_{\text{Ca}} = (2.72 \pm 0.014) \times 10^{-4}$, $(^{39}\text{Ar}/^{37}\text{Ar})_{\text{Ca}} = (6.33 \pm 0.043) \times 10^{-4}$,

Table 10.1   **Interfering Nuclear Reactions Caused by Neutron Irradiation of Mineral Samples (Brereton, 1970)**

ARGON PRODUCED	CALCIUM	POTASSIUM	ARGON	CHLORINE
^{36}Ar	^{40}Ca$(n, n\alpha)$	—	—	—
^{37}Ar	^{40}Ca(n, α)	^{39}K$(n, n\,d)$	^{36}Ar(n, γ)	^{37}Cl(n, γ, β^-)
^{38}Ar	^{42}Ca$(n, n\alpha)$	^{39}K(n, d)	^{40}Ar$(n, n\,d, \beta^-)$	
		^{41}K(n, α, β^-)		
^{39}Ar	^{42}Ca(n, α)	^{39}K$(n, p)^a$	^{38}Ar(n, γ)	—
	^{43}Ca$(n, n\alpha)$	^{40}K(n, d)	^{40}Ar(n, d, β^-)	
^{40}Ar	^{43}Ca(n, α)	^{40}K(n, p)	—	—
	^{44}Ca$(n, n\alpha)$	^{41}K(n, d)		

^{a} This is the principal reaction on which the ^{40}Ar/^{39}Ar method is based.

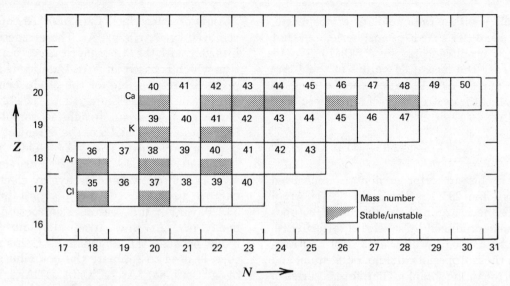

FIGURE 10.1 A segment of the chart of the nuclides showing most of the stable and unstable isotopes of calcium, potassium, argon, and chlorine that participate in nuclear reactions with neutrons. ^{40}K is here treated as a stable nuclide.

and $(^{40}\mathrm{Ar}/^{39}\mathrm{Ar})_\mathrm{K} = 0.0059 \pm 0.00042$. The production ratio of $^{40}\mathrm{Ar}/^{37}\mathrm{Ar}$ due to calcium is 4.6×10^{-3}. By far the most important corrections for Ca-rich samples result from the production of $^{39}\mathrm{Ar}$ by $^{42}\mathrm{Ca}(n, \alpha)^{39}\mathrm{Ar}$ and of $^{36}\mathrm{Ar}$ by $^{40}\mathrm{Ca}(n, n\alpha)^{36}\mathrm{Ar}$. For young samples having high K/Ca ratios, corrections are also necessary to remove $^{40}\mathrm{Ar}$ produced by $^{40}\mathrm{K}(n, p)^{40}\mathrm{Ar}$ and $^{41}\mathrm{K}(n, d)^{40}\mathrm{Ar}$. Brereton (1970) derived an equation that relates the age of an irradiated sample to its $^{40}\mathrm{Ar*}/^{39}\mathrm{Ar}$ ratio corrected for all interfering reactions. Dalrymple and Lanphere (1971) developed a more general expression for their parameter $F\, (=\,^{40}\mathrm{Ar*}/^{39}\mathrm{Ar})$:

$$F = \frac{A - C_1 B + C_1 C_2 D - C_3}{1 - C_4 D} \qquad (10.11)$$

where

$A =$ measured value of the $^{40}\mathrm{Ar}/^{39}\mathrm{Ar}$ ratio,

$B =$ measured value of the $^{36}\mathrm{Ar}/^{39}\mathrm{Ar}$ ratio,

$C_1 = \,^{40}\mathrm{Ar}/^{36}\mathrm{Ar}$ ratio in the atmosphere ($=295.5$),

$C_2 = \,^{36}\mathrm{Ar}/^{37}\mathrm{Ar}$ ratio produced by interfering neutron reactions with calcium ($2.72 \pm 0.014 \times 10^{-4}$),

$C_3 = \,^{40}\mathrm{Ar}/^{39}\mathrm{Ar}$ ratio produced by interfering neutron reactions with potassium (0.0059 ± 0.00042),

$C_4 = \,^{39}\mathrm{Ar}/^{37}\mathrm{Ar}$ ratio produced by interfering neutron reactions with calcium ($6.33 \pm 0.043 \times 10^{-4}$), and

$D = \,^{37}\mathrm{Ar}/^{39}\mathrm{Ar}$ ratio in the sample after correcting for decay of $^{37}\mathrm{Ar}$.

2
The Incremental Heating Technique

We have shown in the preceding section that a date can be calculated solely on the basis of the $^{40}\mathrm{Ar*}/^{39}\mathrm{Ar}$ ratio of a neutron-irradiated sample. Consequently, a series of dates can be obtained for a single sample by releasing argon from it in steps at increasing temperatures. If the sample has been closed to argon and potassium since the time of initial cooling, the $^{40}Ar*/^{39}Ar$ ratios and thus the dates calculated at each step should be constant. However, if radiogenic argon was lost some time after initial cooling, the $^{40}Ar*/^{39}Ar$ ratios of the gas released at different temperatures may vary, and a spectrum of dates will then result from which the time elapsed since initial cooling must be inferred.

Turner (1968, 1969) used a model, illustrated in Figure 10.2, to predict the $^{40}Ar*/^{39}Ar$ spectrum of a K-bearing mineral grain that experienced partial loss of radiogenic ^{40}Ar due to heating by a metamorphic event. Figure 10.2a shows the uniform distribution of the $^{40}Ar*/^{40}Ar$ ratio expected in a closed system. Part (b) illustrates the effect of partial loss of $^{40}Ar*$ due to heating as a result of which the $^{40}Ar*/^{40}K$ ratio at the edge of the grain is reduced to zero, but remains unaffected at its center. Figure 10.2c shows the distribution of the $^{40}Ar*/^{40}K$ ratios in the grain at a later time assuming that it became a closed system again after the episode of metamorphism. When grains of uniform size of a specific K-bearing mineral that has experienced such a history are analyzed by incremental heating, the $^{40}Ar*/^{39}Ar$ ratios of gas fractions released at different temperatures vary in a systematic fashion. The first gas to be released at the lowest temperature originates from the surfaces of the grains and from sites that lose argon readily. This gas has a low $^{40}Ar*/^{39}Ar$ ratio, corresponding to the time elapsed since accumulation of radiogenic ^{40}Ar resumed after metamorphism. Gas fractions released at higher temperatures have higher $^{40}Ar*/^{39}Ar$ ratios because argon is then removed from more retentive sites which lost smaller fractions of $^{40}Ar*$ during metamorphism. Ultimately, the $^{40}Ar*/^{39}Ar$ ratios may reach a plateau corresponding to a date that approaches the

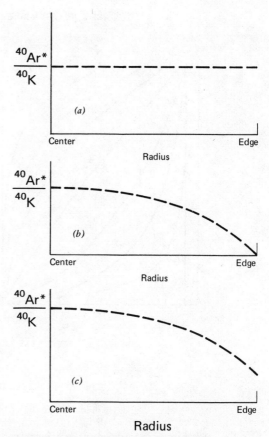

FIGURE 10.2 (a) This is a schematic diagram showing the uniform distribution of the $^{40}Ar*/^{40}K$ ratio in a spherical mineral grain that has been a closed system since the time of its formation. (b) Partial loss of $^{40}A*$ due to diffusion as a result of heating of a spherical mineral grain. Note that the $^{40}Ar*/^{40}K$ ratio is reduced to zero on the grain surface, but remains unchanged at its center. (c) Distribution of $^{40}Ar*/^{40}K$ ratios some time after partial argon loss after the system has again become closed to $^{40}Ar*$. Note that the value of $^{40}Ar*/^{40}K$ on the grain surface reflects the time elapsed since partial outgassing while the value in the center of the grain corresponds to the age of the mineral. (Turner, 1968.)

time elapsed since original cooling of the mineral. Stepwise heating of a mineral sample that has experienced partial loss of radiogenic argon thus permits the calculation of

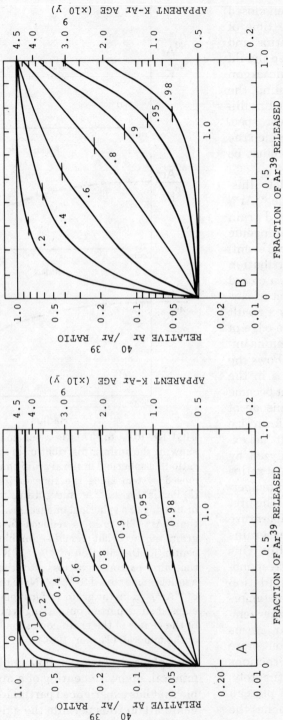

FIGURE 10.3 (a) Theoretical argon release patterns for a sample whose age is 4.55 billion years which was partially outgassed 0.5 billion years ago. The numbers on the curves are the fraction of ^{40}Ar lost and the horizontal bars indicate the average ^{40}Ar/^{39}Ar ratio and the corresponding total-release dates of the sample. Note that the plateau dates for ^{40}Ar loss greater than about 20 percent are less than the age of the sample. The mineral grains were assumed to be spheres of uniform size. (b) Theoretical argon release patterns for the same sample assuming a lognormal distribution of grain radii ($\sigma = 0.33$). Note that the release curves become concave upward in cases where the partial argon loss is about 80 percent or more. The ^{40}Ar/^{39}Ar ratio of gas released at low temperatures from samples with *large* argon loss indicates the date of outgassing, while the high-temperature fraction approaches the age of the sample where argon loss has been *small*. (Reproduced from Figures 4 (p. 391) and 5 (p. 392) of Turner, G. (1968) pp. 387–398, in Origin and Distribution of the Elements, L. H. Ahrens, ed. Pergamon Press Ltd., Oxford, 1178 p., by permission of the publisher.)

170

a series of dates which, in the ideal case, include the time of metamorphism (low-temperature release) and the time of initial cooling (high-temperature plateau). Such theoretical release patterns are shown in Figure 10.3 for a sample that is 4.5 billion years old (cooling age) and that experienced partial argon loss 500 million years ago. The curves are based on calculations of Turner (1968) for uniform spheres (Fig. 10.3a) and for spheres having log-normal size distributions (Fig. 10.3b). Note that the date calculated by *total release* of all gas has a meaningless intermediate value. The principal advantage of the $^{40}Ar*/^{39}Ar$ method of dating lies in this stepwise degassing technique because it permits the determination of a date that approaches, but may underestimate, the original cooling age of K-bearing rocks and minerals that experienced partial loss of radiogenic ^{40}Ar during a metamorphic episode. The incremental heating technique has been used successfully for dating of stony meteorites (Turner et al., 1966; Turner, 1969; Podosek, 1971, 1972; Podosek and Huneke, 1973) and of lunar rocks and minerals (Turner, 1970; Turner et al., 1971; Husain et al., 1971; Husain et al., 1972; Sutter et al., 1971; Husain and Schaeffer, 1973; Husain, 1974).

The incremental heating technique is also being used to date terrestrial rocks and minerals from metamorphic terranes (Fitch et al., 1969; York and Berger, 1970; Bryhni et al., 1971; Brereton, 1972; Lanphere and Albee, 1974; Dallmeyer et al., 1975; Dallmeyer, 1975). Lanphere and Dalrymple (1971) tested the validity of this technique by dating rocks and minerals with known geological histories whose ages had been determined previously by other methods. Their samples included rocks and minerals that had experienced partial argon loss during a postcrystallization heating event, K-feldspar that had not been reheated since crystalliza-

tion, and diabase containing excess ^{40}Ar. A selection of their results is shown in Figure 10.4a and b in the form of age spectra plotted versus the fraction of ^{39}Ar released during a series of heating steps.

Figure 10.4a shows age spectra for biotite and microcline of a sample of granite from the crystalline complex of the Marble Mountains of southeastern California. The age of this complex lies between 1400 and 1450 million years, based on a Rb-Sr date of the biotite (1410 $\pm$ 30 m.y.) and a U-Pb date of zircon (1450 m.y.). The conventional K-Ar date of the biotite is 1152 $\pm$ 30 m.y. and that of the microcline is 992 $\pm$ 26 m.y. The K-Ar dates of both minerals are less than the crystallization age of the complex which indicates that both lost radiogenic argon during a later thermal event. The argon loss of the biotite is about 20 percent. The microcline lost even more ^{40}Ar, consistent with its lower retentivity noted before (Chapter 9). Other age determinations reviewed by Lanphere and Dalrymple suggest that the argon loss occurred 160 to 180 million years ago. The $^{40}Ar*/^{39}Ar$ age spectrum of the biotite fits the pattern for a mineral that has experienced partial argon loss (Turner, 1968). The dates increase from 223 $\pm$ 30 million years for the first argon fraction and attain a plateau after about 30 percent of the ^{39}Ar was released. The average plateau date of 1300 million years is about 150 million years *greater* than the conventional K-Ar date, but is *lower* than the crystallization age by 100 to 150 million years. The total-argon release date has an intermediate value of 1251 $\pm$ 61 million years. The plateau date of this biotite is *less* than its known crystallization age by 5 to 10 percent as predicted by Turner's calculations for mineral grains that lost 20 percent argon.

The spectrum of the microcline differs significantly from that of the biotite. The

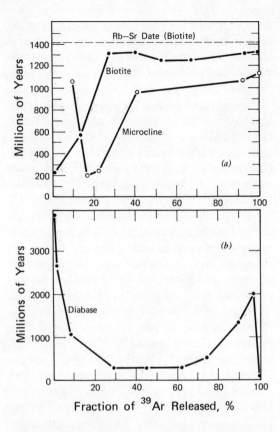

FIGURE 10.4 (*a*) Spectrum of dates calculated from $^{40}Ar^*/^{39}Ar$ ratios of gas fractions released by incremental heating of neutron-irradiated biotite and microcline from the Precambrian basement complex of the Marble Mountains in southeastern California. The age of these rocks lies between 1400 and 1450 million years. The high-temperature gas fractions of the biotite indicate a date of 1300 million years, while its conventional K-Ar date is only 1152 ± 30 million years. The plateau of the microcline dates is not well established which suggests significantly greater losses of ^{40}Ar from this mineral than from the biotite. (Lanphere and Dalrymple, 1971.) (*b*) Spectrum of dates obtained by incremental heating of a neutron-irradiated whole-rock sample of diabase from Liberia. This dike is post-Triassic to pre-Tertiary in age. It was contaminated with ^{40}Ar during intrusion into the granitic basement rocks which range in age from 2700 to 3400 million years. The conventional K-Ar date of this rock is 853 ± 26 million years, which is clearly in excess of its geologic age. The spectrum of dates reveals the presence of excess ^{40}Ar in the low-temperature fractions but does not permit a geologically meaningful interpretation. (Lanphere and Dalrymple, 1971.)

first fraction yielded a date of 1063 million years, while the next two fractions have dates of about 200 million years. The dates then rise to a maximum value of 1133 ± 10 million years without achieving a stable plateau. In such cases, the maximum date at the highest release temperature must be regarded as a lower limit to the crystallization age. The high date obtained for the first fraction of gas may result from a small amount of ^{40}Ar that diffused from crystal lattice to nonlattice sites from which it is readily removed at low temperature. The lowest dates of both the microcline and the biotite are overestimates of the age of the thermal event that caused argon loss from these minerals.

The presence of excess ^{40}Ar in some minerals and whole rock samples causes serious errors in conventional K-Ar dates which cannot be corrected. Lanphere and Dalrymple (1971) therefore analyzed two whole-rock samples of diabase known to contain excess ^{40}Ar to determine whether the stepwise heating technique can overcome this difficulty. The diabase samples were taken from dikes that intrude crystalline rocks of Precambrian age in western Liberia. Dating by the Rb-Sr method has indicated that the basement rocks are from 2700 to 3400 million years old. Conventional K-Ar dates of whole-rock samples and plagioclase from the diabase dikes are highly inconsistent and range from 1200 to 186 million years. K-Ar dates on similar dikes intruding Paleozoic sedimentary rocks elsewhere in Liberia give more concordant dates ranging from 173 to 193 million years, and paleomagnetic pole positions of all of the dikes suggest that they are post-Triassic and pre-Tertiary in age. The discordant K-Ar dates are believed to result from the incorporation of varying amounts of ^{40}Ar from the country rock into the magma during the intrusion of the dikes. Figure 10.4b is the partial-release spectrum of one of the two whole-rock samples of diabase. The conventional K-Ar date of this sample is 853 ± 26 million years and its total-release ^{40}Ar*/^{39}Ar date is 838 ± 42 million years. Both dates are well in excess of the age of this rock. The partial-release patterns of the two specimens are similar, but do not indicate a geologically meaningful result. The high dates of the low-temperature fractions are undoubtedly the result of excess ^{40}Ar. However, the lowest dates (~ 280 m.y.) in the central portion of the spectrum are still significantly above the crystallization age. The dates of the high temperature fractions rise to 1995 ± 16 million years, but the last 2.8 percent of the argon released by complete fusion yields a date of only 79 ± 69 million years. In this case the stepwise heating technique does not solve the problem. This conclusion was later confirmed by Dalrymple et al. (1975). Brereton (1972) reported a similar result for a diabase dike from the Frederikshab area of southwest Greenland, which also contains excess ^{40}Ar.

3
The Argon-Isotope Correlation Diagram

Both the conventional K-Ar and the ^{40}Ar*/^{39}Ar method of dating require a correction for the presence of atmospheric ^{40}Ar. This correction is based on the assumption that ^{36}Ar is of atmospheric origin and that the ^{40}Ar/^{36}Ar ratio of the atmospheric component is 295.5. This assumption may not be strictly valid because it is possible that both

^{40}Ar as well as ^{36}Ar are incorporated into minerals and rocks at the time of initial crystallization (Musset and Dalrymple, 1968). The ^{40}Ar/^{36}Ar ratio of this argon may differ significantly from that of argon in the modern atmosphere. Therefore, the observed amount of ^{36}Ar in a rock or mineral does not necessarily permit an accurate correction to be made for the presence of nonradiogenic ^{40}Ar. If the ^{40}Ar/^{36}Ar ratio of the inherited argon is greater than 295.5, an apparent excess of radiogenic ^{40}Ar will result. If the ^{40}Ar/^{36}Ar ratio of the inherited argon is less than 295.5, there will be an apparent deficiency of radiogenic ^{40}Ar, which is then erroneously attributed to partial loss of ^{40}Ar.

This difficulty can be avoided by use of the argon-isotope correlation diagram which was originally suggested by Merrihue and Turner (1966). The measured ^{40}Ar/^{36}Ar ratios of gas released during the stepwise heating of an irradiated sample (after correction for interfering reaction) is

$$\left(\frac{^{40}\text{Ar}}{^{36}\text{Ar}}\right)_m = \frac{^{40}\text{Ar}_c + {}^{40}\text{Ar}^*}{^{36}\text{Ar}_c} \quad (10.12)$$

where the subscript c identifies contaminant argon which includes both the atmospheric component as well as argon that was either occluded during crystallization or entered the mineral at a later time. Equation 10.12 can be rewritten

$$\left(\frac{^{40}\text{Ar}}{^{36}\text{Ar}}\right)_m = \left(\frac{^{40}\text{Ar}}{^{36}\text{Ar}}\right)_c + \left(\frac{^{40}\text{Ar}^*}{^{39}\text{Ar}}\right)_k \left(\frac{^{39}\text{Ar}}{^{36}\text{Ar}}\right)_m$$

$$(10.13)$$

where the subscript m identifies measured ratios and k refers to argon produced by potassium in the sample. Equation 10.13 represents a family of straight lines in coordinates of $(^{40}\text{Ar}/^{36}\text{Ar})_m$ and $(^{39}\text{Ar}/^{36}\text{Ar})_m$ whose slopes are equal to the ^{40}Ar*/^{39}Ar ratio. It follows that the measured ^{40}Ar/^{36}Ar and ^{39}Ar/^{36}Ar ratios of incremental gas

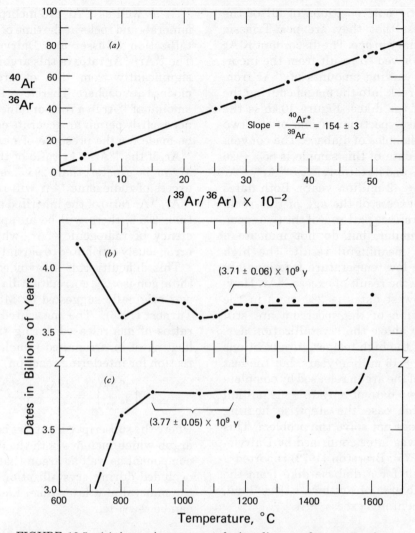

FIGURE 10.5 (a) Argon-isotope correlation-diagram for orange glass
(74220, 39) from the Taurus-Littrow valley of the moon (Apollo 17). The
coordinates of the data points are isotope ratios of argon released by
stepwise degassing of a neutron-irradiated sample weighing only 85 mg.
The slope of the correlation line is the ratio $^{40}Ar^*/^{39}Ar$ and has a value
of 154 $\pm$ 3 that corresponds to an age of 3.70 billion years. (b) Age
spectrum for gas fractions released at different temperatures from the
same orange glass. The $^{40}Ar/^{39}Ar$ ratios were corrected for excess ^{40}Ar
due to ion implantation by the solar wind using the intercept value of
the correlation line (part a) which is $^{40}Ar/^{36}Ar = 3.3 \pm 0.2$. The
average plateau age is 3.71 $\pm$ 0.06 billion years calculated from the gas
fractions released between 1150° and 1350°C. (c) Age spectrum of basalt
(75083,2.5) from the Taurus-Littrow valley. The release pattern is
typical of mare basalts showing evidence of $^{40}Ar^*$ loss in the low-
temperature fractions followed by a stable plateau. The age of this
basalt is 3.77 $\pm$ 0.05 billion years. (Husain and Schaeffer, 1973.)

174

fractions released from undisturbed mineral or rock samples define a series of points that fit a straight line. This line is an isochron whose slope is the $^{40}Ar^*/^{39}Ar$ ratio which is related to the age of the sample by Equation 10.8. The intercept of the isochron is the $^{40}Ar/^{36}Ar$ ratio of the contaminant, that is, the ratio of these isotopes in the nonradiogenic fraction of this gas associated with a given sample. Dalrymple and Lanphere (1974) demonstrated that the isochron dates of 11 undisturbed samples are identical, within experimental errors, with the plateau dates and that in most cases the initial $^{40}Ar/^{36}Ar$ ratio does not differ significantly from the atmospheric value of this ratio, which is equal to 295.5. However, three of the samples they analyzed had somewhat lower initial $^{40}Ar/^{36}Ar$ ratios. These samples may contain inherited argon having a somewhat lower $^{40}Ar/^{36}Ar$ ratio than modern atmospheric argon.

An example of the use of such isotope correlation-diagrams is the work of Husain and Schaeffer (1973) on the dating of the famous orange glass from the Taurus-Littrow valley of the moon. This remarkable material was discovered by the Apollo 17 astronauts in the immediate vicinity of Shorty Crater. At the time it was collected, this material was thought to have resulted from recent volcanic activity on the moon and therefore attracted much attention. Husain and Schaeffer analyzed argon from an 85 mg sample of this glass by the stepwise outgassing procedure and found that its age is 3.71 ± 0.06 billion years. This date is identical to the age of basalt fragments in the soil that covers to floor of Taurus-Littrow valley. Therefore the orange glass is not the product of recent volcanic activity.

One of the problems of dating material from the moon by the $^{40}Ar/^{39}Ar$ method is that both ^{36}Ar and ^{40}Ar are implanted in rocks exposed to the solar wind. In order to calculate the $^{40}Ar^*/^{36}Ar$ ratios of argon released from lunar samples, a correction must therefore be made for the excess ^{40}Ar due to the solar wind. The $^{40}Ar/^{36}Ar$ ratio of this component was determined by Husain and Schaeffer by means of the argon correlation-diagram shown in Figure 10.5a. A least-squares fit of the data points gives an intercept value of 3.3 ± 0.2 and a slope of 154 ± 3. The intercept was used to correct each gas increment for the presence of excess ^{40}Ar. The resulting $^{40}Ar^*/^{39}Ar$ ratios give rise to a spectrum of dates shown in Figure 10.5b. The best estimate of the age of this sample is the mean of the dates obtained for fractions from 1150° to 1350°C. The slope of the isotope-correlation line corresponds to a date of 3.70 billion years, which agrees well with the average $^{40}Ar/^{39}Ar$ plateau age.

4
K-Ar Isochrons and Atmospheric Evolution

The $^{40}Ar/^{36}Ar$ ratio of a K-bearing mineral is given by

$$\frac{^{40}Ar}{^{36}Ar} = \left(\frac{^{40}Ar}{^{36}Ar}\right)_0 + \frac{\lambda_e}{\lambda}\frac{^{40}K}{^{36}Ar}(e^{\lambda t} - 1) \quad (10.14)$$

This equation represents a family of straight lines in coordinates of $^{40}Ar/^{36}Ar$ and $^{40}K/^{36}Ar$ whose slopes are equal to

$$m = \frac{\lambda_e}{\lambda}(e^{\lambda t} - 1) \quad (10.15)$$

Suites of cogenetic K-bearing rocks and minerals, having the same age, are represented by a series of points that fit a straight-line isochron whose slope is related to the age of the samples by Equation 10.15. Such K-Ar isochrons, which are similar to Rb-Sr isochrons discussed in Chapter 6, have been used by McDougall et al. (1969) and Hayatsu and Carmichael (1970) and have been evaluated in detail by Shafiqullah and Damon (1974). It is clear that cogenetic minerals and rocks will form isochrons only when all samples have remained closed to argon and potassium. Moreover, all samples must have the same initial $^{40}Ar/^{36}Ar$ ratio. The K-Ar isochron method of dating therefore appears to offer certain advantages over the conventional K-Ar method in the sense that the goodness of fit of data points to a straight line can be taken as evidence that the samples have remained closed to potassium and argon and that they had identical, or at least similar, initial ratios. The initial $^{40}Ar/^{36}Ar$ ratios reported by Hayatsu and Carmichael (1970) and by Hayatsu and Palmer (1975) differ significantly from the atmospheric value and range up to 5000. These results emphasize that rocks and minerals may contain variable amounts of nonradiogenic ^{40}Ar which is not necessarily of atmospheric origin. In such cases the isochron method may be more appropriate than the conventional K-Ar method because it does not require the assumption that the ratio of nonradiogenic ^{40}Ar to ^{36}Ar is equal to the atmospheric value.

The measurement of initial $^{40}Ar/^{36}Ar$ ratios by means of K-Ar isochrons provides an opportunity to study the isotopic evolution of argon in the Earth. According to Suess (1949), the cosmic $^{40}Ar/^{36}Ar$ ratio is 0.005. Consequently, argon in the terrestrial atmosphere has been enriched by several orders of magnitude as a result of outgassing of radiogenic ^{40}Ar from the mantle and crust of the Earth (Damon and Kulp, 1958). The evolution of the $^{40}Ar/^{36}Ar$ ratio in the crust and mantle depends primarily on the formation of radiogenic ^{40}Ar, but is sensitive also to the possible loss of ^{36}Ar during an episode of catastrophic degassing of the Earth during or shortly after its accretion (Fanale, 1971). Early removal of ^{36}Ar from the interior of the Earth may be reflected by very high initial $^{40}Ar/^{36}Ar$ ratios in rocks derived from the mantle (Schwartzman, 1973; Kaneoka, 1974). On the other hand, Brown et al. (1974) presented initial $^{40}Ar/^{36}Ar$ ratios for rocks and minerals from which they concluded that the $^{40}Ar/^{36}Ar$ ratio of the upper mantle had increased during the past 800 million years from about 130 to its present value of 295.5. Their interpretation was subsequently challenged by Alexander and Schwartzman (1976) on the grounds that most of the samples they considered contain argon of atmospheric origin. It is important to remember that the initial $^{40}Ar/^{36}Ar$ ratio determined from K-Ar isochrons may represent a mixture of atmospheric gas and argon derived from the interior of the Earth. Therefore, the initial $^{40}Ar/^{36}Ar$ ratio is a reliable indicator of the isotopic composition of gas derived from the mantle *only* when it can be shown that the atmospheric component is negligible. Alexander and Schwartzman (1976) suggested that gases trapped in the glassy rinds of submarine basalt flows satisfy this condition and emphasized that the $^{40}Ar/^{36}Ar$ ratios of such glasses are greater than 295.5 (Dalrymple and Moore, 1968; Fisher, 1971; Mellor and Mussett, 1975). Ozima (1975) carried out stepwise degassing experiments of several young, glassy submarine basalts in order to discriminate against the presence of atmospheric argon. The highest $^{40}Ar/^{36}Ar$ ratio he obtained was 1780, which is probably only a lower limit. These and similar results for young submarine basalts and Precambrian ultramafic rocks suggest that the $^{40}Ar/^{36}Ar$ ratio in the mantle is significantly *greater* than that of the atmosphere. This observation, if true, suggests that ^{36}Ar was removed from the interior of the Earth before much radiogenic ^{40}Ar had accumulated. The early removal of ^{36}Ar, in turn, suggests general degassing of the interior and formation of the atmosphere within a few hundred million years after the formation of the Earth.

5 SUMMARY

The ^{40}Ar/^{39}Ar method of dating is based on the production of ^{39}Ar from ^{39}K by an (n, p) reaction during a neutron irradiation. The method requires the use of a flux monitor of known age and corrections for interfering nuclear reactions with isotopes of calcium, potassium, argon, and chlorine. Dates calculated from measured ^{40}Ar*/^{39}Ar ratios after complete release of all gases from irradiated samples are comparable to conventional K-Ar dates. However, they are based only on measurements of isotopic ratios of argon and do not require a separate determination of the potassium concentration.

The principal advantage of the ^{40}Ar/^{39}Ar method of dating is that argon can be released partially by stepwise heating of irradiated samples. In this way, a spectrum of dates can be calculated from the ^{40}Ar*/^{39}Ar ratio of each fraction. Rocks and minerals that have experienced partial argon loss after crystallization may yield age spectra having a plateau formed by argon released from retentive sites at elevated temperatures. Such plateau dates may be equal to or slightly less than the time of crystallization. Rocks or minerals containing excess ^{40}Ar usually yield anomalously old dates at low temperatures but may not achieve reliable plateaus at higher temperatures.

The measured ^{40}Ar/^{36}Ar and ^{39}Ar/^{36}Ar ratios of gas fractions released from undisturbed samples form straight-line isochrons. The slopes of such isochrons are equal to the ^{40}Ar*/^{39}Ar ratio from which the age of the sample can be calculated. The intercept of the isochron is the ^{40}Ar/^{36}Ar ratio of argon that is not associated with potassium. In cases where this contaminant argon is not of atmospheric origin, the initial ^{40}Ar/^{36}Ar ratio should be used to determine the radiogenic ^{40}Ar component in the sample.

Isochron diagrams can also be used to study suites of K-bearing minerals or rocks analyzed by the conventional K-Ar method. Colinearity of points representing cogenetic minerals or rocks in coordinates of ^{40}Ar/^{36}Ar and ^{40}K/^{36}Ar indicates that the samples have remained closed to argon and potassium and had the same initial ^{40}Ar/^{36}Ar ratio. The initial ratios are of considerable interest because they may reflect the isotopic composition of argon in the interior of the Earth. However, in many cases the nonradiogenic argon appears to be dominated by atmospheric argon. Evidence from young volcanic rocks and Precambrian ultramafics suggests that the ^{40}Ar/^{36}Ar ratio of the mantle is about 2000 or higher. Such high values suggest loss of ^{36}Ar early in geologic time and thus have bearing on the origin of the atmosphere of the Earth.

PROBLEMS

1. The ^{40}Ar*/^{39}Ar ratio of a fraction of argon released from a biotite (FMX-25, Bryhni et al., 1971) is 12.31. The J value is 1.925×10^{-2}. Calculate the corresponding ^{40}Ar*/^{39}Ar date. (**ANSWER:** 400 m.y.)

2. The same authors analyzed several argon fractions from another biotite (FM7025) with the following results:

HEATING STEP	^{40}Ar*/^{39}Ar
1	2.27
2	4.97
3	6.68
4	9.58
5	10.25
10	10.10
15	10.26

Calculate dates for each fraction ($J = 2.90 \times 10^{-2}$) and plot them versus the number of the heating step. Estimate the age of this biotite and outline a history for it consistent with the spectrum of dates. (**ANSWER:** about 489 m.y.)

3. A specimen of basalt (15555, 33) from Hadley Rille on the moon (Apollo 15 was analyzed by Husain et al. (1972) with the following results:

CUMULATIVE ^{39}Ar, %	^{40}Ar*/^{39}Ar
3	58.14
10	61.34
27	72.77
61	80.15
79	83.32
100	79.80

Calculate ^{40}Ar*/^{39}Ar dates for each fraction ($J = 9.83 \times 10^{-2}$) and plot them versus the percent of ^{39}Ar released. What is the age of this basalt? (**ANSWER:** 3.28×10^9 y.)

4. The monitor irradiated by Turner et al. (1971) was a hornblende whose age is 1.062×10^9 years based on a conventional K-Ar date. Its ^{40}Ar*/^{39}Ar ratio after irradiation was 29.33. Calculate the value of J. (**ANSWER:** 2.580×10^{-2}.)

5. Podosek and Huneke (1973) measured the following ^{40}Ar*/^{39}Ar ratios in argon fractions released at increasing temperatures from the achondrite meteorite Pasamonte.

EXTRACTION TEMPERATURE °C	$\dfrac{^{40}\text{Ar*}}{^{39}\text{Ar}}$
545	106.0
630	128.0
745	141.2
850	141.5
935	146.9
1040	145.8
1195	168.6
1515	183.8

Their flux monitor had an age of 1.062×10^9 years and its ^{40}Ar*/^{39}Ar ratio was 14.01. Calculate a date for each fraction of argon released and plot them versus the extraction temperature. Observe the resulting spectrum and estimate the age of this meteorite as indicated by these data.

REFERENCES

Alexander, E. C., Jr. and D. W. Schwartzman, 1976, Argon isotopic evolution of upper mantle. Nature, *259*, 104–106.

Brereton, N. R. (1970) Corrections for interfering isotopes in the ^{40}Ar/^{39}Ar dating method. Earth Planet. Sci. Letters, *8*, 427–433.

Brereton, N. R. (1972) A reappraisal of the Ar^{40}/Ar^{39} stepwise degassing technique. Geophys. J. Roy. Astron. Soc., *27*, 449–478.

Brown, J. F., C. T. Harper, and A. L. Odom (1974) Petrogenetic implications of argon isotopic evolution in the upper mantle. Nature, *250*, 130–132.

Bryhni, I., F. J. Fitch, and J. A. Miller (1971) Ar^{40}/Ar^{39} dates from recycled Precambrian rocks in the Gneiss region of the Norwegian Caledonides. Norsk Geologisk Tiddsskrift, *51*, 391–406.

Dallmeyer, R. D. (1975) ^{40}Ar/^{39}Ar ages of biotite and hornblende from a progressively remetamorphosed basement terrane: Their bearing on interpretation of release spectra. Geochim. Cosmochim. Acta, *39*, 1655–1669.

Dallmeyer, R. D., J. F. Sutter, and D. J. Baker (1975) Incremental ^{40}Ar/^{39}Ar ages of biotite and hornblende from the northeastern Reading Prong: Their bearing on late Proterozoic thermal and tectonic history. Geol. Soc. Amer. Bull., *86*, 1435–1443.

Dalrymple, G. B., and J. G. Moore (1968) Argon-40: Excess in submarine pillow basalts from Kilauea volcano, Hawaii. Science, *161*, 1132–1135.

Dalrymple, G. B., and M. A. Lanphere (1971) ^{40}Ar/^{39}Ar technique of K/Ar dating: A comparison with the conventional technique. Earth Planet. Sci. Letters, *12*, 300–308.

Dalrymple, G. B., and M. A. Lanphere (1974) ^{40}Ar/^{39}Ar age spectra of some undisturbed terrestrial samples. Geochim. Cosmochim. Acta, *38*, 715–738.

Dalrymple, G. B., C. S. Grommé, and R. W. White (1975) Potassium-argon age and paleo magnetism of diabase dikes in Liberia: Initiation of central Atlantic rifting. Geol. Soc. Amer. Bull., *86*, 399–411.

Damon, P. E., and J. L. Kulp (1958) Inert gases and the evolution of the atmosphere. Geochim. Cosmochim. Acta, *13*, 280–292.

Fanale, F. P. (1971) A case for catastrophic early degassing of the earth. Chem. Geol., *8*, 79–105.

Fisher, D. E. (1971) Incorporation of Ar in East Pacific basalts. Earth Planet. Sci. Letters, *12*, 321–324.

Fitch, F. J., J. A. Miller, and J. G. Mitchell (1969) A new approach to radioisotopic dating in orogenic belts. In Time and Place in Orogeny, P. E. Kent, G. E. Sattherthwaite, and A. M. Spencer, eds., Geol. Soc. Lond. Spec. Publ. No. 3, 157.

Hayatsu, A., and C. M. Carmichael (1970) K-Ar isochron method and initial argon ratios. Earth Planet. Sci. Letters, *8*, 71–76.

Hayatsu, A., and H. C. Palmer (1975) K-Ar isochron study of the Tudor Gabbro, Grenville Province, Ontario. Earth Planet. Sci. Letters, *25*, 208–213.

Husain, L., J. F. Sutter, and O. A. Schaeffer (1971) Ages of crystalline rocks from Fra Mauro. Science, *173*, 1235–1236.

Husain, L., O. A. Schaeffer, and J. F. Sutter (1972) Age of lunar anorthosite. Science, *175*, 428–430.

Husain, L., and O. A. Schaeffer (1973) Lunar volcanism: Age of the glass in the Apollo 17 orange soil. Science, *180*, 1358–1360.

Husain, L. (1974) ^{40}Ar-^{39}Ar chronology and cosmic ray exposure ages of the Apollo 15 samples. J. Geophys. Res., *79*, 2588–2606.

Kaneoka, I. (1974) Investigation of excess argon in ultramafic rocks from the Kola peninsula by the ^{40}Ar/^{39}Ar method. Earth Planet. Sci. Letters, *22*, 145–156.

Lanphere, M. A., and G. B. Dalrymple (1971) A test of the ^{40}Ar/^{39}Ar age spectrum technique on some terrestrial materials. Earth Planet. Sci. Letters, *12*, 359–372.

Lanphere, M. A., and A. L. Albee (1974) ^{40}Ar/^{39}Ar age measurements in the Worcester Mountains: Evidence of Ordovician and Devonian metamorphic events in northern Vermont. Am. J. Sci., *274*, 545–555.

McDougall, I., H. A. Polach, and J. J. Stipp (1969) Excess radiogenic argon in young subaerial basalts from the Aukland volcanic field, New Zealand. Geochim. Cosmochim. Acta, *33*, 1485–1520.

McDougall, I. (1974) The ^{40}Ar/^{39}Ar method of K-Ar age determination of rocks using HIFAR reactor. Atom. Energy Australia, Aust. Atom. Energy Comm., *17*, No. 3, 3–12.

Mellor, D. W., and A. E. Mussett (1975) Evidence for initial ^{36}Ar in volcanic rocks and some implications. Earth Planet. Sci. Letters, *26*, 312–318.

Merrihue, C. M. (1965) Trace-element determinations and potassium-argon dating by mass spectroscopy of neutron-irradiated samples. Trans. Amer. Geophys. Union, *46*, 125.

Merrihue, C., and G. Turner (1966) Potassium-argon dating by activation with fast neutrons. J. Geophys. Res., *71*, 2852–2857.

Mitchell, J. G. (1968) The argon-^{40}/argon-^{39} method for potassium-argon age determination. Geochim. Cosmochim. Acta, *32*, 781–790.

Musset, A. E., and G. B. Dalrymple (1968) An investigation of the source of air Ar contamination in K-Ar dating. Earth Planet. Sci. Letters, *4*, 422–426.

Ozima, M. (1975) Ar isotopes and Earth-atmosphere evolution models. Geochim. Cosmochim. Acta, *39*, 1127–1134.

Podosek, F. A. (1971) Neutron-activation potassium-argon dating of meteorites. Geochim. Cosmochim. Acta, *35*, 157–173.

Podosek, F. A. (1972) Gas retention chronology of Petersburg and other meteorites. Geochim. Cosmochim. Acta, *36*, 755–772.

Podosek, F. A., and J. C. Huneke (1973) Argon 40-argon 39 chronology of four calcium-rich achondrites. Geochim. Cosmochim. Acta, *37*, 667–684.

Schwartzman, D. W. (1973) On argon degassing models of the Earth. Nature Phys. Sci., *245*, 20.

Shafiqullah, M., and P. E. Damon (1974) Evaluation of K-Ar isochron methods. Geochim. Cosmochim. Acta, *38*, 1341–1358.

Suess, H. E. (1949) Die Häufigkeit der Edelgase auf der Erde und im Kosmos. J. Geol., *57*, 600–607.

Sutter, J. F., L. Husain, and O. A. Schaeffer (1971) $^{40}Ar/^{39}Ar$ ages from Fra Mauro. Earth Planet. Sci. Letters, *11*, 249–253.

Turner, G., J. A. Miller, and R. L. Grasty (1966) The thermal history of the Bruderheim meteorite. Earth Planet. Sci. Letters., *1*, 155–157.

Turner, G. (1968) The distribution of potassium and argon in chondrites, 387–398. In Origin and Distribution of the Elements, L. H. Ahrens, ed. Pergamon Press, London.

Turner, G. (1969) Thermal histories of meteorites by the ^{39}Ar-^{40}Ar method. In Meteorite Research, P. M. Millman, ed., pp. 407–417. Reidel Publ. Co.

Turner, G. (1970) ^{40}Ar-^{39}Ar age determination of lunar rock 12013. Earth Planet. Sci. Letters, *9*, 177–180.

Turner, G. (1971) Argon-40-argon-39 dating: The optimization of irradiation parameters. Earth Planet. Sci. Letters, *10*, 227–234.

Turner, G., J. C. Huneke, F. A. Podosek, and G. J. Wasserburg (1971) ^{40}Ar-^{39}Ar ages and cosmic-ray exposure ages of Apollo 14 samples. Earth Planet. Sci. Letters, *12*, 19–35.

Wänke, H., and H. König (1959) Eine neue Methode zur Kalium-Argon-Altersbestimmung und ihre Anwendung auf Steinmeteorite. Z. Naturforschung, *14a*, 860–866.

York, D., and G. W. Berger (1970) $^{40}Ar/^{39}Ar$ age determinations on nepheline and basic whole rocks. Earth Planet. Sci. Letters, *7*, 333–336.

11

THE Re-Os, Lu-Hf, AND K-Ca METHODS OF DATING

In addition to the Rb-Sr and K-Ar geochronometers, several less well known methods of dating exist which are based on β^- decay of other naturally occurring radioactive isotopes. Most promising among these are geochronometers arising from the decay of ^{187}Re to ^{187}Os, ^{176}Lu to ^{176}Hf, and ^{40}K to ^{40}Ca. Moreover, several of the rare earth elements have naturally occurring isotopes that may be of some interest, particularly the alpha decay of ^{147}Sm to ^{143}Nd and the branched decay of ^{138}La to ^{138}Ce and ^{138}Ba. The ^{147}Sm-^{143}Nd decay scheme is being investigated not only because of its potential for dating terrestrial and extra-terrestrial rocks but also in order to study petrogenesis and the history of the crust and mantle of the Earth (Lugmair et al., 1975; DePaolo and Wasserburg, 1976; Richard et al., 1976). The Re-Os, Lu-Hf, and K-Ca geochronometers have merit primarily because they are applicable to the study of certain rocks and minerals that are not datable by the conventional methods. Moreover, calcium attracts attention because its isotopic composition may be changed by fractionation in the course of certain geological processes.

1

The Re-Os Method of Dating

Rhenium was discovered in 1925 by Ida and Walter Noddack. Its atomic number is 75 and it is a member of Group VIIB together with manganese and technetium. However, rhenium exhibits close geochemical coherence with molybdenum rather than with manganese. Rhenium has two naturally occurring isotopes whose abundances are $^{185}Re = 37.398 \pm 0.016$ percent and $^{187}Re = 62.602 \pm 0.016$ percent giving it an atomic weight of 186.20679 ± 0.00031 (Gramlich et al., 1973). Rhenium 187 is radioactive and decays to stable ^{187}Os by emission of a beta particle.

$$^{187}_{75}Re \rightarrow {}^{187}_{76}Os + \beta^- + \bar{\nu} + Q \quad (11.1)$$

The end point energy of the beta spectrum is only about 2.5 keV which has made it very difficult to determine the half-life of ^{187}Re by direct counting. Hirt et al. (1963a) obtained a value of $(4.3 \pm 0.5) \times 10^{10}$ years by analysis of 10 samples of molybdenite of known age, while Watt and Glover (1962) reported 3×10^{10} years on the basis of direct counting.

Rhenium is a dispersed element occurring almost entirely in minerals of other elements. It is most strongly concentrated in molybdenite (MoS_2) whose rhenium concentrations may vary from a trace up to 1.88 percent. Molybdenites associated with copper sulfide minerals tend to have especially high rhenium concentration. The only mineral in which rhenium may be a major component is dzhezkasganite ($CuReS_4$?), which was discovered in cupriferous sandstone of the Dzhezkasgan deposit in Kazakhstan, U.S.S.R. (Seyfullin et al., 1975). Rhenium is also found in the rare-earth mineral gadolinite, in columbite and tantalite, and in sulfides of copper. Its abundance in igneous rocks is about 0.5 parts per billion.

The most recent summary of the geochemistry and abundance of rhenium is by Morris and Short (1969).

Osmium (atomic number 76) is a member of the platinum metals and belongs to Group VIIIB in the periodic table. It has seven naturally occurring isotopes, all of which are stable. They are: ^{184}Os(0.02 percent), ^{186}Os(1.6 percent), ^{187}Os(1.6 percent), ^{188}Os(13.3 percent), ^{189}Os(16.1 percent), ^{190}Os(26.4 percent), and ^{192}Os(41.0 percent) (Holden and Walker, 1972).

Osmium is strongly siderophile and occurs primarily in the mineral osmiridium which is an alloy of osmium, iridium, and other platinum metals. As a group, the platinum metals are associated with ultramafic rocks and gabbros containing sulfides of copper and nickel. In addition, platinum metals are found in tellurides, selenides, arsenides, and antimonides. The crustal abundance of osmium is of the order of 0.4 ppb. The geochemistry and distribution of the platinum metals has been reviewed by Crocket (1969).

The Re-Os method has been used to date iron meteorites which cannot be dated directly by other methods. Hirt et al. (1963b) determined concentrations of osmium and rhenium in about 30 iron meteorites by neutron activation. They found osmium concentrations ranging from less than 0.025 to 50.4 ppm (average Os = 7.0 ppm) and rhenium concentrations between limits of 0.004 and 4.8 ppm (average Re = 0.58 ppm). The average Re/Os ratio of iron meteorites is 0.083. Whereas the rhenium and osmium *concentrations* of iron meteorites vary by factors of more than 1000, their Re/Os *ratios* vary only by a factor of about four. Morgan and Lovering (1964, 1967) found an average of 0.060 ppm rhenium and 0.71 ppm osmium in several different types of stony meteorites. Therefore, both elements are concentrated in the metallic phases of meteorites by about a factor of 10 relative to the silicate phases.

Nevertheless, the low Re/Os ratios and their limited range contribute significantly to the difficulty of dating iron meteorites by this method.

It is convenient to describe the growth of ^{187}Os in a Re-bearing system by the following equation:

$$\frac{^{187}\text{Os}}{^{186}\text{Os}} = \left(\frac{^{187}\text{Os}}{^{186}\text{Os}}\right)_0 + \frac{^{187}\text{Re}}{^{186}\text{Os}}\left(e^{\lambda t} - 1\right) \quad (11.2)$$

where

$\dfrac{^{187}\text{Os}}{^{186}\text{Os}}$ = ratio of these isotopes at the present time,

$\left(\dfrac{^{187}\text{Os}}{^{186}\text{Os}}\right)_0$ = initial ratio of these isotopes at the time the system became closed to rhenium and osmium,

$\dfrac{^{187}\text{Re}}{^{186}\text{Os}}$ = ratio of these isotopes at the present time,

λ = decay constant of ^{187}Re (1.61×10^{-11} y^{-1}, $T_{1/2} = 4.3 \times 10^{10}$ y), and

t = time elapsed since the system became closed to rhenium and osmium.

Equation 11.2 is analogous to Equation 6.3 for the Rb-Sr method of dating. Re-Os dates can be calculated for single samples whose isotopic composition of osmium and concentrations of osmium and rhenium have been measured. In addition, the initial ^{187}Os/^{186}Os ratio must be assumed. Alternatively, Equation 11.2 can be applied to a suite of samples having a common age and the same initial ^{187}Os/^{186}Os ratio. Such samples form an isochron in coordinates of ^{187}Os/^{186}Os and ^{187}Re/^{186}Os whose slope is proportional to the age of the samples. The zero intercept of the isochron yields the initial ^{187}Os/^{186}Os ratio.

The validity of dates calculated by these methods is subject to the conditions that the samples must have remained closed to rhenium and osmium and that the decay constant of ^{187}Re is known. Under these conditions, and provided that the initial ^{187}Os/^{186}Os ratio is known, Re-Os dates may represent the age of a geological sample. Similarly, the date calculated from the slope of an isochron represents the time elapsed since all specimens had the same ^{187}Os/^{186}Os ratio.

Hirt et al. (1963b) attempted to date a suite of 14 iron meteorites by the isochron method. The data points define an isochron shown in Figure 11.1, whose slope yields a date of 4.0 ± 0.8 billion years. The initial ^{187}Os/^{186}Os ratio is 0.83. The date is not distinguishable from the results obtained by other methods which indicate that meteorites became closed systems between 4.5 and 4.7 billion years ago. The value of the primordial ^{187}Os/^{186}Os ratio derived from the meteorite isochron is less than the ^{187}Os/^{186}Os ratio of any terrestrial sample. The lowest ^{187}Os/^{186}Os ratio of a terrestrial sample is 0.882 ± 0.007 in a sample of

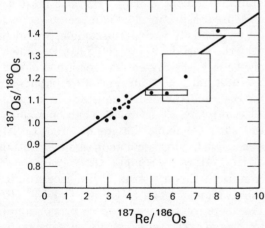

FIGURE 11.1 Re-Os isochron for iron meteorites. The slope of the isochron indicates a date of 4.0 ± 0.8 + 10^9 years assuming a value of 4.3 ± 0.5 × 10^9 years for the half-life of ^{187}Re. The primordial ^{187}Os/^{186}Os ratio is 0.83. (Adapted from Anders, 1962, using data by Herr et al., 1961.)

osmiridium from the Witwatersrand deposits of South Africa. Osmiridium samples from other areas (Australia, the Ural Mountains, South America, and Alaska) have higher ratios (up to 1.086 ± 0.010) indicating that real differences exist in the isotopic composition of terrestrial osmium (Hirt et al., 1963b).

The rhenium contents of *molybdenite* have been measured by Riley (1967), Allegre et al. (1964), and Hirt et al. (1963b). The latter reported concentrations (ranging from 2.0 to 230 ppm) with an average of about 42 ppm for 19 samples. On the other hand, the osmium content of this mineral is less than 0.2 ppm and is due primarily to the presence of radiogenic ^{187}Os formed by in situ decay of ^{187}Re. Consequently, molybdenite has a very high Re/Os ratio and is suitable for dating by the Re-Os method. Such age determinations have been reported by Hirt et al. (1963b) who obtained dates ranging from 105 to 2540 million years. The validity of the dates is difficult to ascertain because the ages of hydrothermal vein deposits are difficult to determine by conventional dating methods. The Re-Os method is of interest *precisely* because it may enable us to date molybdenite-bearing vein deposits. However, the low osmium content, which gives molybdenites their high Re/Os ratios, also causes practical difficulties in the accurate determination of the concentration and isotopic composition of osmium. Nevertheless, the Re-Os method of dating molybdenite and Re-bearing copper sulfide ores deserves additional testing because such deposits are generally not datable by other methods.

2
The Lu-Hf Method of Dating

Lutetium (atomic number 71) is a rare earth that is widely dispersed in igneous, sedimentary and metamorphic rocks. It has two naturally occurring isotopes whose abundances are: $^{175}Lu = 97.4$ percent and $^{176}Lu = 2.6$ percent (Holden and Walker, 1972). The latter is radioactive and is subject to branched decay by beta emission to stable $^{176}_{72}Hf$ and by electron capture to stable $^{176}_{70}Yb$. The ratio of electron captures to beta emissions is of the order of 3 ± 1 percent or less, so that the decay to ^{176}Yb can be neglected (Dixon et al., 1954). The relevant decay scheme is therefore represented by

$$^{176}_{71}Lu \rightarrow {}^{176}_{72}Hf + \beta^- + \bar{\nu} + Q \quad (11.3)$$

The decay is to an excited state of ^{176}Hf and is followed by emission of gamma rays. The half-life of ^{176}Lu has been determined by direct counting experiments and indirectly by analysis of Lu-bearing minerals of known age. The results have been inconsistent, varying from 2 to 7×10^{10} years. Recent direct determinations have narrowed the range considerably: McNair (1961): $3.57 \pm 0.02 \times 10^{10}$; Brinkman et al. (1965): $3.54 \pm 0.05 \times 10^{10}$, $3.50 \pm 0.14 \times 10^{10}$, $3.68 \pm 0.06 \times 10^{10}$; Prodi et al. (1969): $3.27 \pm 0.05 \times 10^{10}$ years. These determinations suggest that the half-life of ^{176}Lu is $3.5 \quad 0.2 \times 10^{10}$ years, in agreement with Boudin and Deutsch (1970) who obtained $3.3 \pm 0.5 \times 10^{10}$ years based on analyses of gadolinite and priorite of known age.

Hafnium (atomic number 72) has six naturally occurring isotopes whose abundances are: $^{174}Hf = 0.17$ percent, $^{176}Hf = 5.2$ percent, $^{177}Hf = 18.5$ percent, $^{178}Hf = 27.2$ percent, $^{179}Hf = 13.8$ percent, and $^{180}Hf = 35.1$ percent (Holden and Walker, 1972). ^{174}Hf is radioactive and decays by alpha emission to stable ^{170}Yb with a half-life of 2.0×10^{15} y (Holden and Walker, 1972). The abundance of ^{176}Hf in Lu-bearing rocks and minerals increases as a function of time due to the decay of ^{176}Lu. The age of Lu-bearing rocks or minerals is obtained by

solution of an equation which is similar to Equation 6.3:

$$\frac{^{176}Hf}{^{177}Hf} = \left(\frac{^{176}Hf}{^{177}Hf}\right)_0 + \frac{^{176}Lu}{^{177}Hf}(e^{\lambda t} - 1) \quad (11.4)$$

where

$\dfrac{^{176}Hf}{^{177}Hf} =$ ratio of these isotopes in the sample at the present time,

$\left(\dfrac{^{176}Hf}{^{177}Hf}\right)_0 =$ initial ratio of these isotopes at the time of formation of the system,

$\dfrac{^{176}Lu}{^{177}Hf} =$ ratio of these isotopes in the sample at the present time,

$\lambda =$ decay constant of $^{176}Lu(1.98 \times 10^{-11} \text{ y}^{-1}$, $T_{1/2} = 3.5 \times 10^{10}$ y), and

$t =$ time elapsed since formation of the rock or mineral.

In order to date a sample by the Lu-Hf method, the concentrations of lutetium and hafnium and the $^{176}Hf/^{177}Hf$ ratio must be measured. Equation 11.4 can then be solved for t which is the age of the sample, provided that the initial $^{176}Hf/^{177}Hf$ ratio is known accurately, the system has remained closed to lutetium and hafnium throughout its history, and the decay constant is known. Alternatively, Equation 11.4 can be applied to a suite of samples having different Lu/Hf ratios but having a common age and initial $^{176}Hf/^{177}Hf$ ratio. Samples that satisfy these conditions lie on a straight-line isochron in coordinates of $^{176}Hf/^{177}Hf$ and $^{176}Lu/^{177}Hf$. The slope of the isochron is proportional to the age of the samples, while its intercept is the initial $^{176}Hf/^{177}Hf$ ratio. This method of dating is similar to the Rb-Sr isochron method.

The rare earth elements (atomic numbers 57 to 71) occur as minor constituents in all types of rocks. Their most common oxidation state is +3, although europium can be reduced to +2 and cerium can be oxidized to +4 in nature. Their ionic radii are similar and decrease slightly with increasing atomic number from 1.13 Å for La^{+3} to 0.94 Å for Lu^{+3}. Because of the similarity of their chemical properties, the rare earths exhibit strong geochemical coherence. However, some fractionation does occur in the course of crystallization of minerals from magma. Reviews of the geochemistry of the rare earths have been published by Haskin et al. (1966), Haskin and Frey (1966), and Herrmann (1970). The concentrations of the rare earths are usually determined by neutron activation or by isotope dilution.

Lutetium is present in virtually all types of rocks in concentrations that rarely exceed one part per million. Table 11.1 is a compilation taken from Herrmann (1970). It shows that the average lutetium concentration of igneous rocks increases with increasing degree of differentiation. However, the rare earths are progressively depleted in lutetium as they are concentrated as a group in the silica-rich rocks. Nevertheless, igneous and sedimentary rocks are generally enriched in lutetium compared to chondritic meteorites by factors of about 20.

Many of the common rock-forming minerals contain detectable concentrations of lutetium. The most important carriers of this element are apatite, zircon, garnet, biotite, and certain rare earth minerals such as the phosphates monazite and xenotime, the oxides euxenite and samarskite, and the silicates allanite and gadolinite. The common rock-forming silicates plagioclase, amphibole, pyroxene, and olivine have lutetium concentrations amounting to fractions of one part per million. In general, the concentration of lutetium in a given mineral in granitic pegmatites is several times greater than the concentration of lutetium in that mineral in basic rocks.

Several fairly common minerals, that is, apatite, garnet, and monazite, may be suitable for dating by the Lu-Hf method. The lutetium concentration of apatite in acid igneous rocks is about 25 ppm, while that of garnet is about 10 ppm. Lutetium concentrations of monazite are probably even higher, but little information is available. Zircon in granites contains an average of 154 ppm of lutetium, but this mineral is not

Table 11.1 **Average Concentrations of Lutetium and Hafnium in Terrestrial and Extraterrestrial Rocks**

ROCK TYPE	Lu ppm	Hf ppm	Lu/Hf
Chondrites	0.031	0.19	0.16
Lunar rocks (Tranquility)	1.85	12.7	0.15
Basalt	0.55	2.44	0.22
Intermed. igneous rocks	0.62	8.09	0.076
Granites	0.68	6.93	0.098
Sedimentary rocks	0.6	2.91	0.21

The Lu data are from Herrmann (1970), those for Hf are from a compilation by Owen (1974).

suitable for dating because of its high hafnium content. Gadolinite contains several thousand parts per million of lutetium and was used by Herr et al. (1958) and by Boudin and Deutsch (1970) to determine the half-life of ^{176}Lu. However, this mineral is not sufficiently abundant to be of much interest for dating.

Hafnium is a member of Group IVB of the periodic table and has a valence of $+4$. Its chemical properties and ionic radius are very similar to those of zirconium with which it is closely associated in nature. Excellent summaries of the physical and chemical properties of zirconium and hafnium and of their analytical chemistry have been published by Elinson and Petrov (1969) and Mukherji (1970).

Hafnium is a dispersed element and occurs in many different minerals by replacement of zirconium and, to a lesser extent, by replacing titanium. Vlasov (1966) listed a large number of minerals that contain appreciable concentrations of hafnium. Among these, zircon is by far the most abundant. In general, alkali-rich igneous rocks have unusually high hafnium concentrations. Brooks (1969) reported average concentrations of 128 and 33 ppm Hf in sodium-rich pyroxene and amphiboles, respectively, from the Kangerdlugssuaq alkaline intrusion of East Greenland. He also found 12.4 ppm Hf in sphene from this intrusive and 12.5 ppm in ilmenite from the Skaergaard complex in Greenland.

Average concentrations of hafnium in terrestrial and extraterrestrial rocks are presented in Table 11.1 together with average lutetium concentrations. Unfortunately, very few rocks or minerals have been analyzed for both elements. Nevertheless, it can be seen that the Lu/Hf ratios of common igneous and sedimentary rocks are of the order of 0.2 or less. These low values suggest that dating of whole-rock samples

by the Lu-Hf method will be difficult. However, the Lu/Hf ratios of igneous rocks do vary by a factor of about two, basic igneous rocks having somewhat higher ratios than acidic igneous rocks.

The principal difficulty in the use of the Lu-Hf method of dating lies in the accurate determination of the isotopic composition of hafnium extracted from geological materials. Boudin and Deutsch (1970) and Owen and Faure (1974) have published procedures that show promise but require further refinement. The extraction of microgram quantities of hafnium is facilitated by the presence of zirconium which acts as a carrier. However, in order to make successful isotope analyses of hafnium, it must be separated from zirconium before being loaded into the mass spectrometer.

While minerals with favorable Lu/Hf ratios offer the greatest promise, dating of suites of cogenetic igneous rocks may be feasible after the analytical difficulties have been overcome. This would not only provide a new method of dating rocks that may not be datable by other methods but would also provide an opportunity to study the isotopic evolution of hafnium in the crust and upper mantle of the Earth. It seems possible that magma formation by partial melting in the upper mantle may have depleted that region in hafnium relative to lutetium, giving it a higher Lu/Hf ratio than the crust. In this case, young basaltic rocks derived from the mantle may have higher ^{176}Hf/^{177}Hf ratios than the sialic crust. No information is currently available to evaluate these suggestions, but further work on the Lu-Hf method seems well worth the effort.

3
The K-Ca Method of Dating

Calcium (atomic number 20) is an important element because of its abundance in the crust of the Earth and because of the existence of a large number of Ca-bearing minerals. It has six stable isotopes whose abundances are: $^{40}Ca = 96.94$ percent, $^{42}Ca = 0.65$ percent, $^{43}Ca = 0.14$ percent, $^{44}Ca = 2.08$ percent, $^{46}Ca = 0.003$ percent, and $^{48}Ca = 0.19$ percent (Holden and Walker, 1972). The isotopic composition of calcium in rocks and minerals may vary because of the formation of ^{40}Ca by beta decay of naturally occurring ^{40}K and because of possible fractionation of Ca-isotopes in the course of certain physical-chemical processes. The possibility of fractionation of calcium isotopes in nature is enhanced by the considerable difference in their masses which amounts to about 20 percent for ^{48}Ca relative to ^{40}Ca. In addition, the isotopic composition of calcium in certain iron meteorites may be changed by the interaction of cosmic rays with the nuclei of iron (Stauffer and Honda, 1962).

The possibility of using the decay of potassium to calcium to study the origin of igneous rocks was first discussed by Holmes (1932) and later by Ahrens (1950). It is now known that ^{40}K is subject to branched decay and forms ^{40}Ar by electron capture and ^{40}Ca by beta (β^-) decay (Chapter 9). The total decay constant of ^{40}K is equal to

$$\lambda = \lambda_e + \lambda_\beta$$
$$= 0.585 \times 10^{-10} + 4.72 \times 10^{-10} \quad (11.5)$$
$$= 5.305 \times 10^{-10} \text{ y}^{-1}$$

Consequently, we can express the $^{40}Ca/^{44}Ca$ ratio of a K-bearing system whose age is equal to t by the equation

$$\frac{^{40}Ca}{^{44}Ca} = \left(\frac{^{40}Ca}{^{44}Ca}\right)_0 + \frac{\lambda_\beta}{\lambda_e + \lambda_\beta}\frac{^{40}K}{^{44}Ca}(e^{\lambda t} - 1)$$
$$(11.6)$$

where

$\dfrac{^{40}Ca}{^{44}Ca} =$ ratio of these isotopes at the present time,

$\left(\dfrac{^{40}Ca}{^{44}Ca}\right)_0 =$ initial ratio of these isotopes at time of formation or last isotopic homogenization of the system,

$\lambda_\beta, \lambda_e =$ decay constants for beta and electron capture decay of ^{40}K, respectively,

$\lambda =$ total decay constant of ^{40}K,

$\dfrac{^{40}K}{^{44}Ca} =$ ratio of these isotopes in the sample at the present time, and

$t =$ time elapsed since formation or last isotopic homogenization.

From a theoretical point of view the K-Ca geochronometer can be used very much like the Rb-Sr geochronometer (Chapter 6). Cogenetic rocks or minerals having the same initial $^{40}Ca/^{44}Ca$ ratio and the same age lie on an isochron in coordinates of $^{40}Ca/^{44}Ca$ and $^{40}K/^{44}Ca$ whose slope is equal to

$$m = \frac{\lambda_\beta}{\lambda_e + \lambda_\beta}(e^{\lambda t} - 1) \quad (11.7)$$

and whose intercept is equal to the initial $^{40}Ca/^{44}Ca$ ratio.

Alternatively, dates may be calculated for single samples by assuming a value for the initial $^{40}Ca/^{44}Ca$ ratio. In actual fact, the K-Ca geochronometer is severely restricted by the high abundance of ^{40}Ca which makes the $^{40}Ca/^{44}Ca$ ratio insensitive to the addition of radiogenic ^{40}Ca. Furthermore, the determination of the isotopic composition of calcium by mass spectrometry is made difficult by the low efficiency of ionization of calcium atoms in a thermionic source and by fractionation of isotopes during that process. For this reason, the K-Ca method of dating has promise only for minerals that are strongly enriched in potassium and depleted in calcium, such as micas in pegmatite and sylvite in evaporite rocks.

Coleman (1971) used the K-Ca method to date a suite of micas from the Scottish Highlands whose ages had been measured previously by the Rb-Sr method. He employed a double-spike technique to correct for mass discrimination of calcium isotopes (Dodson, 1963, 1969) and obtained dates that are in moderately good agreement with the Rb-Sr dates. The radiogenic ^{40}Ca content of these micas ranges from only 0.37 to 1.12 percent of ^{40}Ca. Coleman found considerably more radiogenic ^{40}Ca (up to 16.5 percent) in samples of polylithionite from southwest Greenland whose age is 1086 m.y., based on the Rb-Sr method. However, their calcium contents are variable and smaller grains were found to contain less radiogenic ^{40}Ca than larger grains. Hence their K-Ca dates range from 770 to 1350 m.y. and cannot be interpreted in an unequivocal way.

Polevaya et al. (1958) attempted to date sylvite and other K-bearing evaporite minerals by the K-Ca method. They reported a 9 percent enrichment of radiogenic ^{40}Ca in sylvite of Lower Cambrian age and obtained several K-Ca dates for sylvites consistent with their stratigraphic ages. On the other hand, K-Ar dates of evaporite minerals are invariably too low because of the loss of radiogenic ^{40}Ar. They concluded that radiogenic ^{40}Ca is less mobile than ^{40}Ar and suggested that K-Ca dates of K-bearing evaporite minerals are more reliable than K-Ar dates.

4
Fractionation of Calcium Isotopes

Because of the large mass difference between ^{40}Ca and ^{48}Ca, the isotopes of calcium may be fractionated in nature by diffusion or by isotope exchange reactions in which isotopes are preferentially partitioned among Ca-bearing compounds and solutions. Such fractionation effects would be of considerable interest and could provide information about certain chemical reactions and physical processes that affect calcium in nature. For this reason, efforts have been made to measure the isotopic composition of calcium in various kinds of samples. However, the results have generally been disappointing, in the sense that fractionation of calcium isotopes is not as prevalent in nature as expected.

Nevertheless, two schools of thought persist regarding this question. There are those who claim to have demonstrated significant variations of calcium isotope ratios, while others deny the existence of any measurable fractionation. The difference of opinion may well result from the difficulty in measuring the isotopic composition of calcium on a mass spectrometer. This problem is difficult to control and may depend on the presence of impurities in the sample, the temperatures of the filaments in the source, and on other factors. The problem has been minimized by use of an isotope spike which serves as an internal standard (Hirt and Epstein, 1964) or by rigorous standardization of procedures and operating conditions of the mass spectrometer (Heumann and Lieser, 1973).

Perhaps the strongest advocates of calcium isotope fractionation have been Corless and Winchester (1964) and Corless (1966, 1968) who measured ^{48}Ca/total Ca ratios by neutron activation. Initially, they reported variations of this ratio of up to 13 percent, but later refinement of the technique by Corless indicated that the variations in the relative abundance of ^{48}Ca are of the order of 1 percent or less. Variations in the isotope composition of calcium were also reported by Artemov et al. (1966) and Miller et al. (1966) based on mass spectrometry and by Meshcheryakov and Stolbov (1967) who used a deuteron (deuterium) activation technique.

Table 11.2   **Isotope Ratios of Calcium in Various Samples of Carbonates and Sulfates (Heumann and Lieser, 1973)**

MATERIAL	$\dfrac{^{44}Ca}{^{40}Ca} \times 10^5$	$\dfrac{^{48}Ca}{^{40}Ca} \times 10^6$
Tip of stalactite	2110 ± 6	1869 ± 3
Base of stalactite	2121 ± 3	1880 ± 7
Travertine	2110 ± 6	1866 ± 9
Gypsum tube	2112 ± 5	1870 ± 8
$CaCl_2 \cdot 2H_2O$ reagent[a]	2112 ± 9	1875 ± 14
Gypsum rose		
Fused with Na_2CO_3	2096 ± 7	1847 ± 17
Dissolved in HNO_3	2097 ± 11	1849 ± 25

[a]Prepared by E. Merck Co., Darmstadt, from limestone of Middle Devonian age in Biebertal, Hessen, West Germany.

On the other hand, Heumann and Lieser (1973) analyzed calcium from several samples of calcite and gypsum by carefully controlled mass spectrometry and found no detectable differences in the $^{44}Ca/^{40}Ca$ and $^{48}Ca/^{40}Ca$ ratios. This conclusion is supported by experimental studies carried out by Stahl and Wendt (1968), Möller and Papendorf (1971), and Heumann and Lieser (1972) all of which showed that calcium isotopes are not fractionated during precipitation of calcium carbonate from aqueous solutions. Moreover, Letolle (1968), Stahl (1968), and Heumann and Luecke (1973) all failed to detect measurable differences in the isotopic composition of calcium in natural samples. Backus et al. (1964) reported that the isotopic composition of calcium from the chondritic meteorite Homestead is indistinguishable from that of terrestrial calcium, but claimed that calcium from the achondrite Pasamonte is depleted in ^{40}Ca by up to 3 percent. However, Hirt and Epstein (1964) found that calcium from the chondrite Bruderheim and the achondrite Nuevo Laredo does not differ isotopically from terrestrial calcium.

The weight of the evidence seems to support the conclusion that calcium isotopes are not fractionated by precipitation of solids from aqueous solutions and that variations in the isotope ratios of calcium in nature are small and difficult to detect. However, Heumann and Lieser (1973) did analyze a gypsum rose whose $^{44}Ca/^{40}Ca$ and $^{48}Ca/^{40}Ca$ ratios appear to be significantly less than those of other carbonate and sulfate samples (Table 11.2). This gypsum rose was collected in the northern Sahara at an unspecified depth below the surface corresponding to the deepest level of penetration of meteoric precipitation. They suggested that the isotope fractionation resulted either from ion-exchange reactions with the soil or from diffusion of calcium ions. The calcium in the gypsum rose is depleted in ^{48}Ca by about 1.4 percent and in ^{44}Ca by 0.7 percent compared to reagent calcium. These differences are compatible with isotope fractionation effects caused by diffusion (Senftle and Bracken, 1955). However, further studies are required to substantiate the hypothesis that calcium isotopes are fractionated by diffusion or ion exchange reactions in nature.

5 SUMMARY

The beta decays of ^{187}Re to ^{187}Os, ^{176}Lu to ^{176}Hf and ^{40}K to ^{40}Ca have found only limited application for age determinations of rocks and minerals. The Re-Os method has been used to date iron meteorites and molybdenites from sulfide ore deposits. The Lu-Hf method may be useful for dating certain Lu-bearing minerals such as apatite, garnet, and monazite. Little progress has been made in this direction because of the uncertainty of the half-life of ^{176}Lu and because of analytical problems associated with the measurement of the concentration and isotope composition of hafnium. The K-Ca method is severely handicapped by the high natural abundance of ^{40}Ca and by difficulties with the mass spectrometry of calcium. Some evidence has been presented to suggest that calcium isotopes may be fractionated in nature. However, this effect is less than expected and is of little practical value in geology. Perhaps the most promising applications of these decay schemes are the dating of molybdenite by the Re-Os method and the study of the isotopic evolution of hafnium in the mantle and crust of the Earth. In addition, the recent work on the ^{147}Sm-^{143}Nd decay scheme has yielded promising results.

REFERENCES

THE Re-Os METHOD

Allegre, C., N. Deschamps, J. Faucherre, and M. Hanotel-Monstard (1964) Contribution a la geochronologie par la methode rhenium-osmium. C. R. Sommaire des Seances de la Societe Geologique de France. Fascicule 8, Novembre, 1964, 297–298.

Anders, E. (1962) Meteorite ages. Rev. Mod. Phys., *34*, 287–325.

Crocket, J. H. (1969) Platinum metals. In Handbook of Geochemistry, K. H. Wedepohl exec. ed. Springer-Verlag, Heidelberg and New York.

Gramlich, J. W., T. J. Murphy, E. L. Garner, and W. R. Shields (1973) Absolute isotopic abundance ratio and atomic weight of a reference sample of rhenium. J. Res. Nat. Bur. Stand., A. Phys. and Chem., *77A*, No. 6, 691–698.

Herr, W., W. Hoffmeister, B. Hirt, J. Geiss, and F. G. Houtermans (1961) Versuch zur Datierung von Eisenmeteoriten nach der Rhenium-Osmium Methode. Z. Naturforsch., *16a*, 1053.

Hirt, B., G. R. Tilton, W. Herr, and W. Hoffmeister (1963a) The half-life of ^{187}Re. In Earth Science and Meteorites, J. Geiss and E. D. Goldberg, eds. North-Holland, Amsterdam, pp. 273–280.

Hirt, B., W. Herr, and W. Hoffmeister (1963b) Age determinations by the rhenium-osmium method. In Radioactive Dating. Internat. Atom. Energy Agency, Vienna, 35–44.

Holden, N. E., and F. W. Walker (1972) Chart of the Nuclides (11th ed.) Educational Relations, General Electric Co., Schenectady, N.Y., 12345.

Morgan, J. W., and J. F. Lovering (1964) Rhenium and osmium abundances in stony meteorites. Science, *144*, 835–836.

Morgan, J. W., and J. F. Lovering (1967) Rhenium and osmium abundances in chondritic meteorites. Geochim. Cosmochim. Acta, *31*, 1893–1910.

Morris, D. F. C., and E. L. Short (1969) Rhenium. In Handbook of Geochemistry, K. H. Wedepohl exec. ed., vol. II-1. Springer-Verlag, Heidelberg and New York.

Riley, G. (1967) Rhenium concentrations in Australian molybdenites by stable isotope dilution. Geochim. Cosmochim. Acta, *31*, 1489–1498.

Seyfullin, S. Sh., S. K. Kalinin, A. V. Strutynskiy, et al. (1975) Rhenium in stratified copper deposits and showings in west central Kazakhstan. Geochem. Internat., *11*, No. 2, 414–418.

Watt, D. E., and R. N. Glover (1962) A search for radioactivity among naturally occurring isobaric pairs. Phil. Mag., 7, 105–114.

THE Lu-Hf METHOD

Boudin, A., and S. Deutsch (1970) Geochronology: Recent development in the lutetium-176/hafnium-176 dating method. Science, *168*, 1219–1220.

Brinkman, G. A., A. H. W. Aten, and J. Th. Veenboer (1965) Natural radioactivity of K-40, Rb-87 and Lu-176. Physica, *31*, 1305–1319.

Brooks, C. K. (1969) On the distribution of zirconium and hafnium in the Skaergaard intrusion, East Greenland. Geochim. Cosmochim. Acta, *33*, 357–374.

Dixon, D., A. McNair, and S. C. Curran (1954) The natural radioactivity of lutetium. Phil. Mag. (7), *45*, 683–694.

Elinson S. V., and K. I. Petroy (1969) Analytical chemistry of zirconium and hafnium. Humphrey Science Pub., Ann Arbor, 243 pp.

Haskin, L. A., R. A. Schmitt, and R. H. Smith (1966) Meteoritic, solar, and terrestrial rare earth distributions. In Physics and Chemistry of the Earth, vol. 7, pp. 167–321. Pergamon Press, Oxford.

Haskin, L. A., and F. A. Frey (1966) Dispersed and not-so-rare earths. Science, *152*, 229–314.

Herr, W., E. Merz, P. Eberhardt, and P. Signer (1958) Zur Bestimmung der-Halbwertszeit des Lu-176 durch den Nachweis von radiogenem Hf-176. Z. Naturf., *13a*, 268–273.

Herrmann, A. G. (1970) Yttrium and lanthanides. In Handbook of Geochemistry, K. H. Wedepohl, exec. ed., vol. II-2, pp. 39, 57-71. Springer-Verlag, Berlin.

Holden, N. E., and F. W. Walker (1972) Chart of the Nuclides (11th ed.) Educational Relations, General Electric Co., Schenectady, N.Y. 12345.

McNair, A. (1961) The half-life of long-lived lutetium-176. Phil. Mag., *6*, 851.

Mukherji, A. K. (1970) Analytical Chemistry of Zr and Hf. Pergamon Press, Elmsford, N.Y., 281 p.

Owen, L. B. (1974) Age determinations by the lutetium-176/hafnium-176 method. Unpublished Ph.D. dissertation, Dept. Geol. and Mineral., The Ohio State University, Columbus, Ohio, 300 pp.

Owen, L. B., and G. Faure (1974) Simultaneous determination of hafnium and zirconium in silicate rocks by isotope dilution. Anal. Chem., *46*, 1323–1326.

Prodi, V., K. F. Flynn, and L. E. Glendenin (1969) Half-life and beta spectrum of Lu 176. Phys. Rev., *188*, 1930–1933.

Vlasov, K. A. (1966) Geochemistry and Mineralogy of Rare Elements and Genetic Types of Their Deposits. Israel Program for Scientific Translations, vol. 1, (Geochemistry of rare elements), 688 pp.

THE K-Ca METHOD

Ahrens, L. H. (1950) The feasibility of a calcium method in the determination of geologic age. Geochim. Cosmochim. Acta, *1*, 312–316.

Artemov, Y. M., K. G. Knorre, V. P. Strizkov, and V. I. Ustinov (1966) $^{40}Ca/^{44}Ca$ and $^{18}O/^{16}O$ isotopic ratios in some calcareous rocks. Geochemistry (U.S.S.R.) *3*, 1082–1086.

Backus, M. M., W. H. Pinson, Jr., L. F. Herzog, and P. M. Hurley (1964) Calcium isotope ratios in the Homestead and Pasamonte meteorites and a Devonian limestone. Geochim. Cosmochim. Acta, *28*, 735–742.

Coleman, M. L. (1971) Potassium-calcium dates from pegmatitic micas. Earth Planet, Sci. Letters, *12*, 399–405.

Corless, J. T., and J. W. Winchester (1964) Variations in the ratio $^{48}Ca/(total\ Ca)$ in the natural environment. Pure Appl. Chem., *8*, 317–323.

Corless, J. T. (1966) Determination of calcium-48 in natural calcium by neutron activation analysis. Anal. Chem., *38*, 810–813.

Corless, J. T. (1968) Observations on the isotopic geochemistry of calcium. Earth Planet. Sci. Letters, *4*, 475–478.

Dodson, M. H. (1963) A theoretical study of the use of internal standards for precise isotopic analysis by the surface ionization techniques, Part I. General first order algebraic solutions. J. Sci. Instrum., *40*, 289–295.

Dodson, M. H. (1969) A theoretical study of the use of internal standards for precise isotopic analysis by the surface ionization technique. Part II. Error relationships. J. Sci. Instrum., Ser. 2, *2*, 490–498.

Heumann, K. G., and K. H. Lieser (1972) Untersuchung von Calciumisotopieeffekten bei heterogenen Austauschgleichgewichten. Z. Naturforsch. *276*, 126–133.

Heumann, K. G., and K. H. Lieser (1973) Untersuchung von Isotopenfeinvariationen des Calciums in der Natur an rezenten Karbonaten und Sulfaten. Geochim. Cosmochim. Acta, *37*, 1463–1471.

Heumann, K. G., and W. Luecke (1973) Calcium isotope ratios in natural carbonate rocks. Earth Planet. Sci. Letters, *20*, 341–346.

Hirt, B., and S. Epstein (1964) Die Isotopenzusammensetzung von natürlichem Calcium. Helv. Phys. Acta, *37*, 179.

Holden, N. E., and F. W. Walker (1972) Chart of the Nuclides (11th ed.) Educational Relations, General Electric Co., Schenectady, N.Y., 12345.

Holmes, A. (1932) The origin of igneous rocks. Geol. Mag., *69*, 543–558.

Letolle, R. (1968) Sur la composition isotopique du calcium des echantillons naturels. Earth Planet. Sci. Letters, *5*, 207–208.

Meshcheryakov, R. P., and M. Y. Stolbov (1967) Measurement of the isotopic composition of calcium in natural materials. Geochemistry (U.S.S.R.) *4*, 1001–1003.

Miller, Yu. M., V. I. Ustinov, Yu. M. Artemov, and G. A. Kazakov (1966) Mass spectrometric determination of calcium isotope variations. Geochem. Internat. *3*, 929–933.

Möller, P., and H. Papendorf (1971) Fractionation of calcium isotopes in carbonate precipitates. Earth Planet. Sci. Letters, *11*, 192–194.

Polevaya, N. I., N. E. Titov, V. S. Belyaev, and V. D. Sprintsson (1958) Application of the Ca method in the absolute age determination of sylvites. Geochemistry, *8*, 897–906.

Senftle, F. E., and N. T. J. Bracken (1955) Theoretical effect of diffusion on isotopic abundance in rocks and associated fluids. Geochim. Cosmochim. Acta, *7*, 61.

Stahl, W. (1968) Search for natural variations in calcium isotope abundances. Earth Planet. Sci. Letters, *5*, 171–174.

Stahl, W., and I. Wendt (1968) Fractionation of calcium isotopes in carbonate precipitation. Earth Planet. Sci. Letters, *5*, 184–186.

Stauffer, H., and M. Honda (1962) Cosmic-ray produced stable isotopes in iron meteorites. J. Geophys. Res., *67*, 3503–3512.

THE Sm-Nd METHOD

DePaolo, D. J., and G. J. Wasserburg (1976) Nd isotopic variations and petrogenetic models. Geophys. Res. Letters, *3*, 1–4.

Lugmair, G. W., and N. B. Scheinin (1975) Sm-Nd systematics of the Stannern meteorite. Meteoritics, *10*, 447–448.

Richard, P., N. Shimizu, and C. J. Allegre (1976) ^{143}Nd/^{146}Nd, a natural tracer: An application to oceanic basalts. Earth Planet. Sci. Letters, *31*, 269–278.

12 THE U, Th-Pb METHODS OF DATING

1
Early Methods

The discovery of the radioactivity of uranium and thorium around the turn of the century was soon followed by attempts to use this phenomenon to measure the ages of uranium-bearing minerals. Several methods were developed which depended on the accumulation of helium and lead as decay products of uranium and thorium, including:

1. The chemical Pb-U,Th method
2. The Pb-alpha method
3. The U-He method
4. The U,Th-Pb isotopic method
5. The common-lead method

Only the last two have withstood the test of time, but several others have been added more recently. The chemical Pb-U,Th method, the Pb-alpha method, and the U-He methods are rarely used now because the assumptions on which they are based are largely invalid.

The chemical Ph-U,Th method is based on the premise that all of the lead in a uranium mineral is radiogenic in origin and that its amount increases as a function of time. The Pb-alpha method is a refinement of this approach in the sense that the lead concentration is determined by an optical spectroscopic method and the uranium and thorium concentrations are measured by counting the rate of alpha emission from the sample. The age of the sample is then calculated using an equation of the form

$$t = c \frac{\text{Pb}}{\alpha} \qquad (12.1)$$

where t is the age in millions of years, Pb is the lead concentration in parts per million, α is the observed alpha activity due to the decay of uranium, thorium, and their daughters, and c is a constant that depends on the Th/U ratio of the sample (Gottfried et al., 1959). This method was once used extensively by personnel of the U.S. Geological Survey for dating of igneous rocks using certain accessory minerals such as zircon, monazite, xenotime, and thorite.

The U-He method of dating was originally proposed by Rutherford and received much attention from scientists for many years (Hurley, 1954). It is based on the assumption that certain U, Th-bearing minerals quantitatively retain the helium produced within them by alpha decay of uranium, thorium, and their daughters. This assumption is most nearly satisfied by undisturbed crystals of magnetite, zircon, and sphene. However, even in these minerals the retention of helium is variable depending on the nature of the lattice positions occupied by uranium and thorium, the extent of radiation damage of the crystals, and the geologic histories of the samples. Moreover, it was discovered that significant amounts of uranium and thorium in rocks occur on the surfaces of crystals from where helium can readily escape. For these reasons a U-He date must be regarded as an *underestimate* of the age of a sample unless it is confirmed by an independent method. In spite of these considerable limitations, Fanale and Kulp (1962) reported a reasonable U-He date of 194 ± 4 million years for magnetite from Cornwall, Pennsylvania, which is concordant with K-Ar dates of biotite and muscovite from the Palisade diabase sill of New Jersey. More recently, Fanale and Schaeffer (1965) and Bender (1973) demonstrated the feasibility of dating aragonitic fossil corals and mollusk shells of Tertiary to Pleistocene age by the U-He method. However, Turekian et al. (1970) concluded that fossil bone could not be dated by this method.

While the U-He method is experiencing a modest revival, the chemical Pb-U and the Pb-alpha methods have been superseded by the isotopic U, Th-Pb method of dating which is the topic of this chapter. The common lead method will be presented in Chapter 13.

2
The Geochemistry of Uranium and Thorium

Uranium and thorium are members of the so-called actinide series of elements in which the 5f orbitals are progressively filled with electrons. Because of similar electron configurations, thorium ($Z = 90$) and uranium ($Z = 92$) have similar chemical properties. Both elements occur in nature in the tetravalent oxidation. state and their ions have similar radii ($U^{+4} = 1.05$ Å, $Th^{+4} = 1.10$ Å). Consequently, the two elements can substitute extensively for each other, which explains their geochemical coherence. However, under oxidizing conditions, uranium forms the uranyl ion (UO_2^{+2}) in which uranium has a valence of $+6$. The uranyl ion forms compounds that are soluble in water. Therefore, uranium is a mobile element under oxidizing conditions and is separated from thorium which exists only in the tetravalent state and whose compounds are generally insoluble in water.

The abundances of uranium and thorium in chondritic meteorites are 1×10^{-2} and 4×10^{-2} ppm, respectively. These values may be taken as an indication of the very low abundance of these elements in the man-

tle and crust of the Earth. In the course of partial melting and fractional crystallization of magma, uranium and thorium are concentrated in the liquid phase and become incorporated into the more silica-rich products. For that reason, igneous rocks of granitic composition are strongly enriched in uranium and thorium compared to rocks of basaltic or ultramafic composition. Progressive geochemical differentiation of the upper mantle of the Earth has resulted in the concentration of uranium and thorium into the rocks of the continental crust compared to those of the upper mantle. Table 12.1 contains estimates of the concentrations of uranium, thorium, and lead in some important rock types. The concentrations of all three elements increase from basaltic rocks to low-Ca granites, yet their Th/U and U/Pb ratios remain virtually constant. Low-Ca granites are enriched in thorium relative to uranium, perhaps because some of the uranium enters an aqueous phase as uranyl ion during the final stages of crystallization of granitic magmas. The Th/U ratios of sedimentary rocks are similar to those of igneous rocks with the exception of sedimentary carbonates which are enriched in uranium and have a Th/U ratio of only 0.77. The uranium enrichment of carbonate rocks results from the fact that uranium occurs in the oceans as the uranyl ion that coprecipitates with calcium carbonate, while thorium is associated primarily with water-insoluble sediment.

The concentrations of uranium and thorium in the common rock-forming silicate minerals are uniformly low, on the order of a few parts per million or less. Instead, these two elements occur primarily in certain accessory minerals in which they are either major constituents or replace other elements. The list of these includes uraninite, thorianite (oxides); zircon, thorite, allanite, (silicates); monazite, apatite, xenotime (phosphates); and sphene (titanosilicate).

Table 12.1   **Average Concentrations of U, Th, and Pb in Igneous and Sedimentary Rocks (Turekian and Wedepohl, 1961)**

ROCK TYPE	U ppm	Th ppm	Pb ppm	$\dfrac{Th}{U}$	$\dfrac{U}{Pb}$
Low-Ca granite	3.0	17.	19.	5.7	0.16
High-Ca granite	3.0	8.5	15.	2.8	0.20
Syenite	3.0	13.	12.	4.3	0.25
Basaltic rocks	1.0	4.0	6.	4.0	0.17
Ultramafics	0.001	0.004	1.0	4.0	0.001
Shale	3.7	12.	20.	3.3	0.18
Sandstone	0.45	1.7	7.	3.8	1.73
Carbonates	2.2	1.7	9.	0.77	0.24
Deep sea clay	1.3	7.	80.	5.4	0.016

Many other uranium and thorium minerals exist (Heinrich, 1958). Most of these occur under oxidizing conditions, and uranium is present in these minerals as the uranyl ion. Additional information on the geochemistry or uranium and thorium can be found in review papers by Rogers and Adams (1969a,b) and by Adams et al. (1959).

3
The Decay Series of Uranium and Thorium

Uranium has three naturally occurring isotopes: ^{238}U, ^{235}U, and ^{234}U. All three isotopes are radioactive. Thorium exists primarily as one isotope: ^{232}Th. It, too, is radioactive. In addition, five radioactive isotopes of thorium occur in nature as short-lived intermediate daughters of ^{238}U, ^{235}U, and ^{232}Th. The isotopic abundances, half-lives, and decay constants of the principal isotopes of uranium and thorium are given in Table 12.2. Each of the above isotopes is the parent of a chain of radioactive daughters ending with stable isotopes of lead. The decay of ^{238}U gives rise to the *uranium series* which includes ^{234}U as an interme-

diate daughter and ends in stable ^{206}Pb. The decay of ^{238}U to ^{206}Pb can be summarized as

$$^{238}_{92}U \rightarrow {}^{206}_{82}Pb + 8\,{}^{4}_{2}He + 6\beta^- + Q \qquad (12.2)$$

where $Q = 47.4$ MeV/atom or 0.71 cal/g yr (Wetherill, 1966). Each atom of ^{238}U that decays ultimately produces one atom of ^{206}Pb by emission of eight alpha particles and six beta particles, provided that decay occurs in a closed system with respect to uranium, lead, and all intermediate daughters. Q represents the sum of the decay energies of the entire series in units of million electron volts. The path of this decay chain can be traced in Figure 12.1. Several intermediate daughters in this series undergo branched decay involving the emission of either an alpha particle or a beta particle. The chain therefore splits into separate branches, but ^{206}Pb is the stable end product of all possible decay paths.

The decay of ^{235}U gives rise to the *actinium series* which ends with stable ^{207}Pb after emission of seven alpha and four beta particles:

$$^{235}_{92}U \rightarrow {}^{207}_{82}Pb + 7\,{}^{4}_{2}He + 4\beta^- + Q \qquad (12.3)$$

Table 12.2   **Abundances, Half-Lives, and Decay Constants of the Principal Naturally Occurring Isotopes of Uranium and Thorium**

ISOTOPE	ABUNDANCE %	HALF-LIFE y	DECAY CONSTANT y^{-1}	REFERENCE
^{238}U	99.2739	4.510×10^9	1.537×10^{-10}	1
		4.468×10^9	1.55125×10^{-10}	2
^{235}U	0.7204	0.7129×10^9	9.722×10^{-10}	3
		0.7038×10^9	9.8485×10^{-10}	2
^{234}U	0.0057	2.48×10^5	2.806×10^{-6}	4
^{232}Th	100	13.890×10^9	4.990×10^{-11}	5
		14.008×10^9	4.948×10^{-11}	6

1. Kovarik and Adams (1955)
2. Jaffey et al. (1971)
3. Fleming et al. (1952)
4. Strominger et al. (1958)
5. Picciotto and Wilgain (1956)
6. LeRoux and Glendenin (1963)

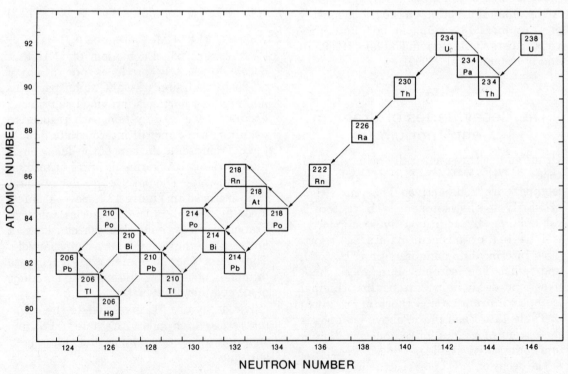

FIGURE 12.1   The decay of ^{238}U to stable ^{206}Pb.

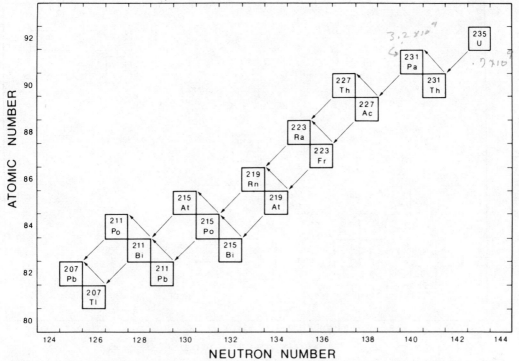

FIGURE 12.2   The decay of ^{235}U to stable ^{207}Pb.

where $Q = 45.2$ MeV/atom or 4.3 cal/g yr (Wetherill, 1966). This series also branches as shown in Figure 12.2.

The decay of ^{232}Th results in the emission of six alpha and four beta particles leading to the formation of stable ^{208}Pb (Fig. 12.3). This decay series therefore can be written as

$$^{232}_{90}\text{Th} \rightarrow {}^{208}_{82}\text{Pb} + 6\,{}^{4}_{2}\text{He} + 4\beta^{-} + Q \quad (12.4)$$

where $Q = 39.8$ MeV/atom or 0.20 cal/g yr (Wetherill, 1966).

In spite of the fact that 43 isotopes of 12 elements are formed as intermediate daughters in these decay series, none is a member of more than one series. In other words, each decay chain always leads to the formation of a specific isotope of lead. The decay of ^{238}U produces ^{206}Pb, ^{235}U produces ^{207}Pb, and ^{232}Th produces ^{208}Pb.

The half-lives of ^{238}U, ^{235}U, and ^{232}Th are all very much longer than those of their respective daughters (Table 12.2). Therefore, these decay series satisfy the prerequisite condition for the establishment of secular equilibrium (Chapter 4). When secular equilibrium exists in a uranium or thorium-bearing mineral, the decay rates of the intermediate daughters are equal to those of their respective parents:

$$N_1\lambda_1 = N_2\lambda_2 = N_3\lambda_3 = \cdots \quad (12.5)$$

If the mineral is a closed system and secular equilibrium has been established, the production rate of the stable daughter at the end of a particular decay chain is equal to the rate of decay of its parent at the head of the chain. Therefore, we can treat the decay of the isotopes of uranium and thorium in minerals in which secular equilibrium has

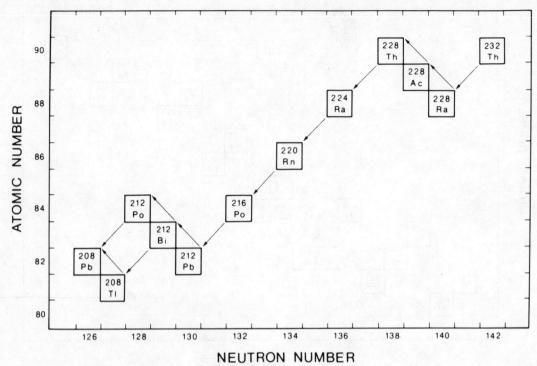

FIGURE 12.3   The decay of ^{232}Th to stable ^{208}Pb.

established itself as though it occurred directly to the respective isotopes of lead:

$$^{238}U \rightarrow {}^{206}Pb$$

$$^{235}U \rightarrow {}^{207}Pb$$

$$^{232}Th \rightarrow {}^{208}Pb$$

As a result, the growth of radiogenic lead in uranium and thorium-bearing minerals can be described by means of simple equations that are similar to those we have already used to treat the decay of ^{87}Rb to ^{87}Sr.

4
The Isotopic U,Th-Pb Methods of Dating

Ordinary lead has four naturally occurring isotopes: ^{208}Pb, ^{207}Pb, ^{206}Pb, and ^{204}Pb. We have seen that the first three are products of decay of uranium and thorium. Only ^{204}Pb is not radiogenic, although it is very weakly radioactive and decays to stable ^{200}Hg by alpha emissions with a half-life of 1.4×10^{17} years (Holden and Walker, 1972). Because of its very long half-life, ^{204}Pb is treated as a stable reference isotope. The isotopic composition of lead in minerals containing uranium and thorium can be expressed in the form of three familiar equations:

$$\frac{^{206}Pb}{^{204}Pb} = \left(\frac{^{206}Pb}{^{204}Pb}\right)_0 + \frac{^{238}U}{^{204}Pb}\left(e^{\lambda_1 t} - 1\right) \quad (12.6)$$

$$\frac{^{207}Pb}{^{204}Pb} = \left(\frac{^{207}Pb}{^{204}Pb}\right)_0 + \frac{^{235}U}{^{204}Pb}\left(e^{\lambda_2 t} - 1\right) \quad (12.7)$$

$$\frac{^{208}Pb}{^{204}Pb} = \left(\frac{^{208}Pb}{^{204}Pb}\right)_0 + \frac{^{232}Th}{^{204}Pb}\left(e^{\lambda_3 t} - 1\right) \quad (12.8)$$

where

$$\frac{^{206}Pb}{^{204}Pb}, \frac{^{207}Pb}{^{204}Pb}, \frac{^{208}Pb}{^{204}Pb}$$

= isotope ratios of lead in the mineral at the time of analysis,

$$\left(\frac{^{206}\text{Pb}}{^{204}\text{Pb}}\right)_0, \left(\frac{^{207}\text{Pb}}{^{204}\text{Pb}}\right)_0, \left(\frac{^{208}\text{Pb}}{^{204}\text{Pb}}\right)_0$$

= initial isotope ratios of lead incorporated into the mineral at the time of its formation,

$$\frac{^{238}\text{U}}{^{204}\text{Pb}}, \frac{^{235}\text{U}}{^{204}\text{Pb}}, \frac{^{232}\text{Th}}{^{204}\text{Pb}}$$

= isotope ratios in the mineral at the time of analysis,

$\lambda_1, \lambda_2, \lambda_3$

= decay constants of ^{238}U, ^{235}U, and ^{232}Th, respectively, and

t = time elapsed since closure of the mineral to uranium, thorium, lead and all intermediate daughters.

The decay constants of uranium and thorium (Table 12.2) have been revised by Jaffey et al. (1971) and by LeRoux and Glendenin (1963) and the newer values have been adopted by many scientists.

In order to date uranium- and thorium-bearing minerals by means of Equations 12.6 to 12.8, the concentrations of uranium, thorium, and lead are usually measured by isotope dilution (Chapter 5) and the isotopic composition of lead is determined on a suitable mass spectrometer. The equations can then be solved for t using reasonable assumed values for the initial lead isotope ratios:

$$t_{206} = \frac{1}{\lambda_1} \ln\left[\frac{\frac{^{206}\text{Pb}}{^{204}\text{Pb}} - \left(\frac{^{206}\text{Pb}}{^{204}\text{Pb}}\right)_0}{\frac{^{238}\text{U}}{^{204}\text{Pb}}} + 1\right] \quad (12.9)$$

The other equations are solved similarly, resulting in three independent *dates* based on the three separate decay series. These dates will be concordant and then represent the *age* of the mineral, provided the following conditions are satisfied.

1. The mineral has remained closed to uranium, thorium, lead, and all intermediate daughters throughout its history.
2. Correct values are used for the initial lead isotope ratios.
3. The decay constants of ^{238}U, ^{235}U, and ^{232}Th are known accurately.
4. The isotopic composition of uranium is normal and has not been modified by isotope fractionation or by the occurrence of a natural chain reaction based on spontaneous fission of ^{235}U, such as the natural reactor at Oklo, Gabon. (Naudet, 1974; Brookins, 1975; Lancelot et al., 1975).
5. All analytical results are accurate and free of systematic errors.

In many instances the dates so calculated for minerals containing uranium and thorium are *not* concordant. The reason seems to be that most minerals are not closed systems, but may lose or gain lead, uranium, thorium, or intermediate daughters after crystallization. The effect of lead loss on U-Pb dates can be minimized by calculating a date based on the $^{207}\text{Pb}/^{206}\text{Pb}$ ratio. This ratio is insensitive to lead loss, especially when the loss occurred recently and when the lead that was lost had the same isotope composition as the lead that remained, that is, when no isotope fractionation occurred. The relationship between the $^{207}\text{Pb}/^{206}\text{Pb}$ ratio and time results from the difference in the half-lives of their respective parents. The equation expressing this relationship can be derived by combining Equations 12.6 and 12.7:

$$\frac{\frac{^{207}\text{Pb}}{^{204}\text{Pb}} - \left(\frac{^{207}\text{Pb}}{^{204}\text{Pb}}\right)_0}{\frac{^{206}\text{Pb}}{^{204}\text{Pb}} - \left(\frac{^{206}\text{Pb}}{^{204}\text{Pb}}\right)_0} = \frac{^{235}\text{U}}{^{238}\text{U}}\left[\frac{e^{\lambda_2 t} - 1}{e^{\lambda_1 t} - 1}\right] \quad (12.10)$$

This equation has several interesting properties. It involves the ratio $^{235}\text{U}/^{238}\text{U}$ which is a constant equal to $1/137.8$ for all uranium of normal isotopic composition in the Earth, the moon and meteorites at the present time. Therefore a "207–206 date" can be calculated without knowing the concentration of uranium in the mineral. Next, we see that

$$\frac{\frac{^{207}Pb}{^{204}Pb} - \left(\frac{^{207}Pb}{^{204}Pb}\right)_0}{\frac{^{206}Pb}{^{204}Pb} - \left(\frac{^{206}Pb}{^{204}Pb}\right)_0} = \left(\frac{^{207}Pb}{^{206}Pb}\right)^* \qquad (12.11)$$

where $(^{207}Pb/^{206}Pb)^*$ is the ratio of radiogenic ^{207}Pb to radiogenic ^{206}Pb. This ratio can be calculated by subtracting assumed initial $^{207}Pb/^{204}Pb$ and initial $^{206}Pb/^{204}Pb$ ratios from the measured values of these ratios. Thus, we can calculate "207–206 dates" solely on the basis of the isotopic composition of lead in the mineral without having to know its concentration. Finally, we discover that the equation

$$\left(\frac{^{207}Pb}{^{206}Pb}\right)^* = \frac{^{235}U}{^{238}U}\left(\frac{e^{\lambda_2 t} - 1}{e^{\lambda_1 t} - 1}\right) \qquad (12.12)$$

cannot be solved for t by algebraic methods because it is transcendental. One possible remedy of this situation is to make a table of $(^{207}Pb/^{206}Pb)^*$ ratios for selected values of t or to use tables such as those of Stieff et al. (1959) and Stacey and Stern (1973) to interpolate t for any desired value of the $(^{207}Pb/^{206}Pb)^*$ ratio. Alternatively, the equation can be solved with a computer by carrying out a series of successive approximations of t until a value is found that solves the equation for a particular $(^{207}Pb/^{206}Pb)^*$ ratio to within the desired level of precision.

A graph for compatible values of $(^{207}Pb/^{206}Pb)^*$ and t is provided in Figure 12.4 based on data compiled in Table 12.3. A difficulty arises in the solution of Equation 12.12 for $t = 0$. In that case,

$$\left(\frac{^{207}Pb}{^{206}Pb}\right)^* = \frac{^{235}U}{^{238}U}\left(\frac{e^{\lambda_2 t} - 1}{e^{\lambda_1 t} - 1}\right) = \frac{^{235}U}{^{238}U} \times \frac{0}{0} \qquad (12.13)$$

which is indeterminate. We avoid this problem by the application of l'Hopital's rule

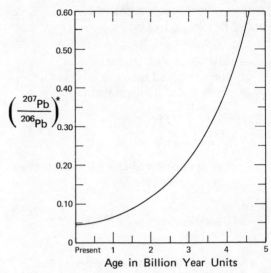

FIGURE 12.4 Plot of Equation 12.12 which relates the ratio of radiogenic $^{207}Pb/^{206}Pb$ of a uranium-bearing system to its age t. The results of the calculations on which this graph is based are compiled in Table 12.3. This curve can be used to determine "207–206 dates" of uranium-bearing minerals based solely on the measured isotopic composition of lead and on an assumed isotopic composition of the lead incorporated into the mineral at the time of its formation.

which states that

$$\lim_{t \to a} \frac{f(t)}{g(t)} = \lim_{t \to a} \frac{f'(t)}{g'(t)}$$

where $f(t)$ and $g(t)$ are functions of the variable t and $f'(t)$ and $g'(t)$ are the first derivatives of these functions with respect to t. Therefore,

$$\lim_{t \to 0} \frac{e^{\lambda_2 t} - 1}{e^{\lambda_1 t} - 1} = \lim_{t \to 0} \frac{\lambda_2 e^{\lambda_2 t}}{\lambda_1 e^{\lambda_1 t}} = \frac{\lambda_2}{\lambda_1} \qquad (12.14)$$

The ratio of radiogenic ^{207}Pb to radiogenic ^{206}Pb produced by decay of ^{235}U and ^{238}U at the present time ($t = 0$) is

$$\left(\frac{^{207}Pb}{^{206}Pb}\right)^* = \frac{^{235}U}{^{238}U}\left(\frac{\lambda_2}{\lambda_1}\right) \qquad (12.15)$$

Table 12.3   **Numerical Values of** $e^{\lambda_1 t} - 1$, $e^{\lambda_2 t} - 1$, **and of the Radiogenic $^{207}Pb/^{206}Pb$ Ratio as a Function of the Age (t)**

$\times 10^9 y$	$e^{\lambda_1 t} - 1$	$e^{\lambda_2 t} - 1$	$\dfrac{^{207}Pb^*}{^{206}Pb}$
0	0.0000	0.0000	0.04607
0.2	0.0315	0.2177	0.05014
0.4	0.0640	0.4828	0.05473
0.6	0.0975	0.8056	0.05994
0.8	0.1321	1.1987	0.06584
1.0	0.1678	1.6774	0.07254
1.2	0.2046	2.2603	0.08017
1.4	0.2426	2.9701	0.08886
1.6	0.2817	3.8344	0.09877
1.8	0.3221	4.8869	0.11010
2.0	0.3638	6.1685	0.12306
2.2	0.4067	7.7292	0.13790
2.4	0.4511	9.6296	0.15492
2.6	0.4968	11.9437	0.17447
2.8	0.5440	14.7617	0.19693
3.0	0.5926	18.1931	0.22279
3.2	0.6428	22.3716	0.25257
3.4	0.6946	27.4597	0.28690
3.6	0.7480	33.6556	0.32653
3.8	0.8030	41.2004	0.37232
4.0	0.8599	50.3878	0.42525
4.2	0.9185	63.5753	0.48651
4.4	0.9789	75.1984	0.55746
4.6	1.0413	91.7873	0.63969

$$\lambda_1(^{238}U) = 1.55125 \times 10^{-10} \, y^{-1};$$
$$\lambda_2(^{235}U) = 9.8485 \times 10^{-10} \, y^{-1}$$
$$\left(\frac{^{207}Pb}{^{206}Pb}\right)^* = \frac{1}{137.8} \frac{e^{\lambda_2 t} - 1}{e^{\lambda_1 t} - 1}$$

$$\left(\frac{^{207}Pb}{^{206}Pb}\right)^* = \frac{1}{137.8} \times \frac{9.8485 \times 10^{-10}}{1.55125 \times 10^{-10}}$$
$$= 0.04607 \qquad (12.15)$$

Equation 12.15 can be obtained also from the consideration that the present-day value of $(^{207}Pb/^{206}Pb)^*$ must be equal to the ratio of the present-day decay rates of the parents. (Remember that the rate of decay of a radioactive isotope is given by λN, where N is the number of atoms present.) The present-day value of the $(^{207}Pb/^{206}Pb)^*$ is a minimum for this ratio, consistent with the low abundance of ^{235}U in modern uranium.

Although uranium and thorium occur in a large number of minerals, only a few are suitable for dating by the U,Th-Pb method. To be useful for dating, a mineral must be retentive with respect to uranium, thorium, lead, and the intermediate daughters, and it should be widely distributed in a variety of rocks. The mineral that most nearly satisfies these conditions is *zircon*. However, several other minerals are also useful: uraninite (pitchblende), monazite, sphene, apatite, and other accessory minerals in igneous rocks listed earlier.

The concentrations of uranium and thorium in zircon range from a few hundred to a few thousand parts per million and average 1330 and 560 ppm, respectively. Zircons in pegmatites may contain more uranium and thorium than those in ordinary igneous rocks. The presence of these elements in zircon can be attributed both to isomorphous replacement of Zr^{+4} (ionic radius = 0.87 Å) by U^{+4} (1.05 Å) and Th^{+4} (1.10 Å) and to the presence of inclusions of thorite. The substitution of Zr^{+4} by U^{+4} and Th^{+4} is limited by differences in their ionic radii. However, equally important to us is the fact that Pb^{+2}, which has an ionic radius of 1.32 Å, is excluded from zircon because of its large radius and lower charge. Therefore, zircon contains very little lead at the time of formation and has very high U/Pb and Th/Pb ratios which enhances its sensitivity as a geochronometer. For this reason, zircon is used most frequently for dating by the U,Th-Pb isotopic method (Silver and Deutsch, 1963; Steiger and Wasserburg, 1969; Doe, 1970; Allegre et al., 1974). The analytical procedure for extracting lead and uranium from zircons has been significantly refined by Krogh (1973) by means of hydrothermal decomposition in a sealed pressure vessel. Age determinations based on apatite and sphene have been reported by Oosthuyzen and Burger (1973), Tilton and Grünenfelder (1968), Tilton and Steiger (1969), Hanson et al. (1971), and others.

Let us examine in detail the model dates calculated by Stern et al. (1971) for a zircon from the Boulder Creek Batholith of Colorado collected near Gross Dam (GP509). They reported the following analytical data for this zircon sample: $U = 792.1$ ppm; $Th = 318.6$ ppm; $Pb = 208.2$ ppm; $^{204}Pb = 0.048$ percent (atom); $^{206}Pb = 80.33$ percent; $^{207}Pb = 9.00$ percent; $^{208}Pb = 10.63$ percent. They assumed that the ordinary lead incorporated into this zircon at the time of crystallization had an isotopic composition given by $204:206:207:208 = 1.00:16.25:15.51:35.73$. This composition corresponds approximately to that of lead in galena formed 1500 million years ago. The atomic weight of this lead is 206.29. We therefore calculate the $^{238}U/^{204}Pb$, $^{235}U/^{204}Pb$, and $^{232}Th/^{204}Pb$ ratios as follows:

$$\frac{^{238}U}{^{204}Pb} = \frac{792.1}{208.2} \times \frac{206.29}{238.03} \times \frac{99.27}{0.048} = 6819.0$$

$$\frac{^{235}U}{^{204}Pb} = \frac{6819.0}{137.8} = 49.48$$

$$\frac{^{232}Th}{^{204}Pb} = \frac{318.6}{208.2} \times \frac{206.29}{232.038} \times \frac{100}{0.048} = 2834.2$$

The values of radiogenic ^{206}Pb, ^{207}Pb, and ^{208}Pb to ^{204}Pb are obtained next:

$$\left(\frac{^{206}Pb}{^{204}Pb}\right)^* = \frac{80.33}{0.048} - 16.25 = 1657.29$$

$$\left(\frac{^{207}Pb}{^{204}Pb}\right)^* = \frac{9.00}{0.048} - 15.51 = 171.99$$

$$\left(\frac{^{208}Pb}{^{204}Pb}\right)^* = \frac{10.63}{0.048} - 35.73 = 185.72$$

We substitute these values into Equations 12.6, 12.7, and 12.8 and calculate the following dates using the new decay constants listed in Table 12.2.

$$t_{206} = \frac{1}{1.55125 \times 10^{-10}} \ln \left[\frac{1657.29}{6819.0} + 1 \right]$$
$$= 1402 \times 10^6 \text{ years}$$

$$t_{207} = \frac{1}{9.8485 \times 10^{-10}} \ln \left[\frac{171.99}{49.48} + 1 \right]$$
$$= 1522 \times 10^6 \text{ years}$$

$$t_{208} = \frac{1}{4.948 \times 10^{-11}} \ln \left[\frac{185.72}{2834.2} + 1 \right]$$
$$= 1283 \times 10^6 \text{ years}$$

The ratio $(^{207}Pb/^{206}Pb)^*$ is found from:

$$\left(\frac{^{207}Pb}{^{206}Pb}\right)^* = \frac{\dfrac{^{207}Pb^*}{^{204}Pb}}{\dfrac{^{206}Pb^*}{^{204}Pb}} = \frac{171.99}{1657.29} = 0.10377$$

This value corresponds to a "207–206 date" of 1688 million years obtained by interpolation in Table 12.3. The model dates calculated above are highly discordant and increase in the order $t_{208} < t_{206} < t_{207} < t_{207-206}$. Such a pattern is typical of discordant zircon dates and must be taken as evidence that one or several of the assumptions of dating are not satisfied. The most likely explanation for the discordance is that the zircon has not been a closed system and that it may have lost radiogenic lead either during an episode of thermal metamorphism or as a result of continuous diffusion. In any case, none of the discordant dates calculated above can be associated with a specific geologic event. If loss of radiogenic lead is the cause of the discordance, the "207–206 date" may approach the age of the zircon most closely, and it is therefore regarded as the most reliable estimate of the age of Boulder Creek Batholith. Although examples of nearly concordant U,Th-Pb dates can be found in the literature (Silver et al., 1963; Ishizaka and Yamaguchi, 1969), in most cases uranium- and thorium-bearing minerals yield discordant dates whose geologic significance is questionable. However, this problem has been overcome by the work of Wetherill (1956, 1963) and Ahrens (1955).

5
The U-Pb Concordia Diagram

The decay of the naturally occurring isotopes of uranium to lead gives rise to two independent geochronometers. When the mineral being dated has remained closed to uranium and all of its daughters and when appropriate corrections are made for the lead incorporated into the mineral at the time of its formation, the two geochronometers give concordant dates.

The decay of ^{238}U to ^{206}Pb as a function of time is described by Equation 12.6, which we rewrite as

$$\frac{^{206}Pb^*}{^{238}U} = e^{\lambda_1 t} - 1 \qquad (12.16)$$

where

$$\frac{^{206}Pb^*}{^{238}U} = \frac{\frac{^{206}Pb}{^{204}Pb} - \left(\frac{^{206}Pb}{^{204}Pb}\right)_0}{\frac{^{238}U}{^{204}Pb}} \qquad (12.17)$$

The decay of ^{235}U to ^{207}Pb is expressed similarly:

$$\frac{^{207}Pb^*}{^{235}U} = e^{\lambda_2 t} - 1 \qquad (12.18)$$

A uranium-bearing mineral that satisfies all of the assumptions of dating yields concordant dates by solution of Equations 12.16 and 12.18. We can also reverse the procedure and use the two equations to calculate compatible sets of $^{206}Pb^*/^{238}U$ and $^{207}Pb^*/^{235}U$ ratios for specified values of t. These are the coordinates of points representing U-Pb systems that have concordant dates in

coordinates of $^{206}Pb^*/^{238}U$ (ordinate) and $^{207}Pb^*/^{235}U$ (abscissa). Equations 12.16 and 12.18 therefore are the parametric equations of a curve that is the locus of all concordant U-Pb systems. Wetherill (1956) called this curve the "concordia." It can be plotted from the data compiled in Table 12.3 which contains values of $e^{\lambda_1 t} - 1$ and $e^{\lambda_2 t} - 1$ for different values of t. Such a concordia curve is shown in Figure 12.5.

Now let us consider the movement of a U-Pb system on the concordia diagram. At the time of crystallization, a uranium-bearing mineral contains no radiogenic lead and therefore such a system plots at the origin. Subsequently, the system will be represented by a point that moves along the concordia curve as long as the system remains closed to uranium and all of its daughters. The age of the system at any time after its formation is indicated by its location on the curve, as shown in Figure 12.5. As long as the system remains closed, it resides on concordia and its U-Pb dates are concordant and indicate the *age* of the system.

We now consider a U-Pb system on concordia of age τ' that experiences an episode of lead loss (or gain of uranium) as a result of metamorphism or chemical weathering. Loss of radiogenic lead changes the coordinates of the point representing the system in such a way that it moves along a chord connecting τ' to the origin. If all of the radiogenic lead that has accumulated in the system is lost, the point representing the system returns to the origin. In this case the U-Pb geochronometers of the system are reset to zero and all memory of the earlier history of the system is lost. If the system loses only a fraction of its radiogenic lead, it will be represented by a point Q somewhere on the chord. All such systems on the chord have discordant dates, and the chord is therefore called *discordia*. The exact

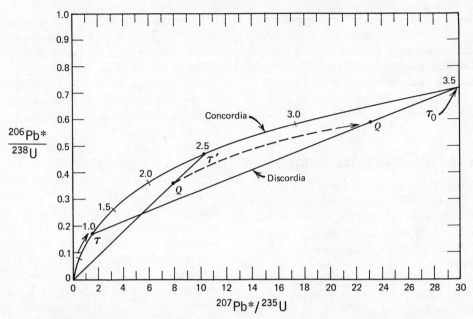

FIGURE 12.5 Concordia diagram illustrating a history of episodic lead loss (or uranium gain). A U-bearing system on concordia lost lead when its age was τ'. As a result, the system moved along a straight line (discordia) directed toward the origin. Point Q represents a system that lost only a fraction of its radiogenic lead. A mineral that had lost all of its radiogenic lead would be represented by a point at the origin. Following the episode of lead loss (or uranium gain), the U-Pb systems reside on discordia that intersects concordia at τ and τ_0, where τ is the time elapsed since closure and τ_0 is the time elapsed since original crystallization of the minerals (or rocks).

position of a U-Pb system on discordia depends on the fraction of radiogenic lead that remains, as we shall see shortly. A system that lost *all* of its radiogenic lead and therefore returned to the origin will once again move along concordia after it becomes closed to uranium and all of its daughters. At the end of an interval of time equal to τ, its position on concordia will be represented by a point whose coordinates yield concordant dates equal to the time elapsed since closure. This date has geologic meaning because it refers to an event in the history of the U-Pb system.

The case we have just considered is an example of extreme behavior that is rarely encountered. Generally, uranium-bearing

minerals such as zircon, sphene, monazite, or apatite lose only a fraction of their radiogenic lead. In fact, zircon crystals in a sample of igneous rock may lose varying fractions of their radiogenic lead even though all of them experienced the same set of conditions. The lead loss seems to be related to the size of the crystals, to their uranium concentrations, and to the radiation damage in the crystals. Smaller grains and those having high uranium concentrations may suffer greater lead losses than larger grains or those containing less uranium. Consequently, it is often possible to recover several zircon fractions from a single sample of rock that plot as a series of points on the discordia line. In practice, this means that we can determine the position of discordia by fitting a straight line to data points representing zircons that have lost varying proportions of their radiogenic lead. By extrapolating discordia, we obtain two points of intersection with concordia. These points are τ_0 and τ where τ_0 is the time elapsed since original crystallization of the minerals and τ is the time elapsed since closure following an episode of lead loss (or uranium gain).

The case we have presented up to this point assumes that lead loss was *episodic* which leads to the interpretation that τ is the time elapsed since closure after loss of lead (or gain of uranium). Actually, as we shall see later, the loss of lead can also be treated as a continuous diffusion process or as an event related to uplift and relaxation of pressure on the zircon crystals. These alternate explanations of the discordance of U-Pb dates lead to different interpretations of the significance of τ. We must therefore qualify our earlier statement with the reminder that τ is the age of an event, such as thermal metamorphism, only in those cases in which its occurrence is confirmed by independent evidence, whereas τ_0 is regarded as the best possible estimate of the age of the uranium-bearing minerals used to define the discordia line. However, the age of zircons is not necessarily identical to the age of the rocks in which they occur, because zircon is a very refractory mineral that may even survive melting of sedimentary rocks to form magma. Zircons in metamorphic rocks commonly contain rounded cores that represent detrital grains derived from older rocks.

The concordia diagram not only permits interpretations of the geologic histories of U-Pb systems but also provides information about the manner in which such systems are disturbed. Consider point Q on the discordia line in Figure 12.6. We define a parameter R

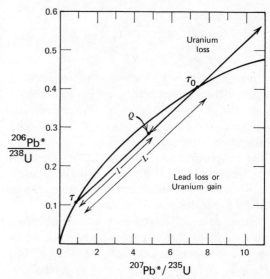

FIGURE 12.6 Concordia diagram illustrating the effects of lead loss and uranium gain or loss on U-Pb systems as discussed in the text.

as

$$R = \frac{l}{L} \qquad (12.19)$$

where l is the distance from τ to Q along discordia and L is the distance τ to τ_0 along discordia.

The parameter R measures the extent to which the daughter to parent ratios of the system represented by Q have been affected either as a result of loss of lead or as a result of gain or loss of uranium. The effect of the addition of lead to the system is not predictable unless the isotopic composition of the new lead is specified. For this reason, gain of lead cannot be treated in this model without additional information. The model includes a further constraint to the effect that lead loss must occur without discrimination against the isotopes of lead on the basis of their masses. With these provisions, the parameter R can be represented as the ratio of daughter to parent immediately after the change divided by the ratio before the change:

$$R = \frac{D_a P_b}{P_a D_b} \qquad (12.20)$$

where D_a, D_b is the amount of radiogenic lead in the system after loss (a) and before loss (b), respectively, and P_a, P_b is the amount of uranium after and before the alteration, respectively.

We can now use Equation 12.20 to predict the response of a U-Pb system on concordia to a variety of situations involving loss of lead or gain/loss of uranium.

Suppose a U-Pb system lost one-half of its radiogenic lead, but experienced no gain or loss of uranium. In this case, $D_a = D_b/2$, but $P_a = P_b$, which results in $R = \frac{1}{2}$, that is, $l = L/2$, and the point Q representing such a system lies halfway between τ and τ_0 on discordia. Now consider another example in which the uranium content is doubled, but the amount of lead remains unchanged. Here we have $D_a = D_b$ and $P_a = 2P_b$, from which it follows that $R = \frac{1}{2}$. Thus the *gain of uranium* by the system has the *same effect* as *loss of lead* and we conclude that any U-Pb system on discordia between τ and τ_0 may have suffered either loss of lead, or gain of uranium, or both. Finally, consider a system that loses one-third of its uranium, but suffers no loss of lead. Here we have $D_a = D_b$ and $P_a = 2P_b/3$, which yields $R = \frac{3}{2}$ and $l = 3L/2$. In other words, loss of uranium causes a U-Pb system to move beyond τ_0 into the area above concordia. Such behavior is sometimes displayed by monazite which tends to plot above concordia on an extension of the discordia line defined by zircons.

We can illustrate the use of the concordia diagram with a study by Catanzaro (1963) of the age of zircons from very old gneisses in southwestern Minnesota. He analyzed three zircon samples collected at Morton and Granite Falls in Minnesota and dated them by the U-Pb method. All three zircons were found to have *discordant* dates ranging from 2640 to 3280 million years ($\lambda_1 = 1.54 \times 10^{-10}$ y^{-1}; $\lambda_2 = 9.72 \times 10^{-10}$ y^{-1}). The 207–206 dates are the oldest in all three cases and the 206–238 dates are the youngest. The pattern of discordance is therefore similar to the case we presented earlier for zircon from the Boulder Creek Batholith of Colorado. Table 12.4 contains the isotopic dates and the corresponding values of the ^{206}Pb*/^{238}U and ^{207}Pb*/^{235}U ratios. Figure 12.7 is a concordia diagram on which we have plotted the three zircons. It is apparent that they

Table 12.4   Discordant U-Pb Dates, ^{206}Pb*/^{238}U, and ^{207}Pb*/^{235}U Ratios of Zircons from the Morton and Montevideo Gneisses of Southwestern Minnesota (Catanzaro, 1963)

| SAMPLE | DATES (10^9 y) | | $\dfrac{207}{206}$ | $\dfrac{^{206}\text{Pb}*}{^{238}\text{U}}$ | $\dfrac{^{207}\text{Pb}*}{^{235}\text{U}}$ |
	206	207			
1	2.64	2.91	3.10	0.5002	15.93
2	2.82	3.07	3.24	0.5434	18.77
3	2.91	3.13	3.28	0.5645	20.00

$\lambda_1(^{238}\text{U}) = 1.54 \times 10^{-10}$ y^{-1}; $\lambda_2(^{235}\text{U}) = 9.72 \times 10^{-10}$ y^{-1}

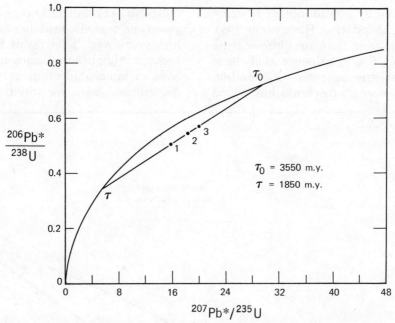

FIGURE 12.7 Concordia diagram for three discordant zircons from gneiss at Morton and Granite Falls in southern Minnesota. The discordia line intersects concordia in two points from whose coordinates the two dates τ_0 and τ can be calculated. The date corresponding to τ_0 is the age of the zircons, while τ may be interpreted in this case as the time elapsed since episodic lead loss. (Data from Catanzaro, 1963, shown also in Table 12.4.)

fit a straight line which intersects concordia at τ and τ_0. Catanzaro (1963) reported a value of 3550 million years for τ_0 and 1850 million years for τ. Evidently these zircons are very old and may have experienced lead loss as a result of metamorphism 1850 million years ago. Age determinations by the K-Ar and Rb-Sr method of biotite and feldspar from Precambrian rocks of this area indicate events at 1800 $\pm$ 100 and 2500 $\pm$ 100 million years. Therefore there is independent evidence in this case for an event about 1800 million years ago that affected the zircons. However, the validity of τ as the age of a metamorphic event is not certain.

Finally, we calculate the fraction of lead remaining in zircon sample 1. From Figure 12.6 we determine that $R = l/L = 0.436$. Assuming that only lead loss occurred, that is, $P_a = P_b$, we find that $D_a = 0.436\ D_b$. In other words, only 43.6 percent of the lead remains, which means that 56.4 percent of the radiogenic lead was lost from this zircon. The other two samples experienced smaller losses.

6
Concordia: Alternative Models

a. CONTINUOUS DIFFUSION

The concordia diagram and the episodic lead-loss model proposed by Wetherill were quickly adopted by geochronologists using the U-Pb method of dating. However, in 1960, Tilton pointed out that uranium-bearing minerals from five continents that have 207–206 dates greater than 2300 million years plot on a single discordia line which indicates a crystallization age of 2800 million years and episodic lead loss around 600 million years ago. This result is remarkable because it implies the occurrence of a worldwide metamorphic event between 500 and 600 million years, for which there is little

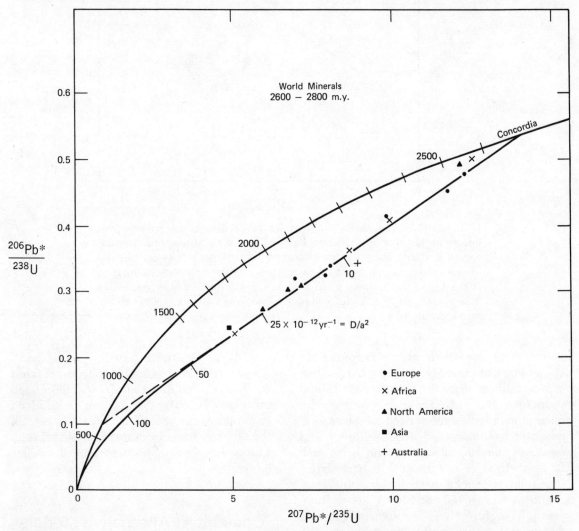

FIGURE 12.8 Concordia plot for minerals from different continents that fit the continuous diffusion model. The position of a U-Pb system on the diffusion trajectory depends on the value of the diffusion parameter D/a^2. A linear extrapolation of the trajectory results in a fictitious value for τ equal to 600 million years. (Reproduced from Figure 1 (p. 2935) of Tilton, G. R., Journal of Geophysical Research, 65, No. 9, 2933–2945, 1960, Copyrighted by American Geophysical Union, by permission of the A.G.U.)

or no evidence. Moreover, the episodic model suggests that negligible lead loss occurred in these minerals from 2800 to 600 million years ago in spite of evidence provided by other dating methods that magmatic activity occurred during this time interval.

Tilton (1960) offered an alternative explanation for the loss of radiogenic lead that is based on continuous diffusion of lead from crystals at a rate governed by a diffusion coefficient D, the effective radius a, and the concentration gradient. He assumed that the crystals are spheres of effective radius a, that uranium is uniformly distributed throughout these spheres, that diffusion of uranium and intermediate daughters is negligible compared to that of lead, that the diffusion coefficient D is a constant independent of time, and that diffusion is governed by Fick's law. With these assumptions, Tilton derived an equation relating the daughter to parent ratio of a mineral to D/a^2 and t, where t is the age of the mineral. The equation is similar to one derived earlier by Wasserburg (1954) for which Nicolaysen (1957) published an extensive table of values.

The solutions of the diffusion equation generate curves on the concordia diagram which are the loci of points representing U-Pb systems of specific ages that have suffered continuous lead loss governed by the parameter D/a^2. Figure 12.8 shows such a curve for U-Pb systems that are 2800 million years old. It is apparent that the curve is very nearly a straight line for systems having $D/a^2 < 50 \times 10^{-12}$ y^{-1}. For values greater than this, the curve deviates from linearity and approaches the origin.

Tilton therefore demonstrated that a linear array of data points representing discordant U-Pb systems on a concordia diagram can be interpreted in two fundamentally different ways. If lead loss was due to an episode of metamorphism, τ obtained by a linear extrapolation indicates the time elapsed since closure after lead loss. On the other hand, if lead loss occurred continuously by diffusion, τ is a fictitious date. The problem is that both episodic loss and continuous loss result in daughter to parent ratios which appear to fit a straight-line discordia. Only U-Pb systems having high D/a^2 values, which therefore lose most of their lead, deviate appreciably from linearity. For this reason, τ should be regarded as a fictitious date (i.e., continuous diffusion), unless independent evidence exists for a metamorphic event at that time (i.e., episodic lead loss).

Tilton's treatment of lead loss by continuous diffusion contained the implicit assumption that the temperature remained constant throughout the history of the zircons and that D/a^2 is therefore invariant with time. The general diffusion case in which the diffusion coefficient is a function of time and both lead and uranium are allowed to diffuse was considered later by Wasserburg (1963) and Wetherill (1963).

b. DILATANCY MODEL

A third interpretation of the discordance of U-Pb dates of uranium-bearing minerals has been proposed by Goldich and Mudrey (1972). They pointed out that such minerals suffer radiation damage as a result of the alpha-decay of uranium, thorium, and their daughters. The extent of radiation damage increases with the age and with the uranium and thorium content of the minerals. The apparent relationship between the radioactivity of zircons and the discordance of U-Pb dates was first demonstrated by Silver and Deutsch (1961) and was used by Wasserburg (1963) who related the diffusion parameter D/a^2 to the radiation damage of zircon crystals. Goldich and Mudrey postulated that radiation damage leads to the formation of microcapillary channels that permit water to enter the crystal. This water is very tightly held until uplift and erosion cause the pressure on the minerals to be released. The resulting dilatance of the zircons allows the water to escape together with dissolved radiogenic lead. Consequently, they proposed that loss of radiogenic lead may be related to uplift and erosion of the crystalline basement complexes of Precambrian shields. Such relatively recent lead loss is consistent with the observation that the 207–206 dates commonly approach the true age of uranium-bearing minerals.

The dilatancy model therefore provides a rational explanation for the observation emphasized by Tilton (1960) that uranium minerals from different continents all seem to have lost radiogenic lead 500 to 600 million years ago, even though no worldwide meta-morphic event is recognized in that period of the Earth's history. According to this model, the value of τ obtained from a linear extrapolation of discordia based on cogenetic suites of uranium-bearing minerals from a particular region may indicate the time of uplift and erosion of the rocks of that region.

c. CHEMICAL WEATHERING

The rocks and minerals used for dating are usually collected from surface outcrops where they have been exposed to chemical weathering. Therefore, we must consider the possibility that discordance of model dates may be the result of disturbances of the daughter to parent ratios caused by chemical weathering. This problem was investigated by Stern et al. (1966) by dating zircons re-

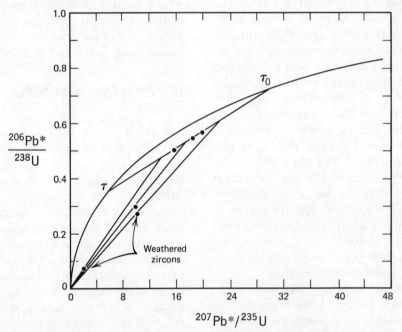

FIGURE 12.9 Concordia diagram showing the effect of lead loss due to chemical weathering of three zircon samples recovered from residual clay derived from the Morton Gneiss near Redwood Falls, Minnesota. (Data from Catanzaro, 1963 and Stern et al., 1966.)

moved from residual clay formed by chemical weathering of the Morton Gneiss near Redwood Falls, Minnesota. Zircons from this rock had been dated previously by Catanzaro (1963) and were discussed elsewhere in this chapter.

The model dates of three weathered zircons from the Morton Gneiss were found to be grossly discordant. However, the 207–206 dates are the oldest in all cases and approach the known age of these zircons. The weathered zircons appear to have lost up to 85 percent of their radiogenic lead, assuming that the uranium content remained unchanged. Figure 12.9 illustrates this interpretation. The weathered zircons lie on straight line chords which start from discordia and lead to the origin.

7
U-Th-Pb Concordias

The concept of U-Pb concordias can be extended to include the decay of ^{232}Th combined with either the decay of ^{238}U or ^{235}U. The systematics of such U-Th-Pb concordias have been discussed by Steiger and Wasserburg (1966) and by Allegre (1967). It turns out that the concordia curve based on ^{208}Pb/^{232}Th and ^{206}Pb/^{238}U ratios is quite straight because of the similarity of the half-lives of ^{232}Th and ^{238}U. Consequently, the intersections of discordias with this concordia have large errors which make this kind of diagram unfavorable for use in geochronometry. However, the concordia based on the decay of ^{232}Th to ^{208}Pb and ^{235}U to ^{207}Pb is shaped very much like the conventional U-Pb concordia when ^{208}Pb/^{232}Th is plotted along the ordinate and ^{207}Pb/^{235}U along the abscissa.

When a U-Th-Pb system experiences loss of parents or daughters or gain of parents, its response in the U-Th-Pb concordias is analogous to that described previously for U-Pb concordias, *provided* that no fractionation of parents or daughters occurs. For example, if the system *gains* uranium and thorium in such a way that its Th/U ratio remains constant, then the point representing that system moves onto a trajectory directed toward the origin. In case of episodic alteration, the lower intersection of discordia with the U-Th-Pb concordia indicates the time of alteration. Similarly, if uranium and thorium are *lost* without fractionation, the point moves into the area *above* concordia and lies on a trajectory passing through the origin. *Loss* of lead, without fractionation, also places the system onto a discordia directed at the origin, as discussed previously. In all such cases, the significance of the date derived from the lower point of intersection is subject to the same ambiguity discussed before regarding the nature of the process, that is, we must distinguish between episodic events and gain or loss by continuous diffusion.

When the alteration of a U-Th-Pb system involves *preferential* gain or loss of parents or disproportionate losses of ^{208}Pb compared to ^{207}Pb or ^{206}Pb, the response is quite different. Because of the differences in the geochemical properties of uranium and thorium, changes in the Th/U ratios are quite likely both during episodic alteration and in the case of continuous gain or loss of parents. Similarly, it is possible that ^{206}Pb and ^{207}Pb may be lost at different rates than ^{208}Pb. As a result of such fractionation, discordant U-Th-Pb systems may lie on discordias that do not pass through the origin. The lower intersections of the resulting discordias with U-Th-Pb concordias have no geological time significance because they reflect the changes that have occurred in the Th/U ratios or lead isotope ratios as a result of preferential gain or loss of parents or loss of lead isotopes.

Both Steiger and Wasserburg (1966) and Allegre (1967) reinterpreted published analysis of zircons by means of U-Th-Pb concordias in coordinates of $^{208}Pb/^{232}Th$ and $^{207}Pb/^{235}U$. Recent examples of the use of this kind of interpretation include studies by Sinha (1972) and Baadsgaard (1973).

8
U-Pb, Th-Pb, and Pb-Pb Isochrons

The decay of uranium and thorium to lead in closed systems is described by equations of the form

$$\frac{^{206}Pb}{^{204}Pb} = \left(\frac{^{206}Pb}{^{204}Pb}\right)_0 + \frac{^{238}U}{^{204}Pb}(e^{\lambda t} - 1) \quad (12.21)$$

where $\left(^{206}Pb/^{204}Pb\right)_0$ is the "initial" ratio incorporated into the system at the time of its formation t years ago. Similar equations describe the decay of ^{235}U to ^{207}Pb and ^{232}Th to ^{208}Pb. Such equations can be used to plot isochrons in coordinates of $^{206}Pb/^{204}Pb$ and $^{238}U/^{204}Pb$ analogous to the Rb-Sr isochrons. In this case, three separate isochrons are obtained whose slopes indicate the age of the suite of samples, provided the samples remained closed systems and had identical initial lead isotopic ratios.

Such U-Pb and Th-Pb isochrons have been used by Sobotovich (1961), Ulrych and Reynolds (1966), Rosholt and Bartel (1969), and by Farquharson and Richards (1970) to date samples of granitic rocks. In general, U-Pb isochrons have not been successful, primarily because rocks exposed to chemical weathering lose a significant fraction of uranium. However, Th-Pb isochrons can be useful because thorium and lead may be retained quantitatively in contrast to uranium. For example, Rosholt and Bartel (1969) obtained a Th-Pb isochron for a suite of whole-rock samples and a feldspar from the Granite Mountains of Wyoming, even

though these rocks have lost up to 93 percent of the uranium they should have contained in order to produce the observed radiogenic ^{206}Pb.

U-Pb systems of cogenetic rocks can also be interpreted in coordinates of $^{207}Pb/^{204}Pb$ and $^{206}Pb/^{204}Pb$ by means of Equation 12.10:

$$\frac{\dfrac{^{207}Pb}{^{204}Pb} - \left(\dfrac{^{207}Pb}{^{204}Pb}\right)_0}{\dfrac{^{206}Pb}{^{204}Pb} - \left(\dfrac{^{206}Pb}{^{204}Pb}\right)_0} = \frac{1}{137.8}\left[\frac{e^{\lambda_2 t} - 1}{e^{\lambda_1 t} - 1}\right]$$

If t is a constant, this equation represents a family of straight lines all of which pass through a point whose coordinates are $(^{207}Pb/^{204}Pb)_0$ and $(^{206}Pb/^{204}Pb)_0$. This condition is satisfied by a suite of cogenetic rocks and minerals whose $^{207}Pb/^{204}Pb$ and $^{206}Pb/^{204}Pb$ ratios must therefore lie along a straight line whose slope is

$$m = \frac{1}{137.8}\left[\frac{e^{\lambda_2 t} - 1}{e^{\lambda_1 t} - 1}\right] \quad (12.22)$$

The goodness of fit of the isotope ratios of lead from any given suite of rocks or minerals is not affected by *recent* losses of uranium or lead, but depends only on the requirement that all samples have the same age, the same initial lead ratios, and that they remained closed until they became exposed to chemical weathering. The theoretical aspects of Pb-Pb isochrons have been presented by Rosholt et al. (1973). Whole-rock Pb-Pb isochron dates have been reported also by Reynolds (1971), Sinha (1972), Moorbath et al. (1973), Oversby (1975), and others. An example of Th-Pb, U-Pb and Pb-Pb isochrons is presented in Figure 12.10 based on the work of Rosholt et al. (1973) on whole-rock samples from the Granite Mountains of Wyoming.

In conclusion, let us consider the interpretation of U-Pb systems of whole-rock samples by means of the concordia diagram.

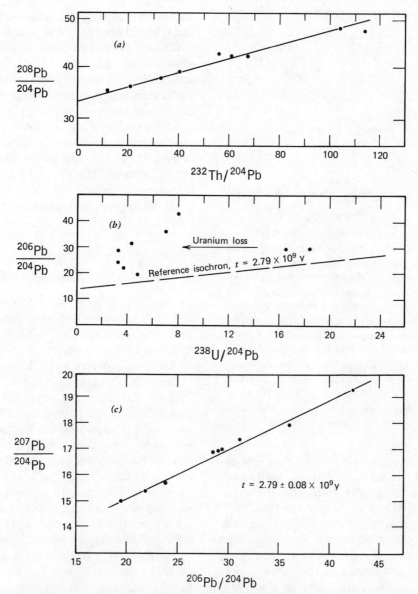

FIGURE 12.10 Th-Pb, U-Pb, and Pb-Pb whole-rock isochrons for granitic rocks from the Granite Mountains of Wyoming. (*a*) These rocks have experienced variable gain or loss of thorium, but on the average the loss of thorium or lead has been small. (*b*) The data points are strongly displaced from the reference isochron drawn for $t = 2.79 \times 10^9$ years, indicating significant losses of uranium ranging from about 37 to 88 percent. (*c*) The isotope ratios of lead define a Pb-Pb isochron with a slope of 0.1911 ± 0.0087 which corresponds to an age of $2.79 \pm 0.08 \times 10^9$y. (The data and interpretations are from Rosholt et al., 1973.)

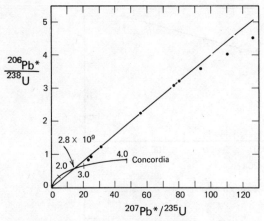

FIGURE 12.11 Interpretation of the U-Pb systems of whole-rock samples from the Granite Mountains of Wyoming. The samples exhibit "reverse" discordance due to variable losses of uranium. The data points form a straight line through the origin consistent with recent loss of uranium due to chemical weathering. The discordia line intersects concordia at 2.8×10^9y in agreement with the Pb-Pb isochron date of these samples (Fig. 12.10). The radiogenic lead/uranium ratios were calculated assuming initial ratios of 13.77 and 14.86 for $^{206}Pb/^{204}Pb$ and $^{207}Pb/^{209}Pb$, respectively. (The data and their interpretation are from Rosholt et al., 1973.)

We showed earlier that all U-Pb systems that remain closed to uranium, lead, and intermediate daughters reside on the concordia curve. When such systems lose lead or gain uranium, they migrate toward the origin along a straight-line discordia. On the other hand, we showed also that the loss of uranium causes them to move along a straight line *away* from the origin. This suggests that rocks that have lost varying amounts of uranium as a result of chemical weathering should lie *above* concordia and should form a straight line that passes through the origin, provided the uranium loss has occurred recently. The age of the rocks is then indicated by the point of intersection on the concordia curve. Such

an interpretation is illustrated in Figure 12.11, based on work of Rosholt et al. (1973).

We conclude that U-Pb and Th-Pb decay schemes of whole-rock samples can be interpreted in several alternate ways and that useful age determinations can be made in spite of the mobility of uranium in rock and mineral systems exposed to chemical weathering. In subsequent chapters we shall extend these interpretive schemes to the dating of lead in galena (Chapter 13) and K-feldspar (Chapter 14), neither of which contains appreciable concentrations of uranium.

9
The U-Xe Method of Dating

Xenon ($Z = 54$) in the atmosphere of the Earth is composed of nine stable isotopes whose abundances are: $^{136}Xe = 8.9$ percent, $^{134}Xe = 10.5$ percent, $^{132}Xe = 27.0$ percent, $^{131}Xe = 21.2$ percent, $^{130}Xe = 3.9$ percent, $^{129}Xe = 26.4$ percent, $^{128}Xe = 1.9$ percent, $^{126}Xe = 0.09$ percent, and $^{124}Xe = 0.10$ percent (Holden and Walker, 1972). Xenon isotopes are also produced by spontaneous fission of ^{238}U with a half-life of 8.03×10^{15} years. Fission xenon consists of ^{136}Xe, ^{134}Xe, ^{132}Xe, ^{131}Xe, and ^{129}Xe. These isotopes may accumulate in U-bearing rocks and minerals and are of potential value for dating by the U-Xe method (Butler et al., 1963; Shukolyukov and Mirkina, 1963; Reynolds, 1963). However, the applicability of this method is limited by the loss of fission xenon even at low temperatures from rocks and most U-bearing minerals.

A significant improvement in the U-Xe method of dating was demonstrated by Teitsma et al. (1975). They irradiated a sample of zircon with thermal neutrons in order to produce xenon by induced fission of ^{235}U and then released the gases by stepwise heating up to a temperature of

1400°C. The xenon released during each step was analyzed on a mass spectrometer and was corrected for atmospheric xenon by means of ^{129}Xe and ^{130}Xe, which are not produced by the fission process in significant amounts. The ratios of any two remaining xenon isotopes can then be related to the number of spontaneous and neutron-induced fission events, designated by S and N, respectively.

The number of spontaneous fission events in the sample is related to the number of ^{238}U atoms:

$$S = \frac{\lambda_f}{\lambda_\alpha} \, ^{238}\text{U}(e^{\lambda_\alpha t} - 1) \qquad (12.23)$$

where λ_f and λ_α are the decay constants of ^{238}U for fission and alpha decay, respectively, and t is the age of the sample. The number of induced fissions in the sample due to irradiation with thermal neutrons is

$$N = \, ^{235}\text{U}\sigma\varphi \qquad (12.24)$$

where $\sigma = 563.7 \times 10^{-24}$ cm^2 is the cross section for neutron fission of ^{235}U and φ is the neutron dose in units of neutrons per square centimeters. Teitsma et al. (1975) measured the neutron dose by means of a 1 percent Co-Al wire which served as the flux monitor. The ratio of spontaneous to neutron induced fissions is

$$\frac{S}{N} = \frac{\lambda_f}{\lambda_\alpha} \times \frac{^{238}\text{U}}{^{235}\text{U}} \times \frac{e^{\lambda_\alpha t} - 1}{\sigma\varphi} \qquad (12.25)$$

The ratio R of any two fission xenon isotopes is related to their absolute fission yields (Y) and to the number of spontaneous (S) and neutron-induced (N) fission events by the following relationship:

$$R = \frac{NY_n^1 + SY_s^1}{NY_s^2 + SY_s^2} \qquad (12.26)$$

where the superscripts identify two fission-xenon isotopes and the subscripts n and s refer to neutron-induced and spontaneous fission, respectively. Therefore the ratio S/N can be calculated from the measured ratio of any two fission xenon isotopes, provided that their fission yields are known. Thus

$$\frac{S}{N} = \frac{Y_n^1 - RY_n^2}{RY_s^2 - Y_s^1} \qquad (12.27)$$

Each fraction of gas released by stepwise heating of the zircon sample yields a value of S/N from which a date can be calculated by solving Equation 12.25:

$$t = \frac{1}{\lambda_\alpha} \ln\left[\left(\frac{\lambda_\alpha}{\lambda_f}\right)\left(\frac{^{235}\text{U}}{^{238}\text{U}}\right)\left(\frac{S}{N}\right)\sigma\varphi + 1\right] \qquad (12.28)$$

This procedure results in a spectrum of dates similar to those generated in the ^{40}Ar/^{39}Ar method of dating. Teitsma et al. (1975) obtained dates of increasing magnitude from gas fractions released at increasing temperatures culminating in a plateau date of 2.82×10^9 years. This date is compatible with the known geologic history of this zircon which originated from the In Ouzzal charnockite in the Sahara. Evidently, the xenon released at the highest temperature originated from sites that had retained most of the xenon produced by spontaneous fission of ^{238}U. This new method of dating requires further testing, but shows considerable promise as a geochronometer of U-bearing minerals. A very similar procedure was proposed by Shukolyukov et al. (1975).

10 SUMMARY

The decay of uranium and thorium to lead permits age determinations of zircon, monazite, sphene, apatite, and certain other minerals containing uranium and thorium.

In many cases the resulting dates are discordant due to loss of lead or uranium. In general, dates based on the radiogenic $^{207}Pb/^{206}Pb$ ratio come closest to the time of crystallization because such dates are not affected by recent losses of lead or uranium.

Discordant U-Pb systems can be interpreted by means of the concordia diagram. The radiogenic $^{206}Pb/^{238}U$ and $^{207}Pb/^{235}U$ ratios of samples whose uranium and lead concentrations have been altered form straight-line discordias that intersect concordia at a point representing the age of the samples. The alteration of U-Pb systems may involve loss of lead as well as gain or loss of uranium and may result from an episode of metamorphism, continuous diffusion of lead, or loss of microcapillary water and chemical weathering at or near the Earth's surface. The concordia concept has been extended to include combinations of Th-Pb with U-Pb decay schemes. The $^{208}Pb/^{232}Th$ versus $^{207}Pb/^{235}U$ concordia is especially useful.

U-Pb and Th-Pb systems can also be interpreted by means of isochron diagrams similar to those used in the Rb-Sr method of dating. In general, U-Pb isochrons for whole-rock samples are not feasible because of extensive loss of uranium. However, Th-Pb isochrons are useful because thorium is not as mobile under near-surface conditions as uranium. Pb-Pb isochrons in coordinates of $^{206}Pb/^{204}Pb$ and $^{207}Pb/^{204}Pb$ are used with increasing frequency because they are not affected by recent losses of either uranium or lead and because only the isotopic ratios of lead are required for dating. Finally, U-Pb systems that have recently lost uranium can also be interpreted by means of concordia diagrams.

The U-Xe method of dating has not been successful in the past because of the loss of xenon at near-surface temperatures. A new method that includes the irradiation of zircon with thermal neutrons followed by the stepwise release of spontaneous fission and neutron-fission xenon may develop into a viable geochronometer.

PROBLEMS

1. Given that the relative abundances of lead isotopes are stated as: $^{204}Pb = 0.143$, $^{206}Pb = 0.100$, $^{207}Pb = 12.95$, $^{208}Pb = 21.96$, recalculate their abundances in terms of atom percent and determine the atomic weight of this sample of lead. The masses of the isotopes in amu are: $^{204}Pb = 203.9730$, $^{206}Pb = 205.9744$, $^{207}Pb = 206.9759$, $^{208}Pb = 207.9766$. (ANSWER: 206.37.) ← wrong

2. The isotopic composition of lead in a zircon is $^{206}Pb/^{204}Pb = 2702.7$, $^{207}Pb/^{204}Pb = 688.92$, and $^{208}Pb/^{204}Pb = 188.92$. The concentration of uranium is 767 ppm, that of lead is 478 ppm. Calculate the radiogenic $^{206}Pb/^{238}U$ and $^{207}Pb/^{235}U$ ratios.* The initial lead isotopic ratios are $(^{206}Pb/^{204}Pb)_0 = 14.2$ and $(^{207}Pb/^{204}Pb)_0 = 15.0$. (ANSWER: $^{206}Pb*/^{238}U = 0.5433$, $^{207}Pb*/^{235}U = 18.7684$.)

3. Calculate three dates from the following data for a zircon: U = 962 ppm, Pb = 548, $^{206}Pb/^{204}Pb = 1960.8$, $^{207}Pb/^{204}Pb = 464.9$, $^{208}Pb/^{204}Pb = 147.4$. The initial lead isotopic data are $(^{206}Pb/^{204}Pb)_0 = 14.2$ and $(^{207}Pb/^{204}Pb)_0 = 15.0$.* Use the new decay constants in Table 12.2. (ANSWER: $t_{206} = 2.61 \times 10^9$ y, $t_{207} = 2.87 \times 10^9$ y, $t_{7/6} = 3.06 \times 10^9$ y.)

4. Nicolaysen et al. (1962) reported the following data for samples of gold-bearing conglomerate from the Witwatersrand district of South Africa:

* Do not neglect to calculate the atomic weight of each lead and the isotopic abundance of ^{204}Pb.

SAMPLE	U %	Pb %	$\frac{^{206}Pb}{^{204}Pb}$	$\frac{^{207}Pb}{^{204}Pb}$	$\frac{^{208}Pb}{^{204}Pb}$
B153	2.46	1.77	571	142	52.6
KCGI	0.201	0.112	249	68.3	62.6
KCGIV	0.520	0.350	326	84.7	69.1

The initial isotope ratios of lead are: $204:206:207:208 = 1:12.4:14.5:32.7$. Calculate the age of these samples by means of a concordia diagram.

5. A suite of whole-rock samples from the Seminoe Mountains of Wyoming (Rosholt et al., 1973) has the following isotope compositions of lead:

$\frac{^{206}Pb}{^{204}Pb}$	$\frac{^{207}Pb}{^{204}Pb}$
30.09	18.03
31.49	18.30
31.86	18.40
32.02	18.14
32.28	18.41
33.02	18.46
34.04	18.88
36.80	19.41

Plot a Pb-Pb isochron and determine the age of these rocks.

REFERENCES

Adams, J. A. S., J. K. Osmond, and J. J. W. Rogers (1959) The geochemistry of thorium and uranium. In Physics and Chemistry of the Earth, vol. 3, pp. 298–348. Pergamon Press, New York, London, Paris, and Los Angeles.

Ahrens, L. H. (1955) Implications of the Rhodesia age pattern. Geochim. Cosmochim. Acta, *8*, 1–15.

Allegre, C. J. (1967) Methode de discussion geochronologique concordia generalisee. Earth Planet. Sci. Letters, *2*, 57–66.

Allegre, C. J., F. Albarede, M. Grünenfelder, and V. Köppel (1974) $^{238}U/^{206}Pb$-$^{235}U/^{207}Pb$-$^{232}Th/^{208}Pb$ zircon geochronology in alpine and non-alpine environment. Cont. Mineral. Petrol. *43*, 163–194.

Baadsgaard, H. (1973) U-Th-Pb dates on zircons from the early Precambrian Amitsoq gneisses, Godthaab district, West Greenland. Earth Planet. Sci. Letters, *19*, 22–28.

Bender, M. L. (1973) Helium-uranium dating of corals. Geochim. Cosmochim. Acta, *37*, 1229–1247.

Brookins, D. G. (1975) Fossil reactor's history probed. Geotimes, *20*, 14–15.

Butler, W. A., P. M. Jeffrey, J. H. Reynolds, and G. J. Wasserburg (1963) Isotopic variations in terrestrial xenon. J. Geophys. Res., *68*, 3283–3291.

Catanzaro, E. J. (1963) Zircon ages in southwestern Minnesota. J. Geophys. Res., *68*, 2045–2048.

Doe, B. R., Lead Isotopes (1970) Springer-Verlag, New York, Heidelberg, Berlin, 137 pp.

Fanale, F. P., and J. L. Kulp (1962) The helium method and the age of the Cornwall Pennsylvania magnetite ore. Econ. Geol., *57*, 735–746.

Fanale, F. P., and O. A. Schaeffer (1965) Helium-uranium ratios for Pleistocene and Tertiary fossil aragonites. Science, *149*, 312–317.

Farquharson, R. B., and J. R. Richards (1970) Whole-rock U-Th-Pb and Rb-Sr ages of the Sybella microgranite and pegmatite, Mount Isa, Queensland. J. Geol. Soc. Aust. *17*, 53–58.

Fleming, G. H., Jr., A. Ghiorso, and B. B. Cunningham (1962) Specific alpha activities and half lives of U234, U235 and U236. Phys. Rev., *88*, 642.

Goldich, S. S., and M. G. Mudrey, Jr. (1972) Dilatancy model for discordant U-Pb zircon ages. In Contributions to Recent Geochemistry and Analytical Chemistry (A. P. Vinogradov volume), A. I. Tugarinov, ed., pp. 415–418. Moscow, Nauka Publ. Office.

Gottfried, D., H. W. Jaffe, and F. E. Senftle (1959) Evaluation of the lead-alpha (Larsen) method for determining ages of igneous rocks. U.S. Geol. Surv. Bull. 1097-A, 63 pp.

Hanson, G. N., E. J. Catanzaro, and D. H. Anderson (1971) U-Pb ages for sphene in a contact metamorphic zone. Earth Planet. Sci. Letters, *12*, 231–237.

Heinrich, E. W. (1958) Mineralogy and geology of radioactive raw materials. McGraw-Hill, New York, 654 pp.

Holden, N. E., and F. W. Walker (1972) Chart of the nuclides. (11th ed.) Educational

Relations, General Electric Co., Schenectady, N.Y.

Hurley, P. M. (1954) The helium age method and the distribution and migration of helium in rocks. In Nuclear Geology, H. Faul, ed., pp. 301–329. John Wiley, New York.

Ishizaka, K., and M. Yamaguchi (1969) U-Th-Pb ages of sphene and zircon from the Hida metamorphic terrain, Japan. Earth Planet. Sci. Letters, 6, 179–185.

Jaffey, A. H., K. F. Flynn, L. E. Glendenin, W. C. Bentley, and A. M. Essling (1971) Precision measurement of the half-lives and specific activities of U^{235} and U^{238} Phys. Rev., C, 4, 1889–1906.

Kovarik, A. F., and N. E. Adams, Jr. (1955) Redetermination of the disintegration constant of ^{238}U. Phys. Rev., 98, 46.

Krogh, T. E. (1973) A low-contamination method for hydrothermal decomposition of zircon and extraction of U and Pb for isotopic age determinations. Geochim. Cosmochim. Acta, 37, 485–494.

Lancelot, J. R., A. Vitrac, and C. J. Allegre (1975) The Oklo natural reactor: Age and evolution studies of U-Pb and Rb-Sr systematics. Earth Planet. Sci. Letters, 25, 189–196.

Moorbath, S., R. K. O'Nions, and R. J. Pankhurst (1973) Early Archaean age for the Isua Iron Formation, West Greenland. Nature, 245, 138–139.

Naudet, R. (1974) Les reacteurs naturels d' Oklo: Bilan des etudes au 1 er mai 1974. Bull. Inform. Scient. Techniques, Comm. a l'Energie Atomique, France, No. 193, 7–45.

Nicolaysen, L. O. (1957) Solid diffusion in radioactive minerals and the measurement of absolute age. Geochim. Cosmochim. Acta, 11, 41–59.

Nicolaysen, L. O., A. J. Burger, and W. R. Liebenberg (1962) Evidence for the extreme age of certain minerals from the Dominion Reef conglomerates and the underlying granite in the western Transvaal. Geochim. Cosmochim. Acta, 26, 15–23.

Oosthuyzen, E. J., and A. J. Burger (1973) The suitability of apatite as an age indicator by the uranium-lead isotope method. Earth Planet. Sci. Letters, 18, 29–36.

Oversby, V. M. (1975) Lead isotopic systematics and ages of Archaean acid intrusives in the Kalgoorlie-Norseman area, Western Australia. Geochim. Cosmochim. Acta, 39, 1107–1125.

Picciotto, E., and S. Wilgain (1956) Confirmation of the period of thorium-232. Nuovo Cimento, 4, 1525.

Reynolds, J. H. (1963) Xenology. J. Geophys. Res., 68, 2939–2956.

Reynolds, P. H. (1971) A U-Th-Pb lead isotope study of rocks and ores from Broken Hill, Australia. Earth Planet. Sci. Letters, 12, 215–223.

Rogers, J. J. W., and J. A. S. Adams (1969a) Thorium. In Handbook of Geochemistry, K. H. Wedepohl, ed., vol. II, No. 1. Springer-Verlag, Berlin, Heidelberg, New York.

Rogers, J. J. W., and J. A. S. Adams (1969b) Uranium. In Handbook of Geochemistry, K. H. Wedepohl, ed., vol. II, No. 1.

Springer-Verlag, Berlin, Heidelberg, New York.

Rosholt, J. N., and A. J. Bartel (1969) Uranium, thorium and lead systematics in Granite Mountains, Wyoming. Earth Planet. Sci. Letters, 7, 141–147.

Rosholt, J. N., R. E. Zartman, and I. T. Nkomo (1973) Lead isotope systematics and uranium depletion in the Granite Mountains, Wyoming. Geol. Soc. Amer. Bull., 84, 989–1002.

Shukolyukov, Y. A., and S. L. Mirkina (1963) Determination of the age of monazites by the Xenon method. Geochemistry, 7, 729–731.

Shukolyukov, Y. A., T. Kirsten, E. K. Jessberger, and G. Sh. Askinadze, 1975, New xenonic neutron-induction method of nuclear geochronology (Xe$_s$–Xe$_n$-method). Geokhimia, 1975, No. 11, 1603–1614.

Silver, L. T., and S. Deutsch (1961) Uranium lead method on zircons. Proc. N.Y. Acad. Sci., 91, 279–283.

Silver, L. T., C. R. McKinney, S. Deutsch, and J. Bolinger (1963) Precambrian age determinations in the western San Gabriel Mountains, California. J. Geol., 71, 196–214.

Silver, L. T., and S. Deutsch (1963) Uranium-lead isotopic variations in zircons: A case study. J. Geol., 71, 721–758.

Sinha, A. K. (1972) U-Th-Pb systematics and the age of the Onverwacht Series, South Africa. Earth Planet. Sci. Letters, 16, 219–227.

Sobotovich, E. V. (1961) Possibility of determining the absolute age of the granites of the Terskey Ala-Tau by the lead included in them. Akad. Nauk SSSR Kom. Opredeleniyu Absolyut. Vozrasta Geol. Formatssii Trudy, sess. 9, 269–280.

Stacey, J. S., and T. W. Stern (1973) Revised tables for the calculation of lead isotope ages. U.S. Dept. Commerce, Natl. Tech. Information Service, Springfield, Virginia, 22151, PB-20919, 35 p.

Steiger, R. H., and G. J. Wasserburg (1966) Systematics in the Pb^{208}-Th^{232}, Pb^{207}-U^{235}, and Pb^{206}-U^{238} systems. J. Geophys. Res., 71, 6065–6090.

Steiger, R. H., and G. J. Wasserburg (1969) Comparative U-Th-Pb systematics in 2.7×10^9 yr plutons of different geological histories. Geochim. Cosmochim. Acta, 33, 1213–1232.

Stern, T. W., S. S. Goldich, and M. F. Newell (1966) Effects of weathering on the U-Pb ages of zircon from the Morton Gneiss, Minnesota. Earth Planet. Sci. Letters, 1, 369–371.

Stern, T. W., G. Phair, and M. F. Newell (1971) Boulder Creek Batholith, Colorado Part II: Isotopic ages of emplacement and morphology of zircon. Geol. Soc. Amer. Bull., 82, 1615–1634.

Stieff, L. R., T. W. Stern, S. Oshiro, and F. E. Senftle (1959) Tables for the calculation of lead isotopic ages. U.S. Geol. Surv. Paper 334-A, 40 p.

Strominger, D., J. M. Hollander, and G. T. Seaborg (1958) Table of isotopes. Rev. Mod. Phys., 30, 585–904.

Teitsma, A., W. B. Clarke, and C. J. Allegre (1975) Spontaneous fission-neutron fission Xenon: A new technique for dating geological events. Science, *189*, 878–879.

Tilton, G. R. (1960) Volume diffusion as a mechanism for discordant lead ages. J. Geophys. Res., *65*, 2933–2945.

Tilton, G. R., and M. H. Grünenfelder (1968) Sphene: Uranium-lead ages. Science, *159*, 1458–1461.

Tilton, G. R., and R. H. Steiger (1969) Mineral ages and isotopic composition of primary lead at Manitouwadge, Ontario. J. Geophys. Res., *74*, 2118–2132.

Turekian, K. K., and K. H. Wedepohl (1961) Distribution of the elements in some major units of the Earth's crust. Geol. Soc. Amer. Bull., *72*, 175–196.

Turekian, K. K., D. P. Kharkar, J. Funkhouser, and O. A. Schaeffer (1970) An evaluation of the uranium-helium method of dating of fossil bones. Earth Planet. Sci. Letters, *7*, 420–424.

Ulrych, T. J., and P. H. Reynolds (1966) Whole-rock and mineral leads from the Llano Uplift, Texas. J. Geophys. Res., *71*, 3089–3094.

Wasserburg, G. J. (1954) Argon[40]: potassium[40] dating. In Nuclear Geology, pp. 341–349, H. Faul, ed. John Wiley, New York.

Wasserburg, G. J. (1963) Diffusion processes in lead-uranium systems. J. Geophys. Res., *68*, 4823–4846.

Wetherill, G. W. (1956) Discordant uranium-lead ages. Trans. Amer. Geophys. Union, *37*, 320–326.

Wetherill, G. W. (1963) Discordant uranium-lead ages—Pt. 2, Discordant ages resulting from diffusion of lead and uranium. J. Geophys. Res., *68*, 2957–2965.

Wetherill, G. W. (1966) Radioactive decay constants and energies. In Handbook of Physical Constants, pp. 514–519, S. E. Clarke, Jr., ed. Geol. Soc. Amer. Mem. 97, 587 p.

13 THE COMMON-LEAD METHOD OF DATING

In the previous chapter we discussed the interpretation of the isotopic composition of lead in uranium-and thorium-bearing minerals and rocks. We now focus attention on "common" lead that occurs in minerals whose U/Pb and Th/Pb ratios are so low that its isotopic composition does not change appreciably with time. The principal common-lead mineral is galena (PbS), but sulfides of other base metals and K-feldspar, in which Pb^{+2} replaces K^{+1}, also contain common lead. Secondary lead minerals such as cerussite ($PbCO_3$) and anglesite ($PbSO_4$) are studied rarely because they are relatively uncommon. However, the isotopic composition of lead in such secondary minerals is usually similar to that of lead in galena with which they are associated.

The isotopic composition of common lead was first determined by Aston (1927). His work established the fact that the atomic weights of the elements reflect the abundances of their naturally occurring isotopes. At that time the atomic weight of lead associated with uranium was already known to be significantly less than that of common lead and the difference was correctly attributed to the presence of radiogenic ^{206}Pb. Common lead, on the other hand, appeared to have a constant atomic weight which suggested that it has a constant isotopic composition.

In 1938, A. O. Nier, working at the University of Minnesota, published the first of a series of papers in which he reported systematic variations of the isotopic compositions of lead in galenas from different sources. These leads had nearly constant atomic weights in spite of significant differences in their isotopic compositions because increases in the $^{206}Pb/^{204}Pb$ ratios were often accompanied by comparable increases of the $^{208}Pb/^{204}Pb$ ratio. The constancy of the atomic weight of common lead is therefore largely fortuitous.

In a subsequent paper Nier et al. (1941) reported isotopic analyses of lead extracted from galenas from different ore deposits. They demonstrated conclusively that such leads have variable isotopic compositions and proposed that these variations result from mixing of radiogenic lead with "primeval" lead prior to the deposition of the galenas. This was a very fruitful suggestion because it opened the way to a quantitative treatment of the isotopic composition of common lead, including a new way to calculate the age of the Earth.

1 The Holmes-Houtermans Model

Nier's work stimulated efforts by other scientists to construct quantitative models for the isotopic evolution of lead in the Earth from which the age of the Earth and the age of common lead minerals can be determined. The first such calculation was made by Gerling (1942) who obtained a value of 3940 million years for the age of the Earth. A few years later, Holmes (1946) and Houtermans (1946) independently formulated a very general model for lead evolution in the Earth that has become known as the Holmes-Houtermans model.

We begin our discussion of this model with a series of statements expressing the assumptions on which it is based.

1. Originally the Earth was fluid and homogeneous.

2. At that time uranium, thorium and lead were uniformly distributed.

3. The isotopic composition of this primeval lead was everywhere the same.

4. Subsequently the Earth became rigid, and small regional differences arose in the U/Pb ratio.

5. In any given region the U/Pb ratio changed only as a result of radioactive decay of uranium to lead.

6. At the time of formation of a common lead mineral, such as galena, the lead was separated from uranium and thorium and its isotopic composition has remained constant since that time.

The Holmes-Houtermans model accounts for the isotopic composition of any given sample of common lead in terms of a single-stage history. It assumes that radiogenic lead is produced by decay of uranium and thorium in the source regions and that the resulting lead (primeval plus radiogenic) is then separated from its parents and incorporated into ore deposits as galena. The isotopic composition of lead in galena does not change because that mineral contains no uranium or thorium.

The $^{206}\text{Pb}/^{204}\text{Pb}$ ratio of a uranium-bearing system of age T that has remained closed to uranium and all of its daughters is

$$\frac{^{206}\text{Pb}}{^{204}\text{Pb}} = \left(\frac{^{206}\text{Pb}}{^{204}\text{Pb}}\right)_0 + \frac{^{238}\text{U}}{^{204}\text{Pb}}(e^{\lambda_1 T} - 1) \quad (13.1)$$

If lead was withdrawn from such a system without isotope fractionation t years ago, the $^{206}\text{Pb}/^{204}\text{Pb}$ ratio of that lead is

$$\left(\frac{^{206}\text{Pb}}{^{204}\text{Pb}}\right)_t = \left(\frac{^{206}\text{Pb}}{^{204}\text{Pb}}\right)_0 + \frac{^{238}\text{U}}{^{204}\text{Pb}}(e^{\lambda_1 T} - 1)$$

$$- \frac{^{238}\text{U}}{^{204}\text{Pb}}(e^{\lambda_1 t} - 1) \quad (13.2)$$

which reduces to

$$\left(\frac{^{206}\text{Pb}}{^{204}\text{Pb}}\right)_t = \left(\frac{^{206}\text{Pb}}{^{204}\text{Pb}}\right)_0 + \frac{^{238}\text{U}}{^{204}\text{Pb}}(e^{\lambda_1 T} - e^{\lambda_1 t}) \quad (13.3)$$

where

$\left(\dfrac{^{206}\text{Pb}}{^{204}\text{Pb}}\right)_t =$ isotope ratio of common lead of age t,

$\left(\dfrac{^{206}\text{Pb}}{^{204}\text{Pb}}\right)_0 =$ isotope ratio of primeval lead in the Earth T years ago,

$\dfrac{^{238}\text{U}}{^{204}\text{Pb}} =$ ratio of these isotopes in a particular source region of common lead in the interior of the Earth at the present time,

$t =$ time elapsed since removal of a common lead sample from its source, and

$T =$ age of the Earth.

Obviously, we can write similar equations for the other two decay schemes. However, before we do, we introduce some new symbols that simplify the writing of these equations. These are:

$$\left(\frac{^{206}\text{Pb}}{^{204}\text{Pb}}\right)_0 = a_0; \qquad \frac{^{238}\text{U}}{^{204}\text{Pb}} = \mu$$

$$\left(\frac{^{207}\text{Pb}}{^{204}\text{Pb}}\right)_0 = b_0; \qquad \frac{^{235}\text{U}}{^{204}\text{Pb}} = \frac{\mu}{137.8}$$

$$\left(\frac{^{208}\text{Pb}}{^{204}\text{Pb}}\right)_0 = c_0; \qquad \frac{^{232}\text{Th}}{^{204}\text{Pb}} = \mu\kappa$$

where $\kappa = {}^{232}\text{Th}/^{238}\text{U}$.

Using these symbols, we can write down the equations for the isotope ratios of common lead according to the Holmes-Houtermans model:

$$\left(\frac{^{206}\text{Pb}}{^{204}\text{Pb}}\right)_t = a_0 + \mu(e^{\lambda_1 T} - e^{\lambda_1 t}) \quad (13.4)$$

$$\left(\frac{^{207}\text{Pb}}{^{204}\text{Pb}}\right)_t = b_0 + \frac{\mu}{137.8}(e^{\lambda_2 T} - e^{\lambda_2 t}) \qquad (13.5)$$

$$\left(\frac{^{208}\text{Pb}}{^{204}\text{Pb}}\right)_t = c_0 + \mu\kappa(e^{\lambda_3 T} - e^{\lambda_3 t}) \qquad (13.6)$$

These equations contain several constants (a_0, b_0, c_0, μ, κ, and T) for which we must provide values before the equations can be used to date samples of common lead. However, we may combine Equations 13.4 and 13.5 and thereby eliminate μ:

$$\frac{\left(\frac{^{207}\text{Pb}}{^{204}\text{Pb}}\right)_t - b_0}{\left(\frac{^{206}\text{Pb}}{^{204}\text{Pb}}\right)_t - a_0} = \frac{1}{137.8}\left[\frac{e^{\lambda_2 T} - e^{\lambda_2 t}}{e^{\lambda_1 T} - e^{\lambda_1 t}}\right] \qquad (13.7)$$

This important equation was first used to estimate the age of the Earth on the basis of the isotopic compositions of lead from galena samples of known age. It was also used to interpret the isotopic composition of lead in meteorites as described in the next section.

2
The Age of Meteorites and the Earth

Meteorites are known to be fragments of larger parent bodies that formed early in the history of the solar system. These parent bodies evolved briefly by partial melting and chemical differentiation and then solidified. During this process, an iron sulfide phase formed, known as troilite (FeS), that contains appreciable concentrations of common lead but is virtually free of uranium and thorium. Therefore, the isotopic composition of lead in troilite has remained very nearly constant since crystallization. It is the least radiogenic lead that is available to us and, for that reason, it is considered to be most nearly representative of the isotopic composition of primeval lead in the Earth.

Use of troilite lead from meteorites as the primeval lead of the Earth implies the assumptions that the lead in the solar nebula was isotopically homogeneous while planetary objects were forming within it, that the Earth and the parent bodies of meteorites formed at the same time, and that the establishment of closed U-Pb and Th-Pb systems in the Earth coincided with the crystallization of troilite in the parent bodies of meteorites. It is unlikely that all of these assumptions are actually satisfied. Nevertheless, the isotopic composition of troilite lead has been used to define primeval terrestrial lead, because any departures from these assumptions are still largely beyond the limits of detection.

The age of meteorites was first established by Patterson (1955, 1956) on the basis of the isotopic composition of lead in three stone and two iron meteorites. For this array of data $t = 0$ and Equation 13.7 therefore reduces to

$$\frac{\frac{^{207}\text{Pb}}{^{204}\text{Pb}} - b_0}{\frac{^{206}\text{Pb}}{^{204}\text{Pb}} - a_0} = \frac{1}{137.8}\left[\frac{e^{\lambda_2 T} - 1}{e^{\lambda_1 T} - 1}\right] \qquad (13.8)$$

This is the equation of a straight line (if T is a constant) in coordinates of $^{206}\text{Pb}/^{204}\text{Pb}$ (x) and $^{207}\text{Pb}/^{204}\text{Pb}$ (y) that passes through a point representing primeval lead whose coordinates are (a_0, b_0). The slope m of this line is

$$m = \frac{1}{137.8}\left[\frac{e^{\lambda_2 T} - 1}{e^{\lambda_1 T} - 1}\right] \qquad (13.9)$$

Patterson used this equation to calculate the age of meteorites from the observed slope of the line, shown in Figure 13.1, formed by the isotope ratios of lead in meteorites. The result was $T = 4.55 \pm 0.07 \times 10^9$ years ($\lambda_1 = 1.537 \times 10^{-10}$ y^{-1}; $\lambda_2 = 9.72 \times 10^{-10}$ y^{-1}). The apparent fit of his data to a straight line confirms the assumptions that (1) all of the meteorites were formed at nearly the same time; (2) they have remained closed to uranium and all of its daughters; and (3) initially they contained lead of very nearly the same isotopic composition.

Patterson's estimate of the age of meteorites has been confirmed recently as follows: Chen and Tilton (1976): $4.565 \pm 0.004 \times 10^9$ y; Tilton (1973): 4.57×10^9 y; Tatsumoto et al. (1973): 4.528×10^9 y for Sioux County and Nuevo Laredo, 4.555×10^9 y for Angra dos Reis; Huey and Kohman (1973): $4.505 \pm 0.008 \times 10^9$ y. All of these authors used the new decay constants measured by Jaffey et al. (1971). These recent studies have disclosed small differences of less than 50 million years in the ages of individual stones, although these differences may also be attributable to differences in their primeval lead ratios.

Patterson (1956) evaluated the assumption that the age of the Earth is similar to that of meteorites. If meteorites and the Earth have a common age, and if they initially contained lead of the same isotopic composition, then terrestrial lead of average isotopic composition must lie on the line formed by the meteorites. Patterson (1956) chose lead from recent oceanic sediment because it is a representative sample of terrestrial lead and showed that it fits the meteorite line within experimental errors (Fig. 13.1).

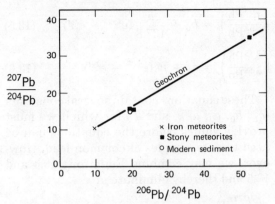

FIGURE 13.1 Lead isochron for meteorites and modern ocean sediment. The slope of this isochron (Equation 13.9) indicates an age of $T = 4.55 \pm 0.07 \times 10^9$ y for meteorites, based on λ_1 (^{238}U) $= 1.537 \times 10^{-10}$ y^{-1} and λ_2 (^{235}U) $= 9.72 \times 10^{-10}$ y^{-1}. (Patterson, 1956.)

This demonstration supports the conclusion that the age of the Earth is essentially the same as that of meteorites and that the isotopic composition of primeval lead of the Earth can be closely approximated by the lead of meteoritic troilite.

The best values for the isotope ratios of primeval lead that are currently available were reported by Tatsumoto et al. (1973) who analyzed lead in troilite from the iron meteorite Canyon Diablo. Their values are in good agreement with those of Tilton (1973) for the chondrite Mezö-Madaras, but differ slightly from those reported earlier by

Table 13.1   **Isotope Ratios of Primeval Lead**

$\left(\dfrac{^{206}\mathrm{Pb}}{^{204}\mathrm{Pb}}\right)_0$	$\left(\dfrac{^{207}\mathrm{Pb}}{^{204}\mathrm{Pb}}\right)_0$	$\left(\dfrac{^{208}\mathrm{Pb}}{^{204}\mathrm{Pb}}\right)_0$	REFERENCE
9.56	10.42	29.71	Murthy and Patterson (1962)
9.307	10.294	29.476	Tatsumoto et al. (1973)
± 0.006	± 0.006	± 0.018	
9.310	10.296	29.57[a]	Tilton, 1973

[a] Estimated assuming ^{232}Th/^{238}U $= 3.80$.

THE COMMON-LEAD METHOD
OF DATING 13

Murthy and Patterson (1962), as shown in Table 13.1.

The study of lead in meteorites has provided estimates of the isotopic composition of primeval lead and of the age of the Earth, which we can now use to continue the development of the common-lead method of dating.

3
Dating of Common Lead

The equations describing the Holmes-Houtermans model (13.4 to 13.7) enable us to determine the ages of common leads that have had single-stage histories. The model assumes that all samples of common lead are mixtures of primeval and radiogenic lead that formed in source regions having differing values of μ and κ. We can use Equations 13.4 and 13.5 to calculate the $^{206}Pb/^{204}Pb$ and $^{207}Pb/^{204}Pb$ ratios of lead removed at different times t from source regions having specified values of μ. Experience has shown that a large number of common leads in ore deposits seem to have evolved in systems having μ values greater than 8, but less than 10. Accordingly, Figure 13.2 shows three lead

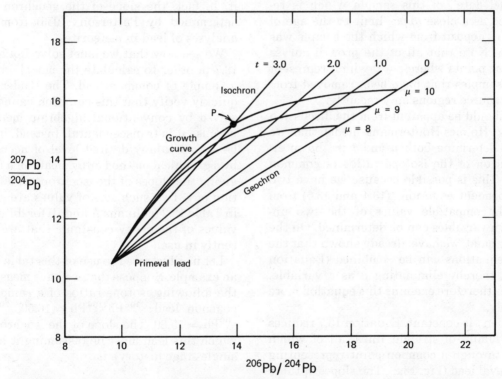

FIGURE 13.2 Graphical representation of the Holmes-Houtermans model. The curved lines are lead growth curves for U-Pb systems having present-day μ values of 8, 9, and 10. The straight lines are isochrons for selected values of t. For example, the coordinates of point P are the $^{207}Pb/^{204}Pb$ and $^{206}Pb/^{204}Pb$ ratios of a galena lead that was withdrawn 3.0×10^9 years ago from a source region whose present $^{238}U/^{204}Pb$ ratio (μ) is 10.0. This diagram was constructed by solving Equations 13.4 and 13.5 assuming that the age of the Earth is $T = 4.55 \times 10^9$ y.

growth curves for $\mu = 8, 9,$ and 10, assuming that the age of the Earth is $T = 4.55 \times 10^9$ y.

We see that the lead growth lines form a fan-shaped array of curved trajectories that spread out from the point representing primeval lead. The position of a point on a particular growth curve indicates the time t when that lead was removed from its source region and was subsequently deposited in the crust as a common lead mineral, such as galena. For example, the coordinates of the point P are the isotope ratios of a common lead sample that evolved in a source region having a value of $\mu = 10$ and that was removed from its source 3.0×10^9 years ago. This is the "model date" of this sample which is regarded as a close *upper* limit to the age of the ore deposit from which the sample was taken. Note that all of the growth curves stop at points where $t = 0$ which represent lead samples that have been removed from their source regions only recently.

It should be apparent from the discussion of the Holmes-Houtermans model that we must determine both μ and t in the interpretation of the isotope ratios of common lead. This is possible because we have two independent equations (13.4 and 13.5) from which compatible values of the two unknown variables can be determined. On the other hand, we have already shown that the two equations can be combined (Equation 13.7), thereby eliminating μ as a variable. Let us therefore examine this equation more closely.

When t is constant, Equation 13.7 reduces to a family of straight lines, all of which pass through a common point representing primeval lead (Fig. 13.2). The slopes of these straight lines are

$$m = \frac{1}{137.8} \left[\frac{e^{\lambda_2 T} - e^{\lambda_2 t}}{e^{\lambda_1 T} - e^{\lambda_1 t}} \right] \quad (13.10)$$

and depend only on t, provided that T is

known. All single-stage leads that were removed from their sources at the *same time t* must lie on these straight lines which are *isochrons*, because all single-stage leads that lie on a particular line have the same age. Common leads that grew in *different* source regions and were removed from them at the *same* time plot at the intersections of their respective growth curves with the isochron corresponding to their age. A series of such isochrons representing different values of t has been drawn in Figure 13.2. The isochron representing leads having $t = 0$ is called the *geochron*, because all modern single-stage leads in the Earth and in meteorites lie on it. In fact, the slope of the geochron was determined by Patterson (1956) from his analyses of lead in meteorites.

We see now that we must solve Equation 13.7 in order to calculate the model date of a sample of common lead. The reader can quickly verify that this equation cannot be solved by conventional algebraic methods because it is transcendental. Instead, it can be solved to any desired level of accuracy by a graphical method or by means of a table giving the slopes of the isochrons as a function of t. Two such sets of values are given in Table 13.2 (columns A and B) for different values of the decay constants that are currently in use.

Let us illustrate the use of this table with an example. Suppose that we have measured the following isotope ratios of a sample of common lead: $^{206}\text{Pb}/^{204}\text{Pb} = 16.50$, $^{207}\text{Pb}/^{204}\text{Pb} = 15.80$. The slope of the isochron on which this lead must lie (assuming it had a single-stage history), is

$$m = \frac{\dfrac{^{207}\text{Pb}}{^{204}\text{Pb}} - b_0}{\dfrac{^{206}\text{Pb}}{^{204}\text{Pb}} - a_0} = \frac{1}{137.8} \left[\frac{e^{\lambda_2 T} - e^{\lambda_2 t}}{e^{\lambda_1 T} - e^{\lambda_1 t}} \right] \quad (13.11)$$

Using the values of Murthy and Patterson

Table 13.2 **Slopes of Isochrons (m) and Corresponding Model Dates (t) of Single-Stage Leads Based on Equation 13.7 (The difference between columns A and B is explained below)**

AGE (t)	SLOPE (m)		AGE (t)	SLOPE (m)	
$\times 10^9$ y	A	B	$\times 10^9$ y	A	B
0	0.5906	0.6180	2.2	0.8909	0.9335
0.2	0.6078	0.6360	2.4	0.9364	0.9816
0.4	0.6264	0.6555	2.6	0.9867	1.0347
0.6	0.6466	0.6766	2.8	1.0424	1.0936
0.8	0.6685	0.6996	3.0	1.1042	1.1590
1.0	0.6924	0.7247	3.2	1.1728	1.2317
1.2	0.7184	0.7520	3.4	1.2492	1.3128
1.4	0.7469	0.7819	3.6	1.3345	1.4034
1.6	0.7780	0.8146	3.8	1.4297	1.5047
1.8	0.8121	0.8505	4.0	1.5364	1.6184
2.0	0.8496	0.8900	4.5	1.8637	1.9681

Column A is based on the "old" decay constants: $\lambda_1(^{238}U) = 1.537 \times 10^{-10}$ y^{-1}, $\lambda_2(^{235}U) = 9.722 \times 10^{-10}$ y^{-1}.

Column B is based on the newer values (Jaffey et al., 1971): $\lambda_1 = 1.55125 \times 10^{-10}$ y^{-1}, $\lambda_2 = 9.8485 \times 10^{-10}$ y^{-1}.

Both sets assume that $T = 4.55 \times 10^9$ y. The calculation of the slope of the isochron from the isotope ratios of a common lead is demonstrated in the text. A table for calculating model dates for single-stage leads has also been published by Doe and Stacey (1974).

(1962) for a_0 and b_0 (Table 13.1), we calculate

$$m = \frac{15.80 - 10.42}{16.50 - 9.56} = 0.7752$$

According to column A (old decay constants), a slope of 0.7752 corresponds to an age t greater than 1.4 but less than 1.6 $\times$ 10^9 years. A linear interpolation yields

$$t = 1.4 \times 10^9 + \frac{0.2 \times 10^9}{0.0311} \times 0.0283$$

$$= 1.581 \times 10^9 \text{ years}$$

If we use the newer values for a_0 and b_0 (Tatsumoto et al., 1973, Table 13.1), we obtain

$m = 0.7654$, from which we calculate a value of $t = 1.518 \times 10^9$ years by means of column A in Table 13.2. Use of the new decay constants (column B) leads to dates of $t = 1.355 \times 10^9$ years and 1.289×10^9 years, respectively. We conclude that the model date of this hypothetical sample of common lead is 1.581×10^9 based on the obsolete constants, but could be as low as 1.289×10^9 according to the most recent values of the decay constants of uranium and the isotope ratios of primeval lead. This demonstration illustrates the need to specify all of the constants used to calculate model lead dates.

After the model date has been calculated (i.e., $t = 1.289 \times 10^9$ y), it can be substituted into Equation 13.4, which is then solved for μ:

$$\mu = \frac{\dfrac{^{206}Pb}{^{204}Pb} - a_0}{e^{\lambda_1 T} - e^{\lambda_1 t}} \qquad (13.12)$$

$$\mu = \frac{16.50 - 9.307}{2.0255 - 1.2213} = 8.94$$

The geological validity of the model date and of the μ value of the source region depend on the assumption that the lead has had a single-stage history. This assumption must be tested before the model date can be used to make statements about the age of the ore deposit from which it was taken. The criteria that can be used for this purpose include (1) the model dates of a representative suite of specimens from a given deposit must be concordant, unless there is evidence of episodic mineralization spanning an interval of time (2) the isotope ratios of lead from a given deposit must be constant within experimental error; (3) the model dates must be positive numbers; (4) the model dates should be in general agreement with isotopic dates of other minerals from the ore and the country rock. However, exact agreement is not to be expected because the model lead date reflects a different event than the isotopic dates of silicate minerals.

Common leads that give meaningful model dates are called "ordinary" and are distinguished from "anomalous" leads whose model dates are meaningless. The number of ore deposits that contain "ordinary" lead and conform to the Holmes-Houtermans model is small, presumably because the lead in most ore deposits has had a more complicated history than is provided for in the Holmes-Houtermans model. Kanasewich (1968), Doe (1972), and Doe and Stacey (1974) have published short lists of the isotopic compositions of lead in ore deposits

that appear to have had single-stage histories.

Anomalous leads were originally subdivided into two categories called J-type and B-type leads (Russell and Farquhar, 1960). J-type leads contain excess radiogenic lead and yield model dates that are less than the age of the ore deposit. They derive their designation from the galena-sphalerite ores at Joplin, Missouri, which contain anomalous leads having negative (i.e., future) model lead dates. The B-type leads are named after ore deposits at Bleiberg, Austria, containing galenas that yield model lead dates that are *older* than the country rock. These leads may actually have single-stage histories but were remobilized later without significant change in their isotopic composition. Therefore, B-type leads are not necessarily anomalous but merely yield geologically anomalous model dates. The distinction between J-type and B-type leads is therefore no longer made. Kanasewich (1968) has proposed an alternate classification of anomalous leads based on several hypothetical, multistage, evolutionary histories.

4
The Th/U Ratio

The isotopic evolution of common lead in the Earth can also be treated with regard to the decay of ^{232}Th to ^{208}Pb. The evolution of the $^{208}Pb/^{204}Pb$ ratio of a single-stage lead is described by Equation 13.6. After the age (t) of a sample of common lead has been determined from the $^{206}Pb/^{204}Pb$ and $^{207}Pb/^{204}Pb$ ratios, the $^{232}Th/^{204}Pb$ ratio of the source region can be calculated from Equation 13.6:

$$\frac{^{232}Th}{^{204}Pb} = \frac{\dfrac{^{208}Pb}{^{204}Pb} - c_0}{e^{\lambda_3 T} - e^{\lambda_3 t}} \qquad (13.13)$$

The $^{232}Th/^{238}U$ ratio (κ) is obtained from

$$\frac{^{232}Th}{^{238}U} = \frac{\dfrac{^{232}Th}{^{204}Pb}}{\dfrac{^{238}U}{^{204}Pb}} \qquad (13.14)$$

The $^{232}Th/^{204}Pb$ ratios calculated in this manner commonly lie between 30 and 40. Since μ generally ranges from 8 to 10, the Th/U ratios of the source regions of ordinary lead vary between limits of 2.90 and 4.84. These values are in satisfactory agreement with the Th/U ratios observed in terrestrial rocks (Table 12.1).

Finally, we shall consider the isotopic evolution of common lead in coordinates of $^{208}Pb/^{204}Pb$ and $^{206}Pb/^{204}Pb$. Such a growth

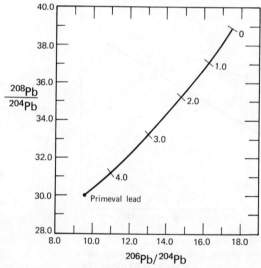

FIGURE 13.3 Isotopic evolution of common lead in coordinates of $^{208}Pb/^{204}Pb$ and $^{206}Pb/^{204}Pb$ according to the Holmes-Houtermans model. The parameters are: $a_0 = 9.56$, $c_0 = 30.0$, $T = 4.55 \times 10^9$ y, $^{232}Th/^{204}Pb = 35.0$, $^{238}U/^{204}Pb$ (μ) = 8.0. The numbers along the growth curve are dates in the past in billions of years when lead was withdrawn from this source.

curve has been plotted in Figure 13.3 by solving Equations 13.4 and 13.6 for the case that $\mu = 8.0$ and $^{232}Th/^{204}Pb = 35.0$ ($\kappa = 4.37$). This diagram could be developed further by plotting several growth curves for different values of κ and by adding isochrons. In addition, a model date can be calculated from the $^{208}Pb/^{204}Pb$ ratio using Equation 13.6 and assuming a $^{232}Th/^{204}Pb$ ratio. Doe and Stacey (1974) recommended a value of 36.50 for this ratio.

5
Ordinary Lead from Conformable Ore Deposits

The existence of anomalous leads is a serious challenge to the validity and usefulness of the Holmes-Houtermans model for dating common leads. Experience has shown that most samples of common lead from ore deposits deviate from the requirements of the model in varying degrees and are to some extent anomalous. Therefore, we need some geological as well as isotopic criteria to distinguish between ordinary and anomalous leads. Some of the relevant characteristics of ordinary leads have already been mentioned.

The problem of identifying ordinary leads that yield meaningful model dates has been at least partly overcome by R. D. Russell and his many collaborators at the University of Toronto and later at the University of British Columbia in Canada. Their contributions have been summarized in very readable form by Russell and Farquhar (1960) and by Kanasewich (1968).

Starting around 1950, the "Toronto group" began dating galenas by means of the isotopic composition of lead using a modified form of the Holmes-Houtermans model. In the course of this work, they were impressed by the observation that most ore leads seemed to have evolved in source regions whose U/Pb and Th/Pb ratios varied only within narrow limits. From this they concluded that such ore leads were probably derived from sources in the upper mantle because only that region of the Earth seemed to be sufficiently homogeneous to account for the apparent uniformity of the observed growth patterns of ore leads of widely varying ages.

The Toronto group was joined in 1958 by R. L. Stanton from Australia. Stanton had made extensive studies of the occurrence of sulfide ore deposits associated with sedimentary and volcanic rocks in old greenstone belts and island arc complexes. These deposits are generally conformable with the structure of the host rock in which they occur, in sharp distinction to the so-called hydrothermal vein deposits that are genetically related to igneous intrusives. Stanton proposed that conformable sulfide deposits in sedimentary/volcanic rocks, or their metamorphic equivalents, formed by syngenetic deposition in sedimentary basins associated with volcanic centers. Consequently, the lead in such conformable deposits could have been derived from sources in the upper mantle without contamination with radiogenic lead derived from the rocks of the continental crust and should satisfy the single-stage evolution model. However, data to be presented in the next chapter indicate that single-stage leads probably did

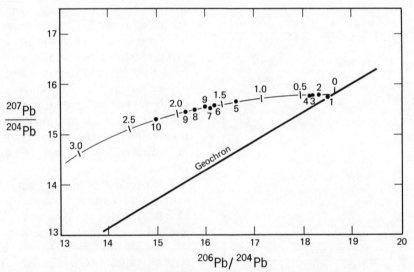

FIGURE 13.4 Isotope ratios of lead from conformable ore deposits that appear to fit a single-stage growth curve having $\mu = 8.99 \pm 0.07$. The constants are: $\lambda_1 = 1.537 \times 10^{-10}\,\text{y}^{-1}$, $\lambda_2 = 9.722 \times 10^{-10}\,\text{y}^{-1}$, $a_0 = 9.56$, $b_0 = 10.42$, $T = 4.55 \times 10^9$ years. The data are from a compilation by Kanasewich (1968) and include samples from the following deposits: (1) Hall's Peak, N.S.W., Australia; (2) Bathurst, N.B., Canada; (3) Cobar, N.S.W., Australia (five mines); (4) Captain's Flat, N.S.W., Australia; (5) Sullivan Mine, B.C., Canada; (6) Mt. Isa, Queensland, Australia (two mines); (7) Broken Hill, N.S.W., Australia (five mines); (8) Finland (five mines including Orijärvi); (9) Errington Mine, Sudbury, Canada (two sets of values); (10) Cobalt, Ontario, Canada.

not evolve in the upper mantle (Richards, 1971; Russell, 1972; Doe and Stacey, 1974).

Nevertheless, Stanton was able to provide a number of geological criteria that could be used to identify ordinary leads. The isotope ratios of these leads were expected to fit a single development line representing lead evolution in the upper mantle. He selected nine ore deposits of varying ages that met his criteria for conformable sulfide deposits. The isotopic ratios of lead from the preselected ore deposits *did* fit closely to a single growth curve (Stanton and Russell, 1959). This was a remarkable convergence of isotope geology and the theory of ore deposits! The Toronto group therefore concluded that all ore leads that do not fit the growth curve for lead from conformable ore deposits are anomalous to some degree and that their model dates are of questionable reliability. Figure 13.4 is a plot of the isotope ratios of lead from conformable ore deposits that appear to fit a single-stage growth curve ($\mu = 8.99 \pm 0.07$), based on a tabulation of data by Kanasewich (1968).

6
The Interpretation of Anomalous Leads

The collaboration between R. L. Stanton and and R. D. Russell resulted in the recognition of a relatively small number of conformable ore deposits containing leads that seem to have had the single-stage histories required by the Holmes-Houtermans model. The isotope ratios of these leads are the coordinates of points whose positions on the growth curve correctly indicate the time of withdrawal of these lead samples from their source region. However, because most common leads are *anomalous* in varying degrees, the interpretation of the isotopic composition of common lead depends largely on our ability to derive useful information from *anomalous* leads.

The most remarkable fact about anomalous leads is that the isotope ratios from a particular ore body or mining district lie close to a straight line in the lead development diagram. Figure 13.5 is a hypothetical

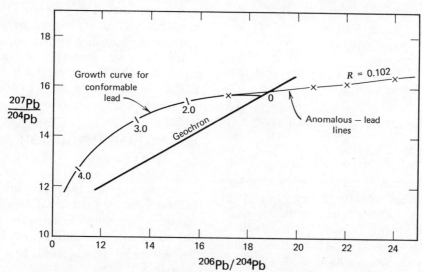

FIGURE 13.5 Hypothetical example of a suite of anomalous leads that form an anomalous-lead line whose slope is $^{207}\text{Pb}/^{206}\text{Pb} = R = 0.102$. The interpretation of these leads is discussed in the text.

example of this phenomenon and will serve as an illustration of the interpretations of anomalous leads that have been proposed, primarily by R. D. Russell and his colleagues. The diagram shows a suite of four anomalous leads that define an "anomalous-lead line" in coordinates of $^{206}Pb/^{204}Pb$ and $^{207}Pb/^{204}Pb$. The slope R of the line in this example is

$$R = \frac{^{207}Pb}{^{206}Pb} = 0.102$$

We see that one of the leads seems to fit the single-stage growth curve ($\mu = 8.99$), while the other three samples lie to the right of the geochron. These three samples yield discordant negative dates that are clearly *not* indicative of the age of this ore deposit. The negative sign presumably means that the dates lie in the future rather than in the past. Let us assume that the isotope ratios of the one sample that does lie on the growth curve are: $^{206}Pb/^{204}Pb = 16.96$ and $^{207}Pb/^{204}Pb = 15.88$. We can calculate a model date for this sample by means of Table 13.2. The slope of its isochron is

$$m = \frac{15.88 - 10.294}{16.96 - 9.307} = 0.7299$$

From Table 13.2 (column B) we find that the model date of this sample is

$$t = 1.0 \times 10^9 + \frac{0.2 \times 10^9}{0.0273} \times 0.0052$$

$$= 1.033 \times 10^9 \text{ y}$$

We may now regard the suite of anomalous leads as the result of the addition of varying amounts of radiogenic lead to ordinary lead that was withdrawn from its source region 1033 million years ago. *Alternatively*, the array of anomalous leads can also be explained in terms of *multistage growth* in two or more source regions having different values of μ. At this point in our discussion we prefer to adopt the mixing model and defer presentation of multistage growth to Chapter 14.

Let us turn therefore to a consideration of the significance of the slope R of the anomalous-lead line. We can represent the slope R as

$$R = \frac{\frac{^{207}Pb}{^{204}Pb} - \left(\frac{^{207}Pb}{^{204}Pb}\right)_0}{\frac{^{206}Pb}{^{204}Pb} - \left(\frac{^{206}Pb}{^{204}Pb}\right)_0}$$

$$= \left(\frac{^{207}Pb}{^{206}Pb}\right)^* \qquad (13.15)$$

where the subscript 0 identifies the coordinates of the point of intersection of the anomalous-lead line with the growth curve for ordinary lead. Thus, we see that R is the $^{207}Pb/^{206}Pb$ ratio of the radiogenic contaminant that was added in varying amounts to the ordinary lead to produce the given array of anomalous leads.

The time at which this radiogenic contaminant was generated by decay of uranium can be estimated in two ways: (1) we can calculate the instant in time in the past when radiogenic lead having the required $^{207}Pb/^{206}Pb$ ratio was formed; and (2) we can assume that the radiogenic lead accumulated during an interval of time which ended when the ore deposit was formed.

a. INSTANTANEOUS-GROWTH MODEL

Let us consider first the "instantaneous-growth" model for formation of the radiogenic contaminant. We have shown (Chapter 12) that the ratio of radiogenic ^{207}Pb to ^{206}Pb being formed at any time t in the past is given by

$$\left(\frac{^{207}Pb}{^{206}Pb}\right)^* = \frac{^{235}U_t\lambda_2}{^{238}U_t\lambda_1} \qquad (13.16)$$

where $^{235}U_t$ and $^{238}U_t$ are the abundances of these isotopes at some time t in the past. We showed that at the present time ($t = 0$)

$$\left(\frac{^{207}Pb}{^{206}Pb}\right)^* = \frac{1}{137.8} \times \frac{9.8485 \times 10^{-10}}{1.55125 \times 10^{-10}}$$

$$= 0.04607$$

This value must be a *minimum* in the history of the solar system because the $^{235}U/^{238}U$ ratio is lower now than it has ever been before. At some time t in the past, the ratio $^{235}U/^{238}U$ was

$$\left(\frac{^{235}U}{^{238}U}\right)_t = \left(\frac{^{235}U}{^{238}U}\right)_p \frac{e^{\lambda_2 t}}{e^{\lambda_1 t}} = \frac{1}{137.8} \frac{e^{\lambda_2 t}}{e^{\lambda_1 t}} \quad (13.17)$$

Therefore, the ratio of radiogenic $^{207}Pb/^{206}Pb$ at time t is

$$R = \left(\frac{^{207}Pb}{^{206}Pb}\right)^*_t = \frac{1}{137.8} \left(\frac{e^{\lambda_2 t}}{e^{\lambda_1 t}}\right)\left(\frac{\lambda_2}{\lambda_1}\right) \quad (13.18)$$

$$R = \frac{e^{t(\lambda_2 - \lambda_1)}}{137.8} \left(\frac{\lambda_2}{\lambda_1}\right) \quad (13.19)$$

Equation 13.19 can be solved for t using the slope R of the anomalous-lead line:

$$t = \frac{1}{\lambda_2 - \lambda_1} \ln\left[\frac{137.8 R \lambda_1}{\lambda_2}\right] \quad (13.20)$$

In our case, $R = 0.102$ and from Equation 13.20 we find that $t = 958$ million years. In other words, the contaminant was formed by decay of uranium 958 million years ago and was subsequently mixed in varying proportions with ordinary lead. If we accept this hypothesis, the age of the ore deposit must be *less* than 958 million years because mixing could have occurred only *after* the radiogenic contaminant was formed.

b. CONTINUOUS-GROWTH MODEL

It is quite unlikely that the radiogenic contaminant was generated during an instant of time. It is far more likely that it originated in the granitic rocks of the continental crust during an interval of time that began with the formation of these granitic rocks and terminated when the radiogenic lead was removed from these rocks and was mixed with other lead just prior to, or in the course of, the formation of the ore deposit. The value of the radiogenic $^{207}Pb/^{206}Pb$ ratio that accumulated during this interval of time is

$$R = \left(\frac{^{207}Pb}{^{206}Pb}\right)^* = \frac{1}{137.8}\left[\frac{e^{\lambda_2 t_r} - e^{\lambda_2 t}}{e^{\lambda_1 t_r} - e^{\lambda_1 t}}\right] \quad (13.21)$$

where t_r is the age of the source rocks in which the radiogenic lead grew and t is the age of the ore deposit, both being measured backward from the present ($t = 0$). Equation 13.21 is analogous to the Holmes-Houtermans equation (13.7) and can be solved for t_r, given the value of R and assuming an appropriate value for t. A limiting condition on possible values of t_r is obtained by setting $t = 0$. In this case Equation 13.21 reduces to

$$R = \left(\frac{^{207}Pb}{^{206}Pb}\right)^* = \frac{1}{137.8}\left[\frac{e^{\lambda_2 t_r} - 1}{e^{\lambda_1 t_r} - 1}\right] \quad (13.22)$$

This equation is identical to Equation 12.12 which we derived in Chapter 12 and used to determine 207–206 dates (Table 12.3, Fig. 12.4). Using that table for the case under consideration ($R = 0.102, t = 0$), we find that $t_r = 1.66 \times 10^9$ years. This date is an *overestimate* of the age of the granitic rocks in which the radiogenic contaminant was generated because, for any desired value of the $(^{207}Pb/^{206}Pb)^*$ ratio, t_r is a maximum when we set $t = 0$. As t *increases*, the value of t_r *decreases* until the two values meet at a date in the past corresponding to the value indicated by the instantaneous-growth model.

In many cases it is possible, on the basis of independent evidence, to estimate the age t of the ore deposit that is being studied.

For example, let us assume that the age of the ore deposit is probably less than 400 million years. We can therefore set $t = 0.4 \times 10^9$ years and then solve Equation 13.22 for t_r. The result of such a calculation indicates that $t_r = 1.42 \times 10^9$ years. The most convenient method of solving Equation 13.22 is to make a table of values of R and t_r for the case $t = 0.4 \times 10^9$ years from which the desired date is obtained by interpolation. Alternatively, Equation 13.22 can be programmed and solved on a computer for a specified value of t.

In conclusion, the foregoing interpretation of the hypothetical suite of anomalous leads suggests that the ordinary component of the lead could have been withdrawn 1033 million years ago from a source region having a value of $\mu = 8.99$. It was then mixed with varying amounts of radiogenic lead having a $^{207}Pb/^{206}Pb$ ratio of $R = 0.102$ before crystallizing as galena. The age of the galena (and of the ore deposit) is therefore 1033 million years or less. Radiogenic lead having the required isotopic composition was generated 958 million years ago (instantaneous-growth model). However, it is more likely that the radiogenic contaminant was formed by decay of uranium and thorium in granitic rocks of the continental crust during an interval of time in the past defined by $t_r - t$, where t_r is the age of these rocks and t is the age of the ore (continuous-growth model). If $t = 0$, we obtain a maximum estimate of the age of the granitic source rocks of 1.66×10^9 years. If we have independent evidence regarding an upper limit to the age of the ore ($t < 0.4 \times 10^9$ y), we obtain a better estimate of the age of the granitic source rocks of 1.42×10^9 years. This treatment of anomalous leads provides information not only about the age of the ore deposit but also about the age of the underlying granitic rocks of the continental crust.

c. THE Th/U RATIO

We now turn briefly to the question of the Th/U ratio of the granitic source rocks of the radiogenic contaminant. By plotting the $^{208}Pb/^{204}Pb$ ratios versus the $^{206}Pb/^{204}Pb$ ratios of a suite of anomalous leads, we can generate a set of points to which we may fit a straight line whose slope is

$$R' = \left(\frac{^{208}Pb}{^{206}Pb}\right)^* \qquad (13.23)$$

where R' can be considered to be the ratio of ^{208}Pb to ^{206}Pb of the radiogenic contaminant.

We then interpret this ratio in terms of the continuous-growth model by means of the following equation:

$$R' = \left(\frac{^{208}Pb}{^{206}Pb}\right)^* = \frac{^{232}Th}{^{238}U}\left[\frac{e^{\lambda_3 t_r} - e^{\lambda_3 t}}{e^{\lambda_1 t_r} - e^{\lambda_1 t}}\right] \qquad (13.24)$$

The values of t_r and t are known from the interpretation of the $^{207}Pb/^{204}Pb$ and $^{206}Pb/^{204}Pb$ ratios of the suite of anomalous leads and R' is determined from the plot. Thus Equation 13.24 can be solved for the ratio $^{232}Th/^{238}U$ from which the Th/U ratio of the granitic source rocks is obtained. We may also solve Equation 13.24 separately for each sample of lead in the suite and thereby derive information about the range of Th/U ratios of the source of the contaminant. In these calculations, a separate value of R' is obtained for each sample by comparing its isotope ratios to the point of intersection of the anomalous-lead line with the growth curve of conformable lead.

7
Lead from the Kootenay Arc of British Columbia, Canada

The Kootenay Arc of southeastern British Columbia is centered about the towns of Trail, Nelson, Kaslo, and Revelstoke and is bordered by the Columbia River which initially flows northwest, turns south at Mount Hooker and flows past Revelstoke and Trail into the state of Washington, in the United States. Within this area exist numerous sulfide deposits, including the very large ore body of the Sullivan Mine at Kimberley, B.C. The deposits were formed by cavity filling and replacement of a wide variety of rocks ranging from limestones of Early Cambrian age to granitic rocks of the Nelson Batholith, whose emplacement began about 170 m.y. ago during the Jurassic period. The Sullivan ore body occurs in the sedimentary Aldridge Formation of Precambrian age, but the origin of the ore is *controversial*. The geologists associated with the Sullivan Mine who have studied the deposit in detail prefer an epigenetic hydrothermal origin (Freeze, 1966). However, the isotopic composition of lead can be interpreted in terms of a single-stage history followed perhaps by subsequent re-

crystallization during deformation and metamorphism of the country rock.

Sinclair (1966) reported isotopic compositions of 21 galena samples originating from 16 deposits of this district, including two samples from the Sullivan Mine. He found that the lead from these deposits is anomalous in varying degrees, even though the isotopic composition of lead in a given deposit is constant. The isotope ratios approach a straight line in coordinates of $^{206}Pb/^{204}Pb$ and $^{207}Pb/^{204}Pb$ as shown in Figure 13.6. The slope of this line is $R = 0.1084 \pm 0.0026$. Note that lead from the Sullivan Mine lies on the growth curve for conformable lead ($\mu = 8.99$, Kanasewich, 1968) which suggests that the anomalous leads of the Kootenay Arc can be interpreted as Sullivan-type leads that have been contaminated with varying amounts of radiogenic lead.

The model date of the Sullivan-type lead can be calculated as before from the slope of

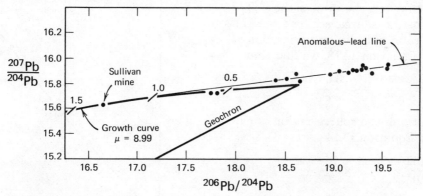

FIGURE 13.6 Anomalous-lead line for galenas from mineral deposits in the Kootenay Arc district of southeastern British Columbia of Canada. The lead from the Sullivan Mine at Kimberley fits a single-stage lead growth curve having a present-day μ value of about 9.0. Its model date is 1.33×10^9 years. The slope of the anomalous-lead line is $R = 0.1084 \pm 0.0026$ and indicates a maximum date of 1.80×10^9 years for the basement rocks in which the radiogenic contaminant lead grew, according to the continuous-growth model. The data are from Sinclair (1966) and the relevant calculations are discussed in the text based on the following constants: $\lambda_1 = 1.537 \times 10^{-10} y^{-1}$, $\lambda_2 = 9.722 \times 10^{-10} y^{-1}$, $\lambda_3 = 4.99 \times 10^{-11} y^{-1}$, $T = 4.55 \times 10^9$, $a_0 = 9.56$, $b_0 = 10.42$, and $c_0 = 30.0$.

its isochron by means of Table 13.2 (column A). The average isotope ratios of the two Sullivan leads are: $^{206}Pb/^{204}Pb = 16.63$ and $^{207}Pb/^{204}Pb = 15.635$. Assuming that $a_0 = 9.56$ and $b_0 = 10.42$, we find that the slope m of the Sullivan isochron is

$$m = \frac{15.635 - 10.42}{16.63 - 9.56} = 0.7376$$

Thus the model date of this lead is

$$t = 1.2 \times 10^9 + \frac{0.2 \times 10^9}{0.0285} \times 0.0192$$

$$= 1.33 \times 10^9 \text{ y}$$

Note that these calculations are based on the old values of the decay constants and on the primeval lead isotope ratios used by Sinclair (1966). The μ value of the source region of Sullivan lead is obtained next from Equation 13.12:

$$\mu = \frac{16.63 - 9.56}{2.0124 - 1.2268} = 9.0$$

We can also calculate the value of the $^{232}Th/^{204}Pb$ ratio of the source region from the $^{208}Pb/^{204}Pb$ ratio of Sullivan lead which is 36.58. Using Equation 13.13, we find that

$$\frac{^{232}Th}{^{204}Pb} = \frac{36.58 - 30.0}{1.2548 - 1.0685} = 35.32$$

The Th/U ratio of the source region is calculated from Equation 13.14:

$$\frac{Th}{U} = \frac{35.32}{9.0} \times \frac{99.28}{100.0} \times \frac{232.038}{238.03} = 3.80$$

Evidently, the Holmes-Houtermans model enables us to obtain a remarkable range of information from the isotope ratios of *ordinary* leads that satisfy the assumptions of the model. In this case we have determined that lead in the ore body of the Sullivan Mine was removed 1.33×10^9 years ago from a source region that is characterized by the following parameter: μ ($^{238}U/^{204}Pb$) = 9.0, $^{232}Th/^{204}Pb = 35.32$, and Th/U = 3.80.

We now turn to the anomalous-lead line whose slope is $R = 0.1084$, according to Sinclair (1966). Using the continuous-growth model (Equation 13.22), we calculate a *maximum* date for the crustal sources of the radiogenic contaminant. Using the old decay constants, we obtain $t_r = 1.80 \times 10^9$ y. If the age of the ore deposits is 170 million years or less, a somewhat lower value of 1.70×10^9 years is obtained for t_r from Equation 13.22. We may conclude, therefore, that the granitic basement underlying the Kootenay Arc region has a probable age of 1.7×10^9 years.

The average Th/U ratio of these basement rocks can be determined from the slope R' of an anomalous-lead line in coordinates of $^{208}Pb/^{204}Pb$ and $^{206}Pb/^{204}Pb$ (Fig. 13.7). We estimate the slope of this line to be $R' = 1.298$ and use this value to calculate the

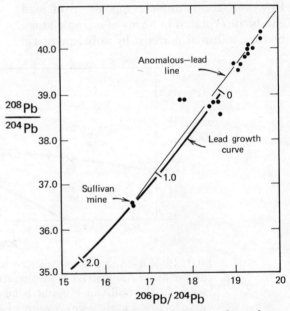

FIGURE 13.7 Anomalous-lead line for galenas from the Kootenay Arc of British Columbia, Canada, in coordinates of $^{208}Pb/^{204}Pb$ and $^{206}Pb/^{204}Pb$. The slope of the line is used to calculate the Th/U ratio of the source regions of the radiogenic contaminant lead as shown in the text. (Sinclair, 1966.)

$^{232}\text{Th}/^{238}\text{U}$ ratio from Equation 13.24 ($t_r = 1.70 \times 10^9$ y, $t = 0.17 \times 10^9$ y):

$$\frac{^{232}\text{Th}}{^{238}\text{U}} = \frac{R'(e^{\lambda_1 t_r} - e^{\lambda_1 t})}{(e^{\lambda_3 t_r} - e^{\lambda_3 t})} \qquad (13.25)$$

$$\frac{^{232}\text{Th}}{^{238}\text{U}} = \frac{1.298(1.2984 - 1.0264)}{(1.0885 - 1.0084)}$$

$$= 4.407$$

The Th/U ratio of the basement rocks is

$$\frac{\text{Th}}{\text{U}} = 4.407 \times \frac{99.2}{100.0} \times \frac{232.038}{238.03} = 4.26$$

The foregoing review of the ore deposits in the Kootenay Arc illustrates the interpretation of suites of anomalous lead. The results not only shed light on the age of the ore deposits but also provide information about the sources of lead and thereby contribute to a better understanding of their origin. The isotopic compositions of lead in many ore deposits around the world have been interpreted using the principles illustrated above. Perhaps most noteworthy are the Pb-Zn deposits in the Mississippi Valley of the central United States which contain anomalous lead first recognized by Nier et al. (1941). The isotopic compositions of these deposits have been interpreted more recently by Heyl et al. (1966), Doe and Delevaux (1972), Richards et al. (1972), and Heyl et al. (1974). Other studies involving the interpretation of anomalous ore leads include Slawson and Austin (1960), Kanasewich and Farquhar (1965), Zartman and Stacey (1971), Rye et al. (1974), Thorpe (1974), Zartman (1974), Kuo and Folinsbee (1974), and Tugarinov et al. (1975).

8 SUMMARY

The isotopic composition of common lead is a mixture of primeval and radiogenic components. According to the Holmes-Houtermans model, common leads evolved by decay of uranium and thorium until the leads were removed from their source regions by geological processes and were deposited as galena in lead-bearing ore deposits. The equations derivable from the model require knowledge of the isotopic composition of primeval lead and of the age of the Earth. Both were determined from a study of lead in stone and iron meteorites that have widely differing U/Pb ratios. The estimates of the age of meteorites and the Earth range from 4.5 to 4.6 $\times 10^9$ years.

The Holmes-Houtermans model can be used to date so-called ordinary common leads that have had single-stage histories. However, only a small number of conformable ore deposits associated with sedimentary and volcanic rocks of island arc complexes satisfy this model. Most common leads appear to be anomalous because they contain excess radiogenic lead. The isotope ratios of suites of anomalous leads from a particular ore body or mining district fit straight lines. The slopes of these anomalous-lead lines can be used to date the granitic basement rocks from which the excess radiogenic lead originated and to determine their Th/U ratios. Under favorable circumstances, the age of the ore bodies can be estimated by means of a model date calculated for the least radiogenic lead in a given suite or from the coordinates of the point of intersection of the anomalous-lead line with the growth curve for ordinary leads in conformable ore deposits. These interpretations of the isotopic composition of common leads can provide much detailed information about their histories and the characteristics of the U, Th-Pb systems through which they have passed. However, the validity of such quantitative interpretations is limited by the provision that the ore leads must satisfy a set of specific assumptions about their evolutionary histories.

PROBLEMS

1. Galena from the ore deposits at Manitouwadge, Ontario, has the following isotopic composition: $^{206}Pb/^{204}Pb = 13.211$, $^{207}Pb/^{204}Pb = 14.401$, $^{208}Pb/^{204}Pb = 33.069$. Calculate the age of this lead and the $^{238}U/^{204}Pb$ (μ), $^{232}Th/^{204}Pb$, $^{232}Th/^{238}U$ (κ), and Th/U ratios of its source region. Use the following constants: λ_1 (^{238}U) = 1.55125×10^{-10} y^{-1}, λ_2 (^{235}U) = 9.8485×10^{-10} y^{-1}, λ_3 (^{232}Th) = 4.948×10^{-11} y^{-1}; primordial lead isotopic ratios: $^{206}Pb/^{204}Pb = 9.307$, $^{207}Pb/^{204}Pb = 10.294$, $^{208}Pb/^{204}Pb = 29.476$, T (age of the Earth) = 4.55×10^9 y. (**ANSWER:** $t = 2.658 \times 10^9$ y, $\mu = 7.97$, $^{232}Th/^{204}Pb = 32.02$, $^{232}Th/^{238}U = 4.02$, Th/U = 4.15.) wrong

2. Lead-bearing fumarolic incrustations of an active andesite volcano on White Island, New Zealand, have the following isotopic composition: $^{206}Pb/^{204}Pb = 18.757$, $^{207}Pb/^{204}Pb = 15.603$, $^{208}Pb/^{204}Pb = 38.644$. Using the constants listed in Problem 1, calculate the age of the Earth. (**ANSWER:** $T = 4.41 \times 10^9$ y.)

3. The average isotopic composition of lead from the Pb-Zn district in southeastern Missouri is $^{206}Pb/^{204}Pb = 20.81$, $^{207}Pb/^{204}Pb = 15.97$, $^{208}Pb/^{204}Pb = 40.08$. Plot a lead-growth curve for $\mu = 9.0$ including the geochron and by means of that diagram determine whether this lead is ordinary or anomalous.

4. Galenas in ore deposits associated with the Tintina Trench in the Selwyn Fold Belt of the Yukon Territory, Canada, have the following isotopic compositions:

DEPOSIT	$\dfrac{^{206}Pb}{^{204}Pb}$	$\dfrac{^{207}Pb}{^{204}Pb}$	$\dfrac{^{208}Pb}{^{204}Pb}$
Faro #1	18.372	15.671	38.416
Fyre Lake	19.360	15.767	39.491
Mt. Hundere	19.450	15.794	39.563
Kathleen Lake	19.701	15.845	40.524
Keno Hill	19.356	15.765	39.651
Mt. Selous	18.521	15.718	38.669
Pike Lake	19.370	15.774	39.511
Nod Claims	19.700	15.861	39.298
Godlin Lake	18.755	15.717	38.769
Metatuff	19.025	15.748	39.044

Make an interpretation of these leads using the slopes of anomalous-lead lines in coordinates of $^{207}Pb/^{204}Pb$, $^{206}Pb/^{204}Pb$ and $^{208}Pb/^{204}Pb$, $^{206}Pb/^{204}Pb$. Estimate the age of the underlying basement assuming that the ore is 500 m.y. old and calculate its Th/U ratio. (See Kuo and Folinsbee, 1974.)

5. Plot a Pb-growth curve for $\mu = 8.79$ using the constants listed in Problem 1 above and superimpose on it the anomalous-lead line for ore samples from the Selwyn Fold Belt (Problem 4). Estimate the time of original separation of the leads from a source having a μ value of 8.79 (See Kuo and Folinsbee, 1974.)

REFERENCES

Aston, F. W. (1927) The constitution of ordinary lead. Nature, *120*, 224.

Chen, J. H., and G. R. Tilton (1976) Isotopic lead investigations on the Allende carbonaceous chondrite. Geochimica Cosmochim. Acta, *40*, 635–643.

Doe, B. R. (1970) Lead Isotopes. Springer-Verlag, Berlin, Heidelberg, and New York, 137 p.

Doe, B. R., and M. H. Delevaux (1972) Source of lead in southeast Missouri galena ores. Econ. Geol., *67*, 409–435.

Doe, B. R., and J. S. Stacey (1974) The application of lead isotopes to the problems of ore genesis and ore prospect evaluation: A review. Econ. Geol., *69*, 757–776.

Fleming, G. H. Jr., A. Ghiorso, and B. B. Cunningham (1962) Specific alpha activities and half-lives of U234, U235, and U236. Phys. Rev., *88*, 642.

Freeze, A. C. (1966) On the origin of the Sullivan ore body, Kimberley, B.C.. In A Symposium on the Tectonic History and Mineral Deposits of the Western Cordillera. Vancouver, B.C., 1964, Can. Inst. Min. and Metall., Spec. vol. 8, 263–294.

Gerling, E. K. (1942) Age of the Earth according to radioactivity data. Compt. Rend. (Doklady) de l'Acad. Sci, de l'URSS, *34*, No. 9, 259–261.

Holmes, A. (1946) An estimate of the age of the earth. Nature, *157*, 680–684.

Heyl, A. V., M. H. Delevaux, R. E. Zartman, and M. R. Brock (1966) Isotopic study of galenas from the Upper Mississippi Valley, the Illinois-Kentucky, and some Appalachian Valley mineral districts. Econ. Geol., *61*, 933–961.

Heyl, A. V., G. P. Landis, and R. E. Zartman (1974) Isotopic evidence for the origin of Mississippi-Valley type mineral deposits: A review. Econ. Geol., *69*, 992–1006.

Houtermans, F. G. (1946) Die Isotopenhäufigkeiten im natürlichen Blei und das Alter des Urans. Naturwissenschaften, *33*, 185–186, 219.

Huey, J. M., and T. P. Kohman (1973) ^{207}Pb-^{206}Pb isochron and age of chondrites. J. Geophys. Res., *78*, 3227–3244.

Jaffey, A. H., K. F. Flynn, L. E. Glendenin, W. C. Bentley, and A. M. Essling (1971) Precision measurement of half-lives and specific activities of U^{235} and U^{238}. Phys. Rev., *C4*, 1889–1906.

Kanasewich, E. R. (1968) The interpretation of lead isotopes and their geological significance. In Radiometric Dating for Geologists, pp. 147–223, E. I. Hamilton and R. M. Farquhar, eds. Interscience, New York, 506 pp.

Kanasewich, E. R., and R. M. Farquhar (1965) Lead isotope ratios from the Cobalt-Noranda area, Canada. Canadian J. Earth Sci., *2*, 361–384.

Kovarik, A. F., and N. E. Adams, Jr. (1955) Redetermination of the disintegration constant of ^{238}U. Phys. Rev., *98*, 46.

Kuo, S-L., and R. E. Folinsbee (1974) Lead isotope geology of mineral deposits spatially related to the Tintina Trench, Yukon Territory. Econ. Geol., *69*, 806–813.

LeRoux, L. J., and L. E. Glendenin (1963) Half-life of thorium-232. Nat. Conf. Nucl. Energy Appl. Isotop. Radiat. Proc. 1963, 77–88.

Murthy, V. R., and C. C. Patterson (1962) Primary isochron of zero age for meteorites and the earth. J. Geophys. Res., 67, 1161.

Nier, A. O. (1938) Variations in the relative abundances of the isotopes of common lead from various sources. J. Amer. Chem. Soc., 60, 1571–1576.

Nier, A. O., R. W. Thompson, and B. F. Murphey (1941) The isotopic constitution of lead and the measurement of geologic time. III. Phys. Rev., 60, 112–116.

Patterson, C. C. (1955) The Pb^{207}/Pb^{206} ages of some stone meteorites. Geochim. Cosmochim. Acta, 7, 151–153.

Patterson, C. C. (1956) Age of meteorites and the earth. Geochim. Cosmochim. Acta, 10, 230–237.

Picciotto, E., and S. Wilgain (1956) Confirmation of the period of thorium-232. Nuovo Cimento, 4, 1525.

Richards, J. R. (1971) Major lead ore-bodies—Mantle origin? Econ. Geol., 66 425–434.

Richards, J. R., A. K. Yonk, and C. W. Keighn (1972) Upper Mississippi Valley lead isotopes re-examined. Mineralium Deposita, 7, 285–291.

Russell, R. D., and R. M. Farquhar (1960) Lead Isotopes in Geology. Interscience, New York, 243 pp.

Russell, R. D. (1972) Evolutionary model for lead isotopes in conformable ores and in ocean volcanics. Rev. Geophys. Space Phys., 10, 529–549.

Rye, D. M., B. R. Doe, and M. H. Delevaux (1974) Homestake mine, South Dakota: II Lead isotopes, mineralization ages, and source of lead in ores of the northern Black Hills. Econ. Geol., 69, 814–822.

Sinclair, A. J. (1966) Anomalous lead from the Kootenay Arc, British Columbia. In A Symposium on the Tectonic History and Mineral Deposits of the Western Cordillera. Vancouver, B.C., 1964. Can. Inst. Min. and Metall. Spec. Vol. 8, 249–262.

Slawson, W. F., and C. F. Austin (1960) Anomalous leads from a selected geological environment in west-central New Mexico. Nature, 187, 400–401.

Stanton, R. L., and R. D. Russell (1959) Anomalous leads and the emplacement of lead sulfide ores. Econ. Geol., 54, 588–607.

Tatsumoto, M., R. J. Knight, and C. J. Allegre (1973) Time differences in the formation of meteorites as determined from the ratio of lead-207 to lead-206. Science, 180, 1279–1283.

Thorpe, R. (1974) Lead isotope evidence on the genesis of the silver-arsenide vein deposits of the Cobalt and Great Bear Lake areas, Canada. Econ. Geol., 69, 777–791.

Tilton, G. R. (1973) Isotopic lead ages of chondritic meteorites. Earth Planet. Sci. Letters, 19, 321–329.

Tugarinov, A. I., E. V. Bibikova, T. V. Gracheva, Z. M. Motorina, and V. A. Makarov (1975) Use of the lead-isotopic method of investigation for solving the genesis of lead deposits of the North-

Caucasus ore province. Geokhimia, *1975*, No. 8, 1156–1163.

Zartman, R. E. (1974) Lead isotopic provinces in the Cordillera of the western United States and their geologic significance. Econ. Geol., *69*, 792–805.

Zartman, R. E., and J. S. Stacey (1971) Lead isotopes and mineralization ages in Belt supergroups rocks, northwestern Montana and northern Idaho. Econ. Geol., *66*, 849–860.

14 THE INTERPRETATION OF MULTISTAGE LEADS

Igneous and metamorphic rocks contain lead whose isotopic compositions reflect multistage histories. This means that the lead may have been associated with several systems having different U/Pb and Th/Pb ratios and it may have resided in these systems for varying lengths of time. The interpretation of the isotopic composition of such multistage leads requires considerable virtuosity, but is based on the interpretive schemes presented in the preceding chapters.

1
Two- and Three-Stage Growth Histories

In the preceding chapter, we showed that the isotope ratios of lead that evolved in a U-bearing source region of age T and that was removed from that system t years ago are given by equations of the form

$$\frac{^{206}\text{Pb}}{^{204}\text{Pb}} = a_0 + \mu(e^{\lambda_1 T} - e^{\lambda_1 t}) \quad (14.1)$$

where a_0 is the primeval $^{206}\text{Pb}/^{204}\text{Pb}$ ratio of the Earth and μ is the $^{238}\text{U}/^{204}\text{Pb}$ ratio in the source region at the present time.

Now let us consider that a sample of lead passes through *two* U-bearing systems characterized by different $^{238}\text{U}/^{204}\text{Pb}$ ratios (μ_1 and μ_2). The $^{206}\text{Pb}/^{204}\text{Pb}$ ratio of that sample of lead will be given by

$$\frac{^{206}\text{Pb}}{^{204}\text{Pb}} = a_0 + \mu_1(e^{\lambda_1 T} - e^{\lambda_1 t_1})$$
$$+ \mu_2(e^{\lambda_1 t_1} - e^{\lambda_1 t_2}) \quad (14.2)$$

where T is the age of the Earth as before, t_1 is the time in the past when the lead was removed from system 1 and was transferred to system 2, and t_2 is the time in the past when the lead was removed from system 2 and when its isotopic composition stopped changing. In other words, our lead sample resides in system 1 for an interval of time equal to $T - t_1$ and in system 2 for a period of time equal to $t_1 - t_2$.

The transfer of the lead from one system to another may involve a geologic process that physically moves the lead from one place to another. For example, lead may be transported by the formation of magma in the mantle and its subsequent intrusion into the crust or extrusion on the surface of the Earth. In this way a sample of lead that originally resided in the upper mantle may become associated with a new system having a different $^{238}\text{U}/^{204}\text{Pb}$ ratio than its former environment. At a later time (t_2) the lead may be withdrawn from the second system and may be recrystallized as galena in an ore deposit.

However, the transition from one system to the next may also occur by changes of the U/Pb ratio of system 1 without requiring movement of the lead. For example, the system may lose or gain uranium or it may *lose* a fraction of its lead. Such events will change the $^{238}\text{U}/^{204}\text{Pb}$ ratio and thus will affect the isotope ratios of the lead that is finally removed at t_2. The *addition* of lead is, of course, also possible. However, in that case the isotopic composition of the lead may be changed because the added lead is likely to have a different isotopic composition.

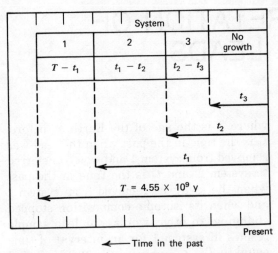

System

| 1 | 2 | 3 | No growth |

| $T - t_1$ | $t_1 - t_2$ | $t_2 - t_3$ | |

t_3

t_2

t_1

$T = 4.55 \times 10^9$ y

Present

←——Time in the past

FIGURE 14.1 Time sequence of events in the evolutionary history of a three-stage lead. System 1 starts T years ago and exists for a period of time equal to $T - t_1$. At t_1 the lead is transferred to system 2 and continues to evolve during interval $t_1 - t_2$. At t_2 the lead is transferred to system 3 and resides there for $t_2 - t_3$ years. At t_3 the lead is withdrawn and placed in an environment containing no uranium or thorium. Between t_3 and the present, no further isotopic evolution occurs.

We can easily extrapolate the two-stage history described above to include three or more stages shown in Figure 14.1. We can also write equations for the other isotope ratios of two- or three-stage leads analogous to Equation 14.2. For example, the $^{207}\text{Pb}/^{204}\text{Pb}$ ratio of a three-stage lead would be

$$\frac{^{207}\text{Pb}}{^{204}\text{Pb}} = b_0 + \frac{\mu_1}{137.8}(e^{\lambda_2 T} - e^{\lambda_2 t_1})$$

$$+ \frac{\mu_2}{137.8}(e^{\lambda_2 t_1} - e^{\lambda_2 t_2})$$

$$+ \frac{\mu_3}{137.8}(e^{\lambda_2 t_2} - e^{\lambda_2 t_3}) \quad (14.3)$$

Clearly, such equations cannot be solved directly because we have no knowledge of

the values of μ_1, μ_2, or μ_3, nor do we have information regarding the residence times $T - t_1$, $t_1 - t_2$, and $t_2 - t_3$. Nevertheless, the growth of lead with such multistage histories generates certain patterns on lead evolution diagrams that can be recognized and interpreted quantitatively.

Let us return to the two-stage lead (Equation 14.2). During the interval $T - t_1$ the lead resided in a source region whose $^{238}\text{U}/^{204}\text{Pb}$ ratio is μ_1 which we take to be equal to about 8.9. In other words, we assume that the first stage of the history of this lead consisted of residence in the source regions of ordinary ore lead. At time t_1 the $^{206}\text{Pb}/^{204}\text{Pb}$ ratio of that lead was

$$\left(\frac{^{206}\text{Pb}}{^{204}\text{Pb}}\right)_{t_1} = a_0 + \mu_1(e^{\lambda_1 T} - e^{\lambda_1 t_1})$$

$$= N_{206}^1 \quad (14.4)$$

where N_{206}^1 is the value of the $^{206}\text{Pb}/^{204}\text{Pb}$ ratio at t_1 that was incorporated initially into system 2. After an interval of time equal to $t_1 - t_2$ the $^{206}\text{Pb}/^{204}\text{Pb}$ ratio of the lead will be

$$\left(\frac{^{206}\text{Pb}}{^{204}\text{Pb}}\right)_{t_2} = N_{206}^1 + \mu_2(e^{\lambda_1 t_1} - e^{\lambda_1 t_2}) \quad (14.5)$$

Similarly, the $^{207}\text{Pb}/^{204}\text{Pb}$ ratio at t_2 will be

$$\left(\frac{^{207}\text{Pb}}{^{204}\text{Pb}}\right)_{t_2} = N_{207}^1 + \frac{\mu_2}{137.8}(e^{\lambda_2 t_1} - e^{\lambda_2 t_2}) \quad (14.6)$$

where N_{207}^1 is the $^{207}\text{Pb}/^{206}\text{Pb}$ ratio of the lead initially incorporated into system 2. Now we combine Equations 14.5 and 14.6 as we did before:

$$\frac{\left(\frac{^{207}\text{Pb}}{^{204}\text{Pb}}\right)_{t_2} - N_{207}^1}{\left(\frac{^{206}\text{Pb}}{^{204}\text{Pb}}\right)_{t_2} - N_{206}^1} = \frac{1}{137.8}\left[\frac{e^{\lambda_2 t_1} - e^{\lambda_2 t_2}}{e^{\lambda_1 t_1} - e^{\lambda_1 t_2}}\right] \quad (14.7)$$

This is the equation of a family of straight

THE INTERPRETATION OF MULTISTAGE LEADS 14

lines passing through a point $P(N_{206}^1, N_{207}^1)$ whose slope m is given by

$$m = \frac{1}{137.8}\left[\frac{e^{\lambda_2 t_1} - e^{\lambda_2 t_2}}{e^{\lambda_1 t_1} - e^{\lambda_1 t_2}}\right] \quad (14.8)$$

In other words, a two-stage lead will lie on a *secondary isochron* that passes through point P on the growth curve of system 1 and whose slope depends only on the interval of time $t_1 - t_2$ during which the lead resided in system 2. Evidently, if we have a suite of leads all of which were removed at the same time t_1 from the same system ($\mu_1 = 8.9$), but which resided in a variety of secondary systems having different values of μ_2 and were removed from these systems at t_2, then they will plot along a straight line (the secondary isochron) in coordinates of $^{206}Pb/^{204}Pb$ and $^{207}Pb/^{204}Pb$. Such leads are, of course, anomalous in the sense of the Holmes-Houtermans model and yield discordant and meaningless model dates. However, they can be interpreted in a meaningful way by means of the slope of the secondary isochron and by means of its point of intersection with the primary growth curve. Note that the secondary isochron is equivalent to the anomalous-lead line discussed in Chapter 13. In fact, Equation 14.8 is identical to Equation 13.21 for the continuous-growth model. We can calculate a maximum value for t_1 from Equation 14.8 using the measured slope of the secondary isochron by setting $t_2 = 0$. If we can estimate t_2 on the basis of independent information, we obtain a lower, but somewhat more meaningful, value for t_1. The date is the best possible estimate of the age of system 2.

Now let us suppose that lead is transferred from system 2 into a third system having a $^{238}U/^{204}Pb$ ratio equal to μ_3 and that it resides in that system for a period of time equal to $t_2 - t_3$. Such a three-stage history is outlined in Figure 14.2. The $^{206}Pb/^{204}Pb$ ratio of a three-stage lead that was removed from

its source t_3 years ago is

$$\left(\frac{^{206}Pb}{^{204}Pb}\right)_{t_3} = N_{206}^{1+2} + \mu_3(e^{\lambda_1 t_2} - e^{\lambda_1 t_3}) \quad (14.9)$$

where N_{206}^{1+2} is given by Equation 14.5. Thus, N_{206}^{1+2} is the $^{206}Pb/^{204}Pb$ ratio of lead initially incorporated into system 3. The superscript merely reminds us that the lead has already passed through systems 1 and 2. We now repeat the procedure of combining equations for $(^{207}Pb/^{204}Pb)_{t_3}$ and $(^{206}Pb/^{204}Pb)_{t_3}$ and obtain

$$\frac{\left(\dfrac{^{207}Pb}{^{204}Pb}\right)_{t_3} - N_{207}^{1+2}}{\left(\dfrac{^{206}Pb}{^{204}Pb}\right)_{t_3} - N_{206}^{1+2}} = \frac{1}{137.8}\left[\frac{e^{\lambda_2 t_2} - e^{\lambda_2 t_3}}{e^{\lambda_1 t_2} - e^{\lambda_1 t_3}}\right]$$

$$(14.10)$$

This equation tells us that all leads that have passed through two previous systems and then enter a third will lie on a straight line (tertiary isochron) on the lead development diagram. The slope of that isochron depends only on the interval of time $t_2 - t_3$ during which the leads reside in system 3. The distribution of points on the tertiary isochron depends on the range of μ_3 values of the third system, just as their distribution on the secondary isochron depends on the range of μ_2 values in the second source. The goodness of fit of lead isotope ratios to a tertiary isochron depends on the *homogenization* of lead generated in the previous environment because only leads having the same values of N_{206}^{1+2} and N_{207}^{1+2} will lie on the isochron. The evolution of lead through a three-stage history similar to that just described is illustrated in Figure 14.2 and can be studied there by referring to the caption. The interpretation of multistage leads has been discussed by Doe (1970), Russell (1972), Gale and Mussett (1973), Doe and Stacey (1974), and Stacey and Kramers (1975).

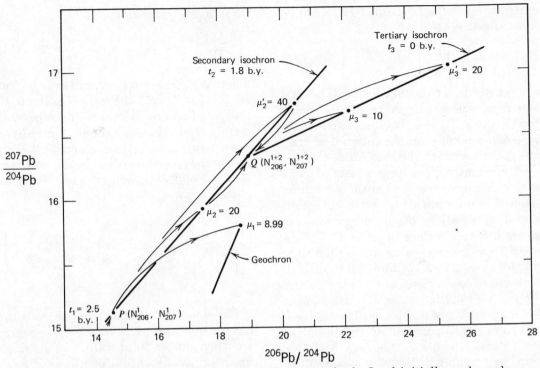

FIGURE 14.2 Isotopic evolution of a suite of three-stage leads. Lead initially evolves along a growth curve corresponding to $\mu_1 = 8.99$. At $t_1 = 2.5$ b.y., when the lead is at point P, it is transferred to secondary systems having $\mu_2 = 20$ and $\mu_2' = 40$. Isotopic evolution continues from point P along diverging growth curves until $t_2 = 1.8$ b.y. At that time the lead is isotopically homogenized to point Q and transferred to tertiary systems having $\mu_3 = 10$ and $\mu_3' = 20$. Again lead evolution continues along diverging growth lines until $t_3 = 0$. Note that all leads in the secondary systems that had initial isotope abundances given by the coordinates of point P lie on the secondary isochron. Similarly, all leads whose initial isotope ratios are given by the coordinates of point Q lie on the tertiary isochron, provided both systems remained closed and provided the leads on the secondary isochron were isotopically homogenized at t_2. Note also that the point P lies on the primary growth curve of leads from conformable ore deposits ($\mu_1 = 8.99$), while the point Q does not.

2
Lead in Young Basaltic Rocks

A very important chapter in the history of the isotope geology of lead (also known as plumbology) began in 1964 with the publication of a paper by Gast, Tilton, and Hedge on the isotopic composition of lead (and strontium) in volcanic rocks from Gough and Ascension islands in the Atlantic Ocean. They reported significant differences among isotopic ratios of lead *within* suites of rocks from each island and also differences *between* the two islands. This was unexpected because the interpretation of lead from conformable ore deposits had given rise to the suggestion that leads derived from the upper mantle should plot close to the point of intersection of the growth curve for conformable ore leads and the geochron. However, the lead from Gough and Ascension Island was found to contain excess radiogenic lead and is therefore anomalous. Many subsequent studies have demonstrated that lead in oceanic and continental basalts *cannot* be accounted for by means of single-stage histories. (Tatsumoto and Knight, 1966a,b; Cooper and Richards, 1966; Gast, 1967, 1969; Doe, 1968; Tatsumoto and Knight, 1969; Armstrong and Cooper, 1971; Oversby and Gast, 1970; Oversby et al., 1971; Oversby, 1971, 1973; Church and Tatsumoto, 1975; Sun and Hanson, 1975). The corollary of these findings is that conformable ore leads do *not* originate in the upper mantle. Their uniform lead-growth pattern may be the result of extensive mixing of lead derived from pelagic sediment subducted under volcanic island arc systems (Russell, 1972; Doe and Stacey, 1974).

In 1967, Gast reported isotope analyses of lead of seven volcanic rocks from St. Helena.

These are shown in Figure 14.3 together with data for Gough, Ascension Island, Tenerife in the Canary Islands, and the Hawaiian Islands. It is evident by inspection of that diagram that most of these leads lie to the right of the geochron and somewhat below the growth curve, that is, they are enriched in ^{206}Pb but deficient in ^{207}Pb. Moreover, the variation of lead ratios for any given island indicates that the lava flows did not originate from a common magma chamber but must have been derived from source regions having different U/Pb ratios. The differences in lead ratios among volcanic rocks from different islands in the Atlantic and Pacific oceans indicate that the apparent heterogeneity of the upper mantle is a worldwide phenomenon and is not restricted to one or two anomalous islands. The isotope ratios of lead from oceanic islands appear

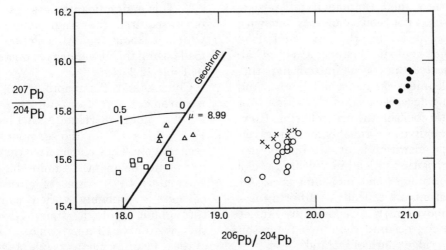

FIGURE 14.3 Isotope ratios of lead from young volcanic rocks. × Ascension Island, △ Gough (Gast et al., 1964); ● St. Helena (Gast, 1967); ○ Tenerife, Canary Islands (Oversby et al., 1971); □ Hawaiian Islands (Tatsumoto, 1966a). The data show both intra- as well as interisland differences in the isotopic composition of lead. These variations are evidence for the heterogeneity of U/Pb ratios in the upper mantle and indicate that the source regions of these volcanic rocks have not remained closed systems throughout geologic time. To put it another way, the data show that leads in volcanic rocks have had multi-stage histories.

to form secondary or higher-order isochrons, thereby implying that such leads have had multistage histories. Consequently, we may conclude that the U/Pb and Th/Pb ratios of the upper mantle have been changed episodically or continuously during geologic time and that these ratios vary significantly from place to place in the mantle on a scale of a few tens of miles.

The variability of the isotope ratios of lead in volcanic rocks, such as those of Gough and Ascension islands, may result from (1) contamination of single-stage, mantle-derived leads with varying amounts of lead originating from old granitic rocks of the continental crust; or (2) evolution of lead in source regions that experienced either episodic or continuous changes of the U/Pb and Th/Pb ratios. The second alternative leads to the conclusion that the upper mantle from which volcanic rocks may be derived is chemically (and isotopically) heterogeneous and has not been a closed system with respect to uranium, thorium, and lead throughout geologic history. Gast et al. (1964) rejected the contamination hypothesis for volcanic rocks in oceanic islands and midocean ridges because of the absence of old granitic rocks in the ocean basins. However, contamination probably must be given serious consideration in explaining the isotopic composition of lead in volcanic rocks on the continents and in island arcs. An example of mixing of leads of differing isotopic compositions was reported by Sun et al. (1975) for oceanic tholeiites from the Reykjanes ridge south of Iceland.

The isotopic compositions of lead in basalts collected along the Juan de Fuca-Gorda ridge and from associated seamounts in the northeast Pacific Ocean are shown in Figure 14.4 (Church and Tatsumoto, 1975). All of these leads are anomalous and plot along straight lines in the lead development diagram. These anomalous-lead lines may be interpreted as secondary isochrons based on the assumption that the leads resided in two systems. The slope of the line in Figure 14.4a is $R = 0.121$, based on all 50 samples analyzed by Church and Tatsumoto. The age of the second system can be calculated from Equation 14.8, assuming that $t_2 = 0$, that is,

$$R = \frac{1}{137.8}\left[\frac{e^{\lambda_2 t_1} - 1}{e^{\lambda_1 t_1} - 1}\right] \quad (14.11)$$

This equation can be solved for t_1 by interpolation in Table 12.3 and yields $t_1 = 1.97 \times 10^9$ years. Using this value, we next calculate the ^{232}Th/^{238}U ratio (κ) from the slope of the line in Figure 14.4b which is $R' = 0.932$. The relevant equation is

$$R' = \frac{^{232}\text{Th}}{^{238}\text{U}}\left[\frac{e^{\lambda_3 t_1} - 1}{e^{\lambda_1 t_1} - 1}\right] \quad (14.12)$$

from which we calculate ^{232}Th/^{238}U (κ) = 3.25. This is significantly less than the value of κ required for single-stage evolution which is about 3.8. All calculations were made using the newer decay constants listed in Table 12.2.

Church and Tatsumoto (1975) obtained isochron dates greater than 2.0×10^9 years for the midocean ridge (MOR) basalts and $1.75 \pm 0.1 \times 10^9$ for the seamount samples, respectively. They emphasized that lead evolution in the mantle cannot be described adequately by a series of closed systems because μ probably varied continuously throughout geologic time. Nevertheless, they were able to approximate the history of these leads by means of five stages involving progressive enrichment of the source regions in uranium and thorium. However, such a model implies instantaneous changes in the parent/daughter ratios which is not compatible with the picture of a dynamic Earth that has emerged from the concept of plate tectonics. Moreover, we must now include the possibility of mixing of leads of

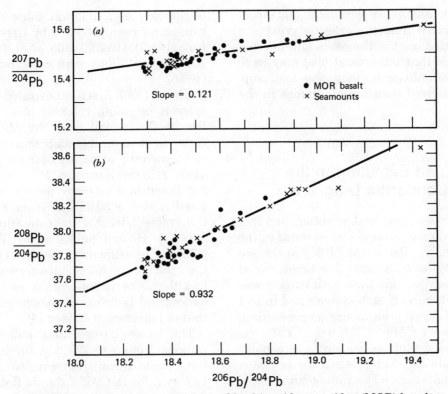

FIGURE 14.4 Isotopic compositions of lead in midocean ridge (MOR) basalts and from seamounts from the Juan de Fuca-Gorda Ridge area of the northeast Pacific Ocean. The data points lie along straight lines in coordinates of $^{207}Pb/^{204}Pb$ versus $^{206}Pb/^{204}Pb$ (a) and $^{208}Pb/^{204}Pb$ vs. $^{206}Pb/^{204}Pb$ (b). By treating the line in (a) as a secondary isochron, a date of 1.97×10^9 years can be calculated. This date represents the time when the μ value of the first source was diversified to produce a set of secondary growth curves for leads that now form the secondary isochron. The date calculated from (a) together with the slope of the isochron in (b) yields an estimate of the $^{232}Th/^{238}U$ ratio (κ) of 3.25. This value is less than that implied by single-stage ore leads of conformable ore deposits. These interpretations are probably no better than crude approximations because of the possibility that the U-Pb and Th-U ratios changed continuously rather than episodically and because lead from different U-Pb systems may have been mixed due to convection in the mantle and subduction of crustal material. (Church and Tatsumoto, 1975.)

differing isotope compositions in the mantle and perhaps also between the mantle and the crust as a result of subduction of crustal material.

The isotopic compositions of lead (and strontium) of igneous rocks of Mesozoic and Cenozoic from the Rocky Mountains of the United States have been reported by Doe (1967, 1968), Doe et al. (1969), Peterman et al. (1970), and Zartman (1974). The results indicate that these rocks either originated from nonuniform crustal or mantle sources or were contaminated with lead (and strontium) derived from the granitic crust. An interesting aspect of the contamination hypothesis is that the crustal lead may have been less radiogenic than the lead supposedly derived from source regions in the mantle.

3
Lead Evolution in the Concordia Diagram

Let us suppose that lead, uranium, and thorium were incorporated into systems in the interior of the Earth 4.55 billion years ago and that these systems remained closed throughout geologic time until magma was generated there. If such systems had in fact remained closed prior to magma generation, their present $^{206}Pb^*/^{238}U$ and $^{207}Pb^*/^{235}U$ ratios would plot on concordia at a point appropriate for U-Pb systems whose age is 4.55 billion years. (The radiogenic component of lead in such systems is calculated by reference to primeval lead of the Earth.) If the U/Pb ratios of young volcanic rocks, produced by crystallization of magma derived from the mantle, were changed by gain or loss of uranium or by loss of lead without isotope fractionation, then such rocks should lie on discordia lines that pass through the origin. On the other hand, if the U/Pb ratios of the source regions had been changed episodically at some time in the past, but were not subsequently altered during magma generation and extrusion of lava, then such systems should lie on a discordia that intersects concordia at a time indicative of the episodic change in the U/Pb ratios of the source regions.

These speculations suggest that the isotope ratios of lead and the U/Pb ratios of young volcanic rocks in the ocean basins can be examined profitably in the context of the concordia diagram. Such interpretations have been proposed by Ulrych (1967), Russel et al. (1968), Welke et al. (1968), and Allegre (1969), but were criticized by Gast (1969).

Ulrych (1967) first demonstrated that U-Pb systems in young volcanic rocks from the Pacific Ocean and from the Mid-Atlantic Ridge fit straight-line discordias that intersect concordia at values of T ranging from 4480 to 4580 million years. His average value for T and best estimate for the age of the Earth is 4530 ± 30 million years. Figure 14.5 illustrates Ulrych's interpretation of the data for Hawaii published by Tatsumoto (1966a). The estimate of the age of the Earth (i.e., the time elapsed since terrestrial lead had the same isotope ratios as troilite in meteorites) is in excellent agreement with that of Patterson (Chapter 13).

The lower intersections indicate dates ranging from 0 to 220 m.y. for leads from the Pacific Ocean, but suggest a date of 1230 m.y. for the Mid-Atlantic Ridge. These dates presumably are the times at which the U/Pb ratio of the upper mantle was altered. The relatively recent dates suggested by volcanic leads from the Pacific Ocean indicate that the heterogeneity of U/Pb ratios in the upper mantle of that region is a fairly recent phenomenon. The diversification of U/Pb ratios under the Mid-Atlantic Ridge, on the other hand, appears to date back to late Precambrian time.

The foregoing interpretation of U-Pb systematics in volcanic rocks implies the assumption that the leads have had two-stage histories and that the observed U/Pb ratios of the rocks are very similar to those of their source regions in the upper mantle. Leads that evolved through more than two stages or rocks whose U/Pb ratios differ significantly from those of their source regions cannot be interpreted by means of the concordia model.

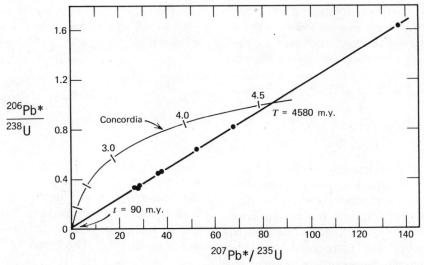

FIGURE 14.5 Isotope evolution of lead in volcanic rocks from Hawaii in the context of the concordia diagram. The coordinates of points were calculated using the measured isotope ratios of lead and the concentrations of uranium and lead. The common lead was assumed to be equal to primeval lead: $a_0 = 9.56$ and $b_0 = 10.42$. The upper intersection (T) indicates the age of the system in which lead evolution occurred. The lower intersection (t) is the time elapsed since the U/Pb ratio of the source region was changed in varying degrees by gain or loss of uranium or loss of lead. This interpretation implies a two-stage mantle history for these leads and suggests that the heterogeneity of the upper mantle (with respect to its U/Pb ratio) under Hawaii is a relatively recent condition. (Interpretation of Ulrych, 1967, based on data by Tatsumoto, 1966a.)

Russell et al. (1968) extended Ulrych's interpretation by demonstrating that it is possible to calculate values for μ_1 from a concordia interpretation of leads in young volcanic rocks, assuming that μ_2 is equal to the observed $^{238}U/^{204}Pb$ ratio of each basalt and that t_1 is a constant for samples from a particular area. They found that μ_1 varies only between narrow limits and has an average value of 8.70 ± 0.12, when T was fixed at 4.55 billion years. In contrast, the values of μ_2 range widely as indicated by the present values of the $^{238}U/^{204}Pb$ ratios in the samples. Russell and his colleagues (1968) pointed out that the average value of μ_1 of the source regions of basalts ($\mu =$

8.70 ± 0.12) is only slightly lower than that indicated by the lead-growth curve of conformable ore deposits ($\mu = 8.94 \pm 0.13$).

We conclude that the concordia diagram provides an alternative method of interpreting isotope ratios of lead in young volcanic rocks. However, this technique is applicable only to two-stage leads in young volcanic rocks whose U/Pb ratios are a close approximation of the U/Pb ratios in the magma sources. The results of this treatment yield estimates of the age of the Earth (T) that are remarkably similar to the age of meteorites. Moreover, the method yields a date (t_1) which presumably refers to the time the U/Pb ratios of a particular mantle region were altered. These events appear to be relatively recent, and this suggests that the apparent heterogeneity of μ values in the upper mantle also is a relatively recent phenomenon.

4
The Whole-Rock Method of Dating

In Chapter 12 we discussed the use of U-Pb and Th-Pb isochrons and pointed out that U-Pb isochrons are rarely successful because of the mobility of uranium in the oxidizing environment of the surface of the Earth. Th-Pb isochrons are more often reliable because thorium is not as mobile as uranium during chemical weathering. We also demonstrated the use of Pb-Pb isochrons which are insensitive to recent losses of uranium or lead.

We now return to a consideration of Pb-Pb isochrons of igneous and metamorphic rocks because the discussion of lead evolution in Chapter 13 has given us better insight into the nature of Pb-Pb isochrons. Let us consider a volume of magma that is isotopically homogeneous and crystallizes to form a suite of rocks having different U/Pb and Th/Pb ratios. Alternatively, we could think of an assemblage of sedimentary and volcanic rocks in which lead is isotopically homogenized as a result of high-grade regional metamorphism. In either case, the lead subsequently evolves along a set of diverging curved trajectories corresponding to different μ values in each specimen, as shown in Figure 13.2. If growth continues uninterrupted to the present time, the leads must lie along an isochron, provided the rocks had the same initial isotopic ratios of lead, formed at the same time, and remained closed to uranium, thorium, and lead after closure. However, recent changes in the concentrations of these elements are permitted,

provided the isotopic compositions of lead are not affected. The equation of such an isochron is

$$\frac{\dfrac{^{207}\text{Pb}}{^{204}\text{Pb}} - \left(\dfrac{^{207}\text{Pb}}{^{204}\text{Pb}}\right)_0}{\dfrac{^{206}\text{Pb}}{^{204}\text{Pb}} - \left(\dfrac{^{206}\text{Pb}}{^{204}\text{Pb}}\right)_0} = \frac{1}{137.8}\left[\frac{e^{\lambda_2 t_1} - e^{\lambda_2 t_2}}{e^{\lambda_1 t_1} - e^{\lambda_1 t_2}}\right]$$

$$(14.13)$$

where the subscript 0 identifies the initial isotope ratios of lead, t_1 is the time since closure, and t_2 is the time since growth of lead isotopes was terminated. If growth continued to the present, $t_2 = 0$ and the slope of the isochron is then given by

$$m = \frac{1}{137.8}\left[\frac{e^{\lambda_2 t} - 1}{e^{\lambda_1 t} - 1}\right] \qquad (14.14)$$

which is identical to Equation 12.22. The age of the suite of samples can be calculated from the slope of the isochron by interpolating in Table 12.3. Note that the initial isotope ratios of the lead and the concentrations of uranium or lead need not be specified.

This method of dating was pioneered by Sobotovich (1961) and was used by him in subsequent publications (Sobotovich and Lovtsyus, 1965). It was also used by Moorbath, Welke, and Gale (1969) who determined isotopic compositions of lead in a suite of metamorphic rocks from the Lewisian basement complex of northwestern Scotland. They determined concentrations of uranium and lead and reported that these rocks are strongly *depleted* in uranium compared to the so-called crustal average. The concentrations of uranium and lead in the Lewisian rocks average 0.24 ppm and 7.9 ppm, respectively, giving them an average $^{238}\text{U}/^{204}\text{Pb}$ ratio (μ) of only 1.76. They found that the isotope ratios of lead from these rocks scatter about a straight line shown in Figure 14.6 which they interpreted as a secondary iso-

THE INTERPRETATION OF MULTISTAGE LEADS 14

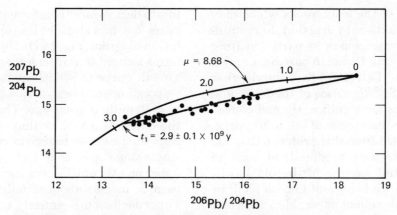

FIGURE 14.6 Plot of isotope ratios of lead from whole-rock samples of the Lewisian Basement Complex of northwestern Scotland. The leads scatter about a secondary isochron that intersects a primary lead growth curve for $\mu = 8.68$ at $t_1 = 2.9 \pm 0.1 \times 10^9$ y and at $t_2 = 0$. The leads in the Lewisian rocks may have been withdrawn about $2.9 \pm 0.1 \times 10^9$ years ago from a source region having $\mu = 8.68$. Since that time, these leads have evolved at different rates determined by their μ values which are significantly *less* than 8.68. The scatter of data points about the isochron may result from the inhomogeneity of the isotopic compositions of initial leads or to alterations of the U/Pb ratios or lead isotope compositions after intial crystallization of these rocks. The constants used are: $\lambda_1 = 1.537 \times 10^{-10}$ y^{-1}, $\lambda_2 = 9.722 \times 10^{-10}$ y^{-1}, $a_0 = 9.56$, $b_0 = 10.42$, $T = 4.55 \times 10^9$ y. (Moorbath et al., 1967.)

chron. However, rather than using the slope of this isochron to calculate the age of these samples, they determined a date from the point of intersection of the isochron with the primary lead growth curve. The primary growth curve was obtained by forcing it to intersect the secondary isochron at $t = 0$ and by assuming values for a_0, b_0, and T listed in the caption to Figure 14.6. The μ value of this growth curve was found to be 8.68, which is close to, but somewhat less than, the value obtained from conformable ore leads. The date indicated by the point of intersection is 2900 ± 100 million years which the authors interpreted as the time of variable uranium loss during pyroxene-granulite metamorphism of the ancestral Lewisian rocks.

However, not all linear patterns of isotope ratios on lead evolution diagrams are necessarily isochrons. Such patterns may also result from mixing of leads of differing isotope composition in the course of petrogenesis. An example of this phenomenon was first reported by Kanasewich and Farquhar (1965) for ore leads from the mines of Cobalt and Noranda in Canada. Another example was found by Moorbath and Welke (1969) in the igneous rocks from the Isle of Skye in northwest Scotland.

This area is well known to petrologists because it contains a wide variety of plutonic and volcanic igneous rocks of early Tertiary age whose petrogenesis has been the subject of much discussion (Anderson and Dunham, 1966). The principal point of the controversy is the origin of the felsic rocks which may have formed either by fractional crystallization of basic magma or by partial melting of the underlying Lewisian basement rocks. Moorbath and Bell (1965) first showed that the initial $^{87}Sr/^{86}Sr$ ratios of the felsic and granitic rocks are significantly and consistently higher than those of the basic rocks. They concluded from this evidence that the granitic rocks were produced, at least in part, by partial melting of the old Lewisian gneiss which is present here at shallow depth. In a subsequent paper, Moorbath and Welke (1969) found that the isotope ratios of lead extracted from the basic and acid igneous rocks of Skye, after correction for

in situ decay since their formation 60 million years ago, fit a straight line on the lead evolution diagram (Fig. 14.7). This straight line forms a chord that intersects a primary lead growth curve ($\mu = 8.92$) at two points corresponding to times of $t_1 = 3100 \pm 50$ and $t_2 = 60$ million years ago. They interpreted the chord as a mixing line of lead derived from the Lewisian basement complex with a single-stage lead that had evolved in the upper mantle until it was incorporated into basaltic magma about 60 million years ago. This conclusion is entirely consistent with their earlier interpretation of the initial $^{87}Sr/^{86}Sr$ ratios. The isotopic compositions of both strontium and lead clearly favor the

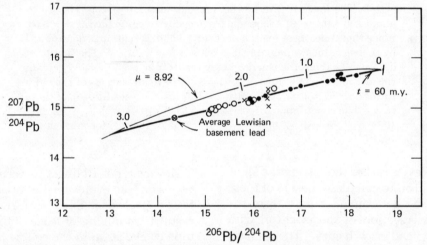

FIGURE 14.7 Isotopic composition of lead in the Tertiary igneous rocks of the Isle of Skye, northwest Scotland. ● mafic and ultramafic rocks, ○ acid rocks, × intermediate and hybrid rocks. The straight line along which these leads plot (corrected for in situ decay) is a mixing line of average lead from the Lewisian Basement Complex and single-stage leads derived from a source region in the upper mantle ($\mu = 8.92$) about 60 million years ago. Note that the lead in the acid rocks is more highly contaminated than that of the mafic and ultramafic rocks. Moreover, the contaminant lead from the Lewisian Complex is *less* radiogenic than the lead derived from the upper mantle 60 million years ago. The constants used in constructing this diagram are: $\lambda_1 = 1.537 \times 10^{-10}$ y^{-1}, $\lambda_2 = 9.722 \times 10^{-10}$ y^{-1}, $a_0 = 9.56$, $b_0 = 10.42$, $T = 4.55 \times 10^9$ y. (Moorbath and Welke, 1969.)

hypothesis that the igneous rocks of Skye contain significant proportions of strontium and lead derived from the Lewisian basement complex and that at least some of the granitic magma could have been derived by partial melting of the basement complex.

5
Dating of K-feldspar by the Common Lead Method

The feldspars of igneous and metamorphic rocks are enriched in lead and depleted in uranium and thorium. Lead concentrations may range from a few tens up to several hundred parts per million, whereas their uranium and thorium concentrations are generally less than one part per million. Consequently, the isotopic composition of lead in feldspars remains virtually unchanged for long periods of time and thus preserves a record of the lead that was initially incorporated into igneous and metamorphic rocks. The isotopic composition of lead in feldspar can therefore be used for dating by the common lead method and to derive information about the petrogenesis of the rocks.

The feasibility of interpreting the isotopic composition of lead in feldspars has been investigated by Tilton et al. (1955), Catanzaro and Gast (1960), Doe et al. (1965), Zartman (1965), Doe (1967), and others. The interpretation of feldspar leads in terms of the common lead method of dating is subject to several assumptions. These are (1) the lead must have had a single-stage history including evolution in a closed U,Th-Pb system prior to its incorporation into the feldspar; (2) it must not be contaminated with other leads having different isotopic compositions during the formation, intrusion and crystallization of magma; (3) the feldspars must remain closed to lead and their Pb/U and Pb/Th ratios must be sufficiently large to

make the isotopic composition virtually invariant with time. Experience has shown that these prerequisite conditions are not satisfied in some cases.

A significant number of examples have come to light in which model lead dates of feldspars are significantly less (and sometimes greater) than dates obtained by other methods. These discrepancies typically imply the presence of excess radiogenic lead that may have been acquired during a multistage history or by contamination prior to crystallization of the feldspar, or may be due to the incorporation of radiogenic lead into the feldspar during episodes of thermal metamorphism subsequent to crystallization. For example, Doe and Hart (1963) found that centimeter-sized crystals of K-feldspar of the Idaho Springs Formation in Colorado had scavenged radiogenic lead during contact metamorphism caused by the intrusion of the Eldora stock (Chapter 6). This effect was detectable up to 300 meters from the contact, whereas no change occurred in the isotopic composition of strontium in the same feldspars beyond a distance of only 15 meters from the contact. This example suggests that lead in feldspars from metamorphosed igneous rocks may be contaminated with varying amounts of radiogenic lead derived from associated U-bearing minerals. In fact, some geochronologists have expressed doubt regarding the reliability of lead isotope compositions of feldspars from any rock containing zircons or other U-bearing minerals that give discordant dates due to loss of radiogenic lead.

6 SUMMARY

The isotope ratios of lead in many rocks and ore deposits are explainable on the basis of multistage histories. Such leads form secondary or higher order isochrons in coordinates of $^{206}Pb/^{204}Pb$ and $^{207}Pb/^{204}Pb$, provided they were isotopically homogenized and provided that the last system with which they were associated was closed to uranium, thorium, and lead.

Lead in young basaltic rocks has variable isotopic ratios which, in many cases, plot to the right of the zero isochron. Moreover, suites of leads from a given region tend to fit lines that have been interpreted as secondary or higher order isochrons. However, such linear arrays may also be the result of contamination of mantle-derived leads with foreign lead from the crust. The isotopic composition of lead in volcanic rocks shows that the upper mantle has not been a closed system with respect to uranium, thorium, and lead and is heterogeneous both chemically and isotopically.

Igneous and metamorphic rocks contain lead whose isotopic composition may fit secondary or higher order isochrons in coordinates of $^{206}Pb/^{204}Pb$ and $^{207}Pb/^{204}Pb$. The history of such leads and the ages of the rocks in which they occur can be determined from the slopes of these isochrons or from the point of intersection with a hypothetical primary lead-growth curve.

Both K-feldspar and plagioclase have high Pb/U ratios and may be dated by the common lead method, provided they have not scavanged radiogenic lead from associated U, Th-bearing minerals and provided the lead has had a single-stage history.

PROBLEMS

1. A sample of lead had the following evolutionary history: Starting with primeval isotope ratios $^{206}Pb/^{204}Pb = 9.307$ and $^{207}Pb/^{204}Pb = 10.294$, it resided from 4.55 to 2.0 billion years in a system characterized by $\mu = 8.9$. At 2.0 billion years in the past, the U/Pb ratio changed such that $\mu_2 = 12.0$. Trace the movement of this lead on a Pb-Pb evolution diagram, draw the growth lines and the isochrons on which it resided at 2.0 b.y. and at the present time, and state the present isotope ratios of this lead. Use the new decay constants: λ_1 $(^{238}U) = 1.5525 \times 10^{-10}$ y^{-1}, λ_2 $(^{235}U) = 9.8485 \times 10^{-10}$ y^{-1}. (**ANSWER:** Final isotopic ratios $^{206}Pb/^{204}Pb = 19.56$, $^{207}Pb/^{204}Pb = 16.07$.)

2. Fifteen samples of basalt from seamounts at the north end of the Juan de Fuca ridge form a line in coordinates of $^{206}Pb/^{204}Pb$ (x) and $^{207}Pb/^{204}Pb$ (y) having a slope equal to $^{207}Pb/^{206}Pb = 0.101$ (Church and Tatsumoto, 1975). Treating it as a secondary isochron, calculate the age of the second U/Pb system in which these leads resided. (**ANSWER:** 1.64×10^9 y.)

3. The same samples (Problem 2 above) form a line in coordinates of $^{206}Pb/^{204}Pb$ (x) versus $^{208}Pb/^{204}Pb$ (y) whose slope is 0.861. Calculate the $^{232}Th/^{238}U$ ratio of the source region of these basalts, assuming that the age of the source is 1.64×10^9 years. (**ANSWER:** $^{232}Th/^{238}U = 2.95$.)

4. The following data were obtained for basalts from Guadelupe Island (Ulrych, 1967):

$\dfrac{^{206}Pb*}{^{238}U}$	$\dfrac{^{207}Pb*}{^{235}U}$
0.518	34.91
0.301	20.56
0.321	22.22
0.401	27.30

Plot these data on a concordia diagram and estimate the age of the Earth by means of a graphical interpretation.

5. A suite of microcline feldspars from the Granite and Seminoe mountains of Wyoming (Rosholt et al., 1973) form a line in coordinates of $^{206}Pb/^{204}Pb$ (x) and $^{207}Pb/^{204}Pb$ (y) whose slope if 0.295. Assume that this line is a secondary isochron for leads that began to evolve 2.79×10^9 years ago. Calculate the time when these leads entered the microcline and lead evolution stopped because of their low U/Pb and Th/Pb ratios. (**ANSWER:** $t = 1.62 \times 10^9$ y.)

REFERENCES

Allegre, C. J. (1969) Comportement des systemes U-Th-Pb dans le manteau superieur et modele d' evolution de ce dernier au cours des temps geologiques. Earth Planet. Sci. Letters, 5, 261–269.

Anderson, F. W., and K. C. Dunham (1966) The geology of northern Skye. Mem. Geol. Surv. Scotland, No. 80 + 90, 216 pp.

Armstrong, R. L., and J. A. Cooper (1971) Lead isotopes in island arcs. Bull. volcanol., 35, 27–63.

Catanzaro, E. J., and P. W. Gast (1960) Isotopic composition of lead in pegmatite feldspars. Geochim. Cosmochim. Acta, 19, 113–126.

Church, S. E., and M. Tatsumoto (1975) Lead isotope relations in oceanic ridge basalts from the Juan de Fuca-Gorda ridge area, N.E. Pacific Ocean. Contrib. Mineral. Petrol., 53, 253–279.

Cooper, J. A., and J. R. Richards (1966) Lead isotopes and volcanic magmas. Earth Planet Sci. Letters, 1, 259–269.

Doe, B. R., and S. R. Hart (1963) The effect of contact metamorphism on lead in potassium feldspars near the Eldora stock, Colorado. J. Geophys. Res., 68, 3521–3530.

Doe, B. R., G. R. Tilton, and C. A. Hopson (1965) Lead isotopes in feldspars from selected granitic rocks associated with regional metamorphism. J. Geophys. Res., 70, No. 8, 1947–1968.

Doe, B. R. (1967) The bearing of lead isotopes on the source of granitic magma. J. Petrol. 8, 51–83.

Doe, B. R. (1968) Lead and strontium isotope studies of Cenozoic volcanic rocks in the Rocky Mountain region—a summary. Colorado School Mines Quart., 63, 149–174.

Doe, B. R., P. W. Lipman, and C. E. Hedge (1969) Primitive and contaminated basalts from the southern Rocky Mountains, U.S.A. Contrib. Mineral. Petrol., 21, 142–156.

Doe, B. R. (1970) Lead Isotopes. Springer-Verlag, New York, Heidelberg, Berlin, 137 pp.

Doe, B. R., and J. S. Stacey (1974) The application of lead isotopes to the problems of ore genesis and ore prospect evaluation. A review. Econ. Geol., 69, 757–776.

Gale, N. H., and A. E. Mussett (1973) Episodic uranium-lead models and the interpretation of variations in the isotopic composition of lead in rocks. Rev. Geophys. Space Sci., 11, 37–86.

Gast, P. W., G. R. Tilton, and C. Hedge (1964) Isotopic composition of lead and strontium from Ascension and Gough Islands. Science, 145, 1181–1185.

Gast, P. W. (1967) Isotope geochemistry of volcanic rocks. In Basalts. The Poldervaart Treatise on Rocks of Basaltic Composition, vol. 1, pp. 325–358, H. H. Hess and A. Poldervaart, eds., Interscience, New York.

Gast, P. W. (1969) The isotopic composition of lead from St. Helena and Ascension Islands. Earth Planet. Sci. Lett., 5, 353–359.

Kanasewich, E. R., and R. M. Farquhar (1965) Lead isotope ratios from the Cobalt-Noranda area, Canada. Can. J. Earth Sci., 2, 361–384.

Moorbath, S., and J. D. Bell (1965) Strontium isotope abundance studies and Rb-Sr age determinations on Tertiary igneous rocks from the Isle of Skye, north-west Scotland. J. Petrol., *6*, 37–66.

Moorbath, S., and H. Welke (1969) Lead isotope studies on igneous rocks from the Isle of Skye, northwestern Scotland. Earth Planet. Sci. Letters, *5*, 217–230.

Moorbath, S., H. Welke, and N. H. Gale (1969) The significance of lead isotope studies in ancient, high-grade metamorphic basement complexes, as exemplified by the Lewisian rocks of northwest Scotland. Earth Planet. Sci. Letters, *6*, 245–256.

Oversby, V. M., and P. W. Gast (1970) Lead from oceanic islands. J. Geophys. Res., *75*, 2097–2114.

Oversby, V. M. (1971) Lead in oceanic islands: Faial, Azores, and Trinidad. Earth Planet. Sci. Letters, *11*, 401–406.

Oversby, V. M., J. Lancelot, and P. W. Gast (1971) Isotopic composition of lead in volcanic rocks from Tenerife, Canary Islands. J. Geophys. Res., *76*, 3402–3413.

Oversby, V. M., and A. Ewart (1973) Lead isotopic compositions of Tonga-Kermadec volcanics and their petrologic significance. Contrib. Mineral. Petrol., *37*, 181–210.

Peterman, Z. E., B. R. Doe, and H. J. Prostka (1970) Lead and strontium isotopes in rocks of the Absaroka volcanic field, Wyoming. Contrib. Mineral Petrol., *27*, 121–130.

Rosholt, J. N., R. E. Zartman, and I. T. Nkomo (1973) Lead isotope systematics and uranium depletion in the Granite Mountains, Wyoming. Geol. Soc. Amer. Bull., *84*, 989–1002.

Russell, R. D., W. F. Slawson, T. J. Ulrych, and P. H. Reynolds (1968) Further applications of concordia plots to rock lead isotope abundances. Earth Planet. Sci. Letters, *3*, 284–288.

Russell, R. D. (1972) Evolutionary model for lead isotopes in conformable ores and in ocean volcanics. Rev. Geophys. Space Phys., *10*, 529–549.

Sobotovich, E. V. (1961) Possibility of determining the absolute age of the granites of the Terskey Ala-Tau by the lead included in them. Akad. Nauk SSSR Kom. Opredeleniyu Absolyut. Vozrasta Geol. Formatsii Trudy, sess. 9, 269–280.

Sobotovich, E. V., and A. V. Lovtsyus (1965) Age of the rocks of the Sharyzhalgay series (Baikal block). Akad. Nauk SSSR Izv. Ser. Geol., No. 9, 38–41.

Stacey, J. S., and T. W. Stern (1973) Revised tables for the calculation of lead isotope ages. U.S. Dept. Commerce, Natl. Tech. Information Service, Springfield, Virginia, 22151, PB-20919, 35 p.

Stacey, J. S., and J. D. Kramers (1975) Approximation of terrestrial lead isotope evolution by a two-stage model. Earth Planet. Sci. Letters, *26*, 207–221.

Sun, S. S., M. Tatsumoto, and J. G. Schilling (1975) Mantle plume mixing along the Reykjanes ridge axis: Lead isotopic evidence. Science, *190*, 143–146.

Sun, S. S., and G. N. Hanson (1975) Origin of Ross Island basanitoids and limitations upon the heterogeneity of mantle sources for alkali basalts and nephelinites. Contrib. Mineral. Petrol., *52*, 77–106.

Tatsumoto, M. (1966a) Isotopic composition of lead in volcanic rocks from Hawaii, Iwo Jima, and Japan. J. Geophys. Res., *71*, 1721–1733.

Tatsumoto, M. (1966b) Genetic relations of oceanic basalts as indicated by lead isotopes. Science, *153*, 1094–1101.

Tatsumoto, M., and R. J. Knight (1969) Isotopic composition of lead in volcanic rocks from central Honshu with regard to basalt genesis. Geochem. J., *3*, 53–86.

Tilton, G. R., C. C. Patterson, H. Brown, M. Inghram, R. Hayden, D. Hess, and E. Larsen, Jr. (1955) Isotopic composition and distribution of lead, uranium, and thorium in a Precambrian granite. Geol. Soc. Amer. Bull., *66*, 1131–1148

Ulrych, T. J. (1967) Oceanic basalt leads: A new interpretation and an independent age for the earth. Science, *158*, 252–256.

Welke, H., S. Moorbath, G. L. Cumming, and H. Sigurdsson (1968) Lead isotope studies on igneous rocks from Iceland. Earth Planet. Sci. Letters, *4*, 221–231.

Zartman, R. E. (1974) Lead isotopic provinces in the Cordillera of the western United States and their geological significance. Econ. Geol., *69*, 792–805.

15 THE FISSION-TRACK METHOD OF DATING

When charged particles travel through a solid medium, they leave a trail of damage resulting from the transfer of energy from the particle to the atoms of the medium. Such tracks were first seen by Silk and Barnes (1959) during the examination of irradiated solids by means of the electron microscope using very high magnification (50,000X). Price and Walker (1962a) discovered that the tracks can be enlarged and made visible under an optical microscope by etching with suitable solutions because of the increased solubility of the damaged regions. The tracks (about 10 microns) are made by the fission fragments of uranium as a result of the large masses of the fragments and because of the large amount of energy liberated by this process (about 200 MeV). Price and Walker (1962b) discovered such fission tracks in natural samples of mica. In the following year, they showed that fission tracks in natural minerals are due primarily to spontaneous fission of ^{238}U (Chapter 3.7) and proposed that the density of such tracks could be used to date samples of mica (biotite, phlogopite, muscovite, and lepidolite) in which they found track densities of up to $5 \times 10^4/cm^2$. Fleischer and Price (1964a) used this method to date tektites and Libyan Desert glass and obtained results that agreed satisfactorily with K-Ar dates for these samples.

The fission-track method is now widely used for dating a variety of minerals and both natural and synthetic glass. It is especially applicable to relatively young samples that have not been reheated since the time of their formation and is therefore of potential interest to archaeology as well as to geology (Fleischer and Price, 1946b). However, the method also provides useful information about the thermal histories of older rocks because the preservation of fission tracks is temperature-dependent, and different minerals lose their tracks at different temperatures (Fleischer et al., 1969).

The study of particle tracks in solids has resulted in many other technical applications including uranium determinations in crystals (Price and Walker, 1963a), deep-sea sediment (Bertine et al., 1970), and natural waters (Reimer, 1975). Other applications have been discussed by Fleischer et al. (1972), Wagner (1974), and Fisher (1975). The literature dealing with this subject has consequently grown rapidly. The principles of track formation in solids and their applications are the subject of a large book by Fleischer et al. (1975).

1
Methodology

In order to date a specimen of glass or mineral by the fission track method, an interior surface is exposed by grinding and is then polished and etched with a suitable solvent under appropriate conditions (see Table 15.1). Small mineral grains may be mounted in clear epoxy resin on a glass slide and can then be ground and polished (Naeser, 1967). After etching, the polished surface is examined with a petrographic microscope (magnification of 800 to 1800X) equipped with a flat-field eyepiece with a graticule to permit counting of tracks in a known area. Fission tracks are readily distinguished by their

Table 15.1   **Typical Etching Procedures for Selected Minerals and Volcanic Glass**

MINERAL	ETCHING SOLUTION	T°C	DURATION	REFERENCE
Apatite	conc. HNO_3	25	10–30 sec	1
	5% HNO_3	20	45 sec	2
Sphene	conc HCl	90	30–90 min	1
	1 $HF:2HCl:3HNO_3:6H_2O$	20	6 min	3
Zircon	100N NaOH	270	1.25 hr	3
Muscovite	HF (48%)	20	20 min	3
Epidote	6gNaOH + 4 ml H_2O	159	150 min	4
	(37.5 N NaOH)			
Volc. glass	HF (24%)	25	60 sec	5

1. Naeser, 1967
2. Wagner and Reimer, 1972
3. Gleadow and Lovering, 1974
4. Bar et al., 1974
5. Lakatos and Miller, 1972

Etching procedures and track-counting methods have also been discussed by Lal et al. (1968) and Burchart et al. (1975).

characteristic tubular shape from other etch pits that result from imperfections or other causes (see Fleischer and Price, 1964b, for many excellent photomicrographs of fission tracks). The track density due to spontaneous fission of ^{238}U is determined by counting a statistically significant number of tracks in a known area. Counting becomes difficult when the track density is less than 10 tracks per square centimeter. However, many minerals and glasses have much higher track densities so that from several hundred to several thousand tracks can be counted. The observed track density is related to the length of time during which tracks have accumulated and to the uranium concentration of the specimen.

The uranium concentration can be measured by a procedure that involves counting fission tracks produced by *induced* fission of ^{235}U due to irradiation of the sample with thermal neutrons in a nuclear reactor (Chap-

ter 4.4). This can be accomplished in several ways. A widely used procedure is to destroy the spontaneous-fission tracks by heating the specimen to cause annealing. The specimen is then exposed to thermal neutrons in a nuclear reactor in order to produce new tracks by induced fission of ^{235}U. After the irradiation, a new surface is polished and etched, and the density of the induced fission tracks is determined. The uranium concentration of the specimen is indicated by the observed density of induced fission tracks, provided the effective thermal neutron flux and the duration of the irradiation are known.

The neutron "dose" (flux density times irradiation time) is determined by means of various flux monitors (Fleischer and Price, 1964; Kaufhold and Herr, 1968). Copper foil is most frequently used. Another monitor is glass of known uranium concentration, such as Corning reference glass U3 or Nat. Bur.

Stds. SRM614, which is irradiated together with a batch of "unknowns" (Fleischer and Hart, 1971; Schreurs et al., 1971; Carpenter and Reimer, 1974). The density of induced tracks in the monitor is a measure of the neutron dose to which the unknowns were exposed.

The principal cause of systematic errors in fission-track dates of natural samples is the fading of tracks due to annealing of the sample at elevated temperatures. For this reason, fission-track dates of natural samples must be interpreted as "cooling ages" and do not necessarily coincide with the times at which the minerals crystallized in an igneous or metamorphic rock. Other sources of error are the uncertainty of the neutron dose, statistical errors in counting tracks, and the possible uneven distribution of uranium in the specimen.

The prerequisite conditions for dating minerals or glasses by the fission track method are (1) the concentration of uranium must be sufficient to produce a track density of greater than 10 tracks per square centimeter in the time elapsed since cooling of the sample; (2) the tracks must be stable at ordinary temperatures for time intervals comparable to the age being measured; (3) the material must be sufficiently free of inclusions, defects, and lattice dislocations to permit identification and counting of etched fission tracks; and (4) the uranium distribution in the specimen must be uniform to permit the concentration of ^{238}U to be determined from the density of induced-fission tracks in a different portion of the sample. Most of the minerals and glasses listed in Table 15.1 have been dated by the fission-track method and therefore are known to satisfy these conditions, at least in some cases. However, a given mineral may not necessarily be suitable for dating in all cases.

2
Derivation of the "Age Equation"

Several naturally occurring isotopes of high atomic number are known to decay by spontaneous fission and therefore produce fission tracks in minerals and glass. However, Price and Walker (1963b) showed that in most cases such tracks are due primarily to spontaneous fission of ^{238}U. The track densities caused by spontaneous fission of ^{235}U and ^{232}Th are negligible. They also concluded that *induced* fission of ^{235}U due to absorption of thermal neutrons produced by *spontaneous* fission of ^{238}U is not important, provided the mineral to be dated was not embedded in uranium ore. For these reasons, fission tracks in natural materials can be safely attributed to spontaneous fission of ^{238}U alone.

The decay constant for spontaneous fission of ^{238}U (λ_f) has been in doubt for some time. Fleischer and Price (1964b) reported a value $(6.85 \pm 0.20) \times 10^{-17}$ y^{-1} based on a comparison of fission track dates of tektites with K-Ar dates. However, when the effect of track fading is taken into consideration, a value of 8.4×10^{-17} y^{-1} is obtained (Gentner et al., 1969, Storzer and Wagner, 1971). The most precise determination is by Galliker et al. (1970), who reported λ_f (^{238}U) = $(8.46 \pm 0.06) \times 10^{-17}$ y^{-1}. This value was confirmed by Storzer (1970) and has been recommended by Wagner et al. (1975) who obtained $(8.7 \pm 0.6) \times 10^{-17}$ y^{-1} on the basis of a study of U-rich glasses of known age of manufacture. Both values (6.85×10^{-17} y^{-1} and 8.46×10^{-17} y^{-1}) are currently in use (Gleadow and Lovering, 1974; Bar et al., 1974). In any case, it is clear that $\lambda_f \ll \lambda_\alpha$ (see Table 12.2) and that the decay of ^{238}U can be attributed entirely to alpha emission.

Let us consider, therefore, a grain of mineral or glass containing ^{238}U atoms distributed evenly throughout its volume. The total number of decays of ^{238}U in a given volume of sample during time t is

$$D = {}^{238}\text{U}(e^{\lambda_\alpha t} - 1) \qquad (15.1)$$

where D is the number of decay events per cubic centimeter of the sample, ^{238}U is the number of ^{238}U atoms per cubic centimeter of the sample at the present time, and λ_α is the decay constant of ^{238}U for alpha emission $= 1.55125 \times 10^{-10}\,\text{y}^{-1}$ (Table 12.2).

The fraction of ^{238}U decays that are due to spontaneous fission and leave a track is

$$F_s = \left(\frac{\lambda_f}{\lambda_\alpha}\right){}^{238}\text{U}(e^{\lambda_\alpha t} - 1) \qquad (15.2)$$

A certain fraction q of these tracks will cross the polished surface and will be counted after etching. The area density of spontaneous-fission tracks in that surface is therefore given by

$$\rho_s = F_s q = \left(\frac{\lambda_f}{\lambda_\alpha}\right){}^{238}\text{U}(e^{\lambda_\alpha t} - 1)q \qquad (15.3)$$

The number of fissions of ^{235}U induced by the thermal neutron irration per cubic centimeter of sample is

$$F_i = {}^{235}\text{U}\varphi\sigma \qquad (15.4)$$

where ^{235}U is the number of ^{235}U atoms per cubic centimeter of sample at the present time, φ is the thermal neutron dose in units of neutrons per square centimeter, and σ is the cross section for induced fission of ^{235}U by thermal neutrons $= 582 \times 10^{-24}\,\text{cm}^2$.

The fraction of induced tracks that cross an interior surface and that will be counted after etching is also equal to q, provided the uranium atoms are evenly distributed throughout the volume of the specimen and provided the etching is performed exactly as before. The density of induced tracks is then equal to

$$\rho_i = F_i q = {}^{235}\text{U}\varphi\sigma q \qquad (15.5)$$

By combining Equations 15.3 and 15.5, we obtain

$$\frac{\rho_s}{\rho_i} = \left(\frac{\lambda_f}{\lambda_\alpha}\right)\frac{I(e^{\lambda_\alpha t} - 1)}{\varphi\sigma} \qquad (15.6)$$

where I is the atomic ratio of ^{238}U/^{235}U. We now solve Equation 15.6 for t and thus derive the "age equation," first formulated by Walker and Price (1963):

$$t = \frac{1}{\lambda_\alpha}\ln\left[1 + \left(\frac{\rho_s}{\rho_i}\right)\left(\frac{\lambda_\alpha}{\lambda_f}\right)\frac{\varphi\sigma}{I}\right] \qquad (15.7)$$

By substituting $\lambda_\alpha = 1.55125 \times 10^{-10}\,\text{y}^{-1}$, $\lambda_f = 8.46 \times 10^{-17}\,\text{y}^{-1}$, $\sigma = 582 \times 10^{-24}\,\text{cm}^2$, and $I = 137.8$, we obtain

$$t = 6.446 \times 10^9 \ln\left[1 + 7.744 \times 10^{-18}\left(\frac{\rho_s}{\rho_i}\right)\varphi\right] \qquad (15.8)$$

Equation 15.8 can be solved for t after ρ_s/ρ_i has been measured and the neutron dose φ has been determined from the monitor.

When t is less than 500 million years, Equation 15.6 can be simplified by setting $e^{\lambda t} - 1 \sim \lambda t$ (Chapter 6.4). In that case

$$\frac{\rho_s}{\rho_i} = \frac{\lambda_f I t}{\varphi\sigma} \qquad (15.9)$$

and

$$t = \left(\frac{\rho_s}{\rho_i}\right)\frac{\varphi\sigma}{\lambda_f I} = 4.992 \times 10^{-8}\varphi\left(\frac{\rho_s}{\rho_i}\right) \qquad (15.10)$$

Dates calculated from the approximate equation are 4 percent low for $t = 500$ m.y. and only 0.7 percent low for $t = 100$ m.y., whereas the analytical uncertainty is rarely less than 5 percent. Equation 15.10 also provides a basis for adjusting the neutron dose so that the density of induced tracks approximately equals that of the spontaneous-fission tracks. If $\rho_s/\rho_i = 1$,

$$\varphi = 2.00 \times 10^7 t \qquad (15.11)$$

where t is in years.

It is important to equalize the densities of the spontaneous and induced tracks in order to reduce the error of measurement of ρ_s/ρ_i.

3
Fading of Fission Tracks

In most materials, fission tracks are stable for long periods of time at room temperature. However, at elevated temperatures the tracks fade as the damage done by the charged particles is healed. Track fading appears to be independent of hydrostatic pressure but is accelerated in muscovite by shear stress. Exposure to shock waves in excess of 20 kilobars causes some erasure of tracks in apatite and complete erasure at shock pressures greater than 400 kilobars. Similarly, sphene loses all fission tracks when exposed to shocks in excess of 100 kilobars (Fleischer et al., 1974). Fission tracks in apatite may also be lost by the action of ionic solutions (Burchart and Reimer, 1972). Lakatos and Miller (1972) showed that the annealing rate of volcanic glass increases with increasing

water content. Nevertheless, *temperature* is the most important parameter controlling the fading of fission tracks in solids.

The rate at which tracks fade at a given temperature varies among different minerals and glasses. Consequently, two different minerals that have been exposed to the same elevated temperature for the same length of time may have differing fission-track dates. Such discordance of dates of cogenetic minerals provides information about the temperature history of the sample.

The differences in the annealing properties of different minerals and glasses are reflected by the temperatures, listed in Table 15.2, that are required to completely eliminate all tracks in one hour. Quartz appears to be the most refractory of the common minerals ($T = 1050°C$), while calcite anneals at the lowest temperature ($T = 320 \pm 25°C$). Actually, neither of these is useful for dating because the uranium concentration of quartz is too low and because calcite anneals at about 20°C in two million years (McDougall and Price, 1974). However, many common minerals that are suitable for dating have

Table 15.2   **Annealing Temperatures Resulting in Complete Loss of Tracks in One Hour at One Atmosphere Pressure (Adapted from McCorkell, 1974)**

SUBSTANCE	TEMPERATURE °C	SUBSTANCE	TEMPERATURE °C
Quartz	1050	Whitlockite	584
Diopside	885	Pigeonite	525 ± 25
Albite	775	Olivine	509
Epidote	715	Hypersthene	475 ± 25
Zircon	700 ± 25	Phlogopite	450
Garnet[a]	685	Enstatite	450
Pollucite	670	Apatite	340
Hornblende	630	Calcite	320 ± 25
Muscovite	680	Tektites	500
Sphene	620	Basalt glass	300

[a] Haack and Potts (1972).

quite different track-fading properties and therefore can record the times in the history of a rock when the temperature last dropped through certain critical values.

In order to describe the fading of fission tracks in minerals as a function of increasing temperature, one must measure the reduction in track density in natural materials as a function of temperature and duration of heating. The results of such studies are conveniently described by equations of the form

$$t = Ae^{U/kT} \qquad (15.12)$$

where t is the annealing time for a specific reduction in track density, A is a constant, U is the activation energy in units of kcal/mole or electron volts, k is Boltzman's constant $= 8.6171 \times 10^{-5}$ eV/°K, and T is the absolute temperature in degrees Kelvin.

By taking natural logarithms of both sides of Equation 15.12 and converting them to $\log_{10}$, we obtain a linear equation:

$$\log t = \log A + \frac{U}{2.303kT} \qquad (15.13)$$

This is the equation of a straight line in coordinates of $\log t$ and $1/T$ having a positive slope equal to $U/2.303k$ and an intercept on the y-axis equal to $\log A$. The linear relationship between $\log t$ and $1/T$ permits the extrapolation of laboratory measurements obtained at elevated temperatures and short time periods to lower temperatures and geological time periods (see also Dakowski et al., 1974).

Naeser and Faul (1969) used this technique to present their experimental results of the fading of fission tracks in apatite and sphene. Their data, shown in Figure 15.1, indicate that the tracks in these two minerals, which are common accessories in igneous and metamorphic rocks, fade at very different rates. We can use the diagram to predict track fading in these minerals in response to an increase in temperature. For a heating period of one million years, apatite begins to lose tracks at about 50°C and is completely annealed at about 175°C, while tracks in sphene do not fade unless the temperature is raised to 250°C and the mineral anneals completely only at 420°C. Therefore, fission-track dates of apatite can be completely reset by episodic heating under conditions implied by its track-fading curve for 100 percent track loss. Under the same conditions, sphene loses no tracks at all, and its fission-track date remains unaffected. Therefore, when fission-track dates of these two minerals extracted from the same rock are concordant, the rock presumably cooled rapidly and was not reheated at a later time. On the other hand, when the dates are discordant, that is, the sphene date is greater than the apatite date, the rock either cooled slowly or was reheated to a temperature at which track fading occurred in apatite but not in sphene.

Apatite is a particularly good indicator of the cooling history of a rock because it retains fission tracks only at temperatures that are significantly less than the "blocking temperatures" for Rb-Sr or K-Ar geochronometers in co-existing micas. The exact temperature at which apatite retains all tracks (the 0 percent loss curve in Fig. 15.1) depends on the cooling rate. When the rate of cooling is high, the cooling time is short, and complete track retention occurs at a higher temperature than when the cooling rate is slow. At any given cooling rate, track retention increases from 0 to 100 percent as the temperature drops so that the observed track density represents approximately the time elapsed since the temperature passed through the value at which 50 percent of the tracks are retained. Consequently, fission track dates of minerals can be interpreted

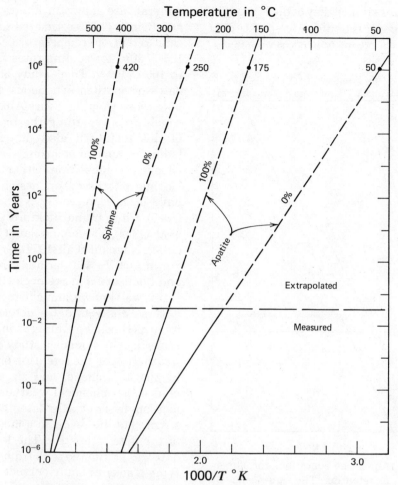

FIGURE 15.1 Fading of fission tracks in apatite and sphene. The lines marked 0 percent indicate temperatures and time periods at which no tracks are lost. The lines labeled 100 percent indicate conditions when all tracks are lost. All fission tracks in apatite are lost when that mineral is heated for one million years at 175°C. Sphene does not begin to lose tracks until the temperature is raised to 250°C for one million years, and it anneals completely only at 420°C. Upon cooling from a high temperature, tracks begin to be accumulated along the line for 100 percent track loss. For example, if apatite cooled from 175°C to 50°C in one million years, some tracks begin to accummulate at 175°C, but complete retention does not occur until the temperature reaches 50°C. The effective temperature recorded by the fission-track date of a mineral is the value at which about 50 percent of the tracks are preserved. (Naeser and Faul, 1969.)

as the time elapsed since they cooled through their 50 percent retention temperatures. Such 50 percent retention curves of several

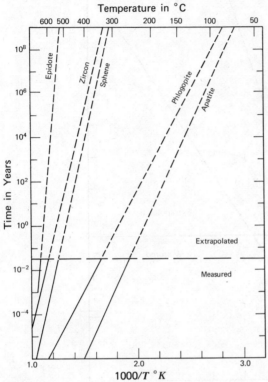

FIGURE 15.2 Annealing conditions for 50 percent track-loss or retention. The significance of the 50 percent track-loss curves is that during cooling from a high temperature, fission-track dates indicate approximately the time elapsed since the temperature dropped through the 50 percent track retention value. The specific value of this temperature depends on the cooling rate. It is evident that epidote is the most refractory of the minerals shown. This means that the fission-track clock in this mineral is activated at the highest temperature and that it is least likely to be disturbed by track fading due to later reheating of the mineral. (Adapted from a compilation of Naeser and Faul, 1969. The epidote annealing curve is from Naeser et al., 1970. The fading of fission tracks in zircon has also been studied by Nishida and Takashima, 1975.)

minerals are shown in Figure 15.2. Wagner and Reimer (1972) estimated that the 50 percent retention temperature for apatite is $125 \pm 20°C$ for cooling rates from $1°C/10^6$ y to $100°C/10^6$ y. Their study of fission-track dates of apatites in igneous and metamorphic rocks from the Swiss Alps led to useful conclusions regarding the tectonic history of this region. It also showed that track fading in apatite is strongly affected by its chemical composition. Other geological investigations based on fission-track dating have been reported by Naeser and Dodge (1969), Church and Bickford (1971), Bigazzi et al. (1972), Naeser and Saul (1974), Hurford (1974), Nagpaul et al. (1974), Crawford (1974), Seward (1975), Naeser and Brookins (1975), and Boellsorff and Steineck (1975).

It is well known that chemical weathering adversely affects dates determined by isotopic methods. For this reason, Gleadow and Lovering (1974) made a study of the effect of weathering on the retention of fission tracks in apatite, sphene, and zircon. They concluded that chemical weathering has no effect on fission-track dates of sphene and zircon, but did result in a modest reduction of the date of apatite. The lowering of the date was attributed to the difficulty in identifying tracks in badly corroded crystals and to partial fading of tracks caused by the action of groundwater. The apparent loss of tracks was partly offset by a lowering of the uranium concentration. The authors drew attention to the fact that zircon crystals in sedimentary rocks might be useful in studies to determine the age of sediment provenance.

4
Alpha-Recoil Tracks

The naturally occurring isotopes of uranium and thorium, and many of their short-lived daughters, decay by emission of alpha particles. The emission of energetic alpha par-

THE FISSION-TRACK METHOD
OF DATING 15

ticles imparts a significant amount of recoil energy to the product nucleus (Chapter 3.5) which causes damage to the surrounding area of the crystal. Huang and Walker (1967) demonstrated that such alpha-recoil tracks can be enlarged by etching and can then be counted with a phase-contrast microscope.

They observed such alpha-recoil tracks in muscovite (etched for two hours at 20°C in 48 percent HF) and phlogopite (etched for 60 seconds). After etching, the alpha-recoil tracks appear in the form of shallow pits with an average depth of about 100 Å. The authors performed a number of experiments which indicated that these pits are produced by the recoil of the nucleus in the course of the sequential emission of alpha particles by the isotopes of uranium, thorium, and their daughters. The alpha-recoil tracks are more abundant than fission tracks by factors ranging from 2.3×10^3 to 4.5×10^3, depending on the Th/U ratio of the sample.

They suggest that alpha-recoil tracks might be useful for dating geological and archaeological objects and pointed out that this method is potentially more sensitive than the conventional fission-track method because of the larger number of tracks produced in unit time. For example, the alpha-recoil method might be successful in dating teeth which are not datable by the fission-track method. It might also permit dating of synthetic glass having low uranium concentrations. The proposed alpha-recoil method of dating requires measurements of the concentrations of uranium and thorium as well as of the track density. However, procedures for dating by this method have not yet been developed.

5
Pleochroic Haloes

When alpha particles interact with the atoms of a crystal lattice, they cause dis-placement of electrons or atoms from their equilibrium positions. The resulting radiation damage manifests itself by a halo of discoloration surrounding uranium and thorium-bearing inclusions in many minerals. The presence of colored aureoles around certain accessory minerals was first described by Michel-Levy (1882), almost 15 years before H. Becquerel announced the discovery of radioactivity. Pleochroic haloes occur not only in colored minerals (biotite, tourmaline, amphibole, and chlorite) but also in colorless minerals (muscovite, cordierite, and fluorite). The inclusions at the centers of the haloes are most often zircon, apatite, and sphene. The connection between pleochroic haloes and radioactivity was first established in 1907 by J. Joly and O. Mügge. Since that time, pleochroic haloes have been studied by many scientists, including Joly (1907, 1923), Poole (1933), Henderson and Bateson (1934), Henderson and Sparks (1939), Deutsch et al. (1956), Deutsch (1960), and Gentry (1974).

Pleochroic haloes contain concentric rings having radii ranging from about 10 to 50 microns. These rings are produced by specific alpha particles emitted by radioactive nuclides residing in the central inclusions. The radii of these rings indicate the ranges and therefore the kinetic energies of the different sets of alpha particles. The rings are formed because the ionizing power of an alpha particle is greatest near the end of its range in a particular medium. The mineral biotite is well suited to the study of the ring structure of pleochroic haloes because of its perfect cleavage and because the discoloration caused by radiation damage is readily visible. Much effort has been expended to identify the nuclides that produce rings with specific radii in biotite. Most of the rings have been successfully identified, based mainly on the work of Joly and Henderson. However, some aspects of this phenomenon remain controversial (Gentry et al., 1974; Moazed et al., 1973).

Pleochroic haloes have also been used for age determinations of rocks and minerals (Joly and Rutherford, 1913; Holmes, 1926; Henderson, 1934; Hayase, 1954; Pasteels, 1960). This possibility arises from the fact that the intensity of the color of pleochroic haloes increases as a function of the alpha dose to which the mineral has been exposed. The dating of minerals by the pleochroic-halo method has been reviewed by Picciotto and Deutsch (1960). The relationship between radiation dose and coloration can be treated in three stages: (1) a normal stage, in which the coloration increases linearly with the radiation dose; (2) a saturated stage, in which the coloration has reached maximum intensity and remains constant with increasing dose; and (3) an inversion stage, in which the coloration decreases in intensity due to extreme radiation damage. Age determinations by the pleochoric-halo method require the assumptions that the coloration is a linear function of the alpha-dose (stage 1) and that the concentrations of radioactive nuclides emitting alpha particles have not changed except by decay processes. In order to date a mineral by this method, one must measure (1) the alpha activity of the inclusion at the center of the pleochroic halo; and (2) the intensity of the color of the halo. In addition, the relationship between alpha-dose and color intensity in the mineral to be dated must be known. This relationship can be established experimentally by irradiating the mineral with alpha particles from an artificial source or by studying haloes in other specimens of that mineral of known age. In principle, dating of minerals (especially biotite) by the pleochroic-halo method is possible. However, the method suffers from several defects: (1) the haloes are partially erased by heating of the mineral during metamorphism; (2) the difficulty in measuring the alpha activity of the inclusions; (3) the uncertainty of the precise relationship between alpha-dose and coloration in biotite. These difficulties limit the accuracy of age determinations based on the study of pleochroic halos. For these reasons, this method has been abandoned in favor of the isotopic dating methods.

6 SUMMARY

Spontaneous fission of ^{238}U in minerals and in natural and synthetic glasses leaves damage tracks that can be enlarged by etching. The number of tracks per unit area is a function of the age of the specimen and its uranium concentration, provided it cooled rapidly and was not reheated later. The uranium concentration can be measured by counting tracks produced by fission of ^{235}U caused by irradiation of the specimen with thermal neutrons in a nuclear reactor. Therefore, fission-track dates of many common minerals (such as micas, apatite, sphene, epidote, and zircon) can be measured without expensive laboratory facilities. In addition, the method can be used to date tektites, volcanic glass, and some archaeological objects.

Fission tracks are known to fade by annealing of solids at elevated temperatures. The annealing rates of different minerals vary within wide limits. Zircon, epidote, and sphene are fairly refractory, while apatite, calcite, and most glasses anneal at temperatures of less than 100°C in less than one million years. Fission-track dates are therefore "cooling ages" and indicate the time elapsed since the temperature dropped below the 50 percent track-retention value.

The fission-track method can be used to date igneous and metamorphic rocks and to study their thermal histories. It may also be useful for dating detrital zircon, and perhaps

THE FISSION-TRACK METHOD OF DATING 15

sphere, in sedimentary rocks and can be used to date archaeological objects, such as glass and ceramics, provided the density of spontaneous-fission tracks is sufficient to permit a determination.

The recoil imparted to the nuclei of atoms by emission of alpha particles causes damage to solids which can be made visible by etching. Such alpha-recoil pits are more numerous than fission tracks and are potentially useful for dating purposes.

Pleochroic haloes are the result of radiation damage caused by alpha particles emitted from small inclusions of radioactive minerals. The ring structure of these haloes reflects the emission of alpha particles of specific energies by radionuclides. Some aspects of this phenomenon are still controversial. Dating of minerals (biotite) by means of pleochroic haloes is possible, but cannot compete with isotopic dating methods.

PROBLEMS

1. Apatite and sphene from the granite at Mt. Ascutney, Vermont, were analyzed by Naeser (1967) with the following results:

	ρ_s (tracks/cm^2)	ρ_i	φ (n/cm^2)
Apatite	4.82×10^5	10.1×10^5	5.88×10^{15}
Sphene	1.95×10^6	4.29×10^6	5.88×10^{15}

Calculate the fission-track dates of both minerals (**ANSWER:** t(apatite) = 138.5×10^6 y, t(sphene) = 132.0×10^6 y.)

2. Calculate the *optimum* irradiation time for a mineral whose age is 100×10^6 y, assuming that the neutron flux density is 5×10^{11} n/cm^2/sec. (**ANSWER:** About 67 minutes.)

3. Church and Bickford (1971) analyzed an apatite from granite of the Sawatch Range in Colorado. The granite has a whole-rock Rb-Sr age of 1650 million years. The rock sample from which the apatite was taken yielded a Rb-Sr mineral-isochron date of 1275 million years. The analytical results for the apatite (A38G-II, 4) are as follows: ρ_s = 20 tracks/0.1 mm^2, ρ_i = 580 tracks/0.1 mm^2, $\varphi = 2.5 \times 10^{16}$ n/cm^2. Calculate a fission-track date and interpret the result.

REFERENCES

FISSION-TRACK DATING

Bar, M., Y. Kolodny, and Y. K. Bentor (1974) Dating faults by fission track dating of epidotes—an attempt. Earth Planet. Sci. Letters, *22*, 157–162.

Bertine, K. K., L. H. Chan, and K. K. Turekian (1970) Uranium determinations in deep-sea sediments and natural waters using fission tracks. Geochim. Cosmochim. Acta, *34*, 641–648.

Bigazzi, G., G. Ferrara, and F. Innocenti (1972) Fission track ages of gabbros from northern Apennines ophiolites. Earth Planet. Sci. Letters, *14*, 242–244.

Boellsorff, J. D., and P. L. Steineck (1975) The stratigraphic significance of fission-track ages on volcanic ashes in the marine Late Cenozoic of southern California. Earth Planet. Sci. Letters, *27*, 143–154.

Burchart, J. R., and G. M. Reimer (1972) Effects of ionic solutions on fission track stability in apatite. Abstract. Am. Nucl. Soc. Trans., *15*, 129.

Burchart, J., M. Dakowski, and J. Galazka (1975) A technique to determine extremely high fission track densities. Bull. Acad. Polon. Sci., Ser. Sci. de la Terre, *23*, 1–8.

Carpenter, B. S., and G. M. Reimer (1974) Calibrated glass standards for fission track use. Nat. Bur. Stds. Spec. Pub. 260, 49 p.

Church, S. E., and M. E. Bickford (1971) Spontaneous fission-track studies of accessory apatite from granitic rocks in the Sawatch range, Colorado. Geol. Soc. Amer. Bull., *82*, 1727–1734.

Crawford, A. R. (1974) Thermal history of Rajasthan pegmatites of India as revealed by fission track studies: Discussion. Can. J. Earth Sci., *11*, 1728.

Dakowski, M., J. Burchart, and J. Galazka (1974) Experimental formula for thermal fading of fission tracks in minerals and natural glasses. Bull. Acad. Polon. Sci., Ser. Sci. de la Terre, *22*, 11–17.

Fisher, D. E. (1975) Geoanalytical applications of particle tracks. Earth-Science Revs., *11*, 291–336.

Fleischer, R. L., and P. B. Price (1964a) Glass dating by fission fragment tracks. J. Geophys. Res., *69*, 331–339.

Fleischer, R. L., and P. B. Price (1964b) Techniques for geological dating of minerals by chemical etching of fission fragment tracks. Geochim. Cosmochim. Acta, *28*, 1705–1714.

Fleischer, R. L., P. B. Price, and R. M. Walker (1969) Nuclear tracks in solids. Scient. American, *220*, 30–39.

Fleischer, R. L., and H. R. Hart, Jr. (1971) Fission track dating: Techniques and problems. In Calibration of Hominoid Evolution: Recent Advances in Isotopic and Other Dating Methods Applicable to the Origin of Man, pp. 135–170, W. W. Bishop and J. A. Miller, eds. Scottish Academic Press, Edinburgh.

Fleischer, R. L., H. W. Alter, S. C. Furman, P. B. Price, and R. M. Walker (1972) Particle track etching. Science, *178*, 255–263.

THE FISSION-TRACK METHOD
OF DATING 15

Fleischer, R. L., R. T. Woods, H. R. Hart, Jr., and P. B. Price (1974) Effect of shock on fission track dating of apatite and sphene crystals from the Hardhat and Sedan underground nuclear explosions. J. Geophys. Res., *79*, 339–342.

Fleischer, R. L., P. Buford, and R. M. Walker (1975) Nuclear Tracks in Solids: Principles and Applications. University of California, Berkeley, 605 p.

Galliker, D., E. Hugentobler, and B. Hahn (1970) Spontane Kernspaltung von U-238 und Am-241. Helv. Phys. Acta, *43*, 593.

Gentner, W., D. Storzer, and G. A. Wagner (1969) Das Alter von Tektiten und verwandten Gläsern. Naturwiss., *56*, 255–260.

Gleadow, A. J. W., and J. F. Lovering (1974) The effect of weathering on fission track dating. Earth Planet. Sci. Letters, *22*, 163–168.

Haack, U. K., and M. J. Potts (1972) Fission track annealing in garnet. Contrib. Mineral. Petrol., *34*, 343–345.

Huang, W. H., and R. M. Walker (1967) Fossil alpha-particle recoil tracks: A new method of age determination. Science, *155*, 1103–1106.

Hurford, A. J. (1974) Fission track dating of a vitric tuff from East Rudolf, North Kenya. Nature, *249*, 236.

Kaufhold, J., and W. Herr (1968) Influence of experimental factors on dating natural and man-made glasses by fission track method. In Radioactive Dating and Methods of Low Level Counting. Int. Atom. Energy Agency, Vienna, 403–413.

Lakatos, S., and D. S. Miller (1972) Evidence for the effect of water content on fission-track annealing in volcanic glass. Earth Planet. Sci. Letters, *14*, 128–130.

Lal, D., A. V. Muralli, R. S. Rajan, A. S. Tammane, J. C. Lorin, and P. Pellas (1968) Techniques for proper revelation and viewing of etch-tracks in meteoritic and terrestrial minerals. Earth Planet. Sci. Letters, *5*, 111–119.

MacDougall, D., and P. B. Price (1974) Attempt to date early South African hominids by using fission tracks in calcite. Science, *185*, 943–944.

McCorkell, R. (1974) Tables for fission track dating. Carleton University, Geol. Paper 74-1, 130 p., Ottawa, Canada.

Naeser, C. W. (1967) The use of apatite and sphene for fission track age determinations. Geol. Soc. Amer. Bull., *78*, 1523–1526.

Naeser, C. W., and F. C. W. Dodge (1969) Fission-track ages of accessory minerals from granitic rocks of the central Sierra Nevada batholith, California. Geol. Soc. Amer. Bull., *80*, 2201–2212.

Naeser, C. W., and H. Faul (1969) Fission track annealing in apatite and sphene. J. Geophys. Res., *74*, 705–710.

Naeser, C. W., J. C. Engels, and F. C. Dodge (1970) Fission-track annealing and age determination of epidote minerals. J. Geophys. Res., *75*, 1579–1584.

Naeser, C. W., and J. M. Saul (1974) Fission track dating of tanzanite. Am. Mineral., *59*, 613–614.

Naeser, C. W., and D. G. Brookins (1975) Comparison of fission-track, K-Ar, and Rb-Sr radiometric age determinations from some granite plutons in Maine. J. Res. U.S. Geol. Serv., *3*, 229–232.

Nagpaul, K. K., P. P. Mehta, and M. L. Gupta (1974) Fission track ages of cogenetic minerals of the Nellore mica belt of India. Pure and Applied Geophys., *112*, 140–148.

Nishida, T., and Y. Takashima (1975) Annealing of fission tracks in zircons. Earth Planet. Sci. Letters, *27*, 257–264.

Price, P. B., and R. M. Walker (1962a) Chemical etching of charged particle tracks in solids. J. Appl. Phys., *33*, 3407.

Price, P. B., and R. M. Walker (1962b) Observation of fossil particle tracks in natural micas. Nature, *196*, 732.

Price, P. B., and R. M. Walker (1963a) A simple method of measuring low uranium concentrations in natural crystals. Appl. Phys. Letters, *2*, 23.

Price, P. B., and R. M. Walker (1963b) Fossil tracks of charged particles in mica and the age of minerals. J. Geophys. Res., *68*, 4847–4862.

Reimer, G. M. (1975) Uranium determination in natural water by the fission-track technique. J. Geochem. Explor., *4*, 425–432.

Schreurs, J. W. H., A. M. Friedman, D. J. Rockop, M. W. Hair, and R. M. Walker (1971) Calibrated U-Th glasses for neutron dosimetry and determination of uranium and thorium concentrations by the fission-track method. Radiation Effects, *7*, 231–233.

Seward, D. (1975) Fission-track ages of some tephras from Cape Kidnappers, Hawke's Bay, New Zealand. N.Z.J. Geol. Geophys., *18*, 507–510.

Silk, E. C. H., and R. S. Barnes (1959) Examination of fission fragment tracks with an electron microscope. Phil. Mag., *4*, 970.

Storzer, D. (1970) Fission-track dating of volcanic glass and the thermal history of rocks. Earth Planet. Sci. Letters, *8*, 55–60.

Storzer, D., and G. A. Wagner (1971) Fission track ages of North American tektites. Earth Planet. Sci. Letters, *10*, 435–440.

Wagner, G. A., and G. M. Reimer (1972) Fission track tectonics: The tectonic interpretation of fission track apatite ages. Earth Planet. Sci. Letters, *14*, 263–268.

Wagner, G. A. (1974) Die Anwendung anätzbarer Partikelspuren zur geochemischen Analyse. Fortschritte Mineral., *51*, 68–93.

Wagner, G. A., G. M. Reimer, B. S. Carpenter, H. Faul, R. Van der Linden, and R. Gijbels (1975) The spontaneous fission rate of U-238 and fission track dating. **Geochim. Cosmochim. Acta.,** *39*, 1279–1286.

PLEOCHROIC HALOES

Deutsch, S., D. Hirschberg, and E. Picciotto (1956) Etude quantitative des halos pleochroiques. Application a l' estimation de l'age des roches granitiques. Bull. Soc. belge Geol., *65*, 267.

Deutsch, S. (1960) Influence de la chaleur sur la coloration des halos pleochroic dans la biotite. Nuovo Cimento, *16*, 269.

Gentry, R. V. (1974) Radiohalos in a radio-chronological and cosmological perspective. Science, *184*, 62–66.

Gentry, R. V., L. D. Hulett, S. S. Cristy, J. F. McLaughlin, J. A. McHugh, and M. Bayard (1974) "Spectacle" array of ^{210}Po halo radiocentres in biotite: A nuclear geophysical enigma. Nature, *252*, 564–565.

Hayase, I. (1954) Relative geologic age measurements on granites by pleochroic haloes and radioactivity of the minerals in their nuclei. Amer. Mineral., *39*, 761.

Henderson, G. H. (1934) A new method of determining the age of certain minerals. Proc. Roy. Soc. (London) *A145*, 591.

Henderson, G. H., and S. Bateson (1934) A quantitative study of pleochroic haloes, I. Proc. Roy. Soc. (London) *A145*, 563–581.

Henderson, G. H., and F. W. Sparks (1939) A quantitative study of pleochroic haloes, IV and V. Proc. Roy. Soc. (London), *A173*, 238–264.

Holmes, A. (1926) Estimates of geologic time with special reference to thorium and uranium haloes. Phil Mag. *1*, 1055.

Joly, J. J. (1907) Pleochroic haloes. Phil. Mag., *13*, 381.

Joly, J. J., and E. Rutherford (1913) The age of pleochroic haloes. Phil. Mag., *25*, 644.

Joly, J. J. (1923) Pleochroic haloes of various geological ages. Proc. Roy. Soc. (London), *A102*, 682–705.

Michel-Levy, A. (1882) Sur les noyaux a polychroisme intense du mica noir. Compt. Rend., *94*, 1196.

Moazed, C., R. M. Spector, and R. F. Ward (1973) Polonium radiohaloes: An alternate interpretation. Science, *180*, 1272–1274.

Mügge, O. (1909) Radioaktivität und pleochroitische Höfe. Centr. Mineral. Geol., *71*, 113–142.

Pasteels, P. (1960) L' age des halos pleochroiques du granite d' Habkern et de quelques roches du Massif de l'Aar. Bull. Swisse Min. et Petr., *40*, 261.

Picciotto, E. E., and S. Deutsch (1960) Pleochroic haloes. Summer course in Nuclear Geology Varenna, Comitato Nazionale per l' Energia Nucleare, Laboratorio de Geologia Nucleare Pisa, 263–310.

Poole, J. H. J. (1933) Radioactivity of samarium and the formation of Hibernium haloes. Nature, *131*, 654.

16 THE U-SERIES DISEQUILIBRIUM METHODS OF DATING

The decay series arising from ^{238}U and ^{235}U (Figs. 12.1 and 12.2) contain radioactive isotopes of many different elements. These daughters of uranium may be separated from their parents and from each other in the course of normal geological processes such as chemical weathering, precipitation of minerals from aqueous solutions by biological and inorganic processes, adsorption on suspended clay minerals, and even during the formation of magma at depth in the Earth and the crystallization of lava flows on its surface. Such processes break the radioactive-decay chains because of differences in the geochemical properties of the daughters of uranium. As a result, two kinds of situations arise that have been exploited by geochronologists: (1) a member of the series is separated from its parent and subsequently decays at a rate determined by its half-life; (2) a daughter nuclide is formed by decay of its parent (previously separated from its daughter) until radioactive equilibrium is reestablished.

These phenomena give rise to several geo-chronometers capable of measuring geologic time ranging from a few years up to one million years or more. These geochronometers therefore provide information about the Earth's history during the last million years and fill an important gap between the carbon-14 and the K-Ar methods of dating. Only the fission-track method competes with them in this respect.

The disequilibrium methods of dating have been most useful in the study of sediment and calcium carbonate deposited in the oceans and in lakes during the Pleistocene epoch. The relevant geochronometers are based on measurements of the activity ratios of $^{230}Th/$ ^{232}Th (the ionium method), $^{234}U/^{238}U$, $^{230}Th/^{238}U$, and $^{230}Th/^{234}U$. All of these nuclides are daughters of ^{238}U. In addition, we shall consider the $^{230}Th/^{231}Pa$ method of dating where ^{231}Pa is a daughter of ^{235}U. The daughters of uranium are also useful in dating snow and ice by the ^{210}Pb method, which has been extended recently to marine and lacustrine sediment deposited in the last 100 years. Table 16.1 contains the half-lives

Table 16.1 Half-Lives and Decay Constants of the Daughters of Uranium Used in Geochronometry

NUCLIDE	HALF-LIFE, YEARS	DECAY CONSTANT, y^{-1}
$^{234}_{92}U$	2.48×10^5	2.794×10^{-6}
$^{230}_{90}Th$	7.52×10^4	9.217×10^{-6}
$^{226}_{88}Ra$	1.622×10^3	4.272×10^{-4}
$^{210}_{82}Pb$	22.26	3.11×10^{-2}
$^{231}_{91}Pa$	3.248×10^4	2.134×10^{-5}

and decay constants of the daughters of uranium that are used most often in geochronometry.

For practical reasons, the abundances of the daughters of uranium are measured in terms of their activities by means of sensitive radiation detectors. The equations to be derived in later sections of this chapter will therefore be phrased in terms of activities rather than in terms of numbers of atoms. The reader is reminded that the activity A of a radioactive nuclide is related to the number of atoms N by the equation

$$A = c\lambda N \qquad (16.1)$$

where λ is the decay constant and c is the counting efficiency. The fundamental equations describing radioactive decay and growth have been presented in Chapter 4.

1
The Ionium Method of Dating Deep-Sea Sediment

Ionium (^{230}Th) is a radioactive isotope of thorium that is produced in the decay series of ^{238}U (Fig. 12.1). Its immediate parent is ^{234}U and its daughter is ^{226}Ra. The ionium method of dating is based on the separation of thorium isotopes from uranium in the oceans as a consequence of differences in their chemical properties. In oxidizing environments on the surface of the Earth, uranium occurs as the uranyl ion UO_2^{+2}, most probably as complexes such as $(UO_2)(CO_3)_3^{-4}$. Thorium, on the other hand, remains in the tetravalent state and is rapidly removed from sea water by adsorption on the surfaces of solids and by incorporation into authigenic minerals such as phillipsite (zeolite) or barite, while uranium is concentrated in the aqueous phase. As a result, the residence time of uranium in the oceans is about 500,000 years, whereas that of thorium is only 300 years. The residence time is defined as

$$t = \frac{A}{dA/dt} \qquad (16.2)$$

where A is the total amount of an element in the dissolved state in the oceans, and dA/dt is the amount of that element added to or precipitated from the oceans per year.

Most of the ionium in sea water forms by decay of ^{234}U in the oceans and the remainder (about 25 percent) is contributed by streams together with ^{232}Th. The preferential removal of ionium (and other thorium isotopes) from sea water separates it from its parent and gives rise to an excess of unsupported ionium in the sediment. The activity resulting from unsupported ionium in the sediment decreases as a function of time as ionium decays to ^{226}Ra with a half-life of 7.52×10^4 years.

The ionium-thorium method of dating deep sea sediment is based on the assumption that ^{230}Th and the common long-lived isotope ^{232}Th are removed simultaneously from sea water (Picciotto and Wilgain, 1954). Since ^{232}Th has a very long half-life (1.40×10^{10} years), the activity ratio of ionium to thorium (^{230}Th/^{232}Th)$_A$ decreases as a function of time only because of the decay of the unsupported ionium. The age of a sample taken at some depth h below the water-sediment interface can then be calculated from its observed ionium/thorium activity ratio, provided the following assumptions are satisfied (Goldberg and Koide, 1962):

1. The ^{230}Th/^{232}Th ratio in the water mass adjacent to the sediment in a given ocean basin has remained constant during the last several hundred thousand years.
2. Ionium and thorium have the same chemical speciation in sea water and there is no isotopic fractionation between sea water and the mineral phases with which the thorium is associated in the sediment.

3. The ionium and thorium occurring in detrital mineral particles are excluded from the analysis.

4. Thorium isotopes do not migrate in the sediment.

The last assumption has been the source of some concern. While ^{230}Th probably does not migrate, there is reason to believe that its parent ^{234}U is mobile (Ku, 1965) and therefore may contribute to the ^{230}Th activity. However, in most instances the ^{230}Th that is supported by migratory ^{234}U is a negligibly small fraction of the total ^{230}Th activity, at least in the upper part of sediment cores.

The analytical procedure for measuring the ionium/thorium activity ratios in sediment samples has been described in detail by Goldberg and Koide (1962). The sediment is treated with hot hydrochloric acid which liberates the thorium that is adsorbed on detrital particles or resides in authigenic phases. The principal detrital minerals (feldspar and quartz) are not dissolved by this procedure which thus permits the recovery of only that fraction of ionium and thorium that entered the sediment from the sea water. After purification, the thorium isotopes are electroplated onto platinum discs, and the alpha activities of ^{230}Th and ^{232}Th are measured with solid-state detectors or ionization chambers.

The total activity of ^{230}Th that is adsorbed on grain surfaces or is present in authigenic minerals can be represented as the sum of the activities due to decay of unsupported ^230Th plus U-supported ^{230}Th:

$$^{230}\text{Th}_A = {}^{230}\text{Th}_{Ax} + {}^{230}\text{Th}_{As} \quad (16.3)$$

where

$^{230}\text{Th}_A$ = total activity of ^{230}Th in the leachable fraction of the sediment at some depth h below the top in units of

disintegrations per unit time per unit weight of dry sediment,

$^{230}\text{Th}_{Ax}$ = the same for the excess, unsupported fraction of ^{230}Th, and

$^{230}\text{Th}_{As}$ = the same for the fraction of ^{230}Th that is supported by decay of ^{234}U. $\}$ *hopefully minor*

Let us consider first the decay of the excess, unsupported fraction of ^{230}Th only. The activity of this fraction at any time t can be expressed as

$$^{230}\text{Th}_{Ax} = {}^{230}\text{Th}^0_{Ax}e^{-\lambda_{230}t} \quad (16.4)$$

where $N = N_0 e^{-\lambda t}$

$^{230}\text{Th}^0_{Ax}$ = activity of excess, unsupported ionium in the sediment at $t = 0$, that is, in freshly deposited sediment at the water-sediment interface,

λ_{230} = decay constant of ^{230}Th (9.217 × 10^{-6} y^{-1}), and

t = time elapsed since deposition of the sediment.

Following the suggestion of Picciotto and Wilgain (1954), we use the activity of long-lived ^{232}Th as a reference and express the activity of unsupported ionium as the activity ratio:

$$\left(\frac{^{230}\text{Th}}{^{232}\text{Th}}\right)_{Ax} = \left(\frac{^{230}\text{Th}}{^{232}\text{Th}}\right)^0_{Ax} e^{-\lambda_{230}t} \quad (16.5)$$

If the contribution of U-supported ionium is negligibly small (Goldberg and Koide, 1962), then we have a very simple relationship between the activity ratio (R) of a given sediment sample and its age:

$$R = R_0 e^{-\lambda t} \quad (16.6)$$

where R is the ionium/thorium activity ratio of the leachable fraction of sediment taken at some depth h below the top of a core, corrected for U-supported ionium, as necessary, R_0 is the same for freshly deposited sediment, λ is the decay constant of ^{230}Th, and t is the time elapsed since deposition. After R has been measured and if R_0 is known, Equation 16.6 can be solved for t:

$$t = \frac{2.303}{\lambda} \log \left(\frac{R_0}{R} \right) \qquad (16.7)$$

Now let us return to a consideration of the activity of U-supported ^{230}Th in the leachable fraction of marine sediment. Its activity as a function of time is derivable from Equation 4.21 (Chapter 4) which we restate here:

$$N_2 = \frac{\lambda_1}{\lambda_2 - \lambda_1} N_1{}^0 (e^{-\lambda_1 t} - e^{-\lambda_2 t}) \quad (16.8)$$

where N_1 and λ_1 refer to the parent and N_2, λ_2 refer to the radioactive daughter. Therefore, the number of U-supported ^{230}Th atoms is

$$^{230}\text{Th}_s = \frac{\lambda_{234}}{\lambda_{230} - \lambda_{234}} {}^{234}\text{U}^0 (e^{-\lambda_{234} t} - e^{-\lambda_{230} t})$$

$$(16.9)$$

If ^{234}U is in secular equilibrium with its great grandparent ^{238}U, its activity is equal to that of ^{238}U (Chapter 4, Equation 4.29):

$$^{234}\text{U}^0 \lambda_{234} = {}^{238}\text{U}_A$$

Because of the very long half-life of ^{238}U ($T_{1/2} = 4.51 \times 10^9$ y), its activity is invariant with time over periods of several million years. The activity of ^{234}U is thus controlled by the decay constant of ^{238}U which is very much smaller than that of ^{230}Th. Therefore

$$\lambda_{230} - \lambda_{234} \simeq \lambda_{230}$$

$$e^{-\lambda_{234} t} \simeq 1$$

With these modifications, we now rewrite Equation 16.9 as

$$^{230}\text{Th}_{As} = {}^{238}\text{U}_A (1 - e^{-\lambda_{230} t}) \quad (16.10)$$

The total activity of ^{230}Th in the leachable fraction of sediment of age t is therefore

$$^{230}\text{Th}_A = {}^{230}\text{Th}_{Ax}^0 e^{-\lambda_{230} t} + {}^{238}\text{U}(1 - e^{-\lambda_{230} t})$$

$$(16.11)$$

Finally, we again introduce the activity of long-lived ^{232}Th as a reference and obtain

$$\left(\frac{^{230}\text{Th}}{^{232}\text{Th}} \right)_A = \left(\frac{^{230}\text{Th}}{^{232}\text{Th}} \right)_{Ax}^0 e^{-\lambda_{230} t}$$

$$+ \left(\frac{^{238}\text{U}}{^{232}\text{Th}} \right)_A (1 - e^{-\lambda_{230} t}) \quad (16.12)$$

This equation describes in a general way the variation of the ^{230}Th/^{232}Th activity ratio as a function of time due to the decay of excess, unsupported ionium and the growth of U-supported ionium. Note that the magnitude of the first term of Equation 16.12 decreases with time, whereas the second term increases. In other words, the contribution by U-supported ionium may well be negligible at the top of a given core, but eventually it becomes dominant for older sediment at some depth sub-bottom.

We now return to the simple case represented by Equations 16.6 and 16.7 where the activity of U-supported activity of ^{230}Th is assumed to be negligible. The ionium-thorium method has been used most often to determine the rate of sedimentation in the open oceans. If the sedimentation rate is a constant equal to $a = h/t$, where h is the depth below the water-sediment interface, then

$$R = R_0 e^{-\lambda h/a} \qquad (16.13)$$

and

$$\log R = \log R_0 - \frac{\lambda h}{2.303 a} \qquad (16.14)$$

This is the equation of a straight line in co-ordinates of $\log R$ and h whose slope is

$$m = -\frac{\lambda}{2.303 a}$$

The sedimentation rate is therefore indicated by the slope of a straight line fitted to data points whose coordinates are log R and h:

$$a = \frac{-2.303m}{\lambda} \qquad (16.15)$$

Goldberg and Koide (1962) recognized four different patterns of variation of ionium/thorium activity ratios in sediment cores from around the world. These are shown in Figure 16.1. Type A illustrates the linear decrease of log R with depth that is expected in sediment deposited at a constant rate. Type B is a pattern consisting of two straight-line segments indicating a change in the rate of deposition. Type C has a constant ionium/thorium ratio near the top, followed by a linear decrease of log R with depth. This pattern results from mixing of sediment at the top by burrowing organisms and/or by currents. The initial activity ratio in this case is obtained by extrapolating the straight-line portion to the water-sediment interface ($h = 0$). In the type D pattern, the activity ratio actually *increases* with depth and eventually levels off at a constant value at greater depth. Such a pattern results from the growth of ionium from uranium in the leachable fraction of the sediment.

An example of a type A pattern is shown in Figure 16.2 for a piston core from the Indian Ocean (Monsoon 49G, 14°27′S and 78°03′E, water depth 5214 meters, Goldberg and Koide, 1963). The data points fit a straight line whose slope yields a sedimentation rate of 2.75 mm/10³ y. This value is

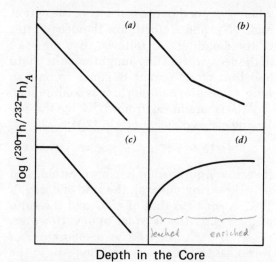

FIGURE 16.1 Variation of the logarithm of the activity of ^{230}Th/^{232}Th with depth in cores of marine sediment. Type A is a normal pattern implying a constant rate of sediment accumulation. Type B is also normal, but indicates a change in the sedimentation rate. Type C has a constant activity ratio at the top of the core due to mixing of sediment by burrowing organisms and/or currents. The rate of deposition is indicated by the slope of the straight-line portion of the pattern. Type D results from the dominance of U-supported ionium that grows in and is then maintained at a constant level. (Goldberg and Koide, 1962.)

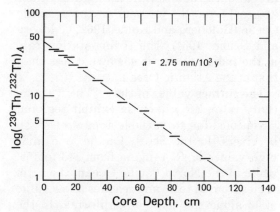

FIGURE 16.2 ^{230}Th/^{232}Th profile of core Monsoon 49G from the Indian Ocean. Note that the activity ratio is constant in the upper 6 cm of the core and that U-supported ionium shows up strongly in the sample from a depth of 130 cm. The remaining points fit a straight line whose slope indicates a constant sedimentation rate $a = 2.75$ mm/10³ y (see Equation 16.14). The surface value of the ^{230}Th/^{232}Th ratio is 42. (Goldberg and Koide, 1963.)

reduced to 1.1 mm/y[3] when biogenic opal (about 57 percent) is removed. The extrapolated Io/Th activity ratio of the sediment at the top of the core is 42. The ratios of the upper 6 cm of the core have been lowered, presumably by burrowing organisms and/or currents. The U-supported component of ionium is significant only at a depth of about 130 cm as indicated by the deviation of that data point from the straight line.

The sedimentation rate in the oceans depends on the input of detrital sediment via rivers, by melting icebergs and glaciers, and by transport through the atmosphere. In addition, biogenic opal and calcium carbonate may contribute significantly to the sedimentation rate. Studies by the ionium/thorium method have indicated rates of accumulation ranging from less than 0.5 mm/10^3 y up to 50 mm/10^3 y on a carbonate and opal-free basis. In general, the lowest deposition rates (0.3 to 0.6 mm/10^3 y) occur in the South Pacific, while the highest values (0.8 to 17 mm/10^3 y) occur in the Atlantic Ocean (Goldberg and Koide, 1962; Goldberg and Griffin, 1964). The sedimentation rate in the Indian Ocean is somewhat less than that of the Atlantic Ocean.

The surface values of the ^{230}Th/^{232}Th activity ratios (R_0) likewise exhibit regional variations due to variable inputs of ^{232}Th by rivers. In the South Pacific, one finds high values of R_0 ranging from 143 to 158, while in the North Pacific Ocean, the ratio varies from 25 to 58 and reaches a low value of 16 along the coast of California. In the Atlantic, R_0 has still lower values ranging from 1.5 to 19, consistent with a larger input of ^{232}Th by the major rivers flowing into this ocean. The R_0 values of the Indian Ocean are intermediate and range from 28 to 80. An alternative interpretation of ionium-thorium ratios in sediment has been proposed by Bernat and Allegre (1974) and is presented in section 6 of this chapter.

2
^{234}U Disequilibrium and the ^{234}U-^{238}U Geochronometer

The decay series arising from ^{238}U, ^{235}U, and ^{232}Th are capable of achieving a state of secular equilibrium, provided that none of the daughters can escape from or enter the system (Chapter 4). When secular equilibrium has been established in a rock or mineral, the decay rates of the daughters of a given decay series are equal to that of their parent:

$$\lambda_1 N_1 = \lambda_2 N_2 = \lambda_3 N_3 = \cdots \lambda_n N_n \qquad (16.16)$$

where $\lambda_1 N_1$ is the rate of decay of the parent and $\lambda_2 N_2$, and so forth are the decay rates of the daughters. It follows, that the *ratio* of the decay rate of any daughter in the chain and that of its parent is equal to one, so long as secular equilibrium is maintained.

^{234}U is produced from ^{238}U by the following series of decays (Fig. 12.1):

$$^{238}\text{U} \xrightarrow{\alpha} {}^{234}\text{Th} \xrightarrow{\beta^-} {}^{234}\text{Pa} \xrightarrow{\beta^-} {}^{234}\text{U}$$

If secular equilibrium has been established in a U-bearing mineral, the rate of decay of ^{234}U is equal to that of ^{238}U and the ratio of their activities is equal to one. However, research begun in 1953 at a laboratory of Kazakhstan University at Alma-Ata in the U.S.S.R. suggested that ^{234}U may not be in secular equilibrium with ^{238}U in samples of groundwater and in secondary minerals deposited from it. Further work in the U.S.S.R. (Starik et al., 1958; Baranov et al., 1958) and later in the United States (Thurber, 1962) revealed wide variations in the activity ratios of ^{234}U/^{238}U of many different kinds of samples collected from the weathering zone of the Earth's surface. A comprehensive discussion of this phenomenon has been published by Cherdyntsev (1969).

In general, the ^{234}U/^{238}U activity ratio of water and of secondary U-bearing minerals on the continents is *greater* than unity due to a significant enrichment of ^{234}U relative to ^{238}U. This enrichment may result from the preferential leaching of ^{234}U from primary U-bearing minerals and from the decay of ^{234}Th dissolved in groundwater. The alpha decay of ^{238}U results in extensive damage to the crystal lattice which permits ^{234}U to migrate into microcapillary fractures in the minerals where it can be oxidized to form the uranyl ion which is soluble in water. Thus ^{234}U is removed into the aqueous phase in preference to ^{238}U which occupies stable lattice positions. In addition, ^{234}Th may be ejected from grain surfaces due to recoil during alpha-decay of ^{238}U. This isotope then decays rapidly to ^{234}U through short-lived ^{234}Pa (Kigoshi, 1971; Kronfeld, 1974). For these reasons, the activity ratios of $^{234}U/^{238}U$ of ground and surface water and of secondary uranium minerals on the continents are commonly greater than one and may reach values up to ten. On the other hand, the activity ratios of primary uranium minerals in rocks exposed to chemical weathering may be less than one because ^{234}U has been preferentially removed from them.

Uranium in solution in groundwater may ultimately enter the oceans where it is isotopically homogenized. It has been shown that the activity ratio of ^{234}U to ^{238}U in sea water varies within fairly narrow limits and has an average value of about 1.15 (Koide and Goldberg, 1965; Miyake et al. 1966). Uranium is removed from the oceans both by adsorption on grain surfaces under reducing conditions and by incorporation into authigenic minerals, including calcium carbonate. After the uranium has been isolated from sea water, the excess ^{234}U decays to ^{230}Th with its characteristic half-life of 2.48×10^5 years until the activity ratio of ^{234}U to ^{238}U approaches the equilibrium value. Ku (1965) evaluated the feasibility of dating pelagic sediment in the oceans by measuring time-dependent changes of the $^{234}U/^{238}U$ activity ratio of leachable uranium. He found that this dating method is not reliable due to postdepositional migration of ^{234}U. However, the decay of excess ^{234}U and the growth of its daughter ^{230}Th can be used for dating marine and nonmarine carbonates of Pleistocene age.

The $^{234}U/^{238}U$ activity ratio can be used to calculate the age of a specimen of calcium carbonate, provided that the initial activity ratio is known and the sample has remained closed to uranium and the daughters between ^{238}U and ^{234}U.

The activity of ^{234}U at any time t after formation of a mineral system is

$$^{234}U_A = {}^{234}U_{As} + {}^{234}U_{Ax} \quad (16.17)$$

where

$^{234}U_A$ = activity of ^{234}U per unit weight of sample at the present time,

$^{234}U_{As}$ = activity of ^{234}U in secular equilibrium with ^{238}U, and

$^{234}U_{Ax}$ = activity of excess ^{234}U per unit weight of sample.

The activity of excess ^{234}U decreases with time according to

$$^{234}U_{Ax} = {}^{234}U_{Ax}^0 e^{-\lambda_{234}t} \quad (16.18)$$

where $^{234}U_{Ax}^0$ is the initial activity of excess ^{234}U, which is expressible as

$$^{234}U_{Ax}^0 = {}^{234}U_A^0 - {}^{234}U_{As} \quad (16.19)$$

where $^{234}U_A^0$ is the total initial activity of ^{234}U and $^{234}U_{As}$ is the component that is supported by ^{238}U. Since $^{234}U_{As} = {}^{238}U_A$, we combine Equations 16.17, 16.18, and 16.19:

$$^{234}U_A = {}^{238}U_A + ({}^{234}U_A^0 - {}^{238}U_A)e^{-\lambda_{234}t}$$

$$(16.20)$$

Dividing by the constant activity of ^{238}U, we obtain

$$\left(\frac{^{234}U}{^{238}U}\right)_A = 1 + \left(\frac{^{234}U_A^0 - {^{238}U_A}}{^{238}U_A}\right)e^{-\lambda_{234}t}$$

$$(16.21)$$

Setting

$$\left(\frac{^{234}U}{^{238}U}\right)_A^0 = \gamma_0$$

$$\left(\frac{^{234}U}{^{238}U}\right)_A = 1 + (\gamma_0 - 1)e^{-\lambda_{234}t}$$

$$(16.22)$$

where γ_0 is the initial $^{234}U/^{238}U$ activity ratio of the sample.

Equation 16.22 has been plotted in Figure 16.3 to show the decrease of $(^{234}U/^{238}U)_A$ as a function of time for the case that $\gamma_0 = 1.15$. The ^{234}U-^{238}U geochronometer has been used to date calcium carbonate of biogenic and inorganic origin deposited both in marine and nonmarine environments (Thompson et al., 1975; Harmon et al., 1975). The reliability of dates of *nonmarine carbonates* is limited by the uncertainty of γ_0 which exhibits both local and time-dependent vari-

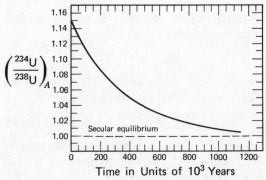

FIGURE 16.3 Decay of excess ^{234}U-activity toward equilibrium with its great-grandparent ^{238}U (Equation 16.22). The initial activity ratio $^{234}U/^{238}U = 1.15$ which is characteristic of uranium in the oceans and in authigenic U-bearing minerals deposited from sea water.

ations. Moreover, it has been found that the shells of both marine and nonmarine mollusks are often unsuitable for dating because they take up uranium after the death of the animal (Kaufman et al., 1971). The method has been most successful in dating corals (Veeh, 1966). In addition, the $^{234}U/^{238}U$ activity ratio of groundwater can be used as a tracer to characterize water masses and to study mixing processes (Osmond et al., 1974; Kronfeld et al., 1975).

3 The ^{230}Th-^{238}U and ^{230}Th-^{234}U Methods of Dating

Calcium carbonate deposited in the oceans and in saline lakes contains appreciable concentrations of uranium (0.1 to 5 ppm) but is virtually free of thorium. As a result, the initial activity of ^{230}Th in some freshly deposited calcium carbonates is essentially equal to zero. After deposition, the activity of ^{230}Th increases as a function of time by decay of ^{234}U.

The activity of ^{230}Th produced by decay of ^{234}U which is in secular equilibrium with ^{238}U was derived earlier (Equation 16.10):

$$^{230}Th_{As} = {^{238}U_A}(1 - e^{-\lambda_{230}t}) \qquad (16.23)$$

A complication arises because of the presence of excess ^{234}U which contributes to the activity of ^{230}Th until the excess has decayed away. The number of ^{230}Th atoms produced by decay of the excess ^{234}U is given by Equation 16.9:

$$^{230}Th_x = \frac{\lambda_{234}}{\lambda_{230} - \lambda_{234}}\,{^{234}U_x^0}(e^{-\lambda_{234}t} - e^{-\lambda_{230}t})$$

$$(16.24)$$

where $^{234}U_x^0$ is the number of excess ^{234}U atoms in a unit weight of carbonate at the time of deposition. Converting to activities as before,

$$^{230}\text{Th}_{Ax} = {}^{230}\text{Th}_x \lambda_{230}$$

$$= \frac{\lambda_{230}}{\lambda_{230} - \lambda_{234}} \, {}^{234}\text{U}^0_{Ax}(e^{-\lambda_{234}t} - e^{-\lambda_{230}t})$$

$$(16.25)$$

The excess activity of ^{234}U is equal to

$$^{234}\text{U}^0_{Ax} = {}^{234}\text{U}^0_A - {}^{234}\text{U}_{As} \qquad (16.26)$$

where $^{234}\text{U}^0_A$ is the total activity of this isotope at the time of deposition and $^{234}\text{U}_{As}$ is the activity of ^{234}U that is in secular equilibrium with ^{238}U. Therefore

$$^{230}\text{Th}_{Ax} = \frac{\lambda_{230}}{\lambda_{230} - \lambda_{234}} \, ({}^{234}\text{U}^0_A - {}^{234}\text{U}_{As})$$

$$\times \, (e^{-\lambda_{234}t} - e^{-\lambda_{230}t}) \qquad (16.27)$$

and the activity of $^{230}\text{Th}_x$ relative to ^{238}U is

$$\left(\frac{^{230}\text{Th}}{^{238}\text{U}}\right)_{Ax} = \frac{\lambda_{230}}{\lambda_{230} - \lambda_{234}} \left(\frac{^{234}\text{U}^0_A - {}^{234}\text{U}_{As}}{^{238}\text{U}_A}\right)$$

$$\times \, (e^{-\lambda_{234}t} - e^{-\lambda_{230}t}) \qquad (16.28)$$

A complete description of the activity ratio of ^{230}Th to ^{238}U as a function of time is therefore given by the equation (Broecker, 1963)

$$\left(\frac{^{230}\text{Th}}{^{238}\text{U}}\right)_A = (1 - e^{-\lambda_{230}t}) + \frac{\lambda_{230}}{\lambda_{230} - \lambda_{234}}$$

$$\times \, (\gamma_0 - 1)(e^{-\lambda_{234}t} - e^{-\lambda_{230}t})$$

$$(16.29)$$

where

$$\gamma_0 = \frac{^{234}\text{U}^0_A}{^{238}\text{U}_A},$$

and

$$^{234}\text{U}_{As} = {}^{238}\text{U}_A.$$

Equations 16.23 and 16.29 have been plotted in Figure 16.4 as curves A and B, respectively. These equations can be used to date corals, calcareous oozes, and mollusk shells

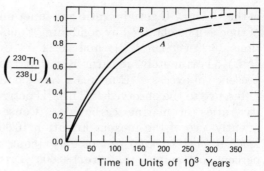

FIGURE 16.4   Growth of ^{230}Th by decay of ^{234}U toward equilibrium with ^{238}U in U-bearing minerals which initially contain no ^{230}Th. Curve A represents the simple case (Equation 16.23) in which the initial activity ratio of $^{234}\text{U}/^{238}\text{U}$ is equal to one, that is, ^{234}U is in secular equilibrium with its parent ^{238}U. Curve B illustrates the growth of ^{230}Th in a mineral whose initial activity ratio $^{234}\text{U}/^{238}\text{U} = 1.15$, (Equation 16.29), that is, ^{234}U is not in secular equilibrium with its parent ^{238}U.

on the basis of the activity ratio of ^{230}Th to ^{238}U, provided the following assumptions are satisfied:

1. The initial $^{230}\text{Th}/^{238}\text{U}$ activity ratio is close to zero.
2. The sample must be a closed system with respect to uranium and intermediate nuclides between ^{238}U and ^{230}Th.
3. The initial activity ratio of $^{234}\text{U}/^{238}\text{U}$ must be known.

Alternatively, the activity of ^{230}Th can be referred to that of its immediate parent ^{234}U. If ^{234}U is in secular equilibrium with ^{238}U, then the $^{230}\text{Th}/^{234}\text{U}$ activity ratio is simply related to time by the equation

$$\left(\frac{^{230}\text{Th}}{^{234}\text{U}}\right)_A = 1 - e^{-\lambda_{230}t} \qquad (16.30)$$

Equation 16.30 is inexact in the sense that it does not allow for the presence of excess ^{234}U. In case $\gamma_0 > 1$, a complete description of the ^{230}Th/^{234}U activity ratio as a function of time can be obtained by combining Equations 16.29 $(^{230}$Th/^{238}U$)_A$ and 16.22 $(^{234}$U/^{238}U$)_A$. Fortunately, it turns out that the presence of excess ^{234}U makes a very small difference to the observed ^{230}Th/^{234}U activity ratios of marine carbonates. Consequently, use of the inexact Equation 16.30 leads to *overestimates* of t of only about 2 percent for samples that are 130,000 years old.

The ^{230}Th-^{234}U method was used by Broecker et al. (1968) to date corals from the three lowest terraces of Barbados in the Lesser Antilles. This island appears to be rising from the sea at a uniform rate and contains a sequence of at least 18 major terraces. As a rule, these terraces represent coral reefs whose ages decrease with decreasing elevation above present sea level and record different stands of sea level in the past. Broecker et al. (1968) determined ^{230}Th/^{234}U activity ratios of corals from the three lowest terraces and calculated dates of 122,000, 103,000, and 82,000 years for them. For example, their sample 1064-C (*Montastria annularis*) from the lowest terrace (Barbados I) has a ^{230}Th/^{234}U activity ratio of 0.53 ± 0.02. In order to calculate the age of this coral, we solve Equation 16.30 for t:

$$t = -\frac{2.303}{\lambda_{230}} \log \left[1 - \left(\frac{^{230}\text{Th}}{^{234}\text{U}} \right)_A \right] \quad (16.31)$$

Setting $\lambda_{230} = 9.217 \times 10^{-6} \text{ y}^{-1}$, we obtain the age:

$$t = -\frac{2.303}{9.217 \times 10^{-6}} (-0.3279) = 81,930 \text{ years}$$

We can use this date to calculate the initial ^{234}U/^{238}U activity ratio by means of Equation 16.22 and the present value of this ratio which was found to be 1.12 ± 0.01. Solving Equation 16.22 for γ_0, we obtain

$$\gamma_0 = (\gamma - 1)e^{\lambda_{234}t} + 1$$

where $\gamma = (^{234}$U/^{238}U$)_A$ and $\lambda_{234} = 2.794 \times 10^{-6} \text{ y}^{-1}$. Thus

$$\gamma_0 = 0.12 \times 1.2572 + 1 = 1.15$$

In other words, the initial activity ratio of ^{234}U/^{238}U in this coral is equal to that of sea water. Broecker et al. (1968) pointed out that the ages of corals on the three lowest terraces in Barbados coincide with times of increased summer radiation at 45°N latitude predicted by Milankovitch on the basis of celestial mechanics.

4
The Ionium-Protactinium Method of Dating

^{231}Pa is a daughter of ^{235}U and is related to it by the following series of decay events (Fig. 12.2)

$$^{235}\text{U} \xrightarrow{\alpha} {}^{231}\text{Th} \xrightarrow{\beta^-} {}^{231}\text{Pa}$$

The geochemical properties of protactinium in the oceans are similar to those of thorium in the sense that the isotopes of both elements are rapidly removed from sea water by adsorption on mineral grains and by incorporation in authigenic minerals. Therefore, sediment deposited in the oceans contains not only an excess of unsupported ^{230}Th (ionium) but also unsupported ^{231}Pa. If these two nuclides are removed from sea water with equal efficiency, the activity ratio of excess ^{230}Th to excess ^{231}Pa changes as a function of time at a rate controlled by the difference in their decay constants:

$$\left(\frac{^{230}\text{Th}}{^{231}\text{Pa}} \right)_{Ax} = \left(\frac{^{230}\text{Th}}{^{231}\text{Pa}} \right)_{Ax}^{0} \frac{e^{-\lambda_{230}t}}{e^{-\lambda_{231}t}} \quad (16.32)$$

where:

$$\left(\frac{^{230}\text{Th}}{^{231}\text{Pa}} \right)_{Ax} = \begin{array}{l} \text{activity ratio of the} \\ \text{unsupported nuclides per} \\ \text{unit weight of sediment} \\ \text{at any time } t, \end{array}$$

$$\left(\frac{^{230}\text{Th}}{^{231}\text{Pa}}\right)^0_{Ax} = \text{the same at the time of deposition,}$$

λ_{231} = decay constant of ^{231}Pa
$(2.134 \times 10^{-5}\,\text{y}^{-1})$, and

λ_{230} = decay constant of
^{230}Th $(9.217 \times 10^{-6}\,\text{y}^{-1})$.

Equation 16.32 implies the assumption that the uranium concentration of the leachable fraction of the sediment is negligible. If that is true, the effective decay constant of the unsupported $^{230}\text{Th}/^{231}\text{Pa}$ activity ratio can be found from

$$\left(\frac{^{230}\text{Th}}{^{231}\text{Pa}}\right)_{Ax} = \left(\frac{^{230}\text{Th}}{^{231}\text{Pa}}\right)^0_{Ax} e^{\lambda' t} \quad (16.33)$$

where $\lambda' = \lambda_{231} - \lambda_{230} = 12.123 \times 10^{-6}\,\text{y}^{-1}$ and the effective half-life is $T_{1/2} = 57{,}100$ years. Therefore, the ratio of the activities of unsupported ^{230}Th and ^{231}Pa of sediment *increases* with depth in the core at a rate proportional to λ'. If the sediment was deposited at a constant rate $a = h/t$, then

$$\log\left(\frac{^{230}\text{Th}}{^{231}\text{Pa}}\right)_{Ax} = \log\left(\frac{^{230}\text{Th}}{^{231}\text{Pa}}\right)^0_{Ax} + \frac{\lambda' h}{2.303 a}$$

$$(16.34)$$

where h is the depth in the core. Equation 16.34 is a straight line in coordinates of $\log(^{230}\text{Th}/^{231}\text{Pa})_{Ax}$ and h whose slope is inversely proportional to the sedimentation rate a and whose intercept is equal to the initial value of the activity ratio of the unsupported nuclides in the sediment. This form of the ionium-protactinium method of dating is analogous to the ionium-thorium method expressed in Equation 16.14. It was used by Ku and Broecker (1967a) to determine a sedimentation rate of about 0.2 cm/10^3 y for a core (T3-63-1) from the Arctic Ocean. The initial activity ratio at the top of this core was found to be 26.

Rosholt et al. (1961) took a different approach that permits dating of single sediment samples without requiring analysis of a suite of samples from a particular core. The activity of ^{231}Pa being produced by decay of ^{235}U in sea water is given by

$$^{231}\text{Pa}_A = {}^{235}\text{U}_A(1 - e^{-\lambda_{231} t}) \quad (16.35)$$

where t is the time interval between the production of ^{231}Pa and its incorporation into the sediment. The growth of ^{230}Th from ^{234}U in sea water is given by a similar equation (16.10) which we derived earlier. Thus the activity ratio of U-supported ^{230}Th to ^{231}Pa in sea water is

$$\left(\frac{^{230}\text{Th}}{^{231}\text{Pa}}\right)_A = \left(\frac{^{234}\text{U}}{^{235}\text{U}}\right)_A \left(\frac{1 - e^{-\lambda_{230} t}}{1 - e^{-\lambda_{231} t}}\right) \quad (16.36)$$

If the relative abundances of the uranium isotopes in sea water are normal, that is, if the atomic abundance ratio $^{238}\text{U}/^{235}\text{U} = 137.8$ and the activity ratio $^{234}\text{U}/^{238}\text{U} = 1.15$, then

$$\left(\frac{^{234}\text{U}}{^{235}\text{U}}\right)_A = 1.15 \left(\frac{\lambda_{238}}{\lambda_{235}}\right)\left(\frac{^{238}\text{U}}{^{235}\text{U}}\right) \quad (16.37)$$

$$\left(\frac{^{234}\text{U}}{^{235}\text{U}}\right)_A = 1.15 \frac{1.55125 \times 10^{-10}}{9.8485 \times 10^{-10}} \times 137.8$$

$$= 25.28$$

(Note that we are using the new decay constants of ^{238}U and ^{235}U, Table 12.2.) Moreover, if the residence times of ^{230}Th and ^{231}Pa in the oceans are both short and nearly equal (about 300 years), then

$$\frac{1 - e^{-\lambda_{230} t}}{1 - e^{-\lambda_{231} t}} \simeq \frac{\lambda_{230} t}{\lambda_{231} t} = \frac{9.217 \times 10^{-6}}{2.134 \times 10^{-5}}$$

$$= 0.4319 \quad (16.38)$$

The activity ratio of U-supported ^{230}Th to ^{231}Pa in sea water is therefore

$$\left(\frac{^{230}\text{Th}}{^{231}\text{Pa}}\right)_A = 25.28 \times 0.4319 = 10.92$$

If both isotopes are removed with equal efficiency from sea water, the initial activity ratio in a freshly deposited sample of sediment should also have this value.

The present activities of ^{230}Th and ^{231}Pa in a sample of sediment deposited t years ago must be corrected for the activities that are supported by decay of uranium in the sample after deposition. Thus,

$$\left(\frac{^{230}\text{Th}}{^{231}\text{Pa}}\right)_{Ax} = \left(\frac{^{230}\text{Th}_t - {}^{230}\text{Th}_s}{^{231}\text{Pa}_t - {}^{231}\text{Pa}_s}\right)_A \quad (16.39)$$

where the subscripts have the following meanings: x = excess unsupported, t = total, and s = supported. Substituting into Equation 16.32,

$$\left(\frac{^{230}\text{Th}}{^{231}\text{Pa}}\right)_{Ax} = \left(\frac{^{230}\text{Th}_t - {}^{230}\text{Th}_s}{^{231}\text{Pa}_t - {}^{231}\text{Pa}_s}\right)_A$$

$$= 10.92 \frac{e^{-\lambda_{230}t}}{e^{-\lambda_{231}t}} \quad (16.40)$$

Solving for t

$$t = \frac{2.303}{\lambda_{231} - \lambda_{230}} \log \left[\frac{\left(\dfrac{^{230}\text{Th}_t - {}^{230}\text{Th}_s}{^{231}\text{Pa}_t - {}^{231}\text{Pa}_s}\right)_A}{10.92}\right]$$

$$(16.41)$$

Finally, the assumption is made that the U-supported activities of ^{230}Th and ^{231}Pa are in secular equilibrium with their parents:

$$^{231}\text{Pa}_{As} = {}^{235}\text{U}_A = \lambda_{235}\,{}^{235}\text{U}$$
$$^{230}\text{Th}_{As} = {}^{234}\text{U}_A = \lambda_{234}\,{}^{234}\text{U} \quad (16.42)$$

where ^{235}U and ^{234}U are the numbers of atoms of these isotopes per unit weight of sample.

An equation very similar to (16.41) was used by Rona and Emiliani (1969) to date sediment from core P6304-8 raised in the Caribbean Sea at 14°59′N, 69°20′W. Their equation differs from (16.41) because they did not include a correction for the enrichment

of ^{234}U in sea water and because they used the old values for the decay constants of ^{238}U and ^{235}U. Thus, they obtained a value of 9.4 for the initial activity ratio of ^{230}Th/^{231}Pa, whereas we calculate a value of 10.92. A sample (321–326 cm), taken from core P6304-8 was found to have a ^{230}Th/^{231}Pa activity ratio of 28.37 (carbonate-free and corrected for U-supported ^{230}Th and ^{231}Pa). Using Equation 16.41, modified as discussed above, we calculate the following date:

$$t = \frac{2.303}{12.123 \times 10^{-6}} \log\left(\frac{28.37}{9.4}\right)$$

$$= 91,134 \text{ years}$$

The ionium-protactinium dates of this and several other cores from the Caribbean are in satisfactory agreement with carbon-14 dates. However, some cores have given erroneous ionium-protactinium dates, presumably because of fractionation of ^{230}Th or ^{231}Pa during adsorption and deposition with the sediment.

5
Dating with ^{210}Pb

The decay series of ^{238}U includes ^{222}Rn which escapes into the atmosphere at a rate of about 42 atoms per minute per square centimeter of land surface. ^{222}Rn subsequently decays through a series of short-lived daughters to ^{210}Pb (see Fig. 12.1). This isotope is rapidly removed from the atmosphere by rain or snow and has a residence time of only about 10 days. After removal from the atmosphere, ^{210}Pb is deposited in the snow and ice of glaciers, in lakes and in the coastal ocean, and may even be incorporated into the woody tissue of trees such as oak and hickory. The activity of the unsupported ^{210}Pb then decreases as a function of time at a rate controlled by its half-life which is 22.26 years. This phenomenon per-

THE U-SERIES DISEQUILIBRIUM METHODS OF DATING 16

mits age determinations of certain geological materials formed within the last 100 years or so. The method was developed by Goldberg (1963) and has been used to study rates of deposition of snow in Antarctica and Greenland (Crozaz et al., 1964; Crozaz and Langway, 1966), and in alpine glaciers (Picciotto et al., 1967; Windom, 1969). It has also been used to date recently deposited lacustrine sediment (Krishnaswamy et al., 1971), and coastal marine sediment (Koide et al., 1972; 1973).

The technique for measuring the activity of ^{210}Pb in geological samples is relatively simple. The isotope is recovered from solutions by anion exchange chromatography using several milligrams of ordinary lead (free of ^{210}Pb) as carrier. The final concentrate is either precipitated as a sulfate or chromate or evaporated directly into a planchet for counting. ^{210}Pb decays to ^{210}Bi by emission of beta particles having a maximum energy of only 0.02 MeV. For this reason, the activity of ^{210}Pb is determined by counting ^{210}Bi$(E_{max} = 1.2 \text{ MeV})$ which grows in with a half-life of about five days. Some workers have counted the alpha particles of ^{210}Po which forms by β^- decay of ^{210}Bi (Fig. 12.1).

a. SNOW AND ICE CHRONOLOGY

The time elapsed since deposition of a sample of snow at a depth h below the surface can be calculated from its activity of ^{210}Pb, provided that the initial activity of this isotope has remained constant:

$$^{210}\text{Pb}_A = {}^{210}\text{Pb}_A^0 e^{-\lambda t} \qquad (16.43)$$

where

$^{210}\text{Pb}_A$ = activity of ^{210}Pb per unit weight of sample at depth h

$^{210}\text{Pb}_A^0$ = activity of ^{210}Pb at the surface $(h = 0)$,

λ = decay constant of ^{210}Pb $(3.11 \times 10^{-2} \text{ y}^{-1})$, and

t = age of the sample of snow.

It follows that

$$t = \frac{2.303}{\lambda} \log \left(\frac{^{210}\text{Pb}^0}{^{210}\text{Pb}} \right)_A \qquad (16.44)$$

When the rate of annual water accumulation is constant such that $a = h/t$, where a is the annual rate of accumulation, then

$$\log {}^{210}\text{Pb}_A = \log {}^{210}\text{Pb}_A^0 - \frac{\lambda h}{2.303a} \qquad (16.45)$$

This is the equation of a straight line in coordinates of $\log {}^{210}\text{Pb}_A$ and h whose slope is

$$m = -\frac{\lambda}{2.303a} \qquad (16.46)$$

and whose intercept on the ordinate axis is $\log {}^{210}\text{Pb}_A^0$. In other words, when the initial activity of ^{210}Pb and the rate of accumulation of water have remained constant at a given location, the ^{210}Pb activities of snow samples taken at different depths below the surface will approach a straight line in coordinates of $\log {}^{210}\text{Pb}_A$ and h. The initial activity $^{210}\text{Pb}_A^0$ is obtained by extrapolating the line to its point of intersection with the ordinate at $h = 0$.

Crozaz et al. (1964) used this method to determine the annual rates of water accumulation by analysis of snow samples collected at Base Roi Baudouin (70°26′S, 24°19′E) and at the South Pole in Antartica. Their data shown in Figure 16.5, fall on straight lines from which average accumulation rates of 45 ± 3 and 6 ± 1 cm of water per year were determined for Base Roi Baudouin and the South Pole, respectively. These results are in good agreement with measurements of the accumulation rates obtained by other methods. The intial activities of ^{210}Pb in

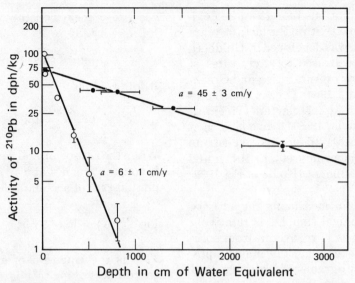

FIGURE 16.5   Plot of the activity of ^{210}Pb (on a logarithmic scale) in snow samples versus depth in water equivalent. ● Base Roi Baudouin, Antarctica (70°26'S, 24°19'E); ○ South Pole station. The data points fit straight lines whose slopes yield the average annual accumulation rates of water from Equation 16.45. (Crozaz et al., 1964.)

Antartica range from about 75 to 110 disintegrations per hour per kilogram of snow (dph/kg).

b.   CHRONOLOGY OF LACUSTRINE AND MARINE SEDIMENT

Goldberg (1963) reported that ^{210}Pb is rapidly removed from river water by inorganic and biochemical reactions. This observation has given rise to a chronometer for dating lacustrine and coastal marine sediment similar to the snow chronometer discussed above. ^{210}Pb enters bodies of water by removal from the atmosphere by rain and by decay of ^{226}Ra within the water. The residence time of ^{210}Pb in the oceans is of the order of only one or two years or less before it is incorporated into the sediment. The activity of ^{210}Pb at some depth h below the water-sediment interface is related to the time elapsed since deposition, provided the initial activity has remained constant. If the rate of sedimentation is also constant, then the activity of ^{210}Pb is linearly related to h as shown in Equation 16.45. A complication arises because of the presence of a small amount of ^{210}Pb that is supported by decay of ^{226}Ra in the deeper layers of sediment. This component can be identified by analyzing sediment in which the unsupported ^{210}Pb has already decayed.

Figure 16.6a is a plot of ^{210}Pb activities of a core from the Santa Barbara basin in California (Koide et al., 1972). The core consists of varved sediment containing annual layers from which the time of deposition of any given sample can be determined. The activity of ^{210}Pb (on a logarithmic scale) decreases linearly with depth in the core. The slope of this line indicates a constant rate of deposition of about 0.39 cm/y in agree-

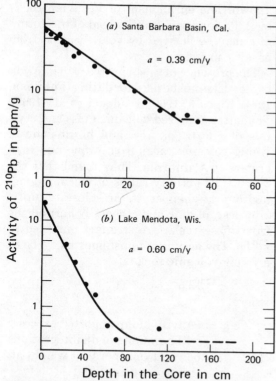

FIGURE 16.6 (a) Variation of the activity of unsupported ^{210}Pb in marine sediment from the Santa Barbara Basin in California as a function of depth in the core. (Koide et al., 1972.) (b) The same for sediment from Lake Mendota in Wisconsin (Koide et al., 1973). The deviation from the straight line of samples at depth in both cores results from the presence of ^{210}Pb supported by decay of ^{226}Ra in the sediment. The parameter a is the rate of sedimentation.

ment with the stratigraphic estimate of 0.4 cm/y. Below a depth of about 30 cm in this core, the ^{210}Pb activity deviates from the straight line because of the increasing importance of ^{210}Pb supported by ^{226}Ra. Figure 16.6b is a similar plot for sediment deposited in Lake Mendota near Madison, Wisconsin (Koide et al., 1973). This core shows the presence of radium-supported ^{210}Pb of about 0.4 dpm/g, which is small compared to the initial activity of about 16.4 dpm/g at the top of

the core. Recent examples of dating lacustrine sediment by the ^{210}Pb method include papers by Petit (1974) and Robbins and Edgington (1975).

When cores are taken by means of conventional piston coring devices the uppermost layers of sediment may be lost. If this condition is not recognized, an erroneous value for the initial activity of unsupported ^{210}Pb may be chosen, resulting in a systematic error in the age of sediment calculated by the ^{210}Pb method. For this reason, special precautions must be taken to assure that the top layers of sediment are not lost. Koide et al. (1972) used a specially constructed box corer and reported an activity of 49.2 dpm/g for sediment deposited from 1960–1962 in the Santa Barbara basin (Fig. 16.6a). Another sample taken at a depth of 21.85 $\pm$ 0.25 cm had an activity of 10.0 dpm/g. According to the authors, the level of supported ^{210}Pb in this core is between 3 and 4 dpm/g. If we assume a value of 3.5 dpm/g, we can calculate the age of this sediment by means of Equation 16.44:

$$t = \frac{2.303}{0.0311} \log\left(\frac{45.8}{6.5}\right) = 62.8 \text{ years}$$

In other words, the sediment was deposited about 62.8 y before 1960–1962. By counting varves, the authors determined that this sample had been deposited between 1900 and 1902, in good agreement with the ^{210}Pb date.

6
U-Series Disequilibrium in Volcanic Rocks

The daughters of ^{238}U in young volcanic rocks are not in radioactive equilibrium with their parent (Somayajulu et al., 1966; Oversby and Gast, 1968). The disequilibrium presumably results from chemical fractionation during magma formation by partial melting in the upper mantle, assuming that secular equilibrium existed in these rocks prior to melting. The radioactive disequilibrium is characteristically indicated by the activity ratios of $^{230}Th/^{238}U$, $^{226}Ra/^{238}U$, and $^{210}Pb/^{238}U$, all of which may be greater than one. On the other hand, no disequilibrium has been found between ^{234}U and ^{238}U because the isotopes of uranium are not fractionated during this process.

Oversby and Gast (1968) studied whole-rock samples of volcanic rocks from Tristan da Cunha, Faial (Azores) and from Mt. Vesuvius. The ages of these rocks ranged from about 500 years to only a few years for lava flows extruded in 1961. The activity ratios of the volcanic rocks from the three volcanic centers are all greater than one, but display significant regional differences shown in Table 16.2. The data clearly show that these young volcanic rocks contain variable amounts of excess unsupported ^{230}Th, ^{226}Ra and ^{210}Pb. The excess ^{210}Pb may be partly due to decay of excess ^{226}Ra.

The specimen from Mt. Vesuvius is especially remarkable because of its very large enrichment in ^{226}Ra, indicated by its $^{226}Ra/^{238}U$ activity ratio of 10.8. The enrichment in ^{210}Pb is equally remarkable, but can be accounted for as a decay product of ^{226}Ra. The high ^{210}Pb activity in the mineral cotunnite ($Pb\ Cl_2$) which forms as a sublimate in the fumaroles around Mt. Vesuvius is well known. The enrichment of Vesuvian rocks in ^{210}Pb suggests that the lead of the cotunnite may be derived by volatilization from basaltic magma.

The presence of unsupported ^{230}Th in volcanic rocks can be used for dating (Fukuoka, 1974; Kigoshi, 1967; Taddeuci et al., 1967). The latter analyzed volcanic glass and crystals of quartz, olivine, and hornblende of young volcanic rocks from Mono and Inyo craters in California. They found that the $^{230}Th/^{238}U$ activity ratios of volcanic glass and quartz phenocrysts are close to unity. However, olivine and especially hornblende from these rocks are strongly enriched in ^{230}Th. The activity of unsupported ^{230}Th in a young volcanic rock is

$$^{230}Th_{Ax} = {}^{230}Th_{Ax}^{0} e^{-\lambda_{230}t} \qquad (16.47)$$

where

$^{230}Th_{Ax}$ = activity of unsupported excess ^{230}Th in units of disintegrations per unit time per unit weight at the present time, and

$^{230}Th_{Ax}^{0}$ = same at the time of crystallization ($t = 0$).

Since

$$^{230}Th_{Ax}^{0} = {}^{230}Th_{A}^{0} - {}^{238}U_{A} \qquad (16.48)$$

where

$^{230}Th_{A}^{0}$ = activity of ^{230}Th at $t = 0$, and

Table 16.2   **Activity Ratios of ^{230}Th, ^{226}Ra, and ^{210}Pb Relative to ^{238}U in Recently Formed Volcanic Rocks (Oversby and Gast, 1968)**

LOCATION	NUMBER OF SAMPLES	$\left(\dfrac{^{230}Th}{^{238}U}\right)_A$	$\left(\dfrac{^{226}Ra}{^{238}U}\right)_A$	$\left(\dfrac{^{210}Pb}{^{238}U}\right)_A$
Tristan da Cunha	5	1.14	1.41	1.04
Faial (Azores)	4	1.45	2.52	1.90[a]
Vesuvius	1	1.05	10.8	8.40

[a] Two samples only.

$^{238}U_A$ = activity of ^{230}Th that is in secular equilibrium with ^{238}U.

$$^{230}\text{Th}_{Ax} = (^{230}\text{Th}_A^0 - {}^{238}\text{U}_A)e^{-230t} \quad (16.49)$$

Dividing by the activity of ^{232}Th, we obtain

$$\left(\frac{^{230}\text{Th}}{^{232}\text{Th}}\right)_{Ax} = \left[\left(\frac{^{230}\text{Th}}{^{232}\text{Th}}\right)_A^0 - \left(\frac{^{238}\text{U}}{^{232}\text{Th}}\right)_A\right]e^{-\lambda_{230}t}$$

$$(16.50)$$

This equation cannot be solved for t directly because the initial ^{230}Th/^{232}Th activity ratio is unknown. However, this parameter can be eliminated by analyzing two cogenetic minerals, provided they formed at nearly the same time and provided they had the same ^{230}Th/^{232}Th activity ratio at the time of crystallization. The dates so calculated by Taddeuci et al. (1967) for glass-hornblende pairs are in moderately good agreement with K-Ar dates of the same rocks.

The activity of ^{230}Th in young volcanic rocks is also described by Equation 16.12 which we derived earlier in connection with the ionium-thorium method of dating sediment. That equation is

$$\left(\frac{^{230}\text{Th}}{^{232}\text{Th}}\right)_A = \left(\frac{^{230}\text{Th}}{^{232}\text{Th}}\right)_{Ax}^0 e^{-\lambda_{230}t}$$
$$+ \left(\frac{^{238}\text{U}}{^{232}\text{Th}}\right)_A (1 - e^{-\lambda_{230}t}) \quad (16.51)$$

The first term of this equation describes the decay of unsupported ^{230}Th, while the second term represents the growth of ^{230}Th that is supported by ^{238}U. Allegre (1968) pointed out that this is the equation of a straight line in coordinates of $(^{230}\text{Th}/^{232}\text{Th})_A$ and $(^{238}\text{U}/^{232}\text{Th})_A$ when t is a constant. Minerals having different $(^{238}\text{U}/^{232}\text{Th})_A$ ratios will satisfy this equation provided they have the same age, the same initial ^{230}Th/^{232}Th activity ratio, and provided they remained closed to uranium and tho-

rium after crystallization. The line formed by a suite of cogenetic minerals that satisfy the above conditions is an isochron whose slope is equal to

$$m = 1 - e^{-\lambda_{230}t} \quad (16.52)$$

The intercept is

$$b = \left(\frac{^{230}\text{Th}}{^{232}\text{Th}}\right)^0 e^{-\lambda_{230}t} \quad (16.53)$$

and can be used to calculate the initial excess activity ratio. Such an isochron is therefore similar to Rb-Sr isochrons, except that the intercept decreases exponentially as a function of time. At the time of crystallization, the minerals form a line whose slope is equal to zero. Subsequently, the slope increases until it reaches a value of one when all excess ^{230}Th has decayed and the remaining ^{230}Th is in secular equilibrium with ^{238}U.

Allegre's interpretation of the data of Taddeuci et al. (1967) is shown in Figure 16.7. It is apparent that hornblende, olivine, and quartz form an isochron whose slope is 0.169 from which we calculate a date of 20,000 years by means of Equation 16.52. The point representing volcanic glass falls below the isochron. Allegre suggested that the isochron indicates the time elapsed since crystallization of the minerals at depth, while the hornblende-glass isochron reflects the eruption of the magma.

In a later study, Bernat and Allegre (1974) used this method to date marine sediment which they split into phillipsite-rich, apatite-rich (fish teeth), and clay-rich fractions. They found that these fractions did indeed form isochrons and that the intercepts of these isochrons decrease with depth in the core. However, the slope of the isochron of the youngest sample was not zero as expected, indicating disequilibrium among the thorium isotopes in the sediment.

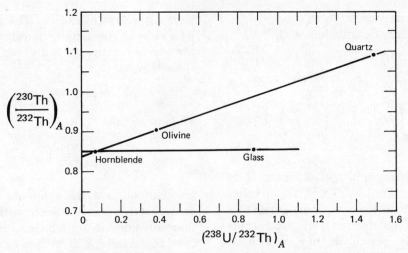

FIGURE 16.7 Ionium-uranium isochron formed by minerals of rhyolitic volcanic rocks from Mono Crater, California. The slope of this isochron is proportional to the time elapsed since crystallization as expressed by Equation 16.52. The date indicated by this line is 20,000 y, which is interpreted as the time elapsed since crystallization at depth. (This diagram is based on the interpretation by Allegre, 1968, of the data of Taddeuci et al., 1967.)

7 SUMMARY

The decay chains formed by ^{238}U and ^{235}U are broken by geological processes because the daughters of uranium are isotopes of different elements and therefore have different geochemical properties. The resulting radioactive disequilibrium can be used for dating over time periods ranging from a few tens of years or less to one million years or more. These geochronometers have been used to measure rates of deposition of sediment in the oceans by the decay of unsupported ionium (^{230}Th) and ^{231}Pa. Biogenically or inorganically precipitated carbonate materials of both marine and nonmarine origin can be dated by the decay of excess ^{234}U or by the growth of ^{230}Th. These methods are very useful for dating corals but may not be reliable when used on the shells of mollusks. Ice and snow of continental ice sheets and of alpine glaciers can be dated by the decay of unsupported

^{210}Pb. This method also permits studies of sedimentation rates in lakes and in the coastal ocean where the effects of our industrialized society are concentrated. Radioactive disequilibrium has even been discovered in young volcanic rocks and may become a useful tool for the study of chemical fractionation during magma production and for dating volcanic rocks of Pleistocene age.

PROBLEMS

1. The following values of the activity ratios $^{230}Th/^{232}Th$ in a manganese nodule were reported by Ku and Broecker (1967b):

DEPTH (mm)	$(^{230}Th/^{232}Th)_A$
0–0.3	43.7 ± 1.8
0.3–0.5	26.2 ± 0.6
0.5–0.85	10.4 ± 0.3

Estimate the rate of accretion of this nodule. (**ANSWER:** ~3.2 mm/10^6 y.)

2. The activity ratios of excess $^{230}Th/^{231}Pa$ in the manganese nodule (Problem 1) are:

DEPTH (mm)	$(^{230}Th/^{231}Pa)_{Ax}$
0–0.3	22.1 ± 0.7
0.3–0.5	44.9 ± 1.3
0.5–0.85	93.8 ± 3.8

Estimate the rate of accretion of this nodule. (**ANSWER:** ~4.3 mm/10^6 y.)

3. The activities of *excess* ^{230}Th and ^{231}Pa in the manganese nodule (Problem 1) are:

DEPTH (mm)	$^{230}Th_{Ax}$ (dpm/g)	$^{231}Pa_{Ax}$ (dpm/g)
0–0.3	1.41	63.6
0.3–0.5	0.76	17.1
0.5–0.85	0.38	4.0

Estimate the accretion rate of this nodule by plotting the logarithms of the activities versus depth and by expressing the slopes of the resulting lines in a manner analogous to Equation 16.13.

4. A coral (*Montastria annularis*) from a coral-reef terrace on Barbados (Broecker et al., 1968, sample 1152-C) has a $^{230}Th/^{234}U$ activity ratio of 0.52. Calculate the age of this coral and its initial $^{234}U/^{238}U$ ratio, given that the present value of this ratio is 1.11. (**ANSWER:** $t = 81,000$ y, $\gamma_0 = 1.15$.)

5. The following activities of ^{210}Pb in samples of firn from a core taken at Camp Century in Greenland were reported by Crozaz and Langway (1966):

CORE DEPTH, METERS WATER EQUIVALENT	^{210}Pb (dph/kg)
0–5.7	170 ± 15
6.9–9.7	55 ± 6
11.8–14.3	58 ± 6
18.1–24.5	21 ± 3
25.5–33.3	10 ± 2
33.3–40.2	5 ± 2
40.2–51.0	3 ± 1

Estimate the average accumulation rate of water per year at this site. (**ANSWER:** ~33 cm/y.)

REFERENCES

Allegre, C. J. (1968) [230]Th dating of volcanic rocks: A comment. Earth Planet. Sci. Letters, *5*, 209–210.

Baranov, V. I., Yu. A. Surkov, and V. D. Vilenskii (1958) Isotopic shift in natural uranium compounds. Geochemistry, *1958*, 591–599.

Bernat, M., and C. J. Allegre (1974) Systematics in uranium-ionium dating of sediments. Earth Planet. Sci. Letters, *21*, 310–314.

Broecker, W. S. (1963) A preliminary evaluation of uranium series inequilibrium as a tool for absolute age measurement on marine carbonates. J. Geophys. Res., *68*, 2817–2834.

Broecker, W. S., D. L. Thurber, J. Goddard, T. L. Ku, R. K. Mathews, and K. J. Mesolella (1968) Milankovitch hypothesis supported by precise dating of coral reefs and deep-sea sediment. Science, *159*, 297–300.

Cherdyntsev, V. V. (1969) Uranium-234. Atomizdat, Moskva. Translated into English by J. Schmorak, Israel Program for Scientif. Translations, 1971, 234 pp.

Crozaz, G., E. Picciotto, and W. De-Breuck (1964) Antartic snow chronology with Pb[210]. J. Geophys. Res., *69*, 2597–2604.

Crozaz, G., and C. C. Langway, Jr. (1966) Dating Greenland firn-ice cores with Pb[210]. Earth Planet. Sci. Letters, *1*, 194–196.

Fukuoka, T. (1974) Ionium dating of acidic volcanic rocks. Geochim. J., *8*, 109–116.

Goldberg, E. D., and M. Koide (1962) Geochronological studies of deep sea sediments by the ionium/thorium method. Geochim. Cosmochim. Acta, *26*, 417–450.

Goldberg, E. D. (1963) Geochronology with Pb-210. In Radioactive Dating. Internat. Atom. Energy Agency, 121–131.

Goldberg, E. D., and M. Koide (1963) Rates of sediment accumulation in the Indian Ocean. In Earth Science and meteoritics, pp. 90–102, J. Geiss and E. D. Goldberg, eds. North-Holland, Amsterdam.

Goldberg, E. D., and J. J. Griffin (1964) Sedimentation rates and mineralogy in the South Atlantic. J. Geophys. Res., *69*, 4293–4309.

Harmon, R. S., P. Thompson, H. P. Schwarcz, and D. C. Ford (1975) Uranium-series dating of speleothems. NSS Bull. Quat. J. Nat. Spel. Soc., *37*, 21–34.

Kaufman, A., W. S. Broecker, T. L. Ku, and D. L. Thurber (1971) The status of U-series methods of dating mollusks. Geochim. Cosmochim. Acta, *35*, 1155–1183.

Kigoshi, K. (1967) Ionium dating of igneous rocks. Science, *156*, 932–934.

Kigoshi, K. (1971) Alpha-recoil thorium-234: Dissolution into water and the uranium-234/uranium-238 disequilibrium in nature. Science, *173*, 47–48.

Koide, M., and E. D. Goldberg (1965) Uranium-234/uranium-238 ratios in sea water. In Progress in Oceanography, vol. 3, pp. 173–178, M. Sears, ed. Pergamon Press. Oxford and New York.

Koide, M., A. Soutar, and E. D. Goldberg (1972) Marine geochronology with Pb^{210}. Earth Planet. Sci. Letters, *14*, 442–446.

Koide, M., K. W. Bruland, and E. D. Goldberg (1973) Th^{228}/Th^{232} and Pb^{210} geochronologies in marine and lake sediments. Geochim. Cosmochim. Acta, *37*, 1171–1188.

Krishnaswamy, S., D. Lal, J. M. Martin, and M. Meybeck (1971) Geochronology of lake sediments. Earth Planet. Sci. Letters, *11*, 407–414.

Kronfeld, J. (1974) Uranium deposition and Th-234 alpha-recoil: An explanation for extreme U-234/U-238 fractionation within the Trinity acquifer. Earth Planet. Sci. Letters, *21*, 327–330.

Kronfeld, J., E. Gradsztajn, H. W. Müller, J. Radin, A. Yaniv, and R. Zach (1975) Excess ^{234}U: An aging effect in confined waters. Earth Planet. Sci. Letters, *27*, 342–344.

Ku, T.-L. (1965) An evaluation of the U^{234}/U^{238} method as a tool for dating pelagic sediments. J. Geophys. Res., *70*, 3457–3474.

Ku, T.-L., and W. S. Broecker (1967a) Rates of sedimentation in the Arctic ocean. In: Progress in Oceanography, vol. 4, 95–104, M. Sears, ed., Pergamon Press, Oxford and New York.

Ku, T.-L., and W. S. Broecker (1967b) Uranium, thorium, and protactinium in a manganese nodule. Earth Planet. Sci. Letters, *2*, 317–320.

Miyake, Y., Y. Sugimura, and T. Uchida (1966) Ratio U^{234}/U^{238} and the uranium concentration in sea water in the western North Pacific. J. Geophys. Res., *71*, 3083–3087.

Osmond, J. K., M. I. Kaufman, and J. B. Cowart (1974) Mixing volume calculations, sources and aging trends of Floridan aquifer water by uranium isotopic methods. Geochim. Cosmochim. Acta, *38*, 1083–1100.

Oversby, V. M., and P. W. Gast (1968) Lead isotope compositions and uranium decay series disequilibrium in Recent volcanic rocks. Earth Planet. Sci. Letters, *5*, 199–206.

Petit, D. (1974) ^{210}Pb et isotopes stables du plomb dans les sediments lacustres. Earth Planet. Sci. Letters, *23*, 199–205.

Picciotto, E., and S. Wilgain (1954) Thorium determination in deep-sea sediments. Nature, *173*, 632–633.

Picciotto, E., G. Crozaz, W. Ambach, and H. Eisner (1967) Lead-210 and strontium-90 in an alpine glacier. Earth Planet. Sci. Letters, *3*, 237–242.

Robbins, J. A., and D. N. Edgington (1975) Determination of recent sedimentation rates in Lake Michigan using Pb-210 and Cs-137. Geochim. Cosmochim. Acta, *39*, 285–304.

Rona, E., and C. Emiliani (1969) Absolute dating of Caribbean cores P6304–8 and P6304-9. Science, *163*, 66–68.

Rosholt, J. N., C. Emiliani, J. Geiss, F. F. Koczy, and P. J. Wangersky (1961) Absolute dating of deep sea-sea cores by the Pa^{231}/Th^{230} method. J. Geol., *69*, 162–185.

Somayajulu, B. L. K., M. Tatsumoto, J. N. Rosholt, and R. J. Knight (1966) Disequilibrium of 238-U series in basalt. Earth Planet. Sci. Letters, *1*, 387–391.

Starik, I. E., F. E. Starik, and B. A. Mikhailov (1958) On the problem of the shift of isotopic ratios in natural formations. Geochemistry, *1958*, 462–464.

Taddeuci, A., W. S. Broecker, and D. L. Thurber (1967) ^{230}Th dating of volcanic rocks. Earth Planet. Sci. Letters, *3*, 338–342.

Thompson, G. M., D. N. Lumsden, R. L. Walker, and J. A. Carter (1975) Uranium series dating of stalagmites from Blanchard Springs Caverns, U.S.A. Geochim. Cosmochim. Acta, *39*, 1211–1218.

Thurber, D. L. (1962) Anomalous $^{234}U/^{238}U$ in nature. J. Geophys. Res., *67*, 4518–4520.

Veeh, H. H. (1966) Th^{230}/U^{238} and U^{234}/U^{238} ages of Pleistocene high sea level stand. J. Geophys. Res., *71*, 3379–3386.

Windom, H. (1969) Atmospheric dust records in permanent snow fields: Implications to marine sedimentation. Geol. Soc. Amer. Bull., *80*, 761–782.

17 THE CARBON-14 METHOD OF DATING

1
The Discovery of Carbon-14

In 1934, A. V. Grosse published a paper in which he reported that the mineral eudialyte (complex silicate of Zr, Fe, Ca, Na, . . .) contains radioactivity in excess of that which can be attributed to uranium and thorium and their daughters. To account for this observation, he offered the inspired suggestion that the excess activity was due to the presence of isotopes produced by the interaction of cosmic rays with elements such as oxygen, silicon, iron, zirconium, and others. More than a decade later, Grosse participated with W. F. Libby and others in a demonstration that ^{14}C, *which is produced by cosmic rays*, occurs in methane derived from sewage gas.

The first indication of the existence of $^{14}_{6}C$ was obtained in 1934 by F. N. D. Kurie at Yale University. He exposed nitrogen to fast neutrons and observed that some nitrogen nuclei emitted a particle which made a very long, thin track in the cloud chamber. Kurie considered several possible explanations of this observation, including the formation of $^{14}_{6}C$ by an (n, p) reaction of $^{14}_{7}N$. The remarkable aspect of Kurie's suggestion was that something other than an alpha particle could emerge from the nucleus of an atom. In the following year, it was found that the reaction also occurred with slow neutrons. Subsequent work by Burcham and Goldhaber (1936) established that the correct reaction was indeed $^{14}_{7}N(n, p)^{14}_{6}C$, thus confirming Kurie's original intuitive suggestion. Additional confirmation of the existence of the radioactive isotope ^{14}C was obtained in 1937

by Kurie and Kamen in experiments with the cyclotron of the Radiation Laboratory of the University of California in Berkley. In fact, Kamen (1963) recalled that in late 1938, E. M. McMillan wanted to produce a separable amount of ^{14}C by irradiating solid ammonium nitrate for several months with an intense neutron beam from the 37-inch cyclotron. Unfortunately the experiment ended when the bottle was accidentally knocked on the floor and was smashed.

There were other hazards in the Radiation Laboratory. Kamen was involved in manufacturing various radionuclides for biological research and in the process became radioactive himself. One day in 1937 when he and Philip Abelson were attempting to measure the radioactivity of a sample, they were frustrated by erratic background readings. Finally, the erratic behavior of the counter was correlated with Kamen's movements about the room, whereupon Abelson had Kamen undress systematically until it was established that the disturbing radiation originated from the front of his pants.

A measurable quantity of ^{14}C was finally prepared by Kamen in early 1940 by means of a (d, p) reaction on ^{13}C in graphite. In the early morning of February 15, 1940, Kamen terminated the irradiation which had lasted nearly a month and staggered home to sleep. According to his own description, he was unshaven, red-eyed and dazed, and was promptly stopped by the police who were looking for an escaped convict. Nevertheless, by the afternoon of February 27, 1940, the last doubt regarding the success of the experiment was eliminated, and on the next day Ruben and Kamen went to see Dr. E. O. Lawrence, the director of the Radiation Laboratory, to tell him that they had produced a measurable quantity of ^{14}C. The initial indication was that the half-life of ^{14}C was about 4×10^3 years, a very good estimate compared to the presently accepted value of 5730 years.

The possibility of producing ^{14}C by an (n, p) reaction on ^{14}N was almost forgotten. Almost, but not quite. Two five-gallon carboys of saturated ammonium nitrate solution were placed along-side the cyclotron where neutrons might interact with the nitrogen atoms. And there they stayed until the cyclotron crew got tired of them and complained. Eventually Kamen and Ruben swept some CO_2 gas out of these bottles and precipitated it as $CaCO_3$. To their amazement, they found that this precipitate was highly radioactive due to the presence of copious amounts of ^{14}C. Thus the (n, p) reaction on ^{14}N turned out to be a far more efficient method of producing ^{14}C than the $^{13}_{6}C(d, p)^{14}_{6}C$ reaction.

The possibility that ^{14}C may be produced in the atmosphere by interactions of cosmic ray neutrons with ^{14}N was first suggested in 1939 by C. G. and D. D. Montgomery, following the earlier discovery of cosmic ray neutrons by Locker in 1933 and by Rumbaugh and Locker in 1936. The search for radiocarbon in nature was led by W. F. Libby and his students and collaborators at the University of Chicago. In 1946, he estimated that the average global production rate of ^{14}C is sufficient to detect it in natural samples. One year later Anderson et al. (1947) demonstrated the presence of ^{14}C in methane extracted from sewage gas and suggested its use for dating of biological specimens. The radio-activity of ^{14}C was measured by means of a screen wall counter, developed by Libby in 1934. In 1949, Arnold and Libby demonstrated the feasibility of the carbon-14

method of dating by analyzing a suite of archaeological samples of known age. Two years later, Anderson and Libby (1951) reported measurements of the specific activity of ^{14}C in the biosphere and showed that it is constant within narrow limits and apparently independent of latitude. Their average value for the disintegration rate of ^{14}C in samples of modern plant tissue was 15.3 ± 0.5 disintegrations per minute per gram (dpm/g) of carbon. They also analyzed one sample of seal oil from Antarctica which had a specific activity of 15.69 ± 0.30 dpm/g, well within the range of activities of plant-tissue samples. In 1952, Libby published his famous book *Radiocarbon Dating*, which was subsequently reissued in 1955 in a second edition. For his pioneering work in developing the carbon-14 dating method, W. F. Libby was awarded the Nobel Prize for chemistry in 1960.

2
Principles of Carbon-14 Dating

It is now known that ^{14}C is produced in the atmosphere by a variety of nuclear reactions based generally on interactions of cosmic-ray produced neutrons with stable isotopes of nitrogen, oxygen, and carbon. By far the most important of these is the reaction between slow cosmic-ray neutrons and the nucleus of stable ^{14}N:

$$^{1}_{0}n + {}^{14}_{7}N \rightarrow {}^{14}_{6}C + {}^{1}_{1}H \qquad (17.1)$$

where $^{1}_{0}n$ is the neutron and $^{1}_{1}H$ is the proton that is emitted by the product nucleus. The atoms of ^{14}C are then incorporated into carbon dioxide molecules by reactions with oxygen or by exchange reactions with stable carbon isotopes in molecules CO or CO_2. The molecules of $^{14}CO_2$ are mixed rapidly throughout the atmosphere and the hydrosphere and attain constant levels of concentration representing a steady-state equilibrium. This equilibrium concentration is maintained by the production of ^{14}C in the atmosphere on the one hand and by its continuous decay on the other. The molecules of $^{14}CO_2$ enter plant tissue as a result of photosynthesis and by absorption through the roots. The concentration of ^{14}C in living green plants is maintained at a constant level by its continuous absorption from the atmosphere and its continuous decay. Animals that feed on plants or absorb carbon-bearing ions or molecules from the atmosphere or hydrosphere also acquire a constant level of radioactivity due to ^{14}C. When the plant or animal dies, the absorption of ^{14}C from the atmosphere stops and its activity due to ^{14}C then declines as a result of radioactive decay. If the activity of ^{14}C in living tissue is known, the activity of ^{14}C of dead plant tissue can be used to calculate the time elapsed since death. This is the "carbon-14 date" of the sample.

The decay of ^{14}C takes place by emission of a negative beta particle and leads to the formation of stable ^{14}N:

$$^{14}_{6}C \rightarrow \, ^{14}_{7}N + \beta^- + \bar{\nu} + Q \qquad (17.2)$$

The end point energy (Q) is 0.156 MeV. The decay is directly to the groundstate of ^{14}N and no gamma ray is emitted. The radioactivity of a specimen of carbon extracted from plant or animal tissue that died t years ago is given by

$$A = A_0 e^{-\lambda t} \qquad (17.3)$$

where A is the measured activity due to ^{14}C in units of disintegrations per minute per gram of carbon and A_0 is the activity of ^{14}C in the same specimen at the time the plant or animal were alive. The best estimate of the specific activity of ^{14}C in equilibrium with the atmosphere (A_0) is 13.56 ± 0.07 dpm/g (Karlèn et al., 1966). The half-life of ^{14}C is 5730 ± 40 years (Godwin, 1962) which was adopted at the Fifth Radiocarbon Dating

Conference held in 1962 in Cambridge, England. However, a value of 5568 ± 30 years is still being used to calculate dates reported in the journal *Radiocarbon* (Box 2161, Yale Station, New Haven, Connecticut, 06520) which has systematically published all carbon-14 age determinations since 1959. The carbon-14 age of a sample containing carbon that is no longer in equilibrium with the ^{14}C of the atmosphere or hydrosphere is obtained by solving Equation 17.3 for t:

$$\ln \left(\frac{A}{A_0} \right) = -\lambda t$$

$$t = \frac{1}{\lambda} \ln \left(\frac{A_0}{A} \right) \qquad (17.4)$$

By changing to logarithms to the base 10 and by substitution of

$$\lambda = \frac{0.693}{5730} = 1.209 \times 10^{-4} \; y^{-1}$$

we obtain the formula

$$t = 19.035 \times 10^3 \log \left(\frac{A_0}{A} \right) \text{ years} \quad (17.5)$$

The relationship between the ^{14}C activity (A) of a sample and the time elapsed since exchange with the reservoir stopped (t) is further illustrated in Figure 17.1.

It is clear that the accuracy of the carbon-14 dating method depends critically on the validity of several important assumptions regarding A_0 and A. It is assumed that the initial activity of ^{14}C in plant and animal tissues (A_0) is a known constant that has been independent of time during the past 70,000 years and that its value is also independent of geographic location and does not depend on the species of plant or animal whose dead tissues are being dated. Moreover, it is assumed that the sample to be dated has not been contaminated with modern ^{14}C and that the observed activity is not affected by radioactive impurities in the

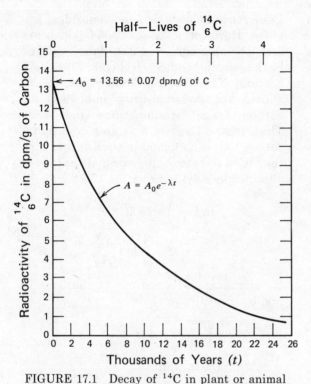

Half-Lives of $^{14}_{6}C$

$A_0 = 13.56 \pm 0.07$ dpm/g of C

$A = A_0 e^{-\lambda t}$

FIGURE 17.1   Decay of ^{14}C in plant or animal tissue that was initially in equilibrium with $^{14}CO_2$ molecules of the atmosphere or hydrosphere. When the plant or animal dies, the exchange stops and the activity due to ^{14}C decreases as a function of time with a half-life of 5730 years. After measuring the remaining ^{14}C activity (A), the carbon-14 age (t) of the specimen can be read from this graph or can be calculated from Equation 17.5.

sample. Libby and his collaborators demonstrated by their measurements that the initial ^{14}C activity of modern plant and animal tissues is constant to a first approximation and that the activity of ^{14}C in archaeological objects of known age can be measured with sufficient precision and accuracy to provide carbon-14 dates that are in reasonably good agreement with their known historical ages. However, more detailed studies have shown that systematic variations of the radiocarbon content of the atmosphere have occurred in the past which give rise to inaccuracies of radiocarbon dates.

3
Variations of the Radiocarbon Content of the Atmosphere

In order to help understand why the radiocarbon content of the atmosphere may exhibit both local and secular variations, let us consider in more detail some of the processes that determine the production rate of ^{14}C and hence its concentration in the atmosphere. The rate of formation of radiocarbon by the (n, p) reaction on ^{14}N depends primarily on the density of the cosmic-ray produced neutron flux. It is known that the neutron flux increases with altitude and reaches a maximum between 40,000 and 50,000 feet above the surface of the Earth and that it is about four times greater in the polar regions than at the equator (Lingenfelter, 1963). Consequently, the production rate of ^{14}C at the poles is significantly greater than it is at the equator, which may be reflected in a higher radiocarbon content of plants and animals in the polar regions compared to those in lower latitudes. Anderson and Libby (1951) originally investigated the latitude effect of the ^{14}C activity by analyzing plant and animal tissue collected at different geomagnetic latitudes. Their data have been plotted in Figure 17.2 from which it can be seen that the specific ^{14}C activity appears to be independent of geomagnetic latitude. The geomagnetic latitude is used because the neutron flux density depends on the proton flux which is modified by the Earth's magnetic field. The apparent uniformity of the ^{14}C activity is due to the rapid mixing (less than two years) that occurs in the atmosphere.

The absolute magnitude of the neutron flux in the upper atmosphere depends on the intensity of the cosmic-ray proton flux impinging upon the Earth. Most of these protons are emitted by the sun. Thus the radiocarbon content of the atmosphere of

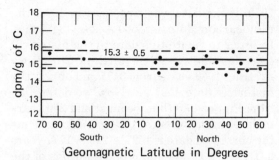

FIGURE 17.2 Measurements of the specific activity of ^{14}C in contemporary samples of plant tissue from different geomagnetic latitudes on the Earth. These data were used by Anderson and Libby (1951) to demonstrate that the ^{14}C activity of biospheric samples is essentially constant and independent of latitude. Their average specific activity was 15.3 ± 0.5 dpm/g of carbon. This value has been superseded by more recent determinations using more efficient detection equipment which indicate a value of 13.56 ± 0.07 dpm/g of carbon. (Libby, 1955.)

the Earth is likely to have varied as a function of time in response to the changing activity of the sun. Moreover, the cosmic-ray proton flux is modulated by changes in the intensity of the Earth's magnetic field. These two factors alone are likely to have produced time-dependent variations in the radiocarbon content of the Earth's atmosphere during the past 30,000 or 40,000 years. The initial work of Arnold and Libby (1949) seemed to suggest that the initial ^{14}C activity of a series of archaeologically dated samples was compatible with the present-day value. However, de Vries (1958) showed later that the radiocarbon content of the atmosphere has varied systematically in the past and that the carbon-14 activity around 1700 and 1500 A.D. was up to 2 percent *greater* than in the nineteenth century. These variations of the carbon-14 content are now known as the "de Vries effect." Another effect was observed by Suess (1955) who found that the activity of twentieth-century wood

is almost 2 percent *lower* than that of nineteenth-century wood which he attributed to the introduction of "dead" CO_2 into the atmosphere by the combustion of fossil fuel since the beginning of the Industrial Revolution. This phenomenon is now recognized as the "Suess effect." Actually, the explosion of nuclear devices in the atmosphere and underground and the operation of nuclear reactors and particle accelerators since about 1945 has greatly increased the level of ^{14}C activity on the surface of the Earth due to the dispersal of artificially produced radiocarbon. For example, at Wellington in New Zealand the radiocarbon content of plant material reached a peak of 69 percent above normal in January of 1965 (Rafter and Stout, 1970).

It is now well established that the radiocarbon content of the atmosphere has varied systematically in the past. These variations introduce serious errors into radiocarbon dates, especially for archaeological purposes. If the initial activity of a sample was less than that indicated by nineteenth-century wood, the date calculated from the observed activity according to Equation 17.5 is too old. If it was greater than the assumed value, the date will be too young. The secular variation of the radiocarbon content of the atmosphere has been studied by analysis of wood from old trees which are dated by dendrochronology (LaMarche and Harlan, 1973). Trees grow by adding a layer of woody tissue to the circumference of their trunks and branches each year. Layers deposited in previous years are then no longer able to absorb ^{14}C from the atmosphere and their radiocarbon content decreases by radioactive decay. The variations in the radiocarbon content of the atmosphere during the past several thousand years can therefore be measured by analysis of wood samples whose age before the present is established by dendrochronology. A great deal of information about the secular variation of the radiocarbon activity has been obtained from studies of the bristlecone pine (*Pinus aristata*) and the sequoia (*Sequoia gigantea*). Michael and Ralph (1970) used the observed age-corrected deviations of the radiocarbon content of such trees to calculate corrections to be applied to conventional radiocarbon dates. Their data are shown in diagrammatic form in Figure 17.3. Similar calibration curves and conversion tables have also been published by Suess (1965) and by Stuiver and Suess (1966). An example will illustrate how these corrections are made. Let us assume that the radiocarbon age of a sample is found to be 3500 years. It is customary to express such dates in terms of years before the present (B.P.), where the present is taken to be

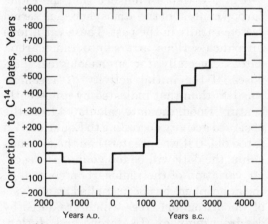

C^{14} Dates of Wood Using $T_{1/2}$ = 5730 Years

FIGURE 17.3 Corrections that must be added to radiocarbon dates to bring them into agreement with tree-ring dates, based on 143 analyses of wood of sequoia and bristlecone which were dated by dendrochronology. These systematic deviations in the radiocarbon dates are due to variations in the radiocarbon content of the atmosphere in the past 6000 years. (Michael and Ralph, 1970). A more detailed and more accurate comparison between carbon-14 and dendrochronology dates for the time interval A.D. 1840 to B.C. 4760 has been published by Ralph et al. (1973).

1950. If the date was measured in 1973, the radiocarbon date is 3500 − 23 = 3477 B.P. In order to relate this date to the historical calendar relative to the birth of Christ, we subtract 1950 years from 3477 and obtain the historical date 1527 B.C. Now we determine from Figure 17.3 that we must add 250 years to dates in the interval 1325 to 1700 B.C. Thus the corrected date is 1527 + 250 = 1777 years B.C. Figure 17.4 is a plot of deviations of the ^{14}C activity of wood samples from all over the world back to 1000 A.D. The curve, which is based on data published by Lerman et al., (1970), shows clearly the departures we described previously as the "de Vries effect" and the "Suess effect."

Dendrochronology can take us only about 7000 years into the past. To explore the variations in the radiocarbon activity in even earlier times, one must rely on varve chronologies of sediment deposited in glacial lakes. In some cases the sediment contains sufficient organic material to be datable by the carbon-14 method. Systematic and continuous differences between the age of the sediment as determined by counting annual varves and the radiocarbon dates are then used to determine the initial ^{14}C activity of the organic matter. In this way Stuiver (1970) showed that radiocarbon dates of organic material from sediment in Lake of the Clouds in Minnesota are consistently lower than varve dates by about 800 years in the interval 3000 to 10,000 years ago. This in turn implies that the radiocarbon content of the atmosphere in this time interval was up to 9 percent greater than during the nineteenth century. The radiocarbon content of the atmosphere prior to 10,000 years B.P. is not well documented. Stuiver (1970) summarized some evidence derived from varve chronologies which indicates that the activity during the interval 10,000 to 12,500 years B.P. was only about 4 percent higher than during the nineteenth century. In general, the existing

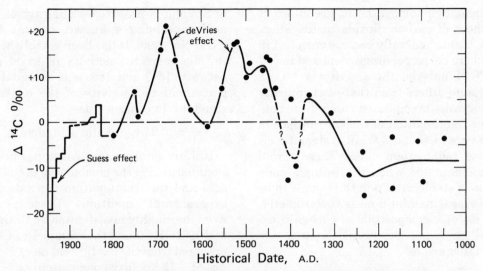

FIGURE 17.4 Deviation of the initial radiocarbon activity in per mil of wood samples of known age relative to 95 percent of the activity of the oxalic acid standard of the National Bureau of Standards. The observed activities were corrected for carbon isotope fractionation and recalculated using a value of 5730 years for the half-life of ^{14}C. The decline in the radiocarbon content starting at about 1900 results from the introduction of fossil CO_2 into the atmosphere by the combustion of fossil fuels (Suess effect). The anomalously high radiocarbon activity around 1710 and 1500 A.D. is known ar the "de Vries effect." Its causes are not understood. (Data from Table 6a and b of Lerman et al., 1970.)

evidence indicates that the radiocarbon content of the atmosphere has varied by 10 percent or more both above and below the nineteenth century value during the past 30,000 years.

The cause or causes for the secular variations are much more difficult to establish. Basically, there are three possible reasons for these variations: (1) variations in the intensity of the cosmic-ray proton flux caused by the behavior of the sun; (2) variations in the strength of the Earth's magnetic field which modulates the proton flux and thereby affects the production rate of ^{14}C; (3) changes in the carbon reservoirs of the Earth due to climatic variations. It seems possible, and perhaps probable, that all three causes have affected the radiocarbon content of the atmosphere in the past. A specific

explanation for the observed secular variation of the carbon-14 content of the atmosphere is not currently available. These problems were thoroughly discussed during the Twelfth Nobel Symposium held in August 1969 in Uppsala, Sweden (Olsson, 1970) and were reviewed by Ralph and Michael (1974).

4
Isotope Fractionation

Another cause for variation of the radiocarbon content of natural samples is the fact that the isotopes of carbon are fractionated by physical and chemical reactions occurring in nature (Chapter 20). As a result, the abundances of the stable isotopes of the carbon ($^{12}_6C$ and $^{13}_6C$), which is incorporated into plants by photosynthesis, are different from those of carbon dioxide in the atmosphere. Plants generally become enriched in ^{12}C and are correspondingly depleted in ^{13}C and ^{14}C. Similarly, the activity of ^{14}C in plant tissue differs from that of chemically or biologically precipitated calcium carbonate. The fractionation of carbon isotopes in nature thus introduces small systematic errors into radiocarbon dates. These errors can be eliminated when the isotopic composition of stable carbon in the sample to be dated is measured on a mass spectrometer.

The isotopic composition of carbon is expressed by means of the $\delta^{13}C$ parameter which is defined as

$$\delta^{13}C = \left[\frac{\left(\frac{^{13}C}{^{12}C}\right)_{spl} - \left(\frac{^{13}C}{^{12}C}\right)_{std}}{\left(\frac{^{13}C}{^{12}C}\right)_{std}} \right] \times 10^3 \permil \quad (17.6)$$

The $\delta^{13}C$ value is simply the per mil difference between the $^{13}C/^{12}C$ ratio of a sample and that of a standard. When the sample is enriched in ^{12}C relative to the standard, $\delta^{13}C$ is negative, and when the sample is enriched in ^{13}C, $\delta^{13}C$ is positive. The standard that is most widely used is carbon dioxide prepared from belemnites (*Belemnitella americana*) collected from the Peedee Formation (Cretaceous) of South Carolina. This is known as the PDB or the University of Chicago standard where this material was first analyzed by Dr. H. C. Urey and his colleagues in 1951 to determine paleotemperatures on the basis of the fractionation of oxygen isotopes.

The problem of correcting radiocarbon activities for isotope fractionation has become entwined with the use of the oxalic acid ^{14}C standard. In order to facilitate interlaboratory calibrations of radiation detectors used in different laboratories to measure the specific decay rate of ^{14}C, the National Bureau of Standards prepared a quantity of oxalic acid containing a known amount of ^{14}C (NBS No. 4990). It has been established that the age-corrected activity of wood grown between 1840 and 1860 A.D. is equal to 95 percent of the activity of the oxalic acid standard. In other words

$$A_o = 0.95 A_{ox} \text{ (oxalic acid) dpm/g} \quad (17.7)$$

Unfortunately, carbon isotopes are fractionated during the combustion of the oxalic acid and the fractionation depends on the experimental conditions. These problems were thoroughly investigated by Craig (1954 and 1961). He found that samples of CO_2 gas prepared from the oxalic acid have $\delta^{13}C$ values of $-19.3\permil$ (dry combustion) or $-19.6\permil$ (wet combustion), relative to the PDB standard. Most laboratories have agreed informally to adopt a standard value of $-19\permil$ and now correct the activity of each batch of CO_2 prepared from oxalic acid according to the measured $\delta^{13}C$ value of that batch. The correction is made by assuming that fractionation depends on mass differences and that ^{14}C is enriched or depleted twice as much as ^{13}C relative to ^{12}C. The formula that has been worked out for this purpose is

$$A_{ox} = A_{ox}' \left[1 - \frac{2(19 + \delta^{13}C')}{1000} \right] \text{ dpm/g} \quad (17.8)$$

where

A_{ox} = activity of the oxalic acid corrected for fractionation to $\delta^{13}C = -19\permil$, relative to the PDB standard,

A_{ox}' = observed activity of a particular batch of CO_2 gas prepared from the oxalic acid standard,

$\delta^{13}C'$ = measured value of this parameter of the CO_2 gas, relative to the PDB standard.

With these corrections the initial activity of nineteenth-century wood, corrected for decay is

$$A_o = 0.95 A'_{ox} \left[1 - \frac{2(19 + \delta^{13}C')}{1000} \right] \text{dpm/g}$$

(17.9)

Further complications arise when it is necessary to measure the radiocarbon activity of samples of wood or shells whose carbon isotope compositions differ from that of average wood for which $\delta^{13}C = -25\permil$. (Wood, on the average, is therefore depleted in ^{13}C relative to the oxalic acid by $6\permil$.) The observed activity of such samples whose $\delta^{13}C$ value is different from $-25\permil$ must be corrected by means of the following formula:

$$A_{corr.} = A_{meas.} \left[1 - \frac{2(25 + \delta^{13}C_{PDB})}{1000} \right] \text{dpm/g}$$

(17.10)

In order to date a carbon-bearing sample by the carbon-14 method, the laboratory must first establish the specific activity of the oxalic acid standard, corrected for isotope fractionation that occurred during its combustion to CO_2 gas (Equation 17.8). It must then measure the specific activity of the sample to be dated and correct it for possible fractionation of carbon isotopes (Equation 17.10). The radiocarbon date is then calculated from the fractionation-corrected activity of the sample and an initial activity equal to $0.95 A_{ox}$. Finally, corrections may have to be made to the calculated date to take into account the secular variations of the radiocarbon content of the atmosphere (Fig. 17.3).

5
Methodology

The principal analytical problem of radiocarbon dating is the detection and accurate measurement of the low level of radioactivity of ^{14}C in natural materials. Originally, Libby and his colleagues burned the sample to be dated to form carbon dioxide which was subsequently reduced to solid carbon by reacting it with hot metallic magnesium. The radioactivity of the solid carbon was then measured by means of a specially designed Geiger counter which was carefully shielded against background radiation. The details of this procedure were described by Libby (1955). The rationale behind the use of solid carbon was that it provided a sufficient concentration of carbon to eliminate the need for isotopic enrichment of the sample in ^{14}C. However, the efficiency of this system was only about 5 percent, primarily because the low-energy beta particles emitted by ^{14}C (average energy 50 keV) are absorbed by the layer of solid carbon, thus reducing the efficiency of the counter. It also became apparent in the 1950s that the method sometimes gave erroneously high counting rates because of contamination of the solid carbon with radioactivity from "fallout" resulting from the atomic weapons testing in the atmosphere that was occurring at that time. For these reasons, most investigators abandoned the solid carbon method by 1956 and changed to gas-counting techniques. (de Vries and Barendsen, 1953). At about the same time, attempts were made to use liquid scintillation techniques. These usually involve the chemical synthesis of an organic compound which can be used as the solvent for the phosphor or which itself acts as the phosphor. The most promising of many schemes that have been proposed is the conversion of the sample carbon dioxide to benzene which has a high carbon content and is a good solvent. (Polach and Stipp, 1967; Noakes et al., 1967).

The procedures that are currently in use have been described by Ralph (1971) and de Vries (1959), among others. After appropriate pretreatment to remove impurities, the carbon is liberated as carbon dioxide by burning organic samples with oxygen or by treating carbonate samples with phosphoric or hydrochloric acid. The gas thus produced is then treated for removal of impurities, such as oxides of nitrogen, oxygen gas, and halogens. In addition, efforts are made to remove radon. Some laboratories convert the carbon dioxide to other gaseous compounds such as methane (Burke and Meinschein, 1955; Fairhall et al., 1961) or acetylene (Suess, 1954). These and other compounds listed by Rankama (1963) are somewhat more difficult to synthesize, involve the risk of an explosion, and are not always formed with 100 percent yields. In the latter case, isotope fractionation may enrich or deplete the product gas in radiocarbon. On the other hand, both methane and acetylene are better counting gases because they are less affected by impurities and require lower voltages in the proportional region of the gas-counting tubes.

roundings. Additional gamma rays originate as a component of cosmic rays. The counting rate due to background gamma radiation is reduced by shielding the counter with steel plates 10 to 30 cm thick. The remaining background is largely due to cosmic ray mesons which are not effectively absorbed by the steel shield. This component of the background is removed by surrounding the radiocarbon counting tube with a ring of Geiger tubes connected to it by an "anticoincidence" circuit. Some additional reduction of the background is achieved by the use of a neutron trap composed of a mixture of paraffin and boric acid which stops locally produced neutrons from entering the counter. Ralph (1971) reported the following counting rates for equipment operated at the Radiocarbon Laboratory at the University of Pennsylvania in Philadelphia:

	COUNTS/MIN
Unshielded counter filled with CO_2 made from anthracite coal	1500
Shielded by iron and mercury	400
Shielded as before with anticoincidence counters turned on	8

The purified carbon dioxide, or whatever gaseous compound is used, is placed inside a counting tube commonly made of purified copper. In order to increase the counting rate, the volume of the counting tube is made large (up to 8 liters) and the gas is compressed (1 to 3 atmospheres), thus increasing the number of ^{14}C atoms present. However, the specific activity of the radiocarbon in the gas can be measured only after the background activity is drastically reduced. The background activity originates from several sources. Most important are gamma rays emitted by naturally occurring or artificially produced radioactive isotopes in the sur-

It is common practice to count all samples for a sufficient number of hours to obtain favorable counting statistics. The standard deviation of 1 percent is achieved by counting 10,000 events, which usually requires from 12 to 24 or more hours. The measurement is then repeated, and a statistical test is made to confirm that the two measured background-corrected counting rates are identical. This check also indicates whether the sample is contaminated with radon which has a half-life of only 3.82 days and whose activity therefore diminishes rapidly within a few days. The various materials suitable for dating are listed in Table 17.1.

Table 17.1   Material Suitable for Dating by the Carbon-14 Method[a] (Ralph, 1971)

MATERIAL	AMOUNT REQUIRED IN GRAMS	COMMENTS
Charcoal and wood	25	Usually reliable, except for finely divided charcoal which may adsorb humic acids, removable by treatment with NaOH. Subject to "post-sample-growth error," i.e., difference in time between growths of tree and use of the wood by humans.
Grains, seeds, nutshells, grasses, twigs, cloth, paper, hide, burned bones	25	Usually reliable. These materials are "short lived" and have negligible post-sample-growth errors.
Organic material mixed with soil	50–300	Should contain at least 1% organic carbon in the form of visible pieces. Efforts should be made to remove as much soil as possible in the field.
Peat	50–200	Often reliable, but intrusive roots of modern plants must be removed. The coincidence of peat formation with the occupation of archaeological sites requires careful consideration.
Ivory	50	Often well preserved and reliable. Interior of tusks is younger than the exterior. Some ivory tools may have been carved from old rather than contemporary material.
Bones (charred)	300	Heavily charred bones are reliable. Lightly charred bones are not, because exchange with modern radiocarbon is possible.
Bones (collagen)	1000 or more	Organic carbon in bones called collagen is reliable. However, the organic carbon content is low and decreases with age to less than 2%.
Shells (inorganic carbon)	100	The carbon in calcite or aragonite of shells may exchange with radiocarbon in carbonate-bearing groundwater. Shell carbon may be initially enriched in ^{14}C relative to wood due to isotope fractionation. It may also be depleted in ^{14}C due to incorporation of "dead" carbon derived by weathering of old carbonate rocks. The reliability of "shell dates" is therefore questionable.
Shells (organic carbon)	Several kilograms	Organic carbon is present in the form of conchiolin which makes up 1 to 2% of modern shells. Dates may be subject to systematic errors due to uncertainty of initial ^{14}C activity of this material.
Lake marl and deep-sea or lake sediment	Variable	Such materials are datable on the basis of the radiocarbon content of calcium carbonate. Special care must be taken to evaluate errors due to special local circumstances.
Pottery and iron	2 to 5 kg	Pottery sherds and metallic iron may contain radiocarbon that was incorporated at the time of manufacture. Reliable dates of such samples have been reported.

[a] The approximate amounts of material required for dating are based on the assumption that 6 grams of carbon should be available which is sufficient to fill an 8-liter counter with CO_2 at a pressure of 1 atmosphere.

6
Dating of Carbonate Samples

The dating of chemically or biologically precipitated calcium carbonate by the radiocarbon method is of great interest to geologists, archaeologists, and oceanographers. However, the radiocarbon content of such samples depends far more on local conditions than does the radiocarbon content of plant material that equilibrated with the atmosphere. Specifically, the initial ^{14}C content of calcium carbonate deposited in water depends on the abundance of this isotope among the ionic and molecular carbonate species in the water. The original work of Libby and his colleagues suggested that the specific activity of ^{14}C in carbonate samples is higher than that of plant material by a factor of about 1.05 because of their enrichment in ^{14}C by isotope fractionation. However, subsequent work by other investigators has revealed a far more complicated picture. It is convenient to discuss the variations of the radiocarbon contents of modern carbonates by distinguishing between freshwater and marine carbonate materials.

The radiocarbon content of the carbonate shells of *modern freshwater* mollusks depends primarily on the source of the bicarbonate ions in solution. In areas underlain by rocks composed of silicate minerals, the bicarbonate ion originates primarily from atmospheric carbon dioxide. For example, the action of aqueous carbon dioxide and water on K-feldspar is represented by the following chemical reaction:

$$2KAlSi_3O_8 + 2CO_2 + 11H_2O \rightarrow 2K^+$$
Orthoclase
$$+ H_4Al_2Si_2O_9 + 4H_4SiO_4 + 2HCO_3^- \quad (17.11)$$
Kaolinite   Silicic Acid Bicarbonate

All of the bicarbonate ions produced by the weathering of K-felspar to kaolinite are derived from atmospheric carbon dioxide. Therefore the specific activity of ^{14}C of such

mollusk shells, corrected for isotope fractionation to $\delta C^{13} = -25‰$, is close to that of plant material in equilibrium with the atmosphere. On the other hand, in bodies of freshwater in contact with old carbonate rocks, up to 50 percent of the bicarbonate ion may consist of "dead" carbon derived from the carbonate rock by the following reaction:

$$CaCO_3 + CO_2 + H_2O \rightarrow$$
$$Ca^{+2} + 2HCO_3^- \quad (17.12)$$

One of the two bicarbonate ions in this reaction originates from the calcium carbonate whose original radiocarbon content has diminished to zero in the time elapsed since deposition. Therefore, the radiocarbon content of modern mollusk shells or of inorganically precipitated calcium carbonate in "hard-water" lakes and rivers is usually less than that predicted on the basis of equilibrium with atmospheric carbon dioxide. As a result, such carbonate samples give fictitious radiocarbon dates that may be too old by several thousand years. The magnitude of this effect varies locally depending on the geology of each watershed and may vary also along the course of a river that traverses different kinds of rocks or that mixes with different water masses. For this reason, the radiocarbon dates of freshwater mollusks have often disagreed with radiocarbon dates of land-based plant or animal remains. However, such samples may be dated reliably when their radiocarbon content is compared to similar material of known age from which the initial activity can be calculated. Even then, the activity of freshwater mollusk shells is subject to the secular variations of the radiocarbon content of the atmosphere discussed before.

Many examples of fictitious radiocarbon dates of mollusk shells have been reported. For example, Keith and Anderson (1963) analyzed shells of *modern* mollusks from rivers, lakes, and from the oceans with the following results: shells from rivers had an average age of 1733 years, while marine shells gave an average age of 155 years. They suggested that the radiocarbon content of freshwater mollusk shells is depleted because of the oxidation of humus which may be as much as 3000 years old. Although this explanation was subsequently disputed by Broecker (1964), it is possible that humus is a contributing cause for the radiocarbon deficiency observed in some freshwater mollusk shells.

The ^{14}C activity of sea water and of *marine mollusk shells* is also less than expected. The reason for this is to be found primarily in the circulation of water in the ocean basins. Cold water sinks in the polar regions because of its higher salinity and then flows toward lower latitudes along the bottom. The rate of flow is sufficiently slow to cause a significant decrease of the radiocarbon content of this cold bottom water. Radiocarbon dates of this water range from several hundred up to about 1000 years. In certain regions of the ocean, such as along the west coasts of North and South America, the bottom water rises to the surface because of the displacement of surface water by the prevailing winds. The surface water in these regions is a mixture of deep water and normal water and is therefore depleted in radiocarbon. Measurements by Berger et al. (1966) and by Taylor and Berger (1967) reveal that the "prebomb" radiocarbon content of marine shells collected from the coastal waters of California to Peru is up to 8.5 percent lower than that of biospheric carbon. The corresponding fictitious radiocarbon dates of modern marine shells from the west coasts of North and South America can be as high as 300 years.

It is clear that dating of mollusk shells collected during archaeological or geological investigations requires a careful consid-

eration of several pertinent factors. Because the radiocarbon contents of freshwater or coastal marine shells reflect local conditions, it is necessary to determine, if possible, where the shells originated from. Once it is known that the shells have not been transported, a study can be made of the radiocarbon content of "prebomb" shells formed in the same or similar environment nearby. Shells originating from the open ocean where the radiocarbon content of the water was not depleted by mixing with old bottom water or by carbonate-rich river water are probably most reliable for dating. However, the possibility remains that the radiocarbon content of any mollusk shell was altered by deposition of secondary calcite or aragonite from percolating groundwater. In spite of these considerable difficulties of dating mollusk shells by the ^{14}C-method, useful information can be obtained from them, and excellent agreement between "wood dates" and "shell dates" has been reported in several instances. (Olson and Broecker, 1958; Berger et al., 1966; Wendorf et al., 1970; Mangerud and Gulliksen, 1975; Donner and Jungner, 1975; Stuiver and Waldren, 1975).

7 SUMMARY

The carbon-14 method of dating, which was developed around 1950 by J. W. Libby, is the culmination of research dating back to the early 1930s. It is based on measurements of the activity of ^{14}C in carbon-bearing materials that originally were in communication with CO_2 gas of the atmosphere. Carbon-14 is generated by the interaction of cosmic-ray produced neutrons and ^{14}N in the atmosphere and forms $^{14}CO_2$ which is rapidly mixed throughout the atmosphere. Living plants (and animals) contain a constant level of ^{14}C; but when the plant (or animal) dies, the activity due to ^{14}C decreases with a characteristic half-life of 5730 years.

Measurements of the ^{14}C activity of wood samples dated by dendrochronology have disclosed significant variations in the initial radiocarbon content. These variations may result from a combination of several factors, including (1) changes in the cosmic-ray flux due to the activity of the sun; (2) changes in the Earth's magnetic field; and (3) changes in the reservoirs of carbon on the Earth. In addition, the radiocarbon content of the atmosphere has decreased in the past 100 years due to combustion of fossil fuel and has increased because of the dispersal of ^{14}C by nuclear explosions in the atmosphere. Small corrections to ^{14}C dates are also required to eliminate the effects of fractionation of carbon isotopes, including ^{14}C.

A wide variety of materials are datable by this method, including calcium carbonate of mollusk shells. However, biogenic or inorganically precipitated calcium carbonate may have anomalously low ^{14}C contents due to the presence of "dead" carbon released by dissolution of limestone or due to aging of oceanic bottom water.

PROBLEMS

1. The specific radiocarbon activity of a sample of wood is 6.25 dpm/g of carbon. The specific activity of the NBS oxalic acid standard is 14.27 dpm/g of carbon. What is the age of the wood sample, assuming that the half-life of ^{14}C is 5730 years? (**ANSWER:** 6367 y.)

2. The specific radiocarbon activity of a sample of wood from the seventeenth-century A.D. that was 310 years old in 1970 when it was analyzed was found to be 15.09 dpm/g of carbon. What was the initial activity of ^{14}C in this sample and how does it differ from that of nineteenth-century wood? (**ANSWER:** 15.67 dpm/g, higher by about 2 percent.)

3. Carbonate from a modern shell of the freshwater mollusk *Elliptio* was found to have a δC^{13} value of $-11.45‰$. Calculate a correction factor by which the observed radiocarbon activity of this shell would have to be multiplied in order to correct it for carbon isotope fractionation to a standard value of $\delta C^{13} = -25‰$. (**ANSWER:** 0.9729.)

4. The δC^{13} value of carbonate from a modern (prebomb) shell of the marine mollusk *Tivela* was found to be $+1.12‰$. Calculate a radiocarbon date for this shell corresponding to its enrichment in radiocarbon as a result of isotope fractionation relative to $\delta C^{13} = -25‰$. (**ANSWER:** -442 y.)

5. A specimen of wood was recovered from an archaeological site and its radiocarbon content and δC^{13} value in 1970 were found to be 10.15 dpm/g and $-28.7‰$, respectively. Calculate the radiocarbon date of this sample corrected for isotope fractionation and express it as a date in terms of years A.D. or B.C. (**ANSWER:** 364 B.C.)

6. A mollusk shell was recovered from some post-Pleistocene lake deposits. The radiocarbon content and δC^{13} value of this shell are 5.62 ± 0.28 dpm/g and $-7.90‰$, respectively. Calculate the radiocarbon date of this shell, corrected for isotope fractionation, and calculate the uncertainty of the date due to the error of the radiocarbon activity. Assume that A_0 was equal to the normal biospheric value. (**ANSWER:** 7570 ± 415 y.)

REFERENCES

Anderson, E. C., W. F. Libby, S. Weinhouse, A. F. Reid, A. D. Kirshenbaum, and A. V. Grosse (1947) Natural radiocarbon from cosmic radiation. Phys. Rev., *72*, 931–936.

Anderson, E. C., and W. F. Libby (1951) World-wide distribution of natural radiocarbon. Phys. Rev., *81*, 64–69.

Arnold, J. R., and W. F. Libby (1949) Age determinations by radiocarbon content: Checks with samples of known age. Science, *110*, 678–680.

Berger, R., R. E. Taylor, and W. F. Libby (1966) Radiocarbon content of marine shells from the California and Mexican west coast. Science, *153*, 864–866.

Broecker, W. S. (1964) Radiocarbon dating: A case against the proposed link between river mollusks and soil humus. Science, *143*, 596–597.

Burcham, W. E., and M. Goldhaber (1936) The disintegration of nitrogen by slow neutrons. Proc. Cambridge Phil. Soc., *32*, 632–636.

Burke, W. H., Jr., and W. G. Meinschein (1955) C^{14} dating with a methane proportional counter. Rev. Sci. Inst., *26*, 1137–1140.

Craig, H. (1954) Carbon-13 in plants and the relationships between carbon-13 and carbon-14 variations in nature. J. Geol., *62*, 115–149.

Craig, H. (1961) Mass-spectrometer analysis of radiocarbon standards. Radiocarbon, *3*, 1–3.

deVries, H., and G. W. Barendsen (1953) A new technique for radiocarbon dating by a proportional counter filled with carbon dioxide. Physica, *19*, 987–1003.

deVries, H. (1958) Variation in concentration of radiocarbon with time and location on earth. Proc. Koninkl. Ned. Akad. Wetenschap, *B61*, 94–102.

deVries, H. (1959) Measurement and use of natural radiocarbon. In Researches in Geochemistry, pp. 169–188, P. H. Abelson, ed. John Wiley, New York.

Donner, J., and H. Jungner (1975) Radiocarbon dating of shells from marine Holocene deposits in the Disko Bugt area, West Greenland. Boreas, *4*, 25–46.

Fairhall, A. W., Schell, W. R., and Y. Takashima (1961) Apparatus for methane synthesis for radiocarbon dating. Rev. Sci. Inst., *32*, 323–325.

Godwin, H. (1962) Half-life of radiocarbon. Nature, *195*, 984.

Grosse, A. V. (1934) An unknown radioactivity. J. Amer. Chem. Soc., *56*, 1922–1923.

Kamen, M. D. (1963) Early history of carbon-14. Science, *140*, 584–590.

Karlen, I., I. U. Olsson, P. Kallberg, and S. Kilicci (1966) Absolute determination of the activity of two C^{14} dating standards. Arkiv Geofysik, *6*, 465–471.

Keith, M. L., and G. M. Anderson (1963) Radiocarbon dating: Fictitious results with mollusk shells. Science, *141*, 634–637.

Kurie, F. N. D. (1934) A new mode of disintegration induced by neutrons. Phys. Rev., *45*, 904–905.

LaMarche, V. C., Jr., and T. P. Harlan (1973) Accuracy of tree-ring dating of bristle-cone pine for calibration of radiocarbon time scale. J. Geophys. Res., *78*, 8849–8858.

Lerman, J. C., W. G. Mook, and J. C. Vogel (1970) C^{14} in tree rings from different localities. In Radiocarbon Variations and Absolute Chronology, pp. 275–299, I. U. Olsson, ed. Wiley-Interscience, New York.

Libby, W. F. (1955) Radiocarbon Dating. (2nd ed.) University of Chicago Press, Chicago, 175 pp.

Lingenfelter, R. E. (1963) Production of carbon 14 by cosmic-ray neutrons. Rev. Geophys., *1*, 35–55.

Locker, G. L. (1933) Neutrons from cosmic-ray stösse. Phys. Rev., *44*, 779–781.

Mangerud, J., and S. Gulliksen (1975) Apparent radiocarbon ages of Recent marine shells from Norway, Spitsbergen and Arctic Canada. Quat. Res., *5*, 263–274.

Michael, H. N., and E. K. Ralph (1970) Correction factors applied to Egyptian radiocarbon dates from the era before Christ. In Radiocarbon Variations and Absolute Chronology, pp. 109–119, I. U. Olsson., ed. Almqvist and Wiksell, Stockholm, and John Wiley, New York.

Montgomery, C. G., and D. D. Montgomery (1939) The intensity of neutrons of thermal energy in the atmosphere at sea level. Phys. Rev., *56*, 10–12.

Noakes, J. E., S. M. Kim, and L. K. Akers (1967) Recent improvement in benzene chemistry for radiocarbon dating. Geochim. Cosmochim. Acta, *31*, 1094–1096.

Olson, E. A., and W. S. Broecker (1958) Sample contamination and reliability of radiocarbon dates. Trans. New York Acad. Sci., Ser. II, *20*, 593–604.

Olsson, I. U., ed. (1970) Radiocarbon variations and absolute chronology. Proc. Twelfth Nobel Symp., Almqvist and Wiksell, Stockholm, and John Wiley, New York, 652 pp.

Polach, H. A., and J. J. Stipp (1967) Improved synthesis techniques for methane and benzene radiocarbon dating. Int. J. Appl. Rad. Isotopes, *18*, 359–364.

Rafter, T. A., and J. D. Stout (1970) Radiocarbon measurements as an index of the rate of turnover of organic matter in forest and grassland ecosystems in New Zealand. In Radiocarbon Variations and Absolute Chronology, pp. 401–415, I. U. Olsson, ed. Almqvist and Wiksell, Stockholm, and John Wiley, New York.

Ralph, E. K. (1971) Carbon-14 dating. In Dating Techniques for the Archaeologist, pp. 1–48, H. N. Michael and E. K. Ralph, eds. The M.I.T. Press, Cambridge, Mass.

Ralph, E. K., H. N. Michael, and M. C. Han (1973) Radiocarbon dates and reality. MASCA Newsletter, *9*, No. 1, 1–18.

Ralph, E. K., and H. N. Michael (1974) Twenty-five years of radiocarbon dating. Amer. Scient., *62*, 553–560.

Rankama, K. (1963) Progress in Isotope Geology. John Wiley, New York, 705 pp.

Rumbaugh, L. H., and G. L. Locker (1936) Neutrons and other heavy particles in cosmic radiation of the stratosphere. Phys. Rev., *49*, 855.

Stuiver, M., and H. E. Suess (1966) On the relationship between radiocarbon dates and true sample ages. Radiocarbon, *8*, 534–540.

Stuiver, M. (1970) Long-term C^{14} variations. In Radiocarbon Variations and Absolute Chronology, pp. 197–213, I. U. Olsson, ed. Almqvist and Wiksell, Stockholm, and John Wiley, New York.

Stuiver, M., and W. H. Waldren (1975) ^{14}C carbonate dating and the age of post-Talayotic lime burials in Mallorca. Nature, *255*, 475–476.

Suess, H. E. (1954) Natural radiocarbon measurements by acetylene counting. Science, *120*, 1–3.

Suess, H. E. (1955) Radiocarbon concentration in modern wood. Science, *122*, 415–417.

Suess, H. E. (1965) Secular variations of the cosmic-ray-produced carbon-14 in the atmosphere and their interpretation. J. Geophys. Res., *70*, 5937–5952.

Taylor, R. E., and R. Berger (1967) Radiocarbon content of marine shells from the Pacific coasts of Central and South America. Science, *158*, 1180–1182.

Wendorf, F., R. Schild, and R. Said (1970) Problems of dating the late Paleolithic age in Egypt. In Radiocarbon Variations and Absolute Chronology, pp. 57–77, I. U. Olsson, ed. Almqvist and Wiksell, Stockholm, and John Wiley, New York.

18 OXYGEN AND HYDROGEN IN THE HYDROSPHERE AND THE ATMOSPHERE

The isotopic compositions of elements having low atomic numbers are variable because their isotopes are fractionated in the course of certain chemical and physical processes occurring in nature. The fractionation is due to slight variations in the physical and chemical properties of isotopes and is proportional to differences in their masses. For this reason, natural isotope fractionation has been detected only up to about mass 40, that is, potassium and calcium. The most important elements in which natural variations of the isotopic composition have been observed include hydrogen, carbon, nitrogen, oxygen, and sulfur. The fractionation of the isotopes of these and other elements has been described by Rankama (1954, 1963) and by Hoefs (1973). The isotope geology of oxygen has been reviewed by Epstein (1959), Epstein and Taylor (1967), and by Garlick (1969).

Isotope fractionation is a consequence of the fact that certain thermodynamic properties of molecules depend on the masses of the atoms of which they are composed. The energy of a molecule in a gas can be described in terms of the interactions among the electrons plus translational, rotational, and vibrational components of the atoms in that molecule. Molecules containing different isotopes of an element in equivalent positions have different energies because of differences in the vibrational components that are mass dependent. In the case of hydrogen and deuterium, differences in the rotational component are

also important. The energy of a molecule decreases with decreasing temperature, and at absolute zero it assumes a certain finite value called the zero-point energy which is greater than the minimum value by $1/2hv$, where h is Planck's constant and v is the vibrational frequency. The vibrational frequency of a molecule is inversely proportional to its mass. Therefore, a given molecule that contains the lighter of two isotopes has a higher vibrational frequency and hence a higher zero-point energy than a similar molecule containing the heavier isotope. One consequence of this phenomenon is that bonds formed by the lighter of two isotopes are weaker and are therefore more easily broken, making the molecule with the lighter isotope more reactive than a similar molecule containing the heavier isotope. The thermodynamic properties of isotopic molecules and the resulting fractionation of isotopes have been treated by Urey (1947), Bigeleisen (1952, 1965), Bigeleisen and Mayer (1947), Broecker and Oversby (1971), and others.

In general, isotopic fractionation occurs during several different kinds of chemical reactions and physical processes:

1. Isotopic exchange reactions involving the redistribution of isotopes of an element among different molecules containing that element.
2. Unidirectional reactions in which reaction rates depend on isotopic compositions of the reactants and products.

3. Physical processes such as evaporation and condensation, melting and crystallization, adsorption and desorption, and diffusion of ions or molecules due to concentration or temperature gradients in which mass differences come into play.

The isotope fractionation that occurs during such processes is indicated by the fractionation factor α which is defined as

$$\alpha = \frac{R_A}{R_B} \qquad (18.1)$$

where R_A is the ratio of the heavy to the light isotope in molecule or phase A and R_B is the same in phase B. In some cases, the fractionation factors can be calculated by means of partition functions derivable from statistical mechanics using experimentally determined vibrational frequencies. Such calculations are only possible for systems composed of gases and liquids and are not directly applicable to systems containing solids. The reason is that in solids one must consider additional energy terms arising from lattice vibrations. In any case, the fractionation factor for any given system is temperature dependent and generally approaches unity at increasing temperatures. This means that isotope fractionation in nature depends on the temperature and is therefore interpretable in terms of environmental temperatures.

Even though fractionation factors are theoretically predictable at any given temperature, the interpretation of observed variations of the isotopic composition of natural materials are largely empirical and rely primarily on generalizations of large bodies of observational data or involve comparisons of natural samples to experimental results obtained in the laboratory. For this reason, we shall refrain from a detailed presentation of the theory of isotope fractiona-

tion and turn our attention first to a description of the isotopic composition of oxygen and hydrogen in the hydrosphere and atmosphere of the Earth.

1
Oxygen and Hydrogen in Water and Water Vapor

Oxygen ($Z = 6$) is the most abundant chemical element in the crust of the Earth and is combined with hydrogen to form water (H_2O). Oxygen has three stable isotopes whose approximate abundances are: $^{16}O = 99.756$ percent, $^{17}O = 0.039$ percent, and $^{18}O = 0.205$ percent. Hydrogen ($Z = 1$) has two stable isotopes whose abundances are $^{1}H = 99.985$ percent and $^{2}H = 0.015$ percent (Holden and Walker, 1972). The heavy isotope of hydrogen, which was discovered by H. C. Urey (Chapter 1), is called deuterium (D). Hydrogen also has a naturally occurring radioactive isotope ^{3}H (tritium, T) which has a half-life of 12.26 y.

Because of the existence of three stable isotopes of oxygen and two stable isotopes of hydrogen, ordinary water molecules have nine different isotopic configurations whose masses are approximately given by their mass numbers: $H_2^{16}O$ (18), $H_2^{17}O$ (19), $H_2^{18}O$ (20), $HD^{16}O$ (19), $HD^{17}O$ (20), $HD^{18}O$ (21), $D_2^{16}O$ (20), $D_2^{17}O$ (21), and $D_2^{18}O$ (22). The vapor pressures of the different isotopic molecules of water are inversely proportional to their masses. Therefore, $H_2^{16}O$ has a significantly higher vapor pressure than $D_2^{18}O$. For this reason, water vapor formed by evaporation of liquid water is enriched in ^{16}O and H while the remaining water is enriched in ^{18}O and D. The fractionation factor at any given temperature is the ratio of the vapor pressures, provided equilibrium between vapor and liquid is maintained.

It is apparent that evaporation of water results in similar isotope fractionation of hydrogen and oxygen such that ^{16}O and H preferentially enter the vapor phase, while ^{18}O and D are concentrated in the liquid phase. For this reason, it is appropriate to consider the fractionation of oxygen and hydrogen together. The fractionation of hydrogen and deuterium in the hydrosphere has been described by Friedman et al. (1964).

The isotopic compositions of oxygen and hydrogen are reported in terms of differences of $^{18}O/^{16}O$ and D/H ratios relative to a standard called SMOW (Standard Mean Ocean Water). This standard was defined by Craig (1961a) with reference to a large volume of distilled water distributed by the National Bureau of Standards in the United States (NBS − 1), such that

$$D/H \, (SMOW) = 1.050 \, D/H \, (NBS - 1) \quad (18.2)$$

$$^{18}O/^{16}O \, (SMOW) = 1.008 \, ^{18}O/^{16}O \, (NBS - 1)$$
$$(18.3)$$

The isotopic compositions of oxygen and hydrogen of a sample are expressed as per mil differences relative to SMOW:

$$\delta^{18}O = \left[\frac{(^{18}O/^{16}O)_{spl} - (^{18}O/^{16}O)_{SMOW}}{(^{18}O/^{16}O)_{SMOW}} \right] \times 10^3 \permil$$
$$(18.4)$$

$$\delta D = \left[\frac{(D/H)_{spl} - (D/H)_{SMOW}}{(D/H)_{SMOW}} \right] \times 10^3 \permil \quad (18.5)$$

Consequently, positive values of $\delta^{18}O$ and δD indicate enrichment of a sample in ^{18}O and D compared to SMOW, while negative values imply depletion of these isotopes in the sample relative to the standard.

The isotopic composition of oxygen is measured on a mass spectrometer (Chapter 5) equipped with a double-collector (Nier et al., 1947) and with additional improvements made by McKinney et al. (1950). The sample to be analyzed is converted into CO_2 gas and the analysis is made by comparing its $^{18}O/^{16}O$ ratio to that of a standard gas (CO_2 equilibrated with SMOW) which is introduced in alternation with the unknown. Carbonate samples are converted to CO_2 by reacting them with 100 percent phosphoric acid at 25°C, while water samples are analyzed by allowing them to equilibrate with CO_2 gas at 25°C (Craig, 1957; Garlick, 1969; Hoefs, 1973). In addition to SMOW, several other standards have been used by different laboratories. One of these is the PDB standard used originally by H. C. Urey and his colleagues at the University of Chicago. The PDB standard is CO_2 gas produced with phosphoric acid from a Cretaceous belemnite (*Belemnitella Americana*) of the Peedee Formation of South Carolina.

Both standards (SMOW and PDB) have been used to express the isotopic compositions of oxygen in geological samples. This has given rise to two sets of numbers which are convertible by means of the following equation given by Craig (1957):

$$\delta^X_{PDB} = \delta^X_{SMOW} + \delta_{SMOW-PDB} + 10^{-3} \, \delta^X_{SMOW} \delta_{SMOW-PDB} \quad (18.6)$$

where δ^X_{PDB} and δ^X_{SMOW} refer to the unknown relative to the two standards and $\delta_{SMOW-PDB}$ is the δ-value of SMOW with respect to PDB. The conversion for carbonate samples (Craig, 1965; Clayton et al., 1968) is

$$\delta^X_{SMOW} = 1.03037 \delta^X_{PDB} + 30.37 \quad (18.6a)$$

The precision of $\delta^{18}O$ values is of the order of 0.2‰ or better including instrumental, analytical, and sampling errors. The reproducibility of δD is ±1‰ (Friedman et al., 1964). Craig and Gordon (1965) achieved a precision of better than 0.02‰ for $\delta^{18}O$ and 0.2‰ for δD.

The fractionation factor α is related to the $\delta^{18}O$ values in the following way: Let R_v, R_l, and R_{sw} be the $^{18}O/^{16}O$ ratios of vapor, liquid water, and SMOW, respectively. It follows from Equation 18.4 that

$$R_v = \frac{\delta^{18}O_v R_{sw}}{1000} + R_{sw} = \frac{R_{sw}(\delta^{18}O_v + 1000)}{1000}$$

(18.7)

$$R_l = \frac{\delta^{18}O_l R_{sw}}{1000} + R_{sw} = \frac{R_{sw}(\delta^{18}O_l + 1000)}{1000}$$

(18.8)

where $\delta^{18}O_v$ and $\delta^{18}O_l$ refer to vapor and liquid, respectively. Substituting into Equation 18.1,

$$\alpha = \frac{R_l}{R_v} = \frac{\delta^{18}O_l + 1000}{\delta^{18}O_v + 1000}$$

(18.9)

When water evaporates from the surface of the ocean, the water vapor is *enriched* in ^{16}O and H because $H_2{}^{16}O$ has a higher vapor pressure than $H_2{}^{18}O$ or water molecules containing D. Consequently, the $\delta^{18}O$ and δD values of water vapor in the atmosphere above the oceans are both negative. The values of the isotopic fractionation factors for evaporation of water under equilibrium conditions at 25°C (Craig and Gordon, 1965)

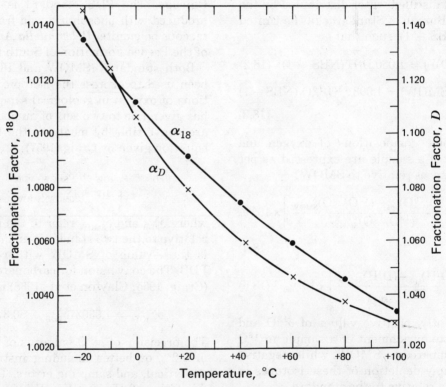

FIGURE 18.1 Temperature variation of isotope fractionation factors for evaporation of water. The fractionation factor for $^{18}O(\alpha_{18})$ is defined as the ratio of $^{18}O/^{16}O$ in the liquid to $^{18}O/^{16}O$ in water vapor in equilibrium with the liquid. The fractionation factor for $D(\alpha_D)$ is similarly defined as the ratio of D/H in the liquid to D/H in the vapor. The graph illustrates the temperature dependence of these isotopic fractionation factors. Plotted from values listed and discussed by Dansgaard (1964).

are

$$\alpha_{18} = \frac{(^{18}O/^{16}O)_l}{(^{18}O/^{16}O)_v} = 1.0092 \quad (18.10)$$

$$\alpha_D = \frac{(D/H)_l}{(D/H)_v} = 1.074 \quad (18.11)$$

The temperature dependence of these fractionation factors is shown in Figure 18.1.

In actual fact, the isotopic composition of water vapor over the oceans does not conform with predictions based on evaporation alone. Craig and Gordon (1965) found that $\delta^{18}O$ values of water vapor in the North Pacific and the North Atlantic Oceans are significantly more negative than the predicted equilibrium values. The difference is due to kinetic effects discussed by Craig and Gordon (1965).

When raindrops form in a cloud by condensation of water vapor, the liquid phase is *enriched* in ^{18}O and D, such that the isotopic composition of the first raindrops is similar to that of ocean water. The continuing preferential removal of ^{18}O and D from a moist air mass causes enrichment of ^{16}O and H in the remaining vapor phase. Therefore, the $\delta^{18}O$ and δD values of water vapor in an air mass become progressively more negative as rain, snow, or hail continue to fall from it. The δ values of the liquid or solid precipitation also become negative because of the depletion of the vapor in ^{18}O and D.

The condensation of water in equilibrium with water vapor and its subsequent removal from a cloud can be described by the Rayleigh distillation equation (Broecker and Oversby, 1971)

$$\frac{R}{R_0} = f^{(\alpha - 1)} \quad (18.12)$$

where R is the $^{18}O/^{16}O$ ratio of the remaining vapor, R_0 is the $^{18}O/^{16}O$ ratio of the vapor before condensation begins, f is the fraction

of the vapor remaining, and α is the isotope fractionation factor $= R_l/R_v$.

We now convert the isotope ratios R and R_0 to the δ-notation by means of Equation 18.9.

$$\frac{R}{R_0} = \frac{\delta^{18}O + 1000}{(\delta^{18}O)_0 + 1000} = f^{(\alpha - 1)}$$

$$\delta^{18}O = [(\delta^{18}O)_0 + 1000]f^{(\alpha - 1)} - 1000 \quad (18.13)$$

Figure 18.2 is a plot of $\delta^{18}O$ of water vapor

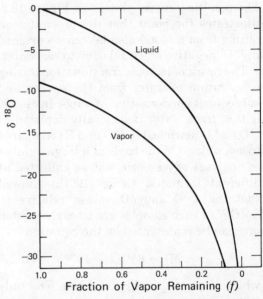

FIGURE 18.2 Fractionation of oxygen isotopes during condensation of water from vapor at 25°C, according to the Rayleigh distillation model ($\alpha = 1.0092$). The initial $\delta^{18}O$ value of the vapor is assumed to be $-9.2‰$. The first condensate in equilibrium with that vapor has $\delta^{18}O = 0$ .. Immediate removal of this condensate from the cloud without reevaporation or isotope exchange continuously enriches the remaining vapor in ^{16}O (i.e., makes its $\delta^{18}O$ value more negative). The condensate that subsequently forms in equilibirum with this vapor also acquires negative $\delta^{18}O$ values. This model accounts in a general way for the fact that $\delta^{18}O$ values of fresh water are negative compared to sea water.

as a function of the fraction of vapor remaining (f), assuming that $\alpha = 1.0092$ ($T = 25°C$) and $(\delta^{18}O)_0 = -9.2‰$. The diagram shows that the $\delta^{18}O$ value of the remaining vapor decreases, that is, is enriched in ^{16}O, as condensation progresses. The $\delta^{18}O$ value of the condensate that is in equilibrium with the vapor at any given instant can be calculated by means of Equation 18.9 by solving for $\delta^{18}O_l$:

$$\delta^{18}O_l = \alpha(\delta^{18}O_v + 1000) - 1000 \quad (18.14)$$

The resulting curve, plotted in Figure 18.2, illustrates the point that the precipitation falling from a cloud also becomes enriched in ^{16}O (negative $\delta^{18}O$) relative to sea water.

The result of isotopic fractionation during evaporation of water from the oceans and subsequent condensation of vapor in clouds is that fresh water is generally depleted in ^{18}O and D (enriched in ^{16}O and H) compared to sea water. On the basis of a large number of analyses of meteoric waters collected at different latitudes, Craig (1961b) showed that the $\delta^{18}O$ and δD values relative to SMOW of such samples are linearly related and can be represented by the equation

$$\delta D = 8\delta^{18}O + 10 \quad (18.15)$$

which is plotted in Figure 18.3a. The only meteoric water samples that do not fit this relationship are from closed basins where excessive evaporation occurs and certain lakes and rivers in East Africa. The range of $\delta^{18}O$ values of meteoric waters is from about 0 to $-60‰$, while δD varies from about $+10$ to less than $-400‰$. The lowest values occur in snow precipitated at high latitudes and/or high elevations at low temperatures. Dansgaard (1964) demonstrated a very striking linear relationship between the $\delta^{18}O$ values of average annual precipitation and the average annual air temperature. This relationship is illustrated in Figure

18.3b as a plot of the equation

$$\delta^{18}O_m = 0.695t - 13.6 \quad (18.16)$$

where $\delta^{18}O_m$ is the annual mean $\delta^{18}O$ value of precipitation and t is the average annual surface air temperature in degrees Centigrade. The relationship between $\delta^{18}O$ and the air temperature reflects the fact that the isotope fractionation factor increases with decreasing temperature.

As a moist air mass moves toward higher latitudes, the $\delta^{18}O$ and δD values of rain or snow falling from it become progressively more negative because of a combination of factors which include: (1) isotope fractionation due to differences in vapor pressure of isotopic water molecules at a given temperature (Fig. 18.2); (2) decrease of the temperature in the air mass and a resulting increase in the fractionation factors (Fig. 18.1); (3) reevaporation of water from rain droplets and from surface water on the ground, both of which enrich the vapor phase in ^{16}O and H; (4) evapo-transpiration of water by plants that also favor ^{16}O and H.

In addition to the nine different isotopic varieties of the water molecule considered so far, we need to mention one more, namely: HTO, where T is tritium. This is the naturally occurring radioactive isotope of hydrogen ($T_{1/2} = 12.26$ y) that is produced in the upper atmosphere by the interaction of fast cosmic-ray neutrons with stable ^{14}N according to the reaction.

$$^{14}_{7}N + n \rightarrow T + ^{12}_{6}C \quad (18.17)$$

The tritium combines with oxygen to form HTO which is subsequently dispersed throughout the hydrosphere. Estimates of the natural production rate of tritium range from 15 to 45 atoms/min/cm^2 of the Earth's surface (Giletti et al., 1958; Libby, 1961; Suess, 1969). The concentrations of tritium in various parts of the hydrosphere have been greatly increased as a result of the

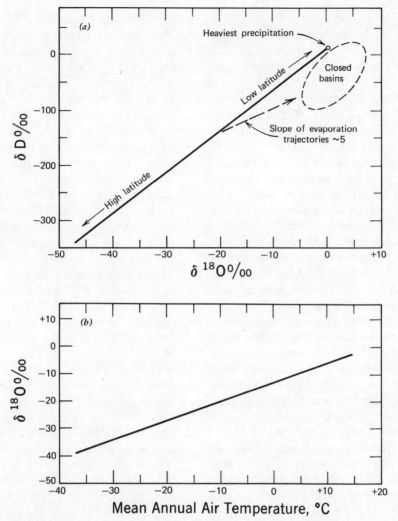

FIGURE 18.3 (a) Relationship between δD and $\delta^{18}O$ (relative to SMOW) in meteoric water samples. The isotopic compositions of water and snow exhibit a pronounced latitude effect which reflects the temperatures of condensation. Water from closed basins and certain northeast African lakes and rivers in which excessive evaporation occurs deviate from the relationship. Such samples lie on trajectories having a slope of about 5. The equation of the relationship between δD and $\delta^{18}O$ in meteoric water is $D = 8\,\delta^{18}O + 10$. (Craig, 1961b.) (b) Relationship between the annual means of $\delta^{18}O$ of meteoric precipitation (relative to SMOW) and the average annual air temperature. The equation of this line is $\delta^{18}O = 0.695t - 13.6$ and is based on analyses of a large number of water samples by Dansgaard (1964).

explosion of hydrogen bombs from 1954 to 1963. The introduction of artificially produced tritium into the atmosphere and hydrosphere has been used to study mixing rates in the atmosphere and in the oceans and to trace the movement of groundwater (Dincer and Davis, 1967). Although water molecules containing tritium are subject to isotopic fractionation, this effect is generally negligible compared to variations in the rate of input and subsequent decay of tritium to stable $_2^3$He by beta emission.

The oxygen of the atmosphere is strongly enriched in ^{18}O and has a δ^{18}O value of $+23.5\%_0$. This is the so-called "Dole effect" (Dole et al., 1954) which may be due to preferential removal of ^{16}O by respiration of plants and animals. The oxygen released during photosynthesis is derived from the water molecule and has a weighted average δ^{18}O value of $+5\%_0$ (Garlick, 1969). Since the fractionation factor for isotope exchange of oxygen between water and oxygen is very nearly equal to one (Urey, 1947), the oxygen of the atmosphere is not in isotopic equilibrium with water of the hydrosphere.

Atmospheric CO_2 has a δ^{18}O value of $+41\%_0$. The fractionation factor for isotope exchange of oxygen between CO_2 and H_2O (*l*) at 25°C is 1.0407 (Garlick, 1969) which indicates that atmospheric CO_2 is in approximate isotopic equilibrium with sea water.

2
Snow and Ice Stratigraphy

The isotopic composition of oxygen (and hydrogen) of snow deposited in the polar regions of the Earth and at high elevations in mountains depends primarily on the temperature. For this reason δ^{18}O values of snow are strongly negative and display variations both in terms of seasonal temperature fluctuations and variations related to altitude and geographic latitude. Systematic measurements of δ^{18}O and δD have therefore been used to study flow patterns of glaciers, snow accumulation rates, and climatic variations in the past 100,000 years.

The seasonal variation of δ^{18}O values of snow can be used to date snow and firn layers by identifying successive summer layers as a function of depth. Snow deposited during the summer has less negative δ^{18}O values than snow deposited during winter when air temperatures are lower. Complications arise in areas of low precipitation, such as the interior of Antarctica, because snow may be removed by deflation and may then be redeposited elsewhere. As a result, the thickness of annual snow layers may be variable and gaps may exist in the record. Nevertheless, δ^{18}O profiles of snow and firn at various places in Antarctica have been used to determine the average rates of accumulation of water. For example, Epstein et al. (1965) found an average annual accumulation rate of 7 cm of water at the South Pole during the time interval 1958–1963, based on a δ^{18}O profile of snow (Fig. 18.4). This estimate is in good agreement with results obtained by identification of annual layers by conventional stratigraphic methods. It also agrees with the accumulation rate obtained by dating snow at the South Pole by means of ^{210}Pb (Chapter 16.5, Fig. 16.5). Similar studies of this kind have been published by Gonfiantini (1965), Deutsch et al. (1966), and others. Moreover, Hamilton and Langway (1967) demonstrated a significant correlation between the concentration of dust particles (0.6 to 3 microns in diameter) in an ice core from Greenland (site 2, close to Camp Century) and δ^{18}O values.

The seasonal fluctuations of δ^{18}O in snow are gradually eliminated due to homogenization of the isotope composition of the oxygen. This results from several causes, including melting and refreezing of water percolating downward through snow or firn,

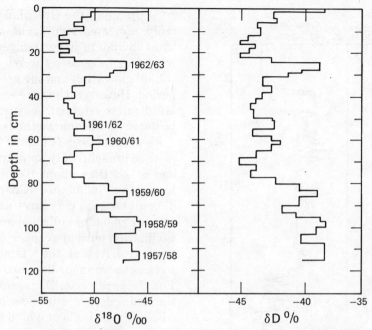

FIGURE 18.4 Seasonal variation of $\delta^{18}O$ and δD in snow and firn at stake 25 of the South Pole accumulation network. The dates identify snow accumulations during austral summers. (Adapted from Figure 2 (p. 1811) of Epstein, S., R. P. Sharp, and A. J. Gow, Journal of Geophysical Research, vol. 70, 1809–1814, 1965, copyrighted by American Geophysical Union, by permission of the A.G.U.)

vertical movement of air trapped in pore spaces, diffusion of water vapor along temperature gradients, and thinning of seasonal ice layers during plastic deformation.

Even though the seasonal variations of $\delta^{18}O$ values of annual layers are gradually obliterated, the absolute values record climatic conditions primarily in terms of the mean air temperatures. Therefore, continuous ice cores recovered from the continental ice sheets in Greenland and Antarctica contain a climatic record extending as far into the past as the age of the oldest ice that is recovered. The first such core was drilled in 1966 at Camp Century, northwest Greenland, about 225 km east of Thule. A similar core was later obtained at Byrd Station in the interior of Antarctica. The Greenland core, which was 1390 meters long and 12 cm in diameter, was analyzed by Dansgaard et al. (1969, 1971). They observed systematic variations of $\delta^{18}O$ values as a function of depth ranging from about -28 to $-40\%_0$. In general, the upper portion of the core has a relatively constant $\delta^{18}O$ value of about $-29\%_0$, except for seasonal and lesser long-term fluctuations. It is succeeded at greater depth by a sharp decrease to about $-40\%_0$, suggesting a significantly lower average air temperature at that time. The $\delta^{18}O$ profile then shows a broad minimum followed by a gradual increase of $\delta^{18}O$ values to about $-26\%_0$ toward the bottom of the core, implying a warmer climate (Fig. 18.5). The

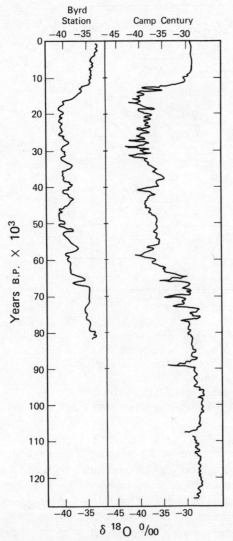

FIGURE 18.5 Variation of $\delta^{18}O$ in ice cores from Byrd Station, Antarctica, and Camp Century, Greenland. The time scale is based on a theoretical ice-flow model. The more negative $\delta^{18}O$ values from about 70,000 to 12,000 years B.P. in both cores reflect colder climatic conditions during the last ice age. More detailed correlations with the Pleistocene chronostratigraphy and a discussion of the ice-flow model can be found in the original. (Modified after Figure 5e and d, (p. 433) of Johnson, S. J., W. Dansgaard, H. B. Clausen and C. C. Langway, Nature, vol. 235, Feb. 25, 429–434, 1972, by permission of Macmillan Journals Ltd.)

$\delta^{18}O$ profile of the Greenland core undoubtedly indicates significant climatic variations including lower temperatures during what may have been the Wisconsin (Weichselian) glaciation and the preceding warmer period. However, the time scale of this core is difficult to establish because the ice cannot be dated by isotopic methods.

As we have seen (Chapter 16.5), the ^{210}Pb method is useful only for dating snow whose age is less than about 100 years. Dating of ice by the ^{14}C method (Chapter 17) is also difficult because very large samples (one ton or more) must be collected and because contamination with atmospheric ^{14}C is difficult to control. At best, the ^{14}C method is limited to samples younger than about 30,000 years. For these reasons, Dansgaard et al. (1969) devised a dynamic ice-flow model (later revised) on the basis of which they calculated the age of the ice as a function of depth in the core. This time scale was questioned by Mörner (1972, 1974) primarily because the model did not allow for changes in ice-flow parameters due to changing climatic conditions. Instead, Mörner proposed a time scale based on correlations of short-term climatic events, displayed by the $\delta^{18}O$ profile, with similar events recognized in the Quaternary time scale. Both time scales indicate ages in excess of about 120,000 years for the oldest ice in the Greenland core, but differ in detail in the dating of significant climatic fluctuations.

The core recovered at Byrd Station in Antarctica has been described by Johnson et al. (1972). Its $\delta^{18}O$ profile is remarkably similar to that of the Camp Century core in the sense that it has the same broad minimum representing colder climatic conditions during the Wisconsin glaciation from about 66,000 to 11,000 years before the Present. However, the $\delta^{18}O$ profile of the Byrd Station core is more subdued, the change from interglacial to glacial conditions being represented by a $\delta^{18}O$ shift from $-34\%_{00}$ to $-41\%_{00}$

(Fig. 18.5). Because of the difficulty in establishing time scales for these cores based on glaciological models, detailed comparisons between the $\delta^{18}O$ records of the ice sheets in Greenland and Antarctica are unreliable. Moreover, the variations of $\delta^{18}O$ values are not necessarily controlled by the air temperature alone, but may also reflect other parameters, such as (1) possible presence of ice that formed elsewhere at different elevations and/or temperatures than the drill site; (2) changes in the temperature at the drill site caused by changes in elevation due to changing thickness of the ice; (3) changes in $\delta^{18}O$ of sea water due to variable storage of water in the continental ice sheets enriched in ^{16}O; (4) changes in atmospheric circulation patterns, such as direction of prevailing winds, that may have caused a change in the $\delta^{18}O$ of water vapor in the clouds over the drill site. Nevertheless, it is clear that the isotope composition of oxygen (and hydrogen) in ice and snow has become an important aspect of glaciological research, especially as it records changing climatic conditions in the past 100,000 years.

3
Isotope Composition of Water in the Oceans

The $\delta^{18}O$ and δD values of sea water are close to zero and vary only within narrow limits. As a consequence of the preferential removal $H_2^{16}O$, the isotopic composition of surface water in the oceans displays relationships between δD and $\delta^{18}O$ of the form

$$\delta D = M\delta^{18}O \qquad (18.18)$$

where M is the slope of linear trajectories and decreases with increasing ratios of evaporation to precipitation in a region. Typical values of M are 7.5 in the North Pacific, 6.5 in the North Atlantic, and 6.0 in the Red Sea.

Preferential loss of $H_2^{16}O$ from the surface of the oceans not only affects the isotopic

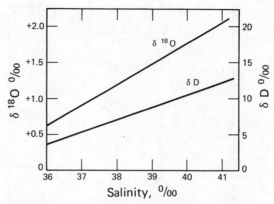

FIGURE 18.6   Relationships between $\delta^{18}O$, δD and salinity of water in the Red Sea due to preferential loss of $H_2^{16}O$ from the surface. The standard is SMOW. (Craig, 1966.)

composition of surface water but also increases its salinity. Although the range in $\delta^{18}O$ values of surface water attributable to this effect is small, careful measurements clearly reveal that $\delta^{18}O$ and δD values of surface water are positively correlated with salinity. This effect is well illustrated by water from the Red Sea (Craig, 1966) shown in Figure 18.6. The salinities of surface, intermediate and deep water in the Red Sea range from about 36‰ to almost 41‰, while the variation of $\delta^{18}O$ and δD is from about +0.6 to +1.9‰ and from +4 to +10‰, respectively. The slopes of the lines are 0.29 for $\delta^{18}O$-S and 1.72 for δD-S.

The $\delta^{18}O$ values and salinities of surface water in high latitudes are also modified by mixing with meltwater derived from ice and snow having very low $\delta^{18}O$ values. This effect is shown in Figure 18.7 by data originally published by Epstein and Mayeda (1953) and later modified by Craig and Gordon (1965). The data points fit a straight line whose equation is

$$\delta^{18}O = -21.2 + 0.61S \qquad (18.19)$$

where S is the salinity in per mil. If we let the salinity of the freshwater contaminant

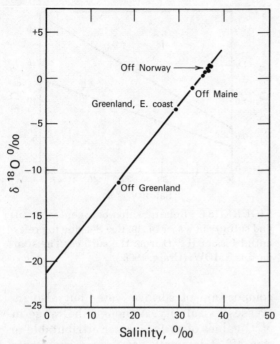

FIGURE 18.7 Relationship between $\delta^{18}O$ and salinity of surface water from the North Atlantic Ocean. (Data from Epstein and Mayeda, 1953, and Craig and Gordon, 1965.) The standard is SMOW.

be about 1‰, Equation 18.19 indicates that its $\delta^{18}O$ value was about -20.6‰, which is quite plausible for meteoric precipitation in high latitudes. Moreover, from Equation 18.15 we calculate a δD value of -154.8‰ for the freshwater contaminant.

Craig and Gordon (1965) presented additional documentation for the existence of functional relationships between $\delta^{18}O$ values and salinities of surface water in the North and South Pacific Ocean. Their data revealed the significant fact that average bottom water in the Pacific Ocean deviates measurably from the relationship displayed by surface water. This observation raises an interesting question about the origin and circulation of bottom water in the world's oceans.

In order to explain this phenomenon, we need to consider the effect of freezing of sea water on its salinity and isotopic composition. The ^{18}O isotope fractionation factor for ice in equilibrium with water is only 1.002, which means that the $\delta^{18}O$ value of sea ice is about $+2$‰ relative to the sea water from which it formed. Therefore, the formation of sea ice leaves the $\delta^{18}O$ value of the water essentially unchanged, but causes a significant increase in its salinity. As a result, the density of such water increases, causing it to sink. It is well known that such dense bottom water forms in the vicinity of Antarctica and thence flows north along the bottom of the ocean. The relationships between surface water in the Pacific and Atlantic oceans, their bottom waters, and Antarctic bottom water are illustrated in Figure 18.8.

The diagram shows trajectories for North Pacific surface water (NPSW) and North Atlantic surface water (NASW). The oxygen isotope composition and salinity of North Atlantic bottom water lie on the surface-water trajectory for the Atlantic Ocean, replotted here from Figure 18.7. However, the bottom waters from the Indian and Pacific oceans are significantly displaced in the direction of higher salinity or lower $\delta^{18}O$ compared to surface water in the oceans. (Trajectories for surface water from the Indian Ocean and for the South Pacific Ocean are not shown here but are given by Craig and Gordon, 1965.) Surface water from the Weddell Sea adjacent to Antarctica is compatible with surface water in the South Pacific and coincides with the trajectory for NASW. Freezing of surface water in the Weddell Sea increases its salinity but does not change its $\delta^{18}O$ value appreciably, as indicated by the arrow pointing toward Antarctic bottom water. The Antarctic bottom water flows north into the Pacific and Indian oceans and is mixed in the process with

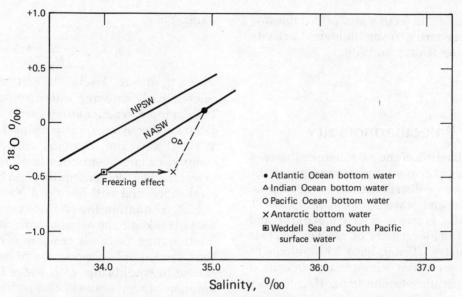

FIGURE 18.8 Relationships between North Pacific surface water (NPSW) and North Atlantic surface water (NASW) to respective bottom waters. The diagram shows that Antarctic bottom water forms by freezing of Weddell Sea surface water. Bottom water in the Pacific and Indian oceans can be regarded as a mixture of Atlantic bottom water and Antarctic bottom water in about equal proportions. However, the displacement of these bottom waters from the mixing line indicates the presence of a third, as yet unidentified, component. (Craig and Gordon, 1965.) The standard is SMOW.

Atlantic bottom water. We can therefore explain the salinities and $\delta^{18}O$ values of bottom waters in the Indian and Pacific oceans as mixtures of about equal proportions of Antarctic bottom water with Atlantic bottom water. However, it is clear that the bottom waters of the Indian and Pacific oceans plot slightly to the left of the mixing line. This indicates the presence of a third component whose salinity is less or whose δ^{18} value is higher, or both, compared to bottom water in the Indian and Pacific oceans. This "third component" has not been fully identified but may be intermediate water in the Indian and Pacific oceans whose approximate composition is $\delta^{18}O = -0.10\%_{00}$ and $S = 34.45\%_{00}$.

In addition to water, the oceans contain other oxygen-bearing ions, including primarily sulfate ($SO_4^=$) and bicarbonate (HCO_3^-). The $\delta^{18}O$ value of marine sulfate ions is constant at about $+9.7\%_{00}$. The fractionation factor for isotope exchange of oxygen between water and sulfate ions is uncertain. Urey (1947) obtained a value of 1.024 based on theoretical calculations, while Lloyd (1968) reported 1.0311 at 25°C derived from experimental determinations. In any case, it is clear that sulfate ions are not in isotopic equilibrium with oxygen in sea water. Lloyd (1968) showed that sulfate-reducing bacteria prefer ^{16}O by about 4.5%$_{00}$ and that periods of time of the order of 10^3 to 10^5 years are required to approach isotopic equilibrium between sulfate and water at surface conditions. The oxygen isotope composition of sulfate ions in fresh water is variable and related primarily to the biological activity in the water (Cortecci, 1973).

4
Paleothermometry

"The estimation of the paleotemperatures of the ancient oceans by measurement of the oxygen isotope distribution between calcium carbonate and water, suggested by H. C. Urey (1947), is one of the most striking and profound achievements of modern nuclear geochemistry" (Craig, 1965). These eloquent words express the widespread admiration of the scientific community for Harold Urey and his many important accomplishments as a scientist, including specifically his study of the fractionation of stable isotopes (Chapter 1).

The paleotemperature scale is based on the fact that the isotope composition of oxygen of calcium carbonate (calcite or aragonite) differs from that of water when this compound precipitates from water under equilibrium conditions. The difference results from an isotope exchange reaction between calcium carbonate and water which can be represented by the equation

$$\tfrac{1}{3}CaCO_3^{16} + H_2O^{18} \rightleftarrows$$
$$\tfrac{1}{3}CaCO_3^{18} + H_2O^{16} \quad (18.20)$$

The equilibrium constant for this reaction is

$$K = \frac{[CaCO_3^{18}]^{1/3}[H_2O^{16}]}{[CaCO_3^{16}]^{1/3}[H_2O^{18}]} \quad (18.21)$$

which can be rewritten as

$$K = \frac{([CaCO_3^{18}]/[CaCO_3^{16}])^{1/3}}{[H_2O^{18}]/[H_2O^{16}]} \quad (18.22)$$

The equilibrium constant in Equation 18.22 is equal to the ratio of $^{18}O/^{16}O$ in the carbonate phase divided by this ratio in the

water. Thus,

$$K = \alpha = \frac{R_c}{R_w} \quad (18.23)$$

where R_c and R_w are the $^{18}O/^{16}O$ ratios of calcium carbonate and water, respectively. The isotopic fractionation factor for the calcite-water system has a value 1.0286 at 25°C (O'Neil and Epstein, 1964). Consequently, calcite is enriched in ^{18}O relative to water when isotopic equilibrium has been established and will have a $\delta^{18}O$ value of +28.6‰ (Equation 18.9) if the δ value of sea water is taken to be equal to zero. Since the fractionation factor is temperature dependent, the isotopic composition of oxygen in calcite in equilibrium with water is also a function of temperature. The isotopic composition of sea water is not affected because the quantity of water is very large compared to the amount of calcium carbonate that is equilibrated with it. The temperature dependence of $\delta^{18}O$ of calcite in equilibrium with water was determined experimentally by Epstein et al. (1953). Their equation expressing the relationship between the temperature of the water and the δ values of calcite and water was later modified by Craig (1965):

$$t°C = 16.9 - 4.2(\delta_c - \delta_w)$$
$$+ 0.13(\delta_c - \delta_w)^2 \quad (18.24)$$

where δ_c is the true δ value of CO_2 gas prepared from the carbonate with 100 percent phosphoric acid at 25°C and δ_w is the true δ value of CO_2 gas equilibrated with the water at 25°C, and both are measured with reference to the same oxygen isotope standard. Equation 18.24 includes the isotopic composition of oxygen in the water as a variable and therefore can be used to determine the temperature of equilibration in any calcite-water system, *provided* that both calcite and water are available for analysis. However, *it is important to note that Equation 18.24 is based on the PDB rather than on the SMOW standard.* Because of the wide variations of $\delta^{18}O$ of fresh water, it is generally not possible to measure paleotemperatures by analysis of calcite deposited in fresh water. Such measurements are possible for *marine* calcite or aragonite (biogenic or inorganically precipitated) only because we may assume that the $\delta^{18}O$ value of sea water is constant, at least to a first approximation. Nevertheless, the accuracy of paleotemperature determinations of skeletal calcium carbonate of marine organisms is limited by several important prerequisite conditions.

First, we must know the $\delta^{18}O$ value of the sea water with which a given specimen of calcium carbonate equilibrated. This is not easy to establish with certainty because the $\delta^{18}O$ values are related to salinity and because the isotope composition of oxygen in the oceans as a whole depends on the amount of water stored on the continents in the form of glacial ice which is strongly enriched in ^{16}O. Using the slope of the $\delta^{18}O$-S relationship for the Red Sea (Fig. 18.6) a difference in salinity of 1‰ implies a difference of 0.29‰ in $\delta^{18}O$ which, according to Equation 18.24, changes the isotope temperature by about 1°C. The importance of continental glaciations on the isotopic composition of oxygen in sea water is widely recognized, although the magnitude of this effect is uncertain. Craig (1965) determined that the present ocean has a mean $\delta^{18}O$ value of $-0.08‰$. If all the continental glaciers were to melt, its δ value would decrease to $-0.60‰$, while at times of maximum glaciation during the Pleistocene epoch the value rose to $+0.90‰$. Shackleton (1968) estimated that the $\delta^{18}O$ of the oceans changed by 1.0 to 1.4‰ during glacial and interglacial ages of the Pleistocene epoch. This problem was again discussed in detail by Emiliani and Shackleton (1974).

A second important limitation on the reliability of the paleotemperature scale arises from the fact that the calcium carbonate secreted by some organisms is not in isotopic equilibrium with water. In such cases the $\delta^{18}O$ value of the carbonate is not a function of temperature alone, and Equation 18.24 therefore does not apply. Some organisms, such as certain mollusks, deposit calcium carbonate in equilibrium with sea water, while others (echinoderma, asteroidea, ophiuroidea, and crinoidea) do not (Weber and Raup, 1966a, b; Weber, 1968). The apparent isotope disequilibrium may be due to the influence of respiratory CO_2 which may exchange oxygen with bicarbonate ions in solution. Other important biological or environmental considerations in the interpretation of paleotemperature data include the seasonality of shell growth, the habitat of the organism, and possible changes in the habitat that occur during the life cycle of the organism.

Last, but not least, we must be concerned with the preservation of the isotopic composition of fossil shells. It is well known that significant mineralogical and chemical changes occur in calcium carbonate shells after the death of the organism. These include the gradual recrystallization of aragonite to calcite, the decay of the organic component of shells, and changes in their trace element composition, such as gain of uranium and loss of strontium and magnesium (Lowenstam, 1961). The end product of such postdepositional alteration is either complete destruction of the shell or its replacement by other minerals. While certain mineralogical and chemical criteria can be useful in avoiding altered shell material, it is difficult to prove that the isotopic composition of oxygen has remained unchanged in shells that appear to be unaltered.

In conclusion, we must recognize that the $\delta^{18}O$ values of unreplaced skeletal calcium carbonates depend not only on the environmental temperature but also on the isotope composition of the water and on possible metabolic effects. For this reason, isotopic temperature determinations are not necessarily accurate in an absolute sense, but can be used to detect *changes* in the temperature of the oceans in the geologic past.

On the basis of $\delta^{18}O$ determinations of the tests of pelagic foraminifers such as *Globigerinoides rubra* and *Globigerinoides sacculifera*, Emiliani (1955, 1966, 1972) has assembled a detailed record of temperature variations in the upper 50 meters of the oceans during the Pleistocene epoch. The paleotemperature curve shows recurring temperature fluctuations having a period of the order of 10^5 years extending for more than 435,000 years into the past. The interpretation of such paleoclimatic records depends on proof of continuity of the sedimentary record. Moreover, the time of deposition of the sediment at different levels in the core must be known. The establishment of a reliable time scale has been a difficult problem and is based largely on ^{14}C-dates of the carbonate fraction (Chapter 17) and on $^{230}Th/^{231}Pa$ dates (Chapter 16.4). The former are usually limited to the last 30,000 years, while the validity of the latter is still in some doubt. A third independent dating method is based on the magnetic-reversal time scale (Chapter 9.2, Fig. 9.2). Emiliani and Shackleton (1974) were able to demonstrate a remarkably good correlation, shown in Figure 18.9, of $\delta^{18}O$ values of *Globigeri-*

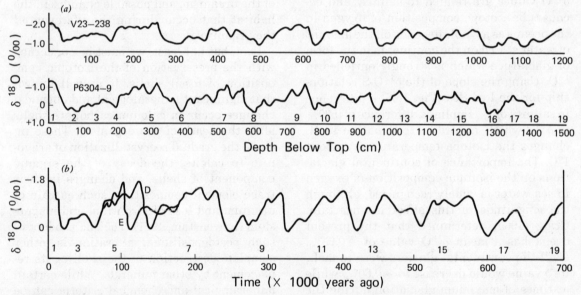

FIGURE 18.9 (*a*) $\delta^{18}O$ values (PDB scale) of tests of the foraminifer *Globigerinoides sacculifera* in cores V28-238 (western equatorial Pacific, 1°1′N, 160°29′E) and P6304-9 (Caribbean Sea, 14°57′N, 69°20′W) during the Brunhes epoch. The horizontal scale of core V28-238 has been adjusted to make the peak of stage 5 coincide with that of core P6304-9. The numbers above the abscissa identify deep-sea core stages. The systematic variations of $\delta^{18}O$ are attributable to temperature fluctuations of surface water in the oceans during the Pleistocene epoch. (*b*) Generalized paleotemperature curve and time scales (D and E) of Emiliani and Shackleton (1974). (Reproduced from Figures 1 and 4 of Emiliani, C. and N. J. Shackleton, Science, vol. 183, No. 4124, 511–514, 1974. Copyright 1974 by the American Association for the Advancement of Science by permission of A.A.A.S.)

noides sacculifera in two cores from the Caribbean Sea (P 6304-9, 14°57′N, 69°20′W) and from the western equatorial Pacific Ocean (V 28-238, 1°1′N, 160°29′E). (Note that the $\delta^{18}O$ values of both cores are on the PDB scale.) The core from the Pacific Ocean encompasses the entire Brunhes normal epoch and crosses the Brunhes/Matuyama boundary which is 700,000 years old.

The study of oxygen isotope compositions of marine carbonates and of glacial ice can provide a detailed record of temperature fluctuations during the Pleistocene epoch and thereby establishes a basis for testing the conflicting hypotheses of the causes of continental glaciation.

Attempts to determine the temperature history of the oceans during the Mesozoic and Cenozoic eras have been made by Spaeth et al. (1971), Tan et al. (1970), Stahl and Jordan (1969), Tourtelot and Rye (1969), Savin et al. (1975), and others. The results have disclosed significant temperature variations during the past 150 million years. However, these interpretations are uncertain because of the alteration of fossil shells, variations of $\delta^{18}O$ due to metabolic effects, and changes in the isotope composition of sea water.

The problem caused by the variation of $\delta^{18}O$ in sea water may be avoided by analysis of two mineral phases, both of which equilibrated with the same water at the same temperature. We would then have two independent equations relating the temperature and the δ value of the water and could thus determine both (Craig, 1965). Knowledge of time-dependent variations of $\delta^{18}O$ in the oceans would be at least as important as knowledge of the temperature. For this reason, much effort has been expended to study oxygen isotope fractionation between water and other phases. Two such phases are biogenic silica (Labeyrie, 1974) and phosphate (Longinelli and Nuti, 1973).

Several organisms (diatoms, radiolarians, and sponges) form siliceous skeletons composed of opal. This material is poorly crystallized and contains significant concentrations of organic matter and water, both as H_2O and as OH^- ions linked to silica tetrahedra. The presence of these impurities interferes with the analysis of the isotopic composition of oxygen (Mopper and Garlick, 1971; Knauth and Epstein, 1975). However, Labeyrie (1974) seems to have solved this problem by heating biogenic silica under vacuum, first at 100°C to remove H_2O and then at 1000°C to drive out OH^-. After this treatment, he obtained $\delta^{18}O$ values for modern sponge spicules and diatom frustules ranging from $+31.40$ to $+40.45\permil$ relative to SMOW. He demonstrated that the fractionation of oxygen between biogenic silica and water is temperature dependent according to the equation

$$t°C = 5 - (4.1 \pm 0.4)(\delta_{Si} - \delta_w - 40) \quad (18.25)$$

where δ_{Si} refers to the oxygen in the silica and δ_w to that in the water. Unfortunately, the slope of this line shown in Figure 18.10 is indistinguishable from that of the calcium carbonate thermometer (Equation 18.24). This means that the difference in the δ values of carbonate and silica in equilibrium with water is constant and independent of the temperature and of the δ value of the water with which both are in equilibrium.

A third paleothermometer arises from the fractionation of oxygen between water and biogenic phosphate. After many years of effort, Longinelli and Nuti (1973) succeeded in measuring $\delta^{18}O$ values of phosphate from the shells of living and fossil marine organisms and from marine phosphorite. Their thermometry equation is

$$t°C = 111.4 - 4.3(\delta_p - \delta_w) \quad (18.26)$$

where δ_p refers to the oxygen in the phosphate and δ_w to that in the water on the

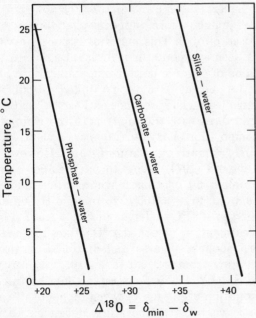

FIGURE 18.10 Temperature dependence of oxygen isotope fractionation between calcium carbonate, phosphate, silica, and water. $\Delta^{18}O$ is the difference between the δ value of a given mineral and that of the water with which it equilibrated. Silica is more strongly enriched in ^{18}O relative to water than are carbonate and phosphate. Nevertheless, the temperature dependence of the fractionation in all three phases is virtually identical. This means that paleotemperatures cannot be determined from the difference in $\delta^{18}O$ of two different phases which equilibrated at the same temperature with water of the same isotopic composition. (The equations for these equilibria are from Craig, 1965, Longinelli and Nuti, 1973, and Labeyrie, 1974.) All three lines are on the SMOW scale.

SMOW scale. Again, the slope of the line (Fig. 18.10) is nearly identical to that of the carbonate-water thermometer (Equation 18.24). One must conclude therefore that it is not possible to determine both the temperature and the isotopic composition of water from measurements of $\delta^{18}O$ in pairs of biogenic carbonate, silica, or phosphate.

5
Geothermal Water and Brines

The origin of water discharged by volcanoes and hot springs is of great interest because of the possibility that some of it is juvenile, that is, it is derived from the upper mantle of the Earth. The amount of juvenile water that is brought to the surface is an important parameter in reconstructing the history of the oceans not only in terms of time-dependent increases in its volume but also because of the potential effect on its isotopic composition.

The isotopic composition of oxygen of juvenile water is probably determined by that of the silicate rocks of the upper mantle. As we shall see in Chapter 19, mafic and ultramafic igneous rocks have $\delta^{18}O$ values ranging from about +6 to +7‰ relative to SMOW. At temperatures of the order of 1000°C, isotopic fractionation effects are negligible so that juvenile water entering the crust should have a $\delta^{18}O$ value of about +7‰ relative to SMOW. However, its isotopic composition may be modified subsequently at lower temperatures by exchange with rocks in the crust. Such exchange reactions generally lead to a depletion of ^{18}O in the aqueous phase and thereby obscure the isotopic composition of juvenile water. Its δD value has been estimated by Sheppard and Epstein (1970) on the basis of D/H ratios of unaltered phlogopites. They suggested $\delta D = -48 \pm 20$‰ for juvenile water.

Whatever its isotopic composition, juvenile water is a philosophical abstraction and probably does not exist in unaltered form in the crust of the Earth. Its isotopic composition is changed not only by exchange with crustal rocks over a wide temperature range but it may also be mixed with meteoric water or sea water trapped in the rocks. It is apparent, therefore, that water discharged by volcanoes and hot springs may have a complex origin that is reflected by its isotopic composition.

Craig (1963) determined $\delta^{18}O$ and δD values of steam or hot waters having a neutral to slightly basic pH. Some of his data are shown in Figure 18.11. It is immediately apparent that the $\delta^{18}O$ values of hot water and steam discharged in these and other geothermal areas are variable, while their δD values remain nearly constant. Craig attributed this "oxygen-isotope shift" to progressive equilibration of oxygen in the water with silicate and carbonate rocks and concluded that the source of the water and steam was local meteoric precipitation. His data for Niland in the geothermal area of the Salton Sea in southern California (Craig, 1966) show that the magnitude of the oxygen isotope shift increases with the temperature and is positively correlated with the solute contents of these brines. The temperature dependence of the oxygen isotope shift is explained by the fact that the fractionation factors decrease toward unity with increasing temperature which allows the $\delta^{18}O$ value of the water to approach that of the rocks more closely at elevated temperatures than is possible at low temperatures. The δD values of the geothermal waters remain

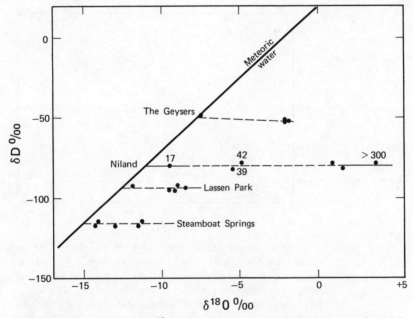

FIGURE 18.11 δD and $\delta^{18}O$ of hot water and steam from geothermal areas. Such samples display oxygen isotope shift due to progressive equilibration of oxygen in the water with oxygen in carbonate and silicate rocks. The δD values of neutral to basic water samples are not changed due to the low hydrogen content of most carbonate and silicate rocks. The magnitude of the shift increases with temperature, but depends also on the $\delta^{18}O$ value of the rocks and on the residence time of water in a given reservoir. (The diagram is based on data presented by Craig, 1963, 1966, but not all of his data points are shown here.) The standard is SMOW.

unchanged primarily because the hydrogen content of silicate and carbonate rocks is low compared to that of water. If these geothermal waters were mixtures of juvenile water ($\delta^{18}O \simeq +7\permil$, $\delta D \simeq -48\permil$) with meteoric waters, the mixing lines should converge toward a small area in the diagram representing the isotopic composition of juvenile water. It is apparent from Craig's work that the lines do *not* converge and that no juvenile water is detectable in these and other geothermal waters that he studied. The $\delta^{18}O$ and δD values of *acid* geothermal waters from Yellowstone Park, Lassen Park, and The Geysers (California) fit parallel lines having positive slopes. Craig (1963) explained these relationships as the result of nonequilibrium evaporation at temperatures of about 70 to 90°C.

Another topic of interest in this context is the origin of oil-field brines and connate water in sedimentary rocks of marine origin. One might expect that the δD and $\delta^{18}O$ values of such brines might converge to the isotopic composition of sea water. Clayton et al. (1966) analyzed suites of brine samples from the Gulf Coast, Illinois, Michigan, and Alberta. Although the data scatter considerably, there is no doubt that brines from a given area fit lines having variable, but posi-

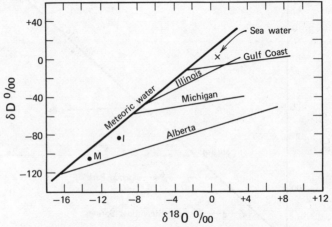

FIGURE 18.12 Relationships between δD and $\delta^{18}O$ in brines from the Gulf Coast, Illinois, Michigan (including southwestern Ontario), and Alberta collected from oil wells in sedimentary rocks of marine origin. The lines that were fitted to experimental data points bear no relationship to the isotopic composition of sea water but intersect the trajectory for meteoric water at compositions close to those of present-day precipitation in these areas. Therefore, these brines have not evolved from sea water but are descendents of modern meteoric precipitation. Points labeled *I* and *M* are exceptional brine samples from the Illinois basin and from southwestern Ontario, respectively. Their δD and $\delta^{18}O$ are anomalously low which suggests that they are meteoric water of Pleistocene age that formed at lower temperatures than those of the Present. (Clayton et al., 1966.) The standard is SMOW.

OXYGEN AND HYDROGEN IN THE HYDROSPHERE AND THE ATMOSPHERE 18

tive, slopes (Fig. 18.12). These lines do not converge toward sea water, but instead intersect the line representing meteoric water. The points of intersections imply isotopic compositions for meteoric water that are quite compatible in most cases with those of meteoric water forming in each respective area under present climatic conditions. A few samples were found to have low isotope compositions suggestive of water that may have precipitated in colder climatic conditions during the Pleistocene epoch (Fig. 18.12). The variation of δD is attributable to isotope exchange with hydrocarbons, hydrogen sulfide, and hydrated minerals such as gypsum and clay. The oxygen isotope shift

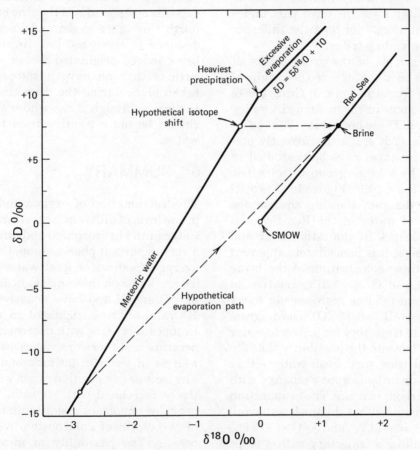

FIGURE 18.13 Isotope diagram for sea water and brine in the Red Sea. The diagram illustrates three possible modes of formation for the brine: (1) evaporation of fresh water along a trajectory having a slope of about 5; (2) oxygen-isotope shift due to exchange between fresh water and silicate and carbonate rocks; (3) deep circulation of sea water without change in isotopic composition. The first two alternatives imply that the fit of the isotopic composition of the brine on the $\delta D - \delta^{18}O$ line for sea water is a coincidence. For that and other reasons Craig (1966), from whom these data were taken, favored the third alternative. The standard is SMOW.

of the brines is very large, increases with temperature and solute concentration, and is due primarily to isotope exchange between water and calcitic limestone having an average $\delta^{18}O$ value of $+23.8 \pm 2\%$ relative to SMOW. The isotopic composition of these brines therefore indicates that they are *not* descendents of sea water but are meteoric water precipitated for the most part under climatic conditions not greatly different from those prevailing today.

Another example of the great utility of isotopic data in the study of the origin of brines arises from the work of Craig (1966) on the brine pools in the median rift valley of the Red Sea. These hot brines ($T = 56°C$) and the metal-rich sediment currently precipitating from them have been studied in great detail by a large group of scientists (Degens and Ross, 1969). Figure 18.13 is a δD versus $\delta^{18}O$ diagram showing the isotope composition of water in the Red Sea and that of the brines in the Atlantis II and Discovery deeps. It is immediately apparent that the isotopic composition of the brine ($\delta D = +7.5\%$, $\delta^{18}O = +1.21\%$ relative to SMOW) lies on the line representing water in the Red Sea ($\delta D = 6.0\ \delta^{18}O$). The diagram also shows the trajectory for meteoric water in order to illustrate the possibility that the brine formed from such fresh water either by evaporation or by isotope exchange with silicate or carbonate rocks. The evaporation hypothesis requires that the fresh water initially had $\delta D = -13.2\%$ and $\delta^{18}O = -2.9\%$ and moved along a trajectory with a slope of about 5 to arrive at the observed isotope composition of the brine. The other alternative is that fresh water with a composition $\delta D = +7.5\%$, $\delta^{18}O = -0.3\%$ experienced oxygen-isotope shift and in the process produced brine of the observed isotope composition. Both alternatives require specific initial isotope compositions for the fresh water and imply that the fit of the δ values of the

brine on the sea water trajectory is a coincidence. Craig (1966) rejected these hypotheses for this reason and concluded instead that the brine originated as present-day surface water near the Straits of Bab el Mandeb in the southern end of the Red Sea. He attributed the high salinity of the brine (257.76‰) to solution of marine evaporite deposits and postulated that the brine flowed north through a system of fractures for a distance of about 800 km. However, if the brine indeed originated as sea water, it is curious that no oxygen-isotope shift has taken place during the deep circulation of the brine through rocks whose $\delta^{18}O$ values should be more positive than that of sea water.

6 SUMMARY

The fractionation of oxygen and hydrogen in the form of different isotopic molecules of water in the hydrosphere and atmosphere is a very important phenomenon. Compared to average ocean water, fresh water and glacial ice and snow on the continents are enriched in ^{16}O and H and have negative $\delta^{18}O$ and δD values. The enrichment in the lighter isotopes increases with decreasing air temperature and therefore varies seasonally as well as in terms of latitude and elevation. The isotope composition of sea water is positively correlated with its salinity and has provided important insight into the problem of the origin of cold bottom water in the oceans. The possibility of measuring the temperature fluctuations of the oceans in the geologic past by analysis of oxygen isotope compositions of calcium carbonate, silica, and phosphate has received a great deal of attention. It must be remembered, however, that such paleotemperature determinations require assumptions regarding the isotope composition of sea water in the past and regarding the existence of isotopic equilibrium. Finally, the characteristic isotopic relationships between oxygen and hydrogen in meteoric water provide a basis for studying the origin of geothermal water and brines of various kinds.

PROBLEMS

1. If the fractionation factor for ^{18}O between liquid and vapor water at $10°C$ is $\alpha = 1.0101$, what is $\delta^{18}O$ of vapor in isotopic equilibrium with water having $\delta^{18}O = -0.80‰$? (**ANSWER:** $\delta^{18}O_v = -10.79‰$.)

2. Calculate δD of water vapor in isotopic equilibrium with fresh water whose δD value is $-65‰$, assuming that α (liquid-vapor) $= 1.090$. (**ANSWER:** $D = -142.2‰$.)

3. Calculate $\delta^{18}O$ of *raindrops* forming in an air mass whose initial $\delta^{18}O$ value was $-9.2‰$, assuming that only 20 percent of the original water vapor remain and that α (liquid-vapor) $= 1.0092$. (**ANSWER:** $\delta^{18}O \simeq -14.75‰$.)

4. What is the δD value of water vapor in an air mass whose initial δD value was $-145‰$ after half of the vapor has condensed, assuming that α (liquid-vapor) $= 1.106$?

5. Show that $(\alpha - 1)\,1000 \simeq \delta_A - \delta_B$, where α is the fractionation factor of an isotope between two phases A and B in isotopic equilibrium. (*Hint:* Let $\delta_B + 1000 \simeq 1000$.)

REFERENCES

Bigeleisen, J., and M. G. Mayer (1947) Calculation of equilibrium constants for isotopic exchange reactions. J. Chem. Phys., *15*, 261–267.

Bigeleisen, J. (1952) The effects of isotopic substitutions on the rates of chemical reactions. J. Phys. Chem., *56*, 823–824.

Bigeleisen, J. (1965) Chemistry of isotopes. Science, *147*, 463–471.

Broecker, W. S., and V. M. Oversby (1971) Chemical Equilibria in the Earth. McGraw-Hill, New York, 318 p.

Clayton, R. N., I. Friedman, D. L. Graf, T. K. Mayeda, W. F. Meents, and N. F. Schimp (1966) The origin of saline formation waters. 1. Isotopic composition. J. Geophys. Res., *71*, 3869–3882.

Clayton, R. N., B. F. Jones, and R. A. Berner (1968) Isotope studies of dolomite formation under sedimentary conditions. Geochim. Cosmochim. Acta, *32*, 415–432.

Cortecci, G. (1973) Oxygen isotope variations in sulfate ions in the water from some Italian lakes. Geochim. Cosmochim. Acta, *37*, 1531–1542.

Craig, H. (1957) Isotopic standards for carbon and oxygen and correction factors for mass spectrometric analysis of carbon dioxide. Geochim. Cosmochim. Acta, *12*, 133–149.

Craig, H. (1961a) Standard for reporting concentrations of deuterium and oxygen-18 in natural waters. Science, *133*, 1833–1934.

Craig, H. (1961b) Isotopic variations in meteoric waters. Science, *133*, 1702–1703.

Craig, H. (1963) The isotopic geochemistry of water and carbon in geothermal areas. In Nuclear Geology on Geothermal Areas. Spoleto, Sept. 9–13, 1963. Consiglio Nazionale delle Ricerche, Laboratorio di Geologia Nucleare, Pisa, 53 p.

Craig, H. (1965) The measurement of oxygen isotope paleotemperatures. In Stable Isotopes in Oceanographic Studies and Paleotemperatures. Spoleto, July 26–27, 1965. Consiglio Nazionale delle Ricerche, Laboratorio di Geologia Nucleare, Pisa, 1–24.

Craig, H., and L. I. Gordon (1965) Deuterium and oxygen-18 variations in the ocean and the marine atmosphere. In Stable Isotopes in Oceanographic Studies and Paleotemperatures. Spoleto, July 26–27, 1965. Consiglio Nazionale delle Ricerche, Laboratorio di Geologia Nucleare, Pisa, 1–122.

Craig, H. (1966) Isotopic composition and origin of the Red Sea and Salton Sea geothermal brines. Science, *154*, 1544–1548.

Dansgaard, W. (1964) Stable isotopes in precipitation. Tellus, *16*, No. 4, 436–468.

Dansgaard, W., S. J. Johnson, J. Moller, and C. C. Langway, Jr. (1969) One thousand centuries of climatic record from Camp Century on the Greenland ice sheet. Science, *166*, 377–381.

Dansgaard, W., S. J. Johnson, H. B. Clausen, and C. C. Langway (1971) Climate record revealed by the Camp Century ice core. In Late Cenozoic Glacial Ages, pp. 37–56, K. K. Turekian, ed., Yale University Press, New Haven.

Degens, E. T., and D. A. Ross, eds. (1969) Hot Brines and Recent Heavy Metal Deposits in the Red Sea. Springer-Verlag, New York, 600 p.

Deutsch, S., W. Ambach, and H. Eisner (1966) Oxygen isotope study of snow and firn on an alpine glacier. Earth Planet. Sci. Letters, *1*, 197–201.

Dincer, T., and G. H. Davis (1967) Some considerations on tritium dating and the estimates of tritium input function. Memoires of the Congress of Istanbul 1967, Internat. Assoc. Hydrogeologists, 276–286.

Dole, M., G. A. Lane, D. P. Rudd, and D. A. Zaukelies (1954) Isotopic composition of atmospheric oxygen and nitrogen. Geochim. Cosmochim. Acta, *6*, 65.

Emiliani, C. (1955) Pleistocene temperatures. J. Geol., *63*, 538–578.

Emiliani, C. (1966) Isotopic paleotemperatures. Science, *154*, 851–857.

Emiliani, C. (1972) Quaternary paleotemperatures and the duration of the high-temperature intervals. Science, *178*, 398–401.

Emiliani, C., and N. J. Shackleton (1974) The Brunhes epoch: Isotopic paleotemperatures and geochronology. Science, *183*, 511–514.

Epstein, S., and T. K. Mayeda (1953) Variation of ^{18}O content of waters from natural sources. Geochim. Cosmochim. Acta, *4*, 213–224.

Epstein, S., R. Buchsbaum, H. A. Lowenstam, and H. C. Urey (1953) Revised carbonate-water isotopic temperature scale. Geol. Soc. Amer. Bull., *64*, 1315–1326.

Epstein, S. (1959) The variation of the O-18/O-16 ratio in nature and some geological applications. In Researches in Geochemistry, vol. 1, pp. 217–240, P. H. Abelson, ed. John Wiley, New York.

Epstein, S., R. P. Sharp, and A. J. Gow (1965) Six-Year record of oxygen and hydrogen isotope variations in South Pole firn. J. Geophys. Res., *70*, 1809–1814.

Epstein, S., and H. P. Taylor, Jr. (1967) Variation of O^{18}/O^{16} in minerals and rocks. In Researches in Geochemistry, vol. 2, pp. 29–62. P. H. Abelson, ed. John Wiley, New York.

Friedman, I., A. C. Redfield, B. Schoen, and J. Harris (1964) The variation of the deuterium content of natural waters in the hydrologic cycle. Rev. Geophys., *2*, 177–224.

Garlick, G. D. (1969) The stable isotopes of oxygen. In Handbook of Geochemistry, vol. II-1, K. H. Wedepohl, ed. Springer-Verlag, Heidelberg and New York.

Giletti, B. J., F. Bazan, and J. L. Kulp (1958) The geochemistry of tritium. Trans. Amer. Geophys. Union, *39*, 807–818.

Gonfiantini, R. (1965) Some results on oxygen isotope stratigraphy in the deep drilling at King Baudouin Station, Antarctica. J. Geophys. Res., *70*, 1815–1819.

Hamilton, W. L., and C. C. Langway, Jr. (1967) A correlation of micro-particle concentrations with oxygen isotope ratios in 700 year old Greenland ice. Earth Planet. Sci. Letters, *3*, 363–366.

Hoefs, J. (1973) Stable Isotope Geochemistry. Springer-Verlag, Heidelberg and New York, 140 p.

Holden, N. E., and F. W. Walker (1972) Chart of the nuclides. Ed. Relations, General Electric Co., Schenectady, N.Y., 12345.

Johnson, S. J., W. Dansgaard, H. B. Clausen, and C. C. Langway (1972) Oxygen isotope profiles through the Antarctic and Greenland ice sheets. Nature, *235*, 429–434.

Knauth, L. P., and S. Epstein (1975) Hydrogen and oxygen isotope ratios in silica from the Joides Deep Sea Drilling Project. Earth Planet. Sci. Letters, *25*, 1–10.

Labeyrie, L. (1974) New approach to surface seawater paleotemperatures using $^{18}O/^{16}O$ ratios in silica of diatom frustules. Nature, *248*, 40–42.

Libby, W. F. (1961) Tritium geophysics. J. Geophys. Res., *66*, 3767–3782.

Lloyd, R. M. (1968) Oxygen isotope behavior in the sulfate-water system. J. Geophys. Res., *73*, 6099–6110.

Longinelli, A., and S. Nuti (1973) Revised phosphate-water isotopic temperature scale. Earth Planet. Sci. Letters, *19*, No. 3, 373–376.

Lowenstam, H. A. (1961) Mineralogy, O-18/O-16 ratios, and strontium and magnesium contents of recent and fossil brachiopods and their bearing on the history of the oceans. J. Geol., *69*, 241–260.

McKinney, C. R., J. M. McCrea, H. A. Allen, and H. C. Urey (1950) Improvements in mass spectrometers for the measurements of small differences in isotope abundance ratios. Rev. Sci. Inst., *21*, 724–730.

Mopper, K., and G. D. Garlick (1971) Oxygen isotope fractionation between biogenic silica and ocean water. Geochim. Cosmochim. Acta, *35*, 1185–1187.

Mörner, N. A. (1972) Time scale and ice accumulation during the last 125,000 years as indicated by the Greenland O^{18} curve. Geol. Mag., *109*, 17–24.

Mörner, N. A. (1974) The Greenland O^{18} curve: time scale and ice accumulation. Geol. Mag., *111*, 431–433.

Nier, A. O., E. P. Ney, and M. G. Inghram (1947) A null method for the comparison of two ion currents in a mass spectrometer. Rev. Sci. Inst., *18*, 294–297.

O'Neil, J. R., and R. N. Clayton (1964) Oxygen isotope geothermometry. In Isotopic and Cosmic Chemistry, pp. 157–168. North-Holland, Amsterdam.

Rankama, K. (1954) Isotope Geology. Pergamon Press, London, 535 p.

Rankama, K. (1963) Progress in Isotope Geology. Interscience, London and New York, 805 p.

Savin, S. M., R. G. Douglas, and F. G. Stehli (1975) Tertiary marine paleotemperatures. Geol. Soc. Amer. Bull., *86*, 1499–1510.

Shackleton, N. J. (1968) Depth of pelagic foraminifera and isotopic changes in Pleistocene oceans. Nature, *218*, 79–80.

Sheppard, S. M. F., and S. Epstein (1970) D/H and O^{18}/O^{16} ratios of minerals of possible mantle or lower crustal origin. Earth Planet. Sci. Letters, *9*, 232–239.

Spaeth, C., J. Hoefs, and U. Vetter (1971) The isotopic composition of belemnites and related paleotemperatures. Geol. Soc. Amer. Bull., *82*, 3139–3150.

Stahl, W., and R. Jordan (1969) General considerations on isotopic paleotemperature determinations and analyses on Jurassic ammonites. Earth Planet Sci. Letters, *6*, 173–178.

Suess, H. E. (1969) Tritium geophysics as an international research project. Science, *163*, 1405–1410.

Tan, F. C., J. D. Hudson, and M. L. Keith (1970) Jurassic (Callovian) paleotemperatures from Scotland. Earth Planet. Sci. Letters, *9*, 421–426.

Tourtelot, H. A., and R. O. Rye (1969) Distribution of oxygen and carbon isotopes in fossils of late Cretaceous age, western interior region of North America. Geol. Soc. Amer. Bull., *80*, 1903–1922.

Urey, H. C. (1947) The thermodynamic properties of isotopic substances. J. Chem. Soc., *1947*, 562–581.

Weber, J. N., and D. M. Raup (1966a) Fractionation of the stable isotopes of carbon and oxygen in marine calcareous organisms—the Echinoidea, Part 1: Variation of ^{13}C and ^{18}O content within individuals. Geochim. Cosmochim. Acta, *30*, 681–704.

Weber, J. N., and D. M. Raup (1966b) Fractionation of the stable isotopes of carbon and oxygen in marine calcareous organisms-the Echinoidea, Part 2: Environmental and genetic factors. Geochim. Cosmochim. Acta, *30*, 705–735.

Weber, J. N. (1968) Fractionation of stable isotopes of carbon and oxygen in calcareous marine invertebrates—the Asteroidea, Ophiuroidea and Crinoidea. Geochim. Cosmochim. Acta, *32*, 33–70.

19 OXYGEN AND HYDROGEN IN THE LITHOSPHERE

Oxygen is an important constituent of most rock-forming minerals, including the silicates, oxides, carbonates, phosphates, and others. These minerals not only form the common igneous, sedimentary, and metamorphic rocks of the crust of the Earth but they occur also in metallic and nonmetallic ore deposits. The isotope composition of oxygen of these minerals contains important information about their origin and conditions of formation. For this reason, the study of the isotope composition of oxygen in the lithosphere has become an important tool in igneous and metamorphic petrology and in the study of ore deposits.

The isotopic composition of oxygen in minerals is expressed in terms of the same delta notation described in the previous chapter. The reference standard is SMOW, except in carbonates for which the PDB standard has been used in the past. Oxygen is liberated from silicate minerals with bromine pentafluoride (BrF_5) or with fluorine gas at 500 to 600°C in nickel tubes (Clayton and Mayeda, 1963).

In general, the $\delta^{18}O$ values of rocks and minerals are positive, that is, they are enriched in ^{18}O compared to SMOW. Most silicate rocks have $\delta^{18}O$ values between $+5$ to $+15$‰. However, the isotopic compositions of igneous, sedimentary, and metamorphic rocks display systematic variations that provide insight into the origin of such rocks and reflect the temperature of final oxygen isotope equilibration. This aspect is particularly important because systematic differences in the $\delta^{18}O$ values of cogenetic minerals may be interpreted in terms of the temperature at which they equilibrated their oxygen with a common reservoir. These interpretations are based on laboratory studies in which different minerals are equilibrated with water over a range of temperatures. For this reason, we shall turn first to a consideration of the experimental equilibration of oxygen between minerals and water at elevated temperature.

1 Oxygen Fractionation of Rock-Forming Minerals

Although the absolute values of isotopic fractionation factors in systems containing solids and liquids cannot be calculated directly from theory, one may assume a temperature dependence of the form

$$\ln \alpha \propto \frac{1}{T^2} \qquad (19.1)$$

where α is the isotopic fractionation factor and T is the absolute temperature. This relationship is true only at high temperatures in excess of 1000°K. However, when the frequencies of molecular vibrations are not too high, the proportionality may be valid at lower temperatures of a few hundred degrees Kelvin. It is generally not valid for compounds containing hydrogen in the form of hydroxyl ions or water of hydration. For such minerals, a more complicated behavior is expected at temperatures below 1000°K. Nevertheless, the proportionality expressed in Equation 19.1 is useful for displaying experimental results and for extrapolating them throughout the temperature range of geological interest, that is, 0 to 1200°C.

The fractionation factors for exchange of oxygen isotopes between two phases under equilibrium conditions have numerical values of the order of $\alpha = 1.00X$, where X is rarely greater than 4. The natural logarithm of 1.004 = 0.00398, which justifies the approximation that

$$1000 \ln 1.00X \simeq X \qquad (19.2)$$

It is therefore convenient to represent the relationships between α and T by means of equations of the form

$$1000 \ln \alpha = A(10^6 T^{-2}) + B \qquad (19.3)$$

in coordinates of $1000 \ln \alpha$ and $10^6 T^{-2}$ where A and B are constants.

Next, let us assume that α_2^1 is the isotope fractionation factor between two phases 1 and 2 such that $\alpha_2^1 = R_1/R_2$ where R stands for the $^{18}O/^{16}O$ ratio of phases 1 and 2, respectively. It follows that

$$\alpha_2^1 - 1 = \frac{R_1 - R_2}{R_2} \qquad (19.4)$$

By using the relationship between δ and isotope ratios (Equation 18.7), we rewrite Equation 19.4:

$$\alpha_2^1 - 1 = \frac{\delta_1 - \delta_2}{\delta_2 + 1000} \qquad (19.5)$$

Since δ_2 is small compared to 1000, we make the approximation that $\delta_2 + 1000 \simeq 1000$ and obtain (Epstein, 1959, p. 332)

$$\alpha_2^1 - 1 \simeq \frac{\delta_1 - \delta_2}{1000} \qquad (19.6)$$

and hence

$$(\alpha_2^1 - 1)1000 \simeq \delta_1 - \delta_2 = \Delta_2^1 \qquad (19.7)$$

Since α is a number of the form $1.00X$ and $1000 \ln 1.00X \simeq X$ (Equation 19.2), we see that $(1.00X - 1)\ 1000$ is also equal to X.

Therefore,

$$1000 \ln \alpha_2^1 \simeq (\alpha_2^1 - 1)1000$$
$$\simeq \delta_1 - \delta_2 = \Delta_2^1 \qquad (19.8)$$

By combining Equation 19.3 with 19.8 we obtain the useful approximation:

$$1000 \ln \alpha_2^1 \simeq \delta_1 - \delta_2 \simeq A(10^6 T^{-2}) + B$$
$$(19.9)$$

In other words, when two solid phases have equilibrated oxygen with a common reservoir at a specific temperature, the difference in their $\delta^{18}O$ values (measured relative to the same standard such as SMOW) is a function of temperature. This is a very important fact that enables us to use the isotope composition of oxygen in rock-forming minerals to determine the temperature of equilibration. This isotope thermometer is based on three assumptions: (1) the exchange reactions must have reached equilibrium; (2) the isotopic compositions were not altered subsequent to the establishment of equilibrium; (3) the temperature dependence of the fractionation factors is known from experimental determinations.

Considerable progress has been made in the experimental determination of fractionation factors for a variety of oxygen-bearing minerals over a range of temperatures. A widely used technique is to equilibrate minerals with water of known isotopic composition and to express such experimental results by means of linear equations of the form of Equation 19.3. Several such mineral-water equilibria are shown in Figure 19.1. The relevant equations are listed in Table 19.1.

These experimental results lead to the important conclusion that the minerals of an igneous or metamorphic rock have different $\delta^{18}O$ values because their respective fractionation factors at any given temperature have different values. This statement is

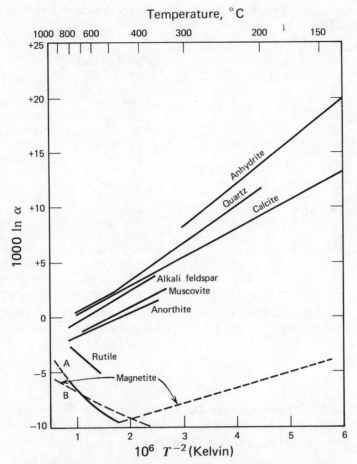

FIGURE 19.1 Temperature variation of the isotope fractionation factor for oxygen between minerals and water. The relationship is expressed by equations of the form: $1000 \ln \alpha = A(10^6 T^{-2}) + B$, where T is the temperature in degrees Kelvin. The equations for the different minerals are listed in Table 19.1. The relationship between α and T for magnetite and water is strongly nonlinear in this temperature range. (Line A is based on O'Neil and Clayton, 1964, and line B is from Hoefs, 1973. For additional information, see Figure 2 of Taylor, 1974.)

true provided the minerals under consideration equilibrated oxygen with the same reservoir at the same temperature and provided the isotope composition was not changed subsequently. If these assumptions are satisfied, then the difference in the $\delta^{18}O$ values of two coexisting minerals in an igneous or metamorphic rock depends only on the temperature of final equilibration. As an example, let us consider the fractionation of oxygen for the quartz-water and calcite-water systems. The relationship between α

Table 19.1 **Temperature Variation of Isotope Fractionation Factors for Mineral-Water Systems**

MINERAL	A	B	TEMPERATURE RANGE (°C)	REFERENCE
Quartz	+3.38	−3.40	200–500	Clayton et al. (1972)
Quartz	+2.51	−1.96	500–750	Clayton et al. (1972)
Quartz	+4.10	−3.70	500–800	Bottinga and Javoy (1973)
Muscovite	+2.38	−3.89	350–650	O'Neil and Taylor (1969)
Muscovite	+1.90	−3.10	500–800	Bottinga and Javoy (1973)
Feldspar	$2.91 - 0.76\beta^a$;	$-3.41 - 0.41\beta^a$		O'Neil and Taylor (1969)
Alk. Feldspar	+2.91	−3.41	350–800	O'Neil and Taylor (1969)
Anorthite	+2.15	−3.82	350–800	O'Neil and Taylor (1969)
Feldspar	$+3.13 - 1.04\beta^a$	−3.70	500–800	Bottinga and Javoy (1973)
Magnetite	−1.59	−3.60	700–800	Anderson et al. (1971)
Magnetite	−1.47	−3.70	500–800	Bottinga and Javoy (1973)
Calcite	+2.78	−3.40	0–800	O'Neil et al. (1969)
Anhydrite	+3.878	−3.40	100–500	Lloyd (1968)
Rutile	−4.1	+0.96	575–775	Addy and Garlick (1974)

[a] β is the mole fraction of anorthite. All equations are of the form $1000 \ln \alpha = A(10^6 T^{-2}) + B$, where α is the isotope fractionation factor between a mineral and water and T is the absolute temperature.

Table 19.2 **Temperature Determinations for Volcanic Rocks Based on Isotope Fractionation of Oxygen Between Plagioclase and Magnetite (Anderson et al., 1971)**

SAMPLE AND LOCATION	Δ_{PM}	ISOTOPE TEMP. °C	INDEPENDENT TEMP. °C
Basalt (HK 1955 AP) Kilauea, Hawaii	2.15	1060	1050[a] 1150[b]
Basalt (KI 67.1, 79.1) Kilauea Iki, Hawaii	2.32	1030	
Alkaline basalt (HM15) Haleakala, Maui	2.08	1070	
Subalkaline basalt (CA5) Mono Craters, Cal.	2.37	990	
Rhyolite (CA3) Mono Craters, Cal.	3.32	840	790–850[b]
Rhyol. ignimbrite (27834) Whakamaru, New Zeal.	3.38	860	

[a] Optical pyrometer determination.
[b] Based on magnetite-ilmenite geothermometer.

and T (200 to 500°C) for quartz is

$$1000 \ln \alpha_{QW} = 3.38(10^6 T^{-2}) - 3.40 \quad (19.10)$$

while for calcite it is

$$1000 \ln \alpha_{CW} = 2.78(10^6 T^{-2}) - 3.40 \quad (19.11)$$

By subtracting Equation 19.11 from 19.10, we obtain

$$1000 \ln \alpha_{QW} - 1000 \ln \alpha_{CW} = 0.60(10^6 T^{-2})$$
$$(19.12)$$

Since $1000 \ln \alpha_{QW} \simeq \delta_Q - \delta_W$ and $1000 \ln \alpha_{CW} \simeq \delta_C - \delta_W$ (Equation 19.8), where δ_Q, δ_C, and δ_W refer to the $\delta^{18}O$ values of quartz, calcite, and water, respectively, we obtain

$$\delta_Q - \delta_C = \Delta_{QC} = 0.60(10^6 T^{-2}) \quad (19.13)$$

In other words, the *difference* in $\delta^{18}O$ values of coexisting quartz and calcite is inversely proportional to the square of the temperature at which these two minerals equilibrated oxygen with the same reservoir. Therefore, Equation 19.13, and others like it for other mineral pairs, are thermometers from which a temperature can be determined. The sensitivity of such thermometers depends on the difference in the slopes of the fractionation lines for each mineral and on the temperature. At increasing temperatures, the Δ value for any given pair of minerals decreases until it approaches the analytical uncertainty. The most sensitive thermometers are therefore based on mineral pairs whose fractionation lines (relative to water) have strongly diverging slopes, such as quartz-magnetite or plagioclase-magnetite (Fig. 19.1).

The plagioclase-magnetite thermometer was discussed by Anderson et al. (1971) using the data for the feldspar-water system of O'Neil and Taylor (1967) and the data for magnetite-water of O'Neil and Clayton (1964) as well as unpublished measurements by O'Neil. From these sources they derived an equation for isotope fractionation be-

tween plagioclase and magnetite at temperatures in excess of 700°C:

$$\Delta_{PM} = (4.72 - 1.19\beta)10^6 T^{-2} \quad (19.14)$$

where β is the mole fraction of anorthite in the plagioclase. They used this relationship to calculate oxygen temperatures for six volcanic rocks shown in the Table 19.2. The crystallization temperatures of two of these rocks had been determined by other methods and are in satisfactory agreement with the isotopic temperatures.

Let us now calculate the temperature of oxygen equilibration for sample HK1955AP (Table 19.2). Anderson et al. (1971) reported that the anorthite content of the plagioclase ranges from 62 to 72 mole percent. We shall assume an average value of 67 percent and set $\beta = 0.67$. Since $\Delta_{PM} = 2.15$ (Table 19.2), we have

$$2.15 = (4.72 - 1.19 \times 0.67)10^6 T^{-2}$$
$$T = 1351°K = 1078°C$$

Anderson et al. (1971) reported a temperature of 1060°C, presumably because they used a slightly higher value for β.

These and other thermometry equations are listed in Table 19.3 and have been plotted in Figure 19.2. We must draw attention to the fact that the validity of these equations is limited by the reliability of the analytical data on which they are based. This problem has been discussed by Bottinga and Javoy (1973) who reinterpreted the experimental results in the light of theoretical considerations and proposed somewhat different fractionation equations for these mineral-water systems.

When three or more oxygen-bearing minerals in a rock equilibrated isotopically at the same temperature, then any pair of these minerals must yield that temperature. Such concordance of the isotopic temperatures is a necessary but not sufficient test that isotopic equilibrium was actually achieved

Table 19.3 **Geothermometry Equations for Mineral Pairs Based on Fractionation of Oxygen Isotopes Under Equilibrium Conditions Above 500°C**

MINERALS	A[a]	B	REFERENCE
Quartz-magnetite	$+5.57$	0	1, 2
Quartz-plagioclase	$+0.97-1.4\beta$[b]	0	1, 2
Quartz-muscovite	$+2.20$	-0.60	1, 2
Plagioclase-muscovite	$+1.23-1.04\beta$	-0.60	1, 2
Plagioclase-pyroxene	$+1.700-1.04\beta$	0	2
Plagioclase-olivine	$+2.940-1.04\beta$	0	2
Plagioclase-garnet	$+1.910-1.04\beta$	0	2
Plagioclase-amphibole	$+2.178-1.04\beta$	-0.30	2
Plagioclase-biotite	$+2.720-1.04\beta$	-0.60	2
Plagioclase-ilmenite	$+4.320-1.04\beta$	0	2
Plagioclase-magnetite	$+4.600-1.04\beta$	0	1, 2
Pyroxene-olivine	$+1.24$	0	2
Pyroxene-garnet	$+0.21$	0	2

[a] A and B are coefficients of the equation:

$$1000 \ln \alpha = A(10^6 T^{-2}) + B$$

where T is the absolute temperature.

[b] β is the mole fraction of anorthite in the feldspar. Note that the coefficients given by Bottinga and Javoy (1975, Table 2) have been recalculated for plagioclase of variable anorthite content.

1. Bottinga and Javoy (1973).
2. Bottinga and Javoy (1975).

in a rock. When any three minerals A, B, and C equilibrated oxygen at different temperatures in a series of different rocks, a plot Δ_{AB} versus Δ_{BC} must yield a smooth curve that passes through the origin and approaches a straight line at high temperature. The goodness of fit of data points to such a line on the concordancy diagram indicates graphically how well minerals actually satisfy the assumptions of oxygen thermometry. Taylor (1968) and others have published such concordancy diagrams for several combinations of mineral pairs taken from igneous and metamorphic rocks. The results indicate that concordance is approached fairly closely in many cases, although minor discrepancies do arise in some rocks.

The concordancy diagram is important also because it can be used to extend the laboratory calibration of oxygen fractionation in one mineral pair to that in other pairs. For example, Taylor (1968) showed that a plot of Δ (quartz-magnetite) versus Δ (quartz-biotite) generates a straight line that passes through the origin and has a slope of 0.59. Shieh and Schwarcz (1974) later used this fact to calibrate a quartz-biotite thermometer based on the experimentally determined quartz-magnetite thermometer.

Δ (quartz-biotite)

$$= 0.59\Delta \text{ (quartz-magnetite)} \quad (19.15)$$

On the basis of the data by O'Neil and

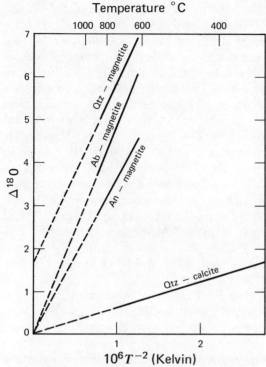

FIGURE 19.2 Variation of $\Delta^{18}O$ with temperature, where $\Delta^{18}O$ is the difference between $\delta^{18}O$ values of two coexisting minerals that have equilibrated oxygen with the same isotope reservoir at the same temperature. The equations for these lines are given in Table 19.3.

Clayton (1964) and Anderson et al. (1971), Shieh and Schwarcz (1974) represented the equilibrium between quartz and magnetite in the temperature range 500 to 750°C as

$$\Delta_{QM} = 4.32(10^6 T^{-2}) + 1.45 \quad (19.16)$$

The quartz-biotite thermometer in the specified temperature range is therefore given by

$$\Delta_{QB} = 2.54(10^6 T^{-2}) + 0.85 \quad (19.17)$$

Taylor (1968) also demonstrated concordance for two additional sets of mineral pairs:

$$\Delta \text{ (quartz-biotite)}$$

$$= 1.5\Delta \text{ (feldspar-biotite)}$$

$$\Delta \text{ (feldspar-magnetite)}$$

$$= 1.47\Delta \text{ (pyroxene-magnetite)}$$

The oxygen isotope concordancy diagram (Fig. 19.3) for the plagioclase-pyroxene-magnetite assemblage (Anderson et al. 1971)

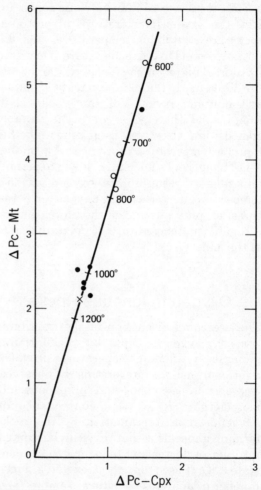

FIGURE 19.3 Oxygen isotope concordancy diagram for plagioclase-pyroxene and pyroxene-magnetite. ○ —plutonic rocks, ● —volcanic rocks, X—Apollo 11 basalts. (Reproduced from Figure 3 of Anderson, A. T., R. N. Clayton, and T. K. Mayeda, Oxygen isotope thermometry of mafic igneous rocks, Journal of Geology, vol. 79, 715–729, 1971, by permission of the University of Chicago Press.)

illustrates yet another important fact. The Δ-values of mineral pairs in plutonic rocks are commonly higher than those of equivalent volcanic rocks. Therefore the oxygen isotope thermometers indicate that plutonic rocks equilibrated their oxygen isotopes at a *lower* temperature than volcanic rocks. In fact, the oxygen temperatures of plutonic rocks are subsolidus temperatures and do not correspond to the temperature of crystallization. This phenomenon is explained by the difference in the cooling rates of volcanic and plutonic rocks. Volcanic rocks cool rapidly and thereby preserve the isotope composition that was established at the liquidus temperature. Plutonic rocks, on the other hand, cool sufficiently slowly to permit minerals to reequilibrate oxygen as the temperature decreases. The oxygen temperatures of plutonic rocks therefore underestimate the temperature of crystallization of the minerals.

2
Oxygen in Igneous Rocks

The preceding discussion of the fractionation of oxygen isotopes by rock-forming minerals at different temperatures provides a rational basis for interpreting the observed oxygen isotope compositions of igneous and metamorphic rocks. One consequence of oxygen-isotope fractionation by the rock-forming minerals is that they can be ranked in order of decreasing tendency to concentrate ^{18}O. Based on a review of a large number of analyses of natural samples and experimental data, Epstein and Taylor (1967) gave the following ranking: quartz, dolomite, alkali feldspar, calcite, intermediate plagioclase, muscovite, anorthite, pyroxene, hornblende, olivine, garnet, biotite, chlorite, ilmenite, and magnetite. Quartz has the strongest tendency to concentrate ^{18}O at a given temperature, while magnetite does so

least of any of the minerals listed above. Moreover, analyses of these minerals from many different kinds of rocks have shown that at a given temperature their $\delta^{18}O$ values vary only within narrow limits. Such uniformity of the isotopic compositions suggests that igneous and metamorphic rocks tend to equilibrate oxygen with a large external reservoir. The tendency of silicate minerals to concentrate ^{18}O is apparently related to their chemical compositions. Minerals in which Si-O-Si bonds predominate are richest in ^{18}O, while those containing Si-O-Mg and Si-O-Fe bonds contain about 2‰ less ^{18}O and those with Si-O-Al bonds have about 4‰ less ^{18}O. This phenomenon has been discussed by Garlick (1966) and by O'Neil and Taylor (1967).

The $\delta^{18}O$ values of most stony meteorites and of lunar rocks are compatible with those of terrestrial rocks and are explainable in terms of isotope fractionation among the constituent minerals at different temperatures. However, Clayton et al. (1973) reported that certain high-temperature phases in several carbonaceous chondrites, including Allende, are anomalously enriched in ^{16}O. The best explanation of this anomaly is that the high-temperature phases include grains containing pure ^{16}O that did not mix with the oxygen of the solar nebula. The exotic oxygen may have been formed by nuclear reactions in a star which exploded and injected matter into the solar nebula. This discovery by Clayton and his colleagues is very significant because it is the first indication that the solar nebula contained matter ejected by more than one star and that it was not homogeneous isotopically. More recently, isotope anomalies have also been reported for magnesium (Lee and Papanastassiou, 1974; Gray and Compston, 1974) and mercury (Jovanovic and Reed, 1976) in carbonaceous chondrites. Another important consequence of Clayton's work is the fact that oxygen in terrestrial and lunar rocks, as well as that of most meteorites, appears to have been derived from the same source. This observation therefore does *not* support the hypothesis that the moon was formed elsewhere in the solar system and was later captured by the Earth.

The regularities in the $\delta^{18}O$ values of rock-forming minerals are reflected in systematic trends of the oxygen-isotope compositions of igneous rocks of different chemical compositions. In general, the $\delta^{18}O$ values of igneous rocks tend to increase with increasing concentration of SiO_2. Ultramafic rocks have the lowest $\delta^{18}O$ values (+5.4 to +6.6‰). Gabbros, basalts and anorthosites have very similar $\delta^{18}O$ values ranging from +5.5 to +7.4‰ and are indistinguishable also from most andesites, trachytes, and syenites. Granitic rocks and pegmatites

have still higher and more variable $\delta^{18}O$ values (+7 to +13‰). We can understand these generalizations, shown graphically in Figure 19.4, by recalling the tendencies of different minerals to concentrate ^{18}O. Ultramafic rocks have relatively low $\delta^{18}O$ values because they are composed of olivine, pyroxene, and magnetite, all of which rank low as ^{18}O concentrators. Similarly, granitic rocks have high $\delta^{18}O$ values because they contain quartz and alkali feldspar, both of which rank high on the scale. These generalizations are useful because they provide a frame of reference within which we can discuss the details of the various processes involved in determining the isotopic composition of oxygen in igneous rocks and because they enable us to recognize anomalous situations. The absolute values of $\delta^{18}O$ of igneous rocks and their minerals are

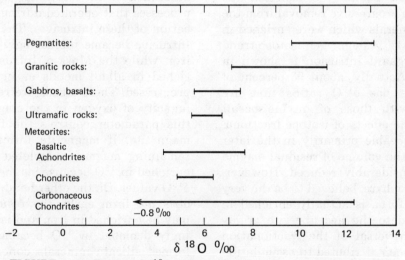

FIGURE 19.4 Range of $\delta^{18}O$ values of igneous rocks and stony meteorites. In general, the $\delta^{18}O$ values increase with increasing SiO_2 content of the rocks. Oxygen in ultramafic rocks has isotopic compositions similar to those of chondrite and achondrite meteorites. The carbonaceous chondrites are very heterogeneous which indicates that they or their parent bodies were never heated sufficiently to achieve isotopic equilibrium. (The meteorite data are from Reuter et al., 1965, and Taylor et al., 1965. The data for igneous rocks are from compilations by Epstein and Taylor, 1967, Taylor, 1968, and Hoefs, 1973.) The standard is SMOW.

determined by several factors, including (1) the temperature of crystallization; (2) the $\delta^{18}O$ of the magma; (3) effects of fractional crystallization; (4) retrograde effects resulting from reequilibration at subsolidus temperatures; and (5) interaction with aqueous solutions.

Let us now turn to a study by Taylor and Epstein (1963) of the oxygen isotope composition of the Skaergaard Intrusion of East Greenland. This has been regarded as one of the best examples of the fractional crystallization of magma of basaltic composition (Wager and Deer, 1939). Taylor and Epstein (1963) originally observed that the $\delta^{18}O$ values of the rocks from this intrusion actually *decrease* with increasing degree of differentiation and that late-stage granophyres are depleted in ^{18}O by up to 5‰ compared to the gabbros of the Lower Zone. They suggested that the ^{18}O-depletion was caused by the progressive removal from the magma of minerals which were enriched in ^{18}O by about 1‰. The oxygen isotope trend of the Skaergaard Intrusion is shown in Figure 19.5. Actually, about 95 percent of the intrusive has $\delta^{18}O$ ratios that are compatible with those of olivine basalt elsewhere. The effects of isotope fractionation are noticeable primarily in the later stages when the volume of residual magma had been considerably reduced. However, the acid granophyre, believed to be the very last rock to form, is actually enriched in ^{18}O compared to the hedenbergite granophyre. This reversal of the fractionation pattern has been attributed to assimilation of granitic country rock having a higher $\delta^{18}O$ ratio than the magma and is compatible with the work of Hamilton (1964) who showed that these granophyres have anomalously high $^{87}Sr/^{86}Sr$ ratios. In a subsequent study, Taylor (1968) presented new data that indicated that the chilled marginal basalt has a $\delta^{18}O$ value of only $+0.4 \pm 0.1$‰

and that anomalously low ^{18}O contents prevail in a marginal zone about 40 meters wide. This effect is not attributable to fractional crystallization nor to assimilation of country rock. Instead, Taylor explained it as the result of isotope equilibration of oxygen in the silicate minerals with meteoric water whose $\delta^{18}O$ value is negative (Chapter 18). We see, therefore, that the $\delta^{18}O$ values of igneous intrusives may be modified extensively by interactions with the country rock and with meteoric water.

The Muskox Intrusive of the Northwest Territories in Canada is similar to the Skaergaard Intrusive in terms of bulk chemical composition and petrologic character. Yet its oxygen isotope composition shown in Figure 19.5 diverges sharply from that of the Skaergaard. A possible explanation for this observation may be based on differences in the fractional crystallization processes that operated during the solidification of these intrusives. The Skaergaard Intrusive became increasingly enriched in iron while the Muskox Intrusive was enriched in alkali metals as crystallization progressed. These trends are related to the fugacity of oxygen in the magma because this parameter controls the formation of magnetite. If magnetite forms early, the remaining magma is depleted in iron but enriched in ^{18}O because magnetite has low $\delta^{18}O$ values. On the other hand, if magnetite does *not* form early, the residual magma may be enriched in iron and is correspondingly depleted in ^{18}O because the early-formed silicate minerals concentrate ^{18}O. In the Skaergaard Complex, magnetite did not form early, and the residual magma was therefore enriched in iron and depleted in ^{18}O. The Muskox Intrusive crystallized under conditions that permitted early formation of magnetite leading to a depletion of iron and enrichment in ^{18}O of the residual magma.

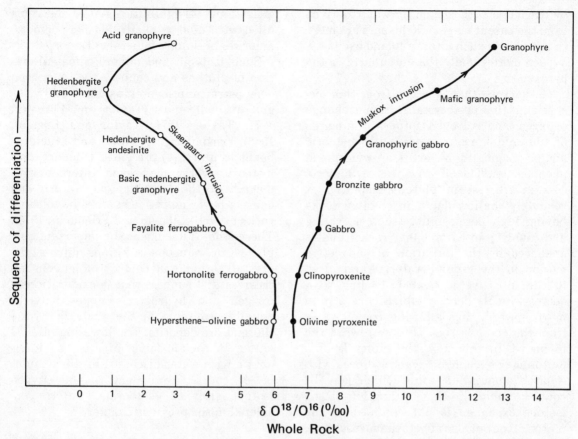

FIGURE 19.5   Variation of $\delta^{18}O$ values in the Skaergaard intrusion of east Greenland and the Muskox intrusion in the Northwest Territory of Canada. The $\delta^{18}O$ values change systematically as a function of increasing stage of magmatic differentiation. The causes for the variation of $\delta^{18}O$ values are discussed in the text. (Reproduced from Figure 5 (p. 39) of Epstein, S. and H. P. Taylor in Abelson, P. H., ed. Researches in Geochemistry, vol. 2, 29–62, John Wiley and Sons, Inc., by permission of the publisher.)

However, this cannot be the whole story of oxygen fractionation in the Muskox Intrusive. Taylor (1968) pointed out that all feldspars in the upper portion (about 10 percent) of the intrusive are anomalously *enriched* in ^{18}O and are not in isotopic equilibrium with coexisting minerals. Such disequilibrium cannot be the result of fractional crystallization of magma as described above. The feldspars in the granophyric gabbros and granophyres are typically turbid and show evidence of late-stage deuteric alteration. Taylor (1968) suggested that the altered state of the feldspars and their high $\delta^{18}O$ values resulted from chemical and isotopic reactions with an aqueous phase (perhaps a brine) that may have been a product of magmatic differentiation and had a positive $\delta^{18}O$ value because of previous equilibration of oxygen with silicate minerals. If this is true, it suggests that the $\delta^{18}O$ values of igneous intrusives may be changed after crystallization not only by interaction with meteoric water but also by exchange with magmatic water derived from the parent magma. Feldspars are more susceptible to such alteration and exchange oxygen more rapidly than do quartz or the pyroxenes.

The picture that emerges from this discussion is that postcrystallization exchange of oxygen and hydrogen isotopes by minerals of plutonic igneous rocks with meteoric and/or magmatic water is a widespread phenomenon. The effectiveness of this process is greatest in plutons emplaced at relatively shallow depth in country rocks having high permeability (such as young and highly jointed volcanic rocks) and is enhanced by the initiation of convective motion in the ground water. Water close to the intrusive is heated, leading to a decrease in its density which causes it to move upward, thus allowing cooler water from the country rock to flow toward the intrusive. In this way, the minerals of a pluton may exchange oxygen isotopes with a large amount of water which may be equal in volume to that of the pluton. The isotope exchange is not confined to the pluton alone but also affects the surrounding country rock. Thus a halo of isotopically altered country rock may be created extending outward from the intrusive contact into the country rock for up to three pluton diameters (Taylor, 1971). The lowering of $\delta^{18}O$ values in the country rock is a form of contact metamorphism that extends far beyond the zone of contact metamorphism recognizable by mineralogical criteria alone. In general, the mineralogical changes that accompany reequilibration of oxygen isotopes in igneous plutons and to a lesser extent in the adjoining country rock include (1) cloudiness of alkali feldspars; (2) alteration of pyroxenes and olivines to uralitic amphibole, chlorite, Fe-Ti oxides, and/or epidote; (3) granophyric (micrographic) intergrowths of turbid alkali feldspar and

quartz; and (4) miarolitic cavities and veins filled with quartz, alkali feldspar, epidote, chlorite, or sulfide minerals (Taylor, 1971).

Such isotopic and mineralogical alteration of plutons and country rock has been demonstrated in several instances by Taylor and his colleagues (Taylor and Forester, 1971; Taylor, 1971; Taylor and Epstein, 1968). Friedman et al. (1974) and Muehlenbachs et al. (1974) presented evidence that meteoric water may even interact with magma at the liquidus temperature and cause a lowering of $\delta^{18}O$ values in volcanic rocks as well (Lipman and Friedman, 1975). These studies document the important role played by aqueous solutions during the crystallization and cooling of igneous intrusives and suggest that consideration of oxygen and hydrogen isotopes may be equally profitable in the study of hydrothermal ore deposits. The lowering of $\delta^{18}O$ values of rock-forming minerals by interaction with water discussed in this chapter is, of course, equivalent to the oxygen isotope shift of geothermal waters considered in the previous chapter.

3
Oxygen and Hydrogen in Hydrothermal Ore Deposits

The evidence for large-scale interactions between cooling magma at shallow depths in the crust of the Earth and large volumes of convecting meteoric water is vitally important to our understanding of the origin of hydrothermal ore deposits and the related alteration of the country rock. The impressive studies of N. L. Bowen and his colleagues at the Geophysical Laboratory on the fractional crystallization of silicate liquids and the work of W. Lindgren and others on hydrothermal ore deposits during the first quarter of this century gave credence to the idea that ore-bearing hydrothermal solutions originate from within cooling igneous intrusives and therefore represent water that was initially dissolved in the magma. This in turn placed a limit on the amount of water that is available to transport metals into fracture systems in the adjoining country rock or within the intrusive itself. A serious problem arose when it was subsequently shown that the common ore-forming sulfide minerals are very insoluble in aqueous solutions and that very large amounts of water are required to transport metals in sufficient quantity to make even a modest ore deposit. This impasse was eventually resolved by the suggestion that metals are transported as complex ions or molecules in chloride brines whose metal-carrying capacity is much greater than that of pure water. The evidence for large-scale convecting systems of meteoric water associated with cooling intrusives means that we can now consider the possibility that very large quantities of water are available to produce wide-spread wall-rock alterations that commonly accompany the formation of hydrothermal ore deposits. Moreover, this insight may give new support to old ideas regarding the source of the metals that are concentrated in such deposits. It is at least conceivable that metals may be removed from the country rock and therefore do not have to be original constituents of the magma.

Many detailed studies of oxygen and hydrogen isotope compositions in rocks and minerals from mining districts have been reported. In fact, an entire issue of *Economic Geology* (vol. 69, no. 6, 1974) was devoted to the isotope geology of ore deposits. It includes an excellent review by Taylor (1974) of the state of the art, including an extensive list of references.

The lowering of $\delta^{18}O$ values of igneous rocks as a result of isotope exchange with meteoric water is well illustrated by a series of Tertiary granodiorite intrusions in the western Cascade Range of Oregon (Taylor, 1971). These include the Brice Creek and Champion Creek stocks of the Bohemia mining district in Lane County, Oregon, which were intruded into volcanic country rock. The Bohemia mining district produced about $1,000,000 in gold from 1870 to 1940 (Taylor, 1974). The stocks, which are less than 3 km in diameter, are surrounded by aureoles of contact metamorphism from 300 to 600 meters wide, consisting of tourmaline hornfels and related rocks grading outward into a zone of propylitic alteration which has a diameter of about 4 km measured east-west across the entire district. The $\delta^{18}O$ values of the stocks and the surrounding country rock have been lowered in a very systematic fashion as indicated by the $\delta^{18}O$ contours in Figure 19.6. The major ore deposit at the south end of the Champion Creek stock is entirely within the $\delta = 0\%$ contour that outlines the area of most intense alteration and highest temperature. The $\delta^{18}O$ values increase outward for about 5 km, more or less parallel to the zone of propylitic alteration of the country rock. A small volume of rock in the center of the Champion Creek stock has positive $\delta^{18}O$ values presumably because it was less accessible to circulating ground water. The area of lowest $\delta^{18}O$ (less than 0‰) is distinctly displaced to the east, especially with reference to the Brice Creek stock, and may indicate the presence of a larger intrusive body at depth.

The isotopic composition of the water associated with the formation of hydrothermal ore deposits and related wall-rock alteration can be estimated in two different ways: (1) by analysis of water trapped in fluid inclusions; and (2) by calculating $\delta^{18}O$ and δD values of the aqueous fluid that equilibrated oxygen and hydrogen with minerals at a specified temperature. The

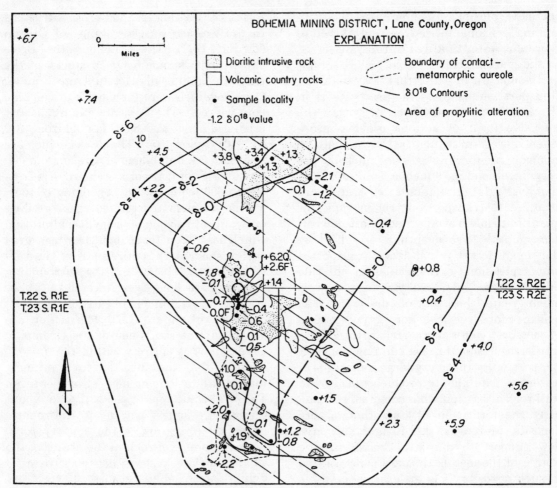

FIGURE 19.6   Lowering of $\delta^{18}O$ values of granodiorite stocks of Tertiary age and volcanic country rock in the Bohemia mining district of Lane County in the western Cascade range of Oregon. This effect is attributable to isotope exchange between rocks and meteoric ground water. (Reproduced from Figure 23 (p. 869) of Taylor, H. P., Jr., Economic Geology, 1974, v. 69, 843–883, by permission of the editor.)

former method is more direct than the latter but suffers from the limitation that the isotopic composition of the water may have been altered by exchange with oxygen and hydrogen of the host mineral. This possibility exists even if the host mineral contains neither oxygen nor hydrogen when that mineral contains fractures or oxygen-bearing inclusions. The second approach relies entirely on the validity of the relevant mineral-water exchange equilibria and on the isotope thermometry equations used to determine the temperature. Although much progress has been made in the study of *oxygen* isotope equilibria, the exchange of *hydrogen* isotopes between minerals and water is still not well known (see Fig. 4 of Taylor, 1974). Nevertheless, it is possible

OXYGEN AND HYDROGEN IN
THE LITHOSPHERE   19

to estimate the isotopic composition of water that participated in ore deposition and wall-rock alteration by these methods and thereby to identify the source of the water (Hall et al., 1974; O'Neil and Silberman, 1974; Sheppard and Taylor, 1974; Ohmoto, 1974; Sheppard et al., 1971; Ohmoto and Rye, 1970; Rye and Rye, 1968).

These studies show that hydrothermal fluids have diverse origins and that different isotopic varieties of water may have been involved in the formation of ore deposits in a given mining district. The hydrothermal fluids may have originated as meteoric water, trapped sea water, geothermal waters variously enriched in ^{18}O, connate formation waters with variable δD and $\delta^{18}O$ values, as well as metamorphic and magmatic waters. The last two can only be studied indirectly by means of the effect they have had on the isotopic composition of minerals with which they equilibrated at a specified temperature. Metamorphic waters are those whose isotopic composition was modified by equilibration with oxygen and hydrogen-bearing minerals during metamorphism at temperatures from 300 to 600°C. According to Taylor (1974), such waters have a relatively restricted range of δD from -20 to -65% but a wide range of $\delta^{18}O$ from $+5$ to $+25\%$. Magmatic waters have isotopic compositions that are controlled by equilibration with magmas at high temperatures ranging from 700 to 1100°C. Because most volcanic and plutonic rocks have relatively uniform $\delta^{18}O$ and δD values, those of magmatic waters are similarly restricted from $+5.5$ to $+10.0\%$ for $\delta^{18}O$ and -50 to -85% for δD. The ultimate source of such metamorphic and magmatic waters is not discernible from their isotopic compositions. However, they may originate in large part as meteoric water or sea water. White (1974) reviewed chemical, isotopic, and physical data of ore deposits from 40 localities and concluded that all of the types of water listed above contributed in varying proportions to the ore-transporting waters. No simple explanations applicable to all hydrothermal ore deposits exist. We must accept diversity.

4
Oxygen and Hydrogen in Sedimentary Rocks

The isotopic compositions of hydrogen and oxygen of detrital sedimentary rocks are controlled by those of the mineral and rock particles of which they are composed. Such rocks are therefore isotopically heterogeneous in the sense that no internal isotopic equilibria exist. Isotopic reequilibration may take place, however, in the course of metamorphism or by interaction with hot aqueous fluids in the vicinity of igneous intrusives. The $\delta^{18}O$ values of shales and deep-sea sediment range from $+5$ to $+25\%$, while their δD values vary between limits of -30 and -100%. Detrital quartz and feldspar do not equilibrate isotopically with sea water. Most of the clay-mineral fraction also appears to be detrital and is not in isotopic equilibrium with sea water.

The isotopic compositions of hydrogen and oxygen of clay minerals formed by chemical weathering of silicate minerals are controlled by (1) the water with which the clay minerals were in contact during their formation; (2) the values of the isotope fractionation factors for oxygen and hydrogen for clay-water systems and the approach to equilibrium; and (3) the temperature of the environment. The isotopic composition of the source rocks is not a controlling factor. Savin and Epstein (1970) estimated isotope fractionation factors for kaolinite, montmorillonite, and glauconite in isotopic equilibrium with water. Their results, shown in Table 19.4, indicate that clay minerals

Table 19.4 Isotope Fractionation Factors for Clay-Water Systems at Earth-Surface Temperatures

MINERAL	OXYGEN	HYDROGEN	REFERENCE
Montmorillonite	1.027	0.94	1
Kaolinite	1.027	0.97	1
Glauconite	1.026	0.93	1
Gibbsite	1.018	0.984	2
Illite	1.0234	—	3

1. Savin and Epstein (1970).
2. Lawrence and Taylor (1972).
3. James and Baker (1976).

are enriched in ^{18}O relative to water but are *depleted* in deuterium at Earth-surface temperatures. Since the isotopic compositions of clay minerals formed at surface temperatures are controlled by the isotopic composition of meteoric water and since δD and $\delta^{18}O$ of meteoric water are related by $\delta D = 8\delta^{18}O + 10$ (Chapter 18, Equation 18.15), it follows that the isotopic compositions of clay minerals formed at a particular temperature should be related by similar equations of the form

$$\delta D = A\delta^{18}O + B \qquad (19.18)$$

where the slope A is determined by the ratios of the fractionation factors and the intercept B by their absolute values (Savin and Epstein, 1970). The equations relating δD and $\delta^{18}O$ for different clay minerals at Earth-surface temperatures are

Montmorillonite:

$$\delta D = 7.3\delta^{18}O - 260 \qquad (19.19)$$

Kaolinite:

$$\delta D = 7.5\delta^{18}O - 220 \qquad (19.20)$$

Glauconite:

$$\delta D = 7.3\delta^{18}O - 260 \qquad (19.21)$$

The kaolinite and montmorillonite lines have been plotted in Figure 19.7 together with the line representing meteoric water. Many clay-rich soils analyzed by Lawrence and Taylor (1971, 1972) lie close to the kaolinite line, while hydrothermal clays formed at higher temperature plot closer to the meteoric water line (Taylor, 1974). The reason is that clays that equilibrated isotopes at elevated temperatures have δD and $\delta^{18}O$ values that approach those of the water because the fractionation factors approach unity with increasing temperatures. Therefore, one can distinguish hypogene clay minerals formed at elevated temperature near hydrothermal ore deposits from supergene clays that equilibrated at Earth-surface temperatures. Moreover, the isotopic composition of clay in soils reflects the climatic conditions that prevailed at the time of their formation because their δD and $\delta^{18}O$ values depend on the isotopic composition of the meteoric water which in turn is controlled by climatic conditions (Chapter 18). The clay minerals in soils over large regions of the United States formed during the Tertiary period under somewhat warmer climatic conditions than those of the present. The fractionation of oxygen and hydrogen isotopes between clay minerals and water at elevated temperatures

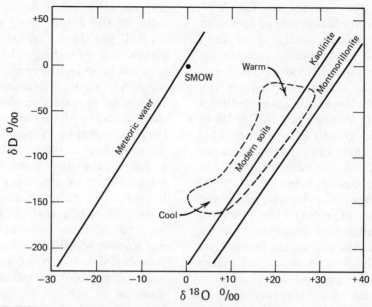

FIGURE 19.7 Relationships between δD and $\delta^{18}O$ of clay minerals formed at Earth-surface temperatures. These relationships result from the isotope equilibration of clay minerals with meteoric water during their formation as products of chemical weathering. The δD and $\delta^{18}O$ values of clay minerals and hydroxides from modern soils overlap the isotope relationships of clay minerals and reflect climatic conditions at the time of formation. Clays formed in cooler climates have lower $\delta^{18}O$ and δD values than those of warmer climates because of differences in the isotopic composition of meteoric water which are climatically controlled. Hypogene clays formed at elevated temperatures in zones of hydrothermal alteration lie closer to the meteoric water line because the isotope fractionation factors approach unity with increasing temperatures. (The lines for kaolinite and montmorillonite are from Savin and Epstein, 1970, the data for modern soil are from Lawrence and Taylor, 1971. The diagram was modified after Taylor, 1974.)

has been studied by O'Neil and Kharaka (1976).

The isotopic composition of oxygen of calcium carbonate deposited in the oceans was discussed in Chapter 18 in connection with paleotemperature determinations. The $\delta^{18}O$ values of marine limestones of Recent age vary from $+28$ to $+30\permil$ relative to SMOW. Keith and Weber (1964) demonstrated that the $\delta^{18}O$ values of such rocks decrease with increasing geologic age to about $+20\permil$ for rocks of Cambrian age. They attributed this effect to postdepositional recrystallization and oxygen isotope exchange with water depleted in ^{18}O compared to sea water. The lowering of $\delta^{18}O$ values of marine limestones is, of course, accompanied by complementary increases of this parameter in the water and is observed in oil field brines and formation water described in Chapter 18. The $\delta^{18}O$ values of freshwater limestones of Mesozoic to Recent Age are generally lower than those of marine limestones of comparable age because fresh water is variably depleted in ^{18}O relative to sea water (Chapter 18). Similarly, the shells of freshwater mollusks have lower $\delta^{18}O$ values than the shells of marine mollusks (Keith et al., 1964). This parameter may therefore be of some use as an indicator of the environment of deposition. The relationship between the $\delta^{18}O$ values of nonmarine carbonate rocks and the concentration of brines formed by evaporative concentration of meteoric water in closed basins on the continents has not yet been studied. The $\delta^{18}O$ values of such brines and of carbonates precipitated from them at a constant temperature may increase with increasing concentration of the brine. Such a relationship, if confirmed, would make it possible to describe the changing salinity of brines in a basin from the $\delta^{18}O$ values of carbonate rocks precipitated as a function of time.

The origin of dolomite has been controversial for many years with no definitive solution in sight (Clayton et al., 1968). Experimental studies at elevated temperatures suggest that dolomite formed at 25°C should be enriched in ^{18}O by 5.0 to 7.0‰ compared to coexisting calcite (O'Neil and Epstein, 1966; Sheppard and Schwarcz, 1970; Tan and Hudson, 1971). The problem is that in most natural samples the difference in $\delta^{18}O$ values is significantly less than predicted. Several explanations have been proposed to account for this discrepancy: (1) the isotope fractionation factor for dolomite relative to calcite may be less than expected; (2) dolomite is deposited as proto-dolomite whose fractionation factor relative to water is less than that of dolomite; (3) dolomite forms by alteration of calcium carbonate

and its $\delta^{18}O$ value depends on the $\delta^{18}O$ value of the dolomitizing solution (Hoefs, 1973). If the third alternative is correct, it means that coexisting calcite and dolomite may not be in isotopic equilibrium and that the $\delta^{18}O$ value of dolomite in nature may vary in an unsystematic fashion. If dolomitization takes place soon after deposition, the enrichment in ^{18}O may be slight because the pore fluid is likely to be similar to sea water. On the other hand, if dolomitization occurs later by reactions with solutions having $\delta^{18}O$ different from those of sea water, the resulting $\delta^{18}O$ values of dolomite may be variable and unpredictable. It is not known whether isotopic equilibria are established between dolomite, calcite, and water during dolomitization. Thus far, the study of oxygen-isotope variations has not brought us much closer to understanding the origin of dolomite.

In conclusion, we turn briefly to the oxygen-isotope composition of calcite in carbonatites. These rocks are believed to be the products of crystallization of carbonatic magmas that are commonly associated with syenitic intrusives, alkali-rich volcanic rocks, and kimberlites. Studies by Conway and Taylor (1969), Deines and Gold (1973), and Pineau et al. (1973) demonstrate that calcites from carbonatites generally have significantly lower $\delta^{18}O$ values than marine limestones. Conway and Taylor (1969) reported $\delta^{18}O$ values for calcite from carbonatites at Oka, Quebec, and from Magnet Cove, Arkansas, ranging from +6.6 to 7.6‰. These low values do not support the hypothesis that carbonatites are recrystallized marine limestones, but are quite consistent with the $\delta^{18}O$ values of mafic and ultramafic rocks. They calculated isotopic temperatures ranging from 720 to 885°C for calcite-magnetite pairs from Oka which are at least qualitatively reasonable. The oxygen-isotope compositions of carbonatites therefore favor an igneous origin for these rocks.

5
Oxygen in Metamorphic Rocks

Metamorphism of sedimentary or igneous rocks encompasses reactions among the constituent minerals of such rocks in response to an increase in the temperature and pressure. These reactions take place without melting but in the presence of a fluid phase composed of water and carbon dioxide. In the course of metamorphism, the isotopic composition of oxygen and hydrogen of the minerals are changed due to equilibration internally or with an external reservoir. If isotopic equilibrium was achieved between coexisting minerals formed during metamorphism, then a temperature can be determined by means of the several geothermometry equations that have been derived from experimental studies in the laboratory. However, it is not always clear whether such isotopic temperatures reflect the maximum temperature attained during metamorphism, because minerals may continue to equilibrate oxygen and hydrogen during subsequent cooling. This effect is especially significant during high-grade regional metamorphism followed by slow cooling and affects primarily those minerals that equilibrate oxygen rapidly, such as calcite. For example, Schwarcz et al. (1970) observed that Δ (quartz-calcite) values of calcareous and pelitic metamorphic rocks from New England decrease with increasing metamorphic grade up to the garnet zone but level off or increase at higher grades presumably because of retrograde equilibration during subsequent cooling. They also showed that even though Δ (quartz-magnetite) values tend to decrease with increasing grade, significant variations of

this value occur in a given metamorphic zone. This effect can be attributed not only to retrograde equilibration but also to variations in the chemical compositions of the minerals and to differences in the fugacity of oxygen or the partial pressure of water. Moreover, anomalous temperatures may result from the fact that some minerals may have formed during a subsequent metamorphic event at a lower temperature which would cause them to be out of equilibrium with previously formed minerals in the rock. It is clear, therefore, that isotopic temperatures of metamorphic rocks are not necessarily the maximum temperatures attained during metamorphism and that discordant temperatures may be observed in polymetamorphic rocks.

Another important consideration is the volume of rock within which oxygen and hydrogen equilibration takes place during metamorphism. If a given mineral existing in different rocks equilibrates isotopically at the same temperature with an isotopically homogeneous fluid that permeates the rocks, then that mineral should have the same $\delta^{18}O$ value regardless of its prior history. One can therefore investigate the extent of oxygen communication in a given area by comparing $\delta^{18}O$ values of a particular mineral. The evidence in this regard shows that in some cases oxygen communication has taken place over thousands of square kilometers while in others isotopic differences have been preserved within a few centimeters. For example, Taylor (1968b) reported $\delta^{18}O$ values of $+9.9 \pm 1.0\%$ for plagioclase from anorthosite, meta-syenite and meta-gabbro from the Adirondack Mountains of New York. This is about 3‰ heavier than $\delta^{18}O$ values of plagioclase from anorthosites elsewhere in the world and may be taken as evidence of large-scale reequilibration of the plagioclase in the

Adirondacks with an oxygen-bearing fluid. On the other hand, Perry et al. (1973) found that differences in $\delta^{18}O$ values of the Biwabik Iron Formation of Minnesota have been preserved even adjacent to the Duluth Gabbro Complex that intrudes the iron formation.

Isotope temperatures of metamorphic rocks can be determined from the differences of $\delta^{18}O$ values of coexisting minerals that equilibrated oxygen at a specific temperature and have not been altered since. One of the most useful mineral pairs for this purpose is quartz-magnetite because quartz is a strong concentrator of ^{18}O, while magnetite does so only weakly. Consequently, Δ_{QM} varies from about 23 to 4‰ between temperatures of 200 and 1000°C. A quartz-magnetite thermometry equation is listed in Table 19.3 based on Bottinga and Javoy (1973). Shieh and Schwarcz (1974) used two slightly different equations:

$$1000 \ln \alpha_{QM} = 4.32(10^6 T^{-2}) + 1.45 \quad (500 - 750°C) \quad (19.22)$$

$$1000 \ln \alpha_{QM} = 5.20(10^6 T^{-2}) \quad (200 - 500°C) \quad (19.23)$$

where $1000 \ln \alpha_{QM} \simeq \delta_Q - \delta_M$ and Q and M refer to quartz and magnetite, respectively. The equations representing oxygen fractionation between quartz-magnetite and other mineral pairs have been plotted in Figure 19.8 versus temperature. Additional thermometry equations for other mineral pairs, not considered in Table 19.3, can be derived from the data in Table 19.1. Isotopic

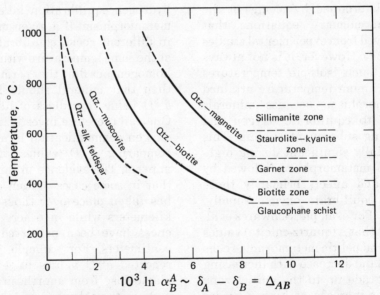

FIGURE 19.8 Oxygen-isotope thermometry equations plotted versus temperature in degrees Centigrade. The quartz-magnetite and quartz-biotite lines are from Shieh and Taylor (1974). The others are from Bottinga and Javoy (1973). The grades of regional metamorphism and their approximate temperatures are from Epstein and Taylor (1967).

temperatures based on $\delta^{18}O$ values of metamorphic rocks have been reported by several investigators, including James and Clayton (1962), Garlick and Epstein (1967), Taylor and Coleman (1968), Shieh and Taylor (1969a, b), Wilson et al. (1970), Eslinger and Savin (1973), and Shieh and Schwarcz (1974). Summaries of the isotope geologies of oxygen and hydrogen of metamorphic rocks have been prepared by Epstein and Taylor (1967), Garlick (1969), and Hoefs (1973).

Garlick and Epstein (1967) reported $\delta^{18}O$ values for minerals from metamorphic rocks of increasing grade. Their data for quartz-mineral pairs of nonretrograded rocks are presented in Table 19.5, averaged for each metamorphic grade. It is apparent that $\Delta^{18}O$ (quartz-mineral) generally decreases with increasing grade, thus implying a higher temperature of equilibration. The corresponding temperatures for oxygen equilibration for quartz-muscovite and quartz-magnetite were calculated using the equations in Table 19.3. Those for quartz-garnet, quartz-biotite, and quartz-ilmenite rely on the following concordancy equations:

$$\Delta \text{ quartz-garnet}$$
$$= 0.51\Delta \text{ (quartz-magnetite)} \quad (19.24)$$

$$\Delta \text{ quartz-biotite}$$
$$= 0.59\Delta \text{ (quartz-magnetite)} \quad (19.25)$$

$$\Delta \text{ quartz-ilmenite}$$
$$= 0.92\Delta \text{ (quartz-magnetite)} \quad (19.26)$$

The quartz-magnetite equation in each case is that of Bottinga and Javoy (1973). It is apparent that each of the isotope thermometers indicates increasing equilibration temperatures with increasing metamorphic grade. Moreover, the temperatures for any given grade of metamorphism are broadly concordant. The only significant exception is the anomalously low $\Delta^{18}O$

Table 19.5   Average $\Delta^{18}O$ Values Between Quartz and Other Minerals for Metamorphic Rocks of Increasing Grade and Corresponding Isotopic Temperatures

METAMORPHIC ZONE	MUSCOVITE	GARNET	BIOTITE	ILMENITE	MAGNETITE
Biotite	3.8	—	7.75	11.35	—
Temp. °C	434		379	399	
Garnet	3.4	5.0	6.25	10.5	10.9
Temp. °C	469	481	452	425	442
Staurol.-kyanite	3.17	4.37?	5.80	8.83	9.8
Temp. °C	491	533	480	488	481
Staurol.-silliman.	2.75	4.75	5.15	8.7	9.3
Temp. °C	537	500	526	494	501
Sillimanite	2.3	4.3	5.0[a]	7.4	9.1
Temp. °C	598	540	538	559	510
Orthocl.-silliman.	—	3.4	4.3	6.8	—
Temp. °C		641	601	595	

[a] Omitted value for AH243.
From Garlick and Epstein, 1967, Figure 19.1.

value and high temperature of the quartz-garnet pair from the staurolite-kyanite zone. Nevertheless, the data in Table 19.5 demonstrate the general validity and wide applicability of this method of thermometry in metamorphic rocks.

6 SUMMARY

Oxygen is a major constituent in most rock-forming minerals. When two minerals *A* and *B* equilibrate oxygen isotopes with a common reservoir at a fixed temperature, the difference in their $\delta^{18}O$ values decreases with increasing temperature. This phenomenon can be used to determine temperatures of final equilibration of cogenetic oxygen-bearing minerals. The reliability of such temperature determinations depends on three assumptions: (1) the minerals achieved isotopic equilibrium at a specific temperature; (2) the isotopic composition of oxygen was not changed subsequently; (3) the isotope fractionation factors are known from experimental studies. Much progress has been made in measuring oxygen and hydrogen fractionation factors between minerals and water at varying temperatures. Equations expressing isotope fractionation of oxygen and hydrogen between minerals and water over a range of temperature can be combined to give mineral isotope-thermometry equations. In general, the minerals of volcanic rocks indicate reasonable temperatures of crystallization while those of plutonic rocks yield lower temperatures due to reequilibration of oxygen isotopes during slow cooling.

The isotopic composition of oxygen in igneous plutons may be modified by interaction with circulating meteoric water. As a result, the $\delta^{18}O$ values of the intrusive and the surrounding country rock are lowered, while those of the water may be raised. Such reequilibration of the oxygen isotopes is accompanied by subtle deuteric alteration of feldspars and ferromagnesian minerals. Water of various origins also plays an important role in the formation of hydrothermal ore deposits and associated wall-rock alteration. The formation of a particular ore deposit may involve different isotopic varieties of water acting at different times and at different temperatures leading to great complexity of the isotopic record.

The isotopic composition of oxygen and hydrogen in clay minerals depends primarily on the composition of water during their formation and on the temperature. As a result, δD and $\delta^{18}O$ of clay minerals can be related by equations that mirror this relationship in meteoric water. The isotopic compositions of oxygen and hydrogen of clay in soils preserve a record of the climate at the time of their formation. Moreover, the isotopic compositions can be used to distinguish between supergene (low temperature) and hypogene (high temperature) clays in ore deposits. Marine carbonates have positive $\delta^{18}O$ values ranging from $+20\%_0$ to $+30\%_0$ relative to SMOW. The $\delta^{18}O$ of marine carbonates of Phanerozoic age decrease with increasing age, presumably due to interaction with meteoric water. Nonmarine carbonates show less enrichment in ^{18}O because they may have formed in isotopic equilibrium with meteoric water. The $\delta^{18}O$ values of dolomite generally resemble those of coexisting calcite and do not exhibit the predicted fractionation of oxygen between these two minerals. The origin of dolomite remains an unresolved problem.

The differences in $\delta^{18}O$ of minerals in metamorphic rocks generally decrease with increasing grade of metamorphism. Temperatures calculated from mineral pairs are concordant and also increase with metamorphic grade. However, they do not necessarily reflect the maximum temperatures because of reequilibration during cooling. This effect is particularly noticeable for high-grade metamorphic rocks and minerals, such as calcite and feldspar, which equilibrate oxygen isotopes rapidly. There is evidence that oxygen equilibration with an external reservoir has taken place in large volumes of rock (Adirondack Mountains). On the other hand, in some cases isotopic differences have been preserved over short distances even at high temperatures.

PROBLEMS

1. Given that the ^{18}O fractionation factor between illite and water is 1.0234 at 22°C (James and Baker, 1976), what is the $\delta^{18}O$ value of illite in equilibrium with water having $\delta^{18}O = -9.00‰$ relative to SMOW? (**ANSWER:** $\delta^{18}O = +14.4‰$.)

2. Given that the fractionation of ^{18}O between the following mineral-water pairs is:

Qtz-water:
$$1000 \ln \alpha_{QW} = 2.51 \times 10^6 (T^{-2}) - 1.96$$

Rutile-water:
$$1000 \ln \alpha_{RW} = 4.1 \times 10^6 (T^{-2}) + 0.96,$$

derive a quartz-rutile geothermometry equation and calculate Δ (quartz-rutile) at 700°C. (**ANSWER:** Δ (quartz-rutile) = $+4.06‰$.)

3. Onuma et al. (1970) reported the following $\delta^{18}O$ values for minerals from lunar basalt 10020:

	$\delta^{18}O‰$
Feldspar (An 85%)	+6.19
Clinopyroxene	+5.74
Olivine	+5.14
Ilmenite	+4.41

Calculate oxygen equilibration temperatures using thermometry equations listed in Table 19.3 for the mineral pairs: feldspar-pyroxene, feldspar-ilmenite, pyroxene-ilmenite (see also Bottinga and Javoy, 1975).

4. Onuma et al. (1972) reported the following $\delta^{18}O$ values for plagioclase, pyroxene, and olivine for several stony meteorites. Plot these data on a concordancy diagram in terms of Δ (feldspar-pyroxene) and Δ (pyroxene-olivine). Is the oxygen in these minerals in isotopic equilibrium and what is the approximate temperature range indicated by these data?
(See also Bottinga and Javoy, 1975.)

METEORITE	FELDSPAR $\delta^{18}O$	PYROXENE $\delta^{18}O$	OLIVINE $\delta^{18}O$	An %
Allegan	+5.12	+4.34	+3.49	12
Estacado	+5.79	+4.38	+3.84	12
Bjurbole	+7.70	+5.87	+4.44	11
Bruderheim	+6.20	+5.10	+4.24	11
Mocs	+6.23	+4.95	+4.10	10
Modoc	+6.17	+5.10	+4.08	10
Shaw	+5.73	+5.03	+4.70	10
Olivenza	+6.27	+5.16	+4.33	10
St. Severin	+6.58	+5.31	+4.27	10

5. Garlick and Epstein (1967) reported the following $\delta^{18}O$ values for a metasedimentary rock (A66a):

	$\delta^{18}O‰$
Quartz	+14.8
Garnet	+11.0
Magnetite	+5.0
Muscovite	+11.4
Biotite	+8.5

Calculate oxygen isotope equilibration temperatures for different mineral pairs. Assuming that these minerals reequilibrated oxygen with water during metamorphism at 480°C, calculate $\delta^{18}O$ of the water using equations listed in Table 19.1 (**ANSWER:** $\delta^{18}O$ water $\simeq$ + 11.5‰ from quartz, muscovite, and magnetite.)

REFERENCES

Addy, S. K., and G. D. Garlick (1974) Oxygen isotope fractionation between rutile and water. Contrib. Mineral. Petrol., *45*, 119–121.

Anderson, A. T., R. N. Clayton, and T. K. Mayeda (1971) Oxygen isotope thermometry of mafic igneous rocks. J. Geol., *79*, 715–729.

Bottinga, Y., and M. Javoy (1973) Comments on oxygen isotope thermometry. Earth Planet. Sci. Letters, *20*, 250–265.

Bottinga, Y., and M. Javoy (1975) Oxygen isotope partitioning among the minerals in igneous and metamorphic rocks. Rev. Geophys. Space Phys., *13*, 401–418.

Clayton, R. N., and T. K. Mayeda (1963) The use of bromine penta-fluoride in the extraction of oxygen from oxide and silicates for isotopic analysis. Geochim. Cosmochim. Acta, *27*, 43–52.

Clayton, R. N., B. F. Jones, and R. A. Berner (1968) Isotope studies of dolomite formation under sedimentary conditions. Geochim. Cosmochim. Acta, *32*, 415–432.

Clayton, R. N., J. R. O'Neil, and T. K. Mayeda (1972) Oxygen isotope exchange between quartz and water. J. Geophys. Res., *77*, 3057–3067.

Clayton, R. N., L. Grossman, and T. K. Mayeda (1973) A component of primitive nuclear composition in carbonaceous chondrites. Science, *182*, 485–487.

Conway, C. M., and H. P. Taylor (1969) $^{18}O/^{16}O$ and $^{13}C/^{12}C$ ratios of coexisting minerals in the Oka and Magnet Cove carbonatite bodies. J. Geol., *77*, 618–626.

Deines, P., and D. P. Gold (1973) The isotopic composition of carbonatite and kimberlite carbonates and their bearing on the isotopic composition of deep-seated carbon. Geochim. Cosmochim. Acta, *37*, 1709–1733.

Epstein, S. (1959) The variation of the $^{18}O/^{16}O$ ratio in nature and some geologic implications. In Researches in Geochemistry, vol. 1, pp. 217–240, P. H. Abelson, ed. John Wiley, New York.

Epstein, S., and H. P. Taylor, Jr. (1967) Variation of O^{18}/O^{16} in minerals and rocks. In Researches in Geochemistry, vol. 2, pp. 29–62, P. H. Abelson, ed. John Wiley, New York, 663 p.

Eslinger, E. V., and S. M. Savin (1973) Oxygen isotope geothermometry of the burial metamorphic rocks of the Precambrian Belt Supergroup, Glacier National Park, Montana. Geol. Soc. Amer. Bull., *84*, 2549–2560.

Friedman, I., P. W. Lipman, J. D. Obradovich, J. D. Gleason, and R. L. Christensen (1974) Meteoric water in magmas. Science, *184*, 1069–1072.

Garlick, G. D. (1966) Oxygen isotope fractionation in igneous rocks. Earth Planet. Sci. Letters, *1*, 361–368.

Garlick, G. D., and S. Epstein (1967) Oxygen isotope ratios in coexisting minerals from regionally metamorphosed rocks. Geochim. Cosmochim. Acta, *31*, 181–214.

Gray, C. M., and W. Compston (1974) Excess ^{26}Mg in the Allende meteorite. Nature, 251, 495–497.

Hall, W. E., I. Friedman, and J. T. Nash (1974) Fluid inclusion and light stable isotope study of the Climax molybdenum deposits, Colorado. Econ. Geol., *69*, 884–901.

Hamilton, E. I. (1964) The isotopic composition of strontium in Skaergaard Intrusion, east Greenland. J. Petrol., *4*, 383–391.

Hoefs, J. (1973) Stable Isotope Geochemistry. Springer-Verlag, New York, Heidelberg and Berlin, 140 p.

James, A. T., and D. R. Baker (1976) Oxygen isotope exchange between illite and water at 22°C. Geochim. Cosmochim. Acta, *40*, 235–240.

James, H. L., and R. N. Clayton (1962) Oxygen isotope fractionation in metamorphosed iron formations of the Lake Superior region and in other iron-rich rocks. In Petrologic Studies (Buddington Volume). Geol. Soc. Amer., pp. 217–239.

Jovanovic, S. and G. W. Reed, Jr. (1976) Interrelations among isotopically anomalous mercury fractions from meteorites and possible cosmological inferences. Science, *193*, 888–890.

Keith, M. L., G. M. Anderson, and R. Eichler (1964) Carbon and oxygen isotopic composition of mollusk shells from marine and fresh-water environments. Geochim. Cosmochim. Acta, *28*, 1757–1786.

Keith, M. L., and J. N. Weber (1964) Carbon and oxygen isotopic composition of selected limestones and fossils. Geochim. Cosmochim. Acta, *28*, 1787–1816.

Lawrence, J. R., and H. P. Taylor, Jr. (1971) Deuterium and oxygen-18 correlation: clay minerals and hydroxides in Quaternary soils compared to meteoric water. Geochim. Cosmochim. Acta, *35*, 993–1003.

Lawrence, J. R., and H. P. Taylor, Jr. (1972) Hydrogen and oxygen isotope systematics in weathering profiles. Geochim. Cosmochim. Acta, *36*, 1377–1393.

Lipman, P. W., and I. Friedman (1975) Interaction of meteoric water with magma: An oxygen isotope study of ash-flow sheets from southern Nevada. Geol. Soc. Amer. Bull., *86*, 695–702.

Lee, T., and D. A. Papanastassiou (1974) Mg isotope anomalies in the Allende meteorite and correlation with O and Sr effects. Geophys. Res. Letters, *1*, 225–229.

Muehlenbachs, K., A. T. Anderson, Jr., and G. E. Sigvaldason (1974) Low-O^{18} basalts from Iceland. Geochim. Cosmochim. Acta, *38*, 588.

Ohmoto, H., and R. O. Rye (1970) The Bluebell mine, British Columbia, I. Mineralogy, paragenesis, fluid inclusions, and the isotopes of hydrogen, oxygen, and carbon. Econ. Geol., *65*, 417–437.

Ohmoto, H. (1974) Hydrogen and oxygen isotopic compositions of fluid inclusions in the Kuroko deposits, Japan. Econ. Geol., *69*, 947–953.

O'Neil, J. R., and R. C. Clayton (1964) Oxygen isotope geothermometry. In Isotopic and Cosmic Chemistry, pp. 157–168, H. Craig, S. L. Miller and G. J. Wasserburg, eds. North-Holland, Amsterdam, 553 p.

O'Neil, J. R., and S. Epstein (1966) Oxygen isotope fractionation in the system dolomite-calcite-carbon dioxide. Science, *152*, 198–201.

O'Neil, J. R., and H. P. Taylor (1967) The oxygen isotope and cation exchange chemistry of feldspars. Amer. Mineralogist, *52*, 1414–1437.

O'Neil, J. R., and H. P. Taylor, Jr. (1969) Oxygen isotope equilibrium between muscovite and water. J. Geophys. Res., 74, 6012–6022.

O'Neil, J. R., R. N. Clayton, and T. K. Mayeda (1969) Oxygen isotope fractionation in divalent metal carbonates. J. Chem. Phys., *51*, 5547–5558.

O'Neil, J. R., and M. L. Silberman (1974) Stable isotope relations in epithermal Au-Ag deposits. Econ. Geol., *69*, 902–909.

O'Neil, J. R., and Y. K. Kharaka (1976) Hydrogen and oxygen isotope exchange reactions between clay minerals and water. Geochim. Cosmochim. Acta, *40*, 241–246.

Onuma, N., R. N. Clayton, and T. K. Mayeda (1972) Oxygen isotope temperature of equilibrated ordinary chondrites. Geochim. Cosmochim. Acta, *36*, 157–168.

Onuma, N., R. N. Clayton, and T. K. Mayeda (1970) Apollo 11 rocks: Oxygen isotope fractionation between minerals and an estimate of the temperature of formation. Geochim. Cosmochim. Acta, Suppl. 1, *2*, 1429–1434.

Perry, E. C., Jr., F. C. Tan, and G. B. Morey (1973) Geology and stable isotope geochemistry of the Biwabik Iron Formation, northern Minnesota. Econ. Geol., *68*, 1110–1125.

Pineau, F., M. Javoy, and C. J. Allegre (1973) Etude systematique des isotopes de l' oxygene, du carbone et du strontium dans les carbonatites. Geochim. Cosmochim. Acta, *37*, 2363–2377.

Reuter, J. H., S. Epstein, and H. P. Taylor, Jr. (1965) O^{18}/O^{16} ratios of some chondritic meteorites and terrestrial ultramafic rocks. Geochim. Cosmochim. Acta, *29*, 481–488.

Rye, R. O., and J. R. O'Neil (1968) The O^{18} content of water in primary fluid inclusions from Providencia, north-central Mexico. Econ. Geol., *63*, 232–238.

Savin, S. M., and S. Epstein (1970) The oxygen and hydrogen isotope geochemistry of clay minerals. Geochim. Cosmochim. Acta, *34*, 25–42.

Schwarcz, H. P., R. N. Clayton, and T. K. Mayeda (1970) Oxygen isotopic studies of calcareous and pelitic metamorphic rocks, New England. Geol. Soc. Amer. Bull., *81*, 2299–2316.

Sheppard, S. M. F., and H. P. Schwarcz (1970) Fractionation of carbon and oxygen isotopes and magnesium between coexisting metamorphic calcite and dolomite. Contrib. Mineral. and Petrol., *26*, 161–198.

Sheppard, S. M. F., R. L. Nielsen, and H. P. Taylor, Jr. (1971) Oxygen and hydrogen isotope ratios in minerals from porphyry copper deposits. Econ. Geol., *66*, 515–542.

Sheppard, S. M. F., and H. P. Taylor, Jr. (1974) Hydrogen and oxygen isotope evidence for the origins of water in the Butte ore deposits, Montana. Econ. Geol., *69*, 926–946.

Shieh, Y. N., and H. P. Taylor (1969a) Oxygen and hydrogen isotope studies of contact metamorphism in the Santa Rosa Range, Nevada and other areas. Contrib. Mineral. Petrol., *20*, 306–356.

Shieh, Y. N., and H. P. Taylor (1969b) Oxygen and carbon isotope studies of contact metamorphism of carbonate rocks. J. Petrol., *10*, 307–331.

Shieh, Y. N., and H. P. Schwarcz (1974) Oxygen isotope studies of granite and migmatite, Grenville province of Ontario, Canada. Geochim. Cosmochim. Acta, *38*, 21–45.

Tan, F. C., and J. D. Hudson (1971) Carbon and oxygen isotopic relationships of dolomites and coexisting calcites, Great Estuarine Series (Jurassic) Scotland. Geochim. Cosmochim. Acta, *35*, 755–768.

Taylor, H. P., and S. Epstein (1963) O^{18}/O^{16} ratios in rocks and coexisting minerals of the Skaergaard Intrusion, East Greenland. J. Petrol., *4*, 51–74.

Taylor, H. P., Jr., M. B. Duke, L. T. Silver, and S. Epstein (1965) Oxygen isotope of minerals in stony meteorites. Geochim. Cosmochim. Acta, *29*, 489–512.

Taylor, H. P., Jr. (1968a) The oxygen isotope geochemistry of igneous rocks. Contrib. Mineral. Petrol., *19*, 1–71.

Taylor, H. P. (1968b) Oxygen isotope studies of anorthosites with special reference to the origin of bodies in the Adirondack Mountains, New York. In Origin of Anorthosites. N.Y. Museum Special Publication.

Taylor, H. P., Jr., and S. Epstein (1968) Hydrogen isotope evidence for influx of meteoric ground water into shallow igneous intrusions. Geol. Soc. Amer. Spec. Paper 121, 294.

Taylor, H. P., and R. G. Coleman (1968) O^{18}/O^{16} ratios of coexisting minerals in glaucophane-bearing metamorphic rocks. Geol. Soc. Amer. Bull., *79*, 1727–1756.

Taylor, H. P., Jr., and S. Epstein (1970) O^{18}/O^{16} ratios of Apollo 11 lunar rocks and minerals. Geochim. Cosmochim. Acta, Supplement 1, *34*, 1613–1626.

Taylor, H. P., Jr., and R. W. Forester (1971) Low-O^{18} igneous rocks from the intrusive complexes of Skye, Mull and Ardnamurchan, western Scotland. J. Petrol., *12*, 465–497.

Taylor, H. P., Jr. (1971) Oxygen isotope evidence for large-scale interaction between meteoric ground waters and Tertiary granodiorite intrusions, western Cascade range, Oregon. J. Geophys. Res., *76*, 7855–7874.

Taylor, H. P., Jr. (1974) The application of oxygen and hydrogen isotope studies to problems of hydrothermal alteration and ore deposition. Econ. Geol., *69*, 843–883.

Wager, L. R., and W. A. Deer (1939) Geological investigations in East Greenland. Part III. The petrology of the Skaergaard intrusion, Kangerdlussuaq. Medd. Gronland, *105*, 1–353.

White, D. E. (1974) Diverse origins of hydrothermal ore fluids. Econ. Geol., *69*, 954–973.

Wilson, A. F., D. C. Green, and R. L. Davidson (1970) The use of oxygen isotope geothermometry on the granulites and related intrusives, Musgrave Ranges, Central Australia. Contrib. Mineral. Petrol., *27*, 166–178.

20 CARBON

Carbon ($Z = 6$) has two stable isotopes: $^{12}C = 98.89$ percent and $^{13}C = 1.11$ percent. In addition, radioactive ^{14}C occurs in nature due to its formation in the upper atmosphere by an (n, p) reaction on stable $^{14}_7N$ (Chapter 17). Carbon is one of the most abundant elements in the universe and is the basis for the existence of life on the Earth. Consequently, it is the most important element of the biosphere, but occurs also in the crust and mantle of the Earth and in its hydrosphere and atmosphere. Carbon occurs in the reduced form in organic compounds and in coal. It also occurs in the oxidized state primarily as carbon dioxide, carbonate ions in aqueous solution, and as carbonate minerals. In addition, it is found as native element in the form of graphite and diamond.

The isotopes of carbon are fractionated by a variety of natural processes, including photosynthesis and isotope exchange reactions among carbon compounds. Photosynthesis leads to enrichment of ^{12}C in biologically synthesized organic compounds. On the other hand, isotope exchange reactions between CO_2 gas and aqueous carbonate species tend to enrich carbonates in ^{13}C. As a result, the isotopic abundance of ^{13}C in terrestrial carbon varies by about 10 percent.

The isotopic composition of carbon is expressed in terms of the delta notation used also for oxygen and hydrogen (Chapters 18 and 19). We therefore define the parameter:

$$\delta^{13}C = \left[\frac{(^{13}C^{12}C)_{spl} - (^{13}C/^{12}C)_{std}}{(^{13}C/^{12}C)_{std}} \right] \times 10^3 \%_0$$

(20.1)

The reference standard is CO_2 gas obtained by reacting belemnites of the Peedee formation with 100 percent phosphoric acid, that is, the PDB standard of the University of Chicago (*Belemnitella Americana*, Peedee Formation, Cretaceous, South Carolina). In actual practice, a working standard is used which has been compared to PDB. Other standards that have been used and their $\delta^{13}C$ values relative to PDB include (Hoefs, 1973a):

	$\delta^{13}C\%_0$ (PDB)
Solenhofen limestone, NBS No. 20	−1.1
$BaCO_3$, Stockholm standard	−10.3
Graphite, NBS No. 21	−27.8
Petroleum standard, NBS No. 22	−29.4

Carbon is always analyzed as CO_2 gas using mass spectrometers equipped with double collectors (Craig, 1957). Carbonate samples are reacted with 100 percent phosphoric acid at 25°C, while organic compounds are oxidized to CO_2 at 900 to 1000°C. The isotope geology of carbon has been reviewed by Craig (1953), Rankama (1963), Schwarcz (1969), Degens (1969), Hoefs (1973a), and Junge et al. (1975).

1
Carbon in the Modern Biosphere

Photosynthesis is a complex process by means of which atmospheric carbon dioxide is incorporated into green plants. The process can be represented by the equation:

$$6CO_2 + 6H_2O \rightarrow C_6H_{12}O_6 + 6O_2 \quad (20.2)$$

In the course of photosynthesis, the carbon that is fixed in plant tissue is significantly enriched in ^{12}C relative to atmospheric carbon dioxide, whose $\delta^{13}C$ value is $-7.0‰$. Photosynthetic plants can be subdivided into two large categories on the basis of their $\delta^{13}C$ values (Smith and Epstein, 1971). Most terrestrial plants have $\delta^{13}C$ values ranging from -24 to $-34‰$, while those of aquatic plants, desert and salt marsh plants, and tropical grasses vary from -6 to $-19‰$. Algae and lichens form an intermediate group with $\delta^{13}C$ values from -12 to $-23‰$. In general, different species of plants growing in a given environment may have slightly different $\delta^{13}C$ values, but individuals of the same species growing in the same area have similar values. The same is also true of animals whose $\delta^{13}C$ values reflect that of their food supply.

The isotope fractionation of carbon in plants takes place in three steps (Park and Epstein, 1960; Whelan et al., 1973). The first step involves the preferential incorporation of $^{12}CO_2$ from the atmosphere across the walls of cells and its dissolution in the cytoplasm. Fractionation results from kinetic effects and can vary widely depending on the concentration of carbon dioxide in the air around the plant, and other factors. Fractionation is inhibited at low concentrations of CO_2, perhaps because of increased back-diffusion and the depletion of ^{12}C in atmospheric carbon dioxide.

The second step in the fractionation of carbon isotopes occurs during the preferential conversion of $^{12}CO_2$ dissolved in cytoplasm into phosphoglyceric acid by the action of enzymes. This leads to enrichment of ^{13}C in the dissolved carbon dioxide which inhibits further fractionation unless the residual carbon dioxide is removed. Vascular plants may translocate carbon dioxide and may expel it through the roots. It is also expelled through the leaves by respiration during dark periods. The respiratory carbon dioxide is initially enriched in ^{13}C relative to the total plant (Abelson and Hoering, 1961) presumably because of the removal of "heavy" carbon dioxide from the cytoplasm. Keeling (1958, 1961) demonstrated a diurnal cycle in the $\delta^{13}C$ value of atmospheric carbon dioxide that is attributable to the respiration of plants at night and photosynthesis during the day. He showed that the concentration of carbon dioxide in rural air reaches a peak around midnight when its $\delta^{13}C$ value is lowest due to release of respiratory carbon dioxide having $\delta^{13}C$ from -21 to $-26‰$. Conversely, during the daytime, the CO_2 concentration decreases and $\delta^{13}C$ increases as plants preferentially take up $^{12}CO_2$. Parker (1964) observed a similar effect in the $\delta^{13}C$ values for carbonate ions in sea water over an oyster-green flagellate community. Lowdon and Dyck (1974) reported a seasonal variation of $\delta^{13}C$ for maple leaves ranging from $-22‰$ in early spring to $-28‰$ in late fall.

The third step in the fractionation of carbon isotopes by plants occurs during the subsequent synthesis of a variety of organic compounds from phosphoglyceric acid. Measurements by Degens et al. (1968) of $\delta^{13}C$ values in organic compounds extracted from marine plankton show a progressive *enrichment* of ^{12}C in cellulose, "lignin," and lipids (oils, fats, and waxes) compared to total organic matter and a relative *depletion* of ^{12}C in total carbohydrates, hemicellulose, proteins, and pectin. These authors also discussed the isotopic diversification of carbon in the course of metabolic reactions in marine plankton.

Organic matter in Recent sediment has $\delta^{13}C$ values ranging from -10 to $-30‰$

with a maximum between -20 and $-27\%_0$ (Eckelman et al., 1962; Shultz and Calder, 1976). Sackett and Thompson (1963) found that $\delta^{13}C$ values of sediment from the Gulf Coast changed landward from about -21 to $-26\%_0$. They attributed this to the presence of plant debris derived from land since land plants characteristically have more negative $\delta^{13}C$ values than aquatic plants. This result suggests that the $\delta^{13}C$ value of organic matter in sediment might be used to distinguish between sedimentary rocks of marine origin and those deposited in terrestrial or near-shore environments. However, this criterion does not work because the $\delta^{13}C$ value of organic matter in sedimentary rocks changes as a function of time due to the preferential destruction of carbohydrates and proteins and the resulting enrichment of the remaining organic matter in lipids, lignin, and cellulose. In this way the organic matter is progressively enriched in ^{12}C so that sedimentary rocks of pre-Tertiary age have average $\delta^{13}C$ values of $-28\%_0$ regardless of the origin of the organic matter they contain (Hoefs, 1973a).

2
Fossil Fuels

The carbon contained in fossil fuels (coal, petroleum, natural gas, as well as oil shale and tar sands) is strongly enriched in ^{12}C, which is consistent with the view that these materials were derived from biogenic material. The isotopic composition of carbon in coal is similar to that of modern land plants and apparently does not vary significantly with increasing rank of the coal or its geologic age. The $\delta^{13}C$ value of coal averages about $-25\%_0$ relative to the PDB standard. The constancy of the isotopic composition indicates that no appreciable isotope fractionation takes place during coalification, and that the plants from which coal has

formed had an isotopic composition similar to modern plants.

The $\delta^{13}C$ values of petroleum are also remarkably uniform with an average of about $-28\%_0$. Petroleum is believed to be derived from plant or animal remains deposited in marine or nonmarine basins. The processes by which such material is converted into petroleum are not well understood. In general, petroleum appears to be enriched in ^{12}C by about $10\%_0$ relative to modern aquatic plants. The enrichment may be due to isotopic fractionation during the formation of petroleum from organic material or to selective accumulation of certain fractions that are enriched in ^{12}C relative to the total organic matter. We know that the lipids are enriched in ^{12}C by up to $8\%_0$ compared to other biogenic compounds. In addition, this fraction is the most stable of all plant constituents, is insoluble in water, and contains hydrocarbons that most closely resemble petroleum in composition (Silverman, 1964). Moreover, it has been demonstrated that lipids can be converted to mixtures of hydrocarbons resembling petroleum by the catalytic action of clay minerals such as montmorillonite. For these reasons, Silverman (1964) concluded that petroleum is derived from the lipid fraction, whereas coal forms from the cellulose fraction of plants.

The carbon in natural gas is even more strongly enriched in ^{12}C than the petroleum with which it is associated (Sackett, 1968). Rosenfeld and Silverman (1959) reported that bacterially produced methane is enriched in ^{12}C by as much as $75\%_0$ relative to the source material. Colombo et al. (1966) found that the $\delta^{13}C$ values of methane from Italian natural gas fields varied widely, even within a single field, between limits of about -40 to $-70\%_0$ relative to PDB. Heavier fractions were found to be enriched in ^{13}C relative to methane in the order: ethane, propane, butane +, listed in order of increasing ^{13}C-enrichment. These results agree with the experiments of Silverman (1964) who demonstrated that the most volatile fraction of crude oil from the West Coyote field of California is enriched in ^{12}C relative to heavier fractions. He suggested that the light hydrocarbon fraction of petroleum is produced by thermal cracking of heavier molecules and that the light fraction is enriched in ^{12}C during this process.

The burning of fossil fuel has significantly increased the carbon dioxide content of the atmosphere (Broecker, 1975) and has caused a measurable decrease in its $\delta^{13}C$ value because fossil fuels are enriched in ^{12}C compared to atmospheric CO_2. In addition, the [14]C- content of the atmosphere has been lowered because fossil fuels contain no [14]C. This is known as the Suess effect, which was discussed in Chapter 17. According to Farmer and Baxter (1974), the combustion of fossil fuels during the twentieth century has increased the carbon dioxide concentration of the atmosphere by about 10 percent from 290 parts per million by volume in 1900 to 323 ppm in 1972. The addition of carbon dioxide to the atmosphere due to burning of fossil fuels can be studied by means of the $\delta^{13}C$ values of wood grown since the onset of the Industrial Revolution, provided that isotope fractionation during photosynthetic fixation of carbon has remained constant. Farmer and Baxter (1974) found that $\delta^{13}C$ values of an oak and a larch grown in England declined by 1.5 and 1.1‰, respectively, from 1900 to 1920. The $\delta^{13}C$ of the oak continued to decline and reached a minimum of about −27.3‰ around 1935, while that of the larch fluctuated irregularly. Between 1944 and 1958, the $\delta^{13}C$ values of both trees actually increased and then declined rapidly until 1964. They attributed these variations not only to the combustion

of fossil fuels but also to increased oxidation of plant debris caused by increased cultivation of arable land. The increase in $\delta^{13}C$ values between 1944 and 1958 may reflect increased uptake of atmospheric carbon dioxide by the oceans in response to a drop of 0.1 to 0.2°C in the mean global temperature or may be due to an increase in the biomass of the Earth stimulated by the previous increase in the CO_2- content of the atmosphere. These results suggest that the isotopic composition of carbon in plants may be interpretable in terms of large-scale processes affecting the carbon reservoirs of the Earth. A decrease in $\delta^{13}C$ values of organic matter in sediment deposited recently in the Baltic Sea has been reported by Erlenkeuser et al. (1974). However, this phenomenon may not only reflect an increase in the ^{12}C content of the atmosphere but may also be due to local discharge of sewage sludge.

3
Sedimentary Rocks of Precambrian Age

One of the most important discoveries in the Earth Sciences during the 1960s was that sedimentary rocks of Precambrian age contain evidence for the existence of life in the oceans (Schopf, 1970). This conclusion is generally based on three lines of evidence: (1) recognition of morphologically preserved microscopic plants such as blue-green algae and bacteria (Schopf, 1968); (2) extraction of complex mixtures of organic compounds believed to be the products of decomposition of biogenic matter (Oró et al., 1965; Hoering, 1967); (3) significant enrichment of the reduced carbon in ^{12}C similar to that found in modern sediment (Oehler et al., 1972). The presence of morphologically preserved plant fossils in Precambrian cherts that are more than three billion years old (Onverwacht Series, South Africa) indicates that life existed in the oceans in early Precambrian time (Engel et al., 1968; Barghoorn and Schopf, 1966). However, the presence of complex organic molecules, including amino acids, in Precambrian rocks must be interpreted with caution. The reason is that even chert is sufficiently permeable to ground water to permit organic compounds derived from modern plants to be deposited within them (Nagy, 1970).

Isotopic compositions of reduced carbon in sedimentary rocks of Precambrian age have been reported by Hoefs and Schidlowsky (1967), Barker and Friedman (1969), Oehler et al. (1972), McKirdy and Powell (1974), and others. The $\delta^{13}C$ values scatter rather widely between limits of about -15 to $-40‰$ relative to PDB. Nevertheless, it is clear that the reduced carbon in these rocks is enriched in ^{12}C and therefore may be of photosynthetic origin. For example, the sedimentary rocks of the Witwatersrand of South Africa contain appreciable quantities of carbonaceous matter called thucholite, which was probably formed by polymerization of hydrocarbons by the action of alpha and gamma-radiation emitted by uraninite. This material is strongly enriched in ^{12}C, according to Hoefs and Schidlowsky (1967), and may have experienced isotope fractionation during photosynthesis.

The wide scatter of $\delta^{13}C$ values may largely result from diagenesis and progressive thermal alteration. McKirdy and Powell (1974) as well as Barker and Friedman (1969) noted that $\delta^{13}C$ values become less negative (implying enrichment in ^{13}C) with increasing metamorphic grade. The enrichment in ^{13}C may result from the formation of methane by thermal cracking of kerogen. We know that this process enriches methane in ^{12}C and leaves the residue enriched in ^{13}C.

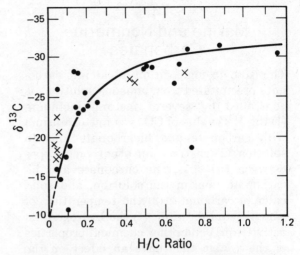

FIGURE 20.1 Plot of $\delta^{13}C$ of reduced carbon in sedimentary rocks of Precambrian to Middle Cambrian age versus the rank of the organic fraction as indicated by the atomic H/C ratio. The solid circles represent samples from Australia while the crosses are cherts from the Onverwacht Series of South Africa. These data suggest that organic carbon in sedimentary rocks is enriched in ^{13}C by the formation and subsequent loss of methane (which concentrates ^{12}C) during increasing thermal modification of kerogen as indicated by its atomic H/C ratio. (The data are from McKirdy and Powell, 1974.) The standard is PDB.

McKirdy and Powell were able to demonstrate this effect by means of the H/C ratio of the organic fraction. Their data, presented in Figure 20.1, clearly show that the reduced carbon of sedimentary rocks of Precambrian to Middle Cambrian age from Australia is progressively enriched in ^{13}C with increasing rank (decreasing H/C ratio) of the organic matter. Cherts from the Onverwacht Series of South Africa also follow this pattern which may explain the anomalous enrichment in ^{13}C of some samples from these rocks originally reported by Oehler et al. (1972).

4
Marine and Nonmarine Carbonates

The isotopic composition of calcium carbonate precipitated from aqueous solutions is controlled by several factors, including (1) the $\delta^{13}C$ value of CO_2 gas in equilibrium with carbonate and bicarbonate ions in solution; (2) fractionation of carbon isotopes between CO_2 gas, the carbonate and bicarbonate ions in the solution, and solid calcium carbonate; (3) the temperature of isotopic equilibration; (4) the hydrogen ion activity (pH) and other chemical properties of the system that have an effect on the abundance of carbonate and bicarbonate ions in the system (Deines et al., 1974). The fractionation of carbon isotopes between CO_2 gas and carbonate species in aqueous solution has been studied by Deuser and Degens (1967), Wendt (1968), Vogel et al. (1970), and Emrich et al. (1970), among others.

The carbonate equilibria can be represented by the following equations:

The temperature variation of these fractionation factors from 20° to 60°C is shown in Figure 20.2 in terms of the parameters $1000 \ln \alpha$ and $10^3 T^{-1}$, where T is the absolute temperature. These results indicate that calcium carbonate precipitated in isotopic and chemical equilibrium with CO_2 gas is enriched in ^{13}C by about 10‰. For example, if $\delta^{13}C$ of the CO_2 gas is -7.0‰, the $\delta^{13}C$ value of calcium carbonate precipitated in equilibrium with CO_2 at 20°C can be calculated from Equation 18.9:

$$\alpha_{AB} = \frac{\delta_A + 1000}{\delta_B + 1000}$$

Therefore,

$$\delta^{13}C \text{ (carbonate)}$$
$$= 1.01017(-7.0 + 1000) - 1000$$
$$= +3.09\text{‰}.$$

Deines et al. (1974) used all available data and derived a series of linear equations representing the variation of the fractionation factors in coordinates of $1000 \ln \alpha$ and $10^6 T^{-2}$. Their equations give fractionation

---

$$CO_{2(g)} \rightleftarrows CO_{2(aq)}$$
$$CO_{2(aq)} + H_2O \rightleftarrows H^+ + HCO_{3(aq)}^-$$
$$CaCO_{3(s)} + H^+ \rightleftarrows Ca_{(aq)}^{+2} + HCO_{3(aq)}^-$$

$$\overline{CO_{2(g)} + H_2O + CaCO_{3(s)} \rightleftarrows Ca_{(aq)}^{+2} + 2HCO_{3\,aq}^-}$$

(20.3)

The fractionation factors for carbon isotopes in this system at 20°C reported by Emrich et al. (1970) are:

factors at 20°C that are in good agreement with those of Emrich et al. (1970). The equations are listed in Table 20.1.

	α, 20°C
Calcium carbonate-bicarbonate:	1.00185 ± 0.00023
Bicarbonate-carbon dioxide gas:	1.00838 ± 0.00012
Calcium carbonate-carbon dioxide gas:	1.01017 ± 0.00018

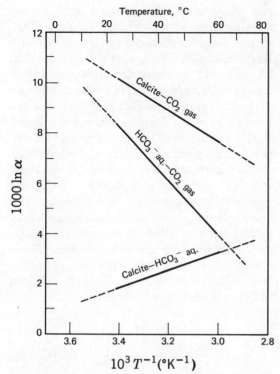

FIGURE 20.2 Variation of carbon isotope fractionation in the system: CO_2 (gas) – HCO_3^- (aqueous) – $CaCO_3$ (solid) as a function of temperature. Note that $1000 \ln \alpha \simeq (\alpha - 1) 1000 \simeq \Delta AB$, where A and B are two carbon-bearing phases in isotopic equilibrium and Δ is the difference between their δ values. If $\Delta = 8.8\permil$, the fractionation factor $\alpha = 1.0088$. (Replotted from experimental data of Emrich et al., 1970.)

The $\delta^{13}C$ values of carbonate rocks of marine origin of Cambrian to Tertiary age are virtually constant and have values close to zero on the PDB scale. Keith and Weber (1964) obtained an average value of $+0.56 \pm 1.55\permil$ for 321 selected samples of marine carbonate rocks. On the other hand, freshwater carbonates are enriched in ^{12}C compared to marine carbonates and have more variable $\delta^{13}C$ values. Keith and Weber (1964) reported an average value of $-4.93 \pm 2.75\permil$ for 183

Table 20.1   **Variation of Carbon Isotope Fractionation Factors of Various Carbonate Species Relative to CO_2 Gas (Deines et al., 1974)**

CARBONATE SPECIES	A	B
H_2CO_3 aq	0.0063	-0.91
HCO_3^- aq	1.099	-4.54
$CO_3^=$ aq	~ 0.87	~ -3.4
$CaCO_3$ solid	1.194	-3.63

The equations are of the form:

$$1000 \ln \alpha = A(10^6 T^{-2}) + B$$

selected freshwater carbonates. The relative enrichment in ^{12}C and greater variability of the $\delta^{13}C$ values of freshwater carbonates is attributable in large measure to the presence of CO_2 gas derived by oxidation of plant debris in soils and by plant respiration. The isotopic composition of carbon in the shells of modern marine and nonmarine mollusks is similar to that of carbonate rocks. Keith et al. (1964) found $\delta^{13}C$ values ranging from $+4.2$ to $-1.7\permil$ in the shells of marine mollusks and -0.6 to $-15.2\permil$ in freshwater species. Consequently, the $\delta^{13}C$ (as well as the $\delta^{18}O$) values of carbonates have some value as environmental indicators. However, the $\delta^{13}C$ (as well as $\delta^{18}O$) values of carbonate rocks may be altered after deposition (Choquette, 1968), making such interpretations uncertain.

Several occurrences of calcium carbonate having unusual isotopic compositions of carbon have been reported. Hathaway and Degens (1969) found $\delta^{13}C$ values ranging from -23.1 to $-60.6\permil$ relative to PDB in aragonite taken from sand deposits of Pleistocene age at the edge of the continental shelf off the northeast coast of the United States. They attributed the origin of these carbonates and their unusual enrichment in

^{12}C to the oxidation of methane. Similar enrichment in ^{12}C was reported by Cheney and Jensen (1965, 1967) for calcium carbonate from cap rocks of salt domes in the Gulf Coast area of the United States. The average δ^{13}C value for such rocks is $-36.2 \pm 6.2\permil$ relative to PDB. Calcites associated with organic material in the uranium deposits of the Wind River Formation (Lower Eocene) in the Gas Hills, Wyoming, have an average δ^{13}C $= -22.5 \pm 4.0\permil$, and sulfur-bearing limestones from Sicily have δ^{13}C $= -28.7 \pm 8.9\permil$. In all three cases the calcium carbonate may have formed from CO_2 derived by bacterial or inorganic oxidation of methane, which is strongly enriched in ^{12}C relative to total organic matter in sedimentary rocks.

A very different isotopic composition was reported by Clayton (1963) for carbonates in carbonaceous chondrites. The δ^{13}C values range from $+58.6$ to $+64.4\permil$ relative to PDB. This is the most extreme example of ^{13}C-enrichment on record. The reduced carbon of carbonaceous chondrites is depleted in ^{13}C and has δ^{13}C values of -7 to $-30\permil$ (Belsky and Kaplan, 1970). Lancet and Anders (1970) demonstrated experimentally that such differences in the isotopic composition of carbon may occur during inorganic synthesis of hydrocarbons by reactions between CO, H_2, and NH_3 in the presence of a nickel-iron or magnetite catalyst.

The isotopic composition of carbon in carbonate rocks of Precambrian age seems to be similar to that in carbonate rocks deposited in modern times. However, Schidlowski et al. (1975, 1976) have reported that dolomite samples from the middle Precambrian Lomagundi Formation of Rhodesia have an average δ^{13}C value of $+8.2 \pm 2.6\permil$ relative to PDB. They concluded that the anomalous enrichment of these rocks in ^{13}C resulted from the preferential removal of ^{12}C by organic material in a closed basin. Becker

and Clayton (1972) reported δ^{13}C values ranging from $+2$ to $-2\permil$ for dolomitic limestones from the Hamersley Group of northwest Australia which was deposited about 2.1 ± 0.1 billion years ago. However, ankerite and siderite from the Dales Gorge member of the Brockman Iron Formation (Hamersley Range) have δ^{13}C values between -9 and $-11\permil$ on the PDB scale, indicating significant enrichment in ^{12}C. They attributed the ^{12}C-enrichment to the presence of bicarbonate ions formed by oxidation of organic matter of biologic origin in freshwater. The dolomitic limestones, on the other hand, may have been deposited during marine transgressions into the Hamersley basin. These conclusions make an important contribution toward solving the problem of the origin of Precambrian iron formations.

5
Carbonatites and Diamonds

The isotopic composition of carbon in carbonatites and diamonds is of interest because both originate from subcrustal sources in the Earth. Carbonatites are commonly associated with alkali-rich igneous rocks and crystallize from carbonate-rich magmas. The δ^{13}C values of most carbonate minerals of carbonatites range from -2.0 to $-8.0\permil$ relative to PDB (Conway and Taylor, 1969; Pineau et al., 1973; Galimov et al. 1975). Calcites generally are more strongly depleted in ^{13}C than dolomites and ankerites. There may be small but significant differences in the average δ^{13}C values among the carbonatites that have been studied in sufficient detail. Deines and Gold (1973) gave the following average values from a review of the literature: Oka (Canada): $-5.08 \pm 0.49\permil$; Laacher See (Germany): $-7.14 \pm 0.37\permil$; Kaiserstuhl (Germany): $-6.21 \pm 0.48\permil$; Chadobets (U.S.S.R.): $-3.36 \pm 0.81\permil$. Although the validity of these data is somewhat limited by inadequate sampling, it is apparent that real differences may exist in the isotopic composition of carbon in carbonatites. Deines and Gold (1973) obtained an average $\delta^{13}C$ value of $-5.1 \pm 1.4\permil$ for 22 selected carbonatites and $-4.7 \pm 1.2\permil$ for carbonates from 13 kimberlites. The regional differences in the $\delta^{13}C$ values of carbonatites may be due either to the isotopic inhomogeneity of carbon in the subcrustal sources of carbonatites or to isotope fractionation during the evolution of carbonatite magmas, or both. In any case, it appears that carbonatites are not reliable indicators of the isotopic composition of deep-seated carbon. Moreover, the variability of $\delta^{13}C$ values makes it difficult to distinguish carbonate minerals of igneous carbonatites from marine or nonmarine sedimentary carbonates.

Diamonds occur in kimberlite pipes and certain other rocks believed to originate from below the crust of the Earth. They require high temperatures and pressures (1200°C, 45 kilobars equivalent to a depth of 150 km) to form in thermodynamic equilibrium with graphite. For this reason, rocks containing diamonds are thought to originate in the mantle of the Earth. However, Mitchell and Crocket (1971) have suggested that diamonds may also form metastably at shallower depths during the ascent of kimberlite magma by crystallization around diamond nuclei formed previously within their stability field.

The isotopic composition of carbon in diamonds appears to be variable within narrow limits with $\delta^{13}C$ between -2.0 to $-10.0\permil$ relative to PDB. The carbon isotope fractionation factors in the system graphite-diamond-carbon dioxide were calculated by Bottinga (1969a) from theoretical considerations. The results, shown in Figure 20.3, suggest that diamond should be appreciably depleted in ^{13}C relative to CO_2 but slightly enriched in ^{13}C with respect to graphite. Apparently, isotope fractionation of carbon is possible even at temperatures of 1200°C. If diamonds can form metastably at lower temperatures, isotope fractionation effects may be even more important. Therefore, the observed isotopic compositions of terrestrial diamonds are not necessarily identical to that of carbon present in the mantle of the Earth, but may depend on the temperature of formation and on the isotopic composition of whatever carbon compound they equilibrated with.

Koval'skiy and Cherskiy (1973) pointed out that the apparent isotopic homogeneity of carbon in diamonds is based primarily on analyses of colorless crystals. They analyzed a variety of diamonds having different colors from deposits in Yakutia (eastern Siberia) and found $\delta^{13}C$ values ranging from -5.0 to $-32.3\permil$ relative to PDB. The average $\delta^{13}C$ value of these colored Yakutian diamonds is $-11.8\permil$ and appears to be significantly less than those of colorless diamonds from these deposits whose average $\delta^{13}C$ value is $-7.2\permil$. These results indicate that the isotopic composition of carbon in diamonds may be more variable than previously stated and that it overlaps with that of carbon of biogenic origin. The greater variability of the $\delta^{13}C$ values of diamonds is compatible with the hypothesis of Mitchell and Crocket (1971) that diamonds may form metastably in nature at shallower depth and lower temperature than indicated by their stability field. We conclude that the $\delta^{13}C$ values of carbonatites and diamonds overlap but vary significantly and do not indicate unequivocally the isotopic composition of carbon in the mantle. The basic source of the difficulty of determining the isotopic composition of carbon in the upper mantle is that carbon isotopes are fractionated appreciably even at elevated temperatures of 1000°C or more.

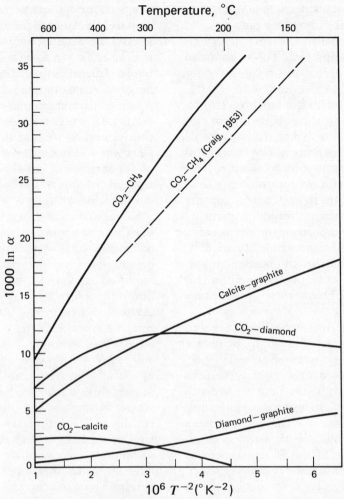

FIGURE 20.3 Calculated isotope fractionation factors for carbon in the system calcite-CO_2-diamond-graphite-CH_4. (Bottinga, 1969a, b; Craig, 1953.)

6
Carbon in Igneous Rocks and Volcanic Gases

Igneous rocks contain a variety of carbon compounds: (1) carbonate minerals and CO_2 gas in fluid or gaseous inclusions; (2) elemental carbon in the form of graphite or diamonds; (3) mixtures of organic molecules and carbides. The isotopic composition of carbon in the oxidized form (carbonates and CO_2 gas) differs markedly from that in the reduced state (graphite, carbides, and organic compounds). Both Hoefs (1973b) and Fuex and Baker (1973) showed that the *reduced* carbon (30 to 360 ppm) is strongly enriched in ^{12}C and has $\delta^{13}C$ values between about −20 and −28‰. The *oxidized* carbon

(0 to 20,000 ppm or more) has distinctly different $\delta^{13}C$ values ranging from $+2.9$ to $-18.2‰$ relative to PDB.

The concentrations as well as the isotopic compositions of *oxidized* carbon in igneous rocks are highly variable and clearly exceed the range of $\delta^{13}C$ values of carbonatites. The carbonate minerals therefore cannot be primarily of magmatic origin. Fuex and Baker (1973) observed that the concentration of oxidized carbon correlates positively with the degree of alteration of feldspars. This relationship suggests that the oxidized carbon is of secondary origin and may have been introduced by circulating hydrothermal fluids or by groundwater. The $\delta^{13}C$ values of carbonates and CO_2 in igneous rocks may therefore reflect the isotopic compositions of carbonate species in aqueous solutions that passed through a given rock during its history after crystallization. However, some of the oxidized carbon may be of magmatic origin. Fuex and Baker (1973) observed that oxidized carbon in the interior of a camptonite dike (4 feet thick) is enriched in ^{12}C by about 1.3‰ relative to that in its chilled margin. They attributed the difference to the effects of isotope fractionation during the crystallization of the magma which implies that the carbon is of magmatic origin.

The *reduced* carbon in igneous rocks has a more uniform isotopic composition than the oxidized carbon and has $\delta^{13}C$ values that overlap those of biogenic carbon. It is possible, therefore, that the reduced carbon is also of secondary origin or that it was incorporated into magma by assimilation of sedimentary rocks containing biogenic carbon. The work of Nagy (1970) certainly supports the hypothesis that organic compounds of modern origin can be infiltrated into relatively impermeable rocks given enough time and a sufficient quantity of water. However, it is doubtful whether such a process can account for the relatively constant and appreciable concentrations of reduced carbon (about 200 ppm) in igneous rocks. Fuex and Baker (1973) drew attention to the fairly constant difference in $\delta^{13}C$ values between oxidized and reduced carbon which implies isotope fractionation. Similarly, Hoefs (1973b) referred to the possibility of inorganic synthesis of hydrocarbons by the Fischer-Tropsch process investigated by Lancet and Anders (1970) and suggested that such compounds may have been preserved in the interior of the Earth. Therefore, the ^{12}C-enrichment of reduced carbon in igneous rocks is not necessarily attributable to the presence of hydrocarbons of biogenic origin.

Volcanic gases emanating from fumaroles and cooling lava flows contain CO_2 and much smaller concentrations of CH_4 (about 1 percent of the gas). The isotopic compositions of carbon in these gases have been studied by Craig (1963), Wasserburg et al. (1963), Ferrara et al. (1963), and Hulston and McCabe (1962). In general, the results show that the CH_4 is significantly enriched in ^{12}C relative to CO_2. The $\delta^{13}C$ values of CO_2 collected in geothermal areas generally range from -2 to $-6‰$. However, CO_2 taken from cooling lava flows has $\delta^{13}C$ values between -14 and $-28‰$. Most geothermal methanes have $\delta^{13}C$ values between -20 and $-30‰$, indicating that geothermal methanes may be *enriched* in ^{13}C relative to methane derived from hydrocarbons of biogenic origin. The difference in $\delta^{13}C$ values of coexisting CO_2 and CH_4 of fumarolic gases may be due to isotope fractionation associated with the equilibrium:

$$CO_2 + 4H_2 \rightleftarrows CH_4 + 2H_2O \quad (19.4)$$

Bottinga (1969b) has calculated fractionation factors for the system $CO_2 - CH_4$ that indicate that methane is strongly enriched in ^{12}C relative to CO_2 even at elevated temperatures of up to 600°C or more (Fig. 20.3). Therefore, the $\delta^{13}C$ values of co-existing CO_2 and CH_4 gas can be used to calculate the temperature of isotope equilibration, provided that isotope equilibration actually occurred and that the gases were not contaminated subsequently. For example, Ferrara et al. (1963) analyzed mixtures of CO_2 and CH_4 from steam jets in Larderello, Tuscany (Italy), and observed the following average $\delta^{13}C$ values in 19 samples: $\delta^{13}C(CO_2) = -3.74\permil, \delta^{13}C(CH_4) = -26.74\permil$ relative to PDB. The difference is: $\Delta(CO_2 - CH_4) = 23.00\permil$ and indicates an average equilibration temperature of 258°C, based on the fractionation factors of Craig (1953) shown in Figure 20.3. This is a reasonable value similar to temperatures reported by Craig (1953) for the geysers and hot springs of Yellowstone Park and by Hulston and McCabe (1962) for the geothermal areas on the North Island of New Zealand. However, note that use of Bottinga's fractionation factors leads to a somewhat higher temperature estimate of 328°C. Nevertheless, the difference in the $\delta^{13}C$ values of CO_2 and CH_4 in fumarolic gases appears to result from isotope fractionation at elevated temperatures. The same mechanism may also play a role in controlling the isotopic composition of oxidized and reduced carbon of igneous rocks.

The CO_2, and associated CH_4, discharged by fumaroles is not necessarily juvenile, that is, derived from the upper mantle. Craig (1963) noted that in areas where marine carbonate rocks are present in subsurface, the isotopic composition of CO_2 gas is similar to that of the carbonate rocks (see also section 7 of this chapter). However, the CO_2 gas emanating from lava flows in Hawaii is probably *not* derived from marine carbonates because it is strongly enriched in ^{12}C. Moreover, limestones are not known at Steamboat Springs (Nevada), Lassen Park (California), and The Geysers (California) which discharge CO_2 gas having isotope compositions between $\delta^{13}C = -8$ and $-11\permil$.

The isotopic composition of terrestrial carbon may have to be estimated on the basis of geochemical balance calculations and knowledge of the isotopic compositions of carbon in the major reservoirs. Sedimentary carbonate rocks contain about 73 percent of the total carbon in the crust of the Earth. Most of the remaining 27 percent is in the form of fossil fuel and disseminated amorphous carbon of sedimentary rocks. The atmosphere, hydrosphere, and biosphere contain less than 0.2 percent of the total amount of carbon in the crust. The isotopic composition of carbon in carbonate rocks is close to $0\permil$ while that of biogenic carbon is $-25\permil$. Using these figures, Fuex and Baker (1973) estimated that the average $\delta^{13}C$ value of carbon in the crust of the Earth is about $-7.0\permil$. If most of the carbon that is now in the crust was derived by outgassing of the upper mantle, then carbon in the mantle should also have this composition. This estimate is compatible with the known $\delta^{13}C$ values of diamonds and carbonatites, provided that allowance is made for isotope fractionation at the high temperatures required for the formation of diamonds and the generation of carbonatite magmas. The differences in $\delta^{13}C$ values among carbonatites, diamonds of different colors, and between CO_2 and CH_4 of fumarolic gases all emphasize the importance of high-temperature fractionation of carbon isotopes between different carbon compounds.

7
Marble and Graphite

When marine carbonate rocks are subjected to contact metamorphism, the isotopic compositions of carbon as well as oxygen are changed. In general, both $\delta^{13}C$ (relative to PDB) and $\delta^{18}O$ (relative to SMOW) are lowered close to the intrusive contact. Deines and Gold (1969) studied the isotopic composition of calcite in an argillaceous limestone of Late Ordovician age (Rosemount member of the Montreal Formation of the Trenton Group) in a contact metamorphic aureole caused by the intrusion of the Mount Royal Pluton (Early Cretaceous) in Montreal, Quebec. The rocks in the contact zone, which is several hundred feet wide, show the effects of metamorphism by bleaching, by an increase in the grainsize of calcite, and by the presence of calc-silicate minerals near the contact. Deines and Gold (1969) found that both $\delta^{13}C$ and $\delta^{18}O$ values of calcite decrease as follows:

	$\delta^{13}C‰$	$\delta^{18}O‰$
Unaltered limestone	+1.13	+24.12
Marble, contact aureole	−0.58	+18.4
Calcite within intrusive	−3.91	+11.06

They concluded that the isotopic composition of both carbon and oxygen had been altered to the point that the metamorphosed calcite could no longer be distinguished from igneous calcite in carbonatites.

Shieh and Taylor (1969) reported similar results from a contact between the Inyo Batholith and Cambrian sedimentary rocks near Deep Springs Valley, Inyo County, California. They found that dolomite coexisting with calcite is consistently enriched in ^{13}C and that the $\delta^{13}C$ values of both decrease irregularly over a distance of about 200 feet as the contact is approached. They also noted that marbles containing calc-silicates appear to be depleted in ^{13}C and that calcite in skarn has the lowest $\delta^{13}C$ values ($-3.0‰$). Shieh and Taylor (1969)

attributed the decrease in $\delta^{13}C$ values of the carbonate minerals to decarbonation reactions and estimated that the CO_2 released during this process was enriched in ^{13}C by about 6‰ relative to calcite. Some representative decarbonation reactions are

$$\text{Calcite} + \text{Quartz} \rightarrow \\ \text{Wollastonite} + CO_2 \quad (20.5)$$

$$\text{2 Dolomite} + \text{Quartz} \rightarrow \\ \text{2 Calcite} + \text{Forsterite} + 2CO_2 \quad (20.6)$$

$$\text{3 Dolomite} + \text{K-feldspar} + \text{Water} \rightarrow \\ \text{Phlogopite} + \text{3 Calcite} + 3CO_2 \quad (20.7)$$

$$\text{5 Dolomite} + \text{8 Quartz} + \text{Water} \rightarrow \\ \text{3 Calcite} + \text{Tremolite} + 7CO_2 \quad (20.8)$$

The preferential *enrichment in* ^{13}C of CO_2 gas produced during decarbonation reactions, if true, presents a problem because fumarolic CO_2 has *negative* $\delta^{13}C$ values even in areas where marine sedimentary rocks exist in subsurface. In addition, Bottinga (1969b) has shown that CO_2 gas in equilibrium with calcite at temperatures above about 200°C is enriched in ^{13}C by *only* 2.8‰ (Fig. 20.3). Therefore, decarbonation reactions may not be the dominant source of fumarolic CO_2. On the other hand, the ^{13}C-enrichment inferred by Shieh and Taylor (1969) for CO_2 evolved during decarbonation reactions is *greater than predicted* by theoretical calculations of Bottinga (1969b). The importance of these discrepancies depends on the question to what extent $\delta^{13}C$ values of carbonate rocks are lowered by increasing grades of metamorphism. The evidence compiled by Shieh and Taylor (1969), based in part on studies by Sheppard and Schwarcz (1968) and Schwarcz (1966), does not support the conclusion that $\delta^{13}C$ values of carbonate minerals correlate with metamorphic grade. The source of fumarolic CO_2 and the isotope fractionation of carbon by decarbonation reactions are as yet unresolved.

Graphite occurs not only in metamorphosed sedimentary rocks but also in some igneous rocks. It is generally assumed that graphite is produced from organic compounds by progressive loss of hydrogen leading to the recrystallization of the residual atomic carbon as graphite. Analyses by Craig (1953, 1954), Landergren (1961), and by Hahn-Weinheimer (1966) indicate that graphites in metamorphosed sedimentary rocks have $\delta^{13}C$ values that are compatible with a biogenic origin of the carbon. However, "heavy" graphites having $\delta^{13}C$ values from -2.0 to $-8.0‰$ have been found in meteorites, ultramafic igneous rocks, and skarn. It is doubtful that these graphites could have been produced directly from biogenic carbon compounds. Hoefs (1973b) suggested that an abiogenic origin of such heavy graphite may be possible by reduction of carbonates or CO_2 with molecular hydrogen. According to Giardini and Salotti (1969), carbonate minerals can be reduced by H_2 in the temperature range 200 to 600°C according to the following simplified reaction:

$$CaCO_3 + 4H_2 \rightarrow CH_4 + H_2O + Ca(OH)_2$$

$$(20.9)$$

The methane produced in this way may then decompose at elevated temperature to form graphite and H_2.

Regardless of its origin, graphite and calcite commonly occur together in marbles and similar rocks and may be in isotopic equilibrium with each other. Bottinga (1969b) calculated the fractionation factors for carbon isotopes in this system (Fig. 20.3). The results suggest that graphite is depleted in ^{13}C relative to calcite by up to 18‰ between 100° and 700°C. However, the temperatures calculated from analyses of

Landergren (1961) and Hahn-Weinheimer (1966) scatter widely, indicating that isotopic equilibrium between graphite and calcite was not reached in many samples.

8
Hydrothermal Ore Deposits

The carbon content of hydrothermal ore deposits is represented primarily by carbonates of calcium, magnesium, iron, and manganese and by CO_2 and CH_4 gas in fluid and gaseous inclusions in ore and gangue minerals. The principal sources of carbon in hydrothermal fluids include marine limestones ($\delta^{13}C \simeq 0‰$), deep-seated or average crustal sources ($\delta^{13}C \simeq -7‰$), and organic compounds of biogenic origin ($\delta^{13}C \simeq -25‰$). The $\delta^{13}C$ values of CO_2 in fluid inclusions range from -4 to $-12‰$ with respect to PDB. Most early-formed carbonate minerals in hydrothermal veins have $\delta^{13}C$ values in the range from -6 to $-9‰$. However, carbonates deposited late in the paragenetic sequence are often enriched in ^{13}C compared to early carbonates and may even have positive $\delta^{13}C$ values, as for example the Pine Point deposits, Northwest Territories, Canada (Fritz, 1969). The low $\delta^{13}C$ values of CO_2 gas and carbonate minerals in hydrothermal ore deposits generally suggest a deep-seated source for the carbon.

The isotopic composition of carbon in hydrothermal carbonates depends not only on the $\delta^{13}C$ of the total carbon in the ore-bearing fluid but also on the fugacity of oxygen, the pH, the temperature, the ionic strength of the fluid, and on the total concentration of carbon. These relationships have been investigated by Ohmoto (1972) and were reviewed by Rye and Ohmoto (1974). The dependence arises from the fact that various physical and chemical parameters control the concentrations of different carbon-bearing ions and molecules which fractionate carbon isotopes depending on the temperature. A change in any one of the above physical or chemical parameters affects the chemical equilibria by which the ions and molecules are related and thereby also changes their isotopic compositions.

The principal carbon species that may become important in a hydrothermal fluid below about 600°C are: $CO_2(aq)$, H_2CO_3, HCO_3^-, CO_3^{-2}, and $CH_4(aq)$. The concentrations of the carbonate species are directly controlled by the hydrogen ion activity (pH) and indirectly by the ionic strength and temperature of the fluid. The ionic strength (I) affects their activity coefficients, while the temperature determines the magnitudes of the equilibrium constants. At low pH (less than about 6), H_2CO_3 is dominant over HCO_3^-, while CO_3^{-2} is negligible. As the pH increases, HCO_3^- becomes dominant up to pH $\simeq$ 12. At still higher values, in strongly basic solutions, the carbonate ion dominates. The fugacity of oxygen affects the oxidation states of the carbon species. If the oxygen fugacity is about 10^{-38} atmospheres, most of the carbon is oxidized and $CH_4(aq)$ is negligible. At lower oxygen fugacities $CH_4(aq)$ rapidly increases in abundance.

Changes in the abundance of $CH_4(aq)$ have a dramatic effect on the $\delta^{13}C$ values of coexisting carbonate species because CH_4 is strongly enriched in ^{12}C (Fig. 20.3, $CO_2 - CH_4$ line). For example, the $\delta^{13}C$ value of $CO_2(aq) + H_2CO_3$ (At pH = 6, T = 250°C, I = 1.0, $\delta^{13}C_{\Sigma C} = 0.0\%_0$) increases from about $0\%_0$ at $fO_2 = 10^{-36}$ atmospheres to $+29\%_0$ at $fO_2 = 10^{-41}$ atmospheres (Ohmoto, 1972, Fig. 11). The enrichment of the oxidized carbon species in ^{13}C is due to the preferential partitioning of ^{12}C into $CH_4(aq)$ which becomes important at low oxygen fugacity. Similarly, the $\delta^{13}C$ value of (CO_2 aq + H_2CO_3) at a fixed value of fO_2 varies with pH because of the changing abundances of HCO_3^- and CO_3^{-2}. Thus, at $fO_2 = 10^{-39}$ the value of $\delta^{13}C$ for CO_2 aq + H_2CO_3 is about $+22\%_0$ from pH = 2 to 6 and then decreases to about $+4\%_0$ as the pH rises from 6 to 12.

Evidently, both pH and fO_2 may have a pronounced effect on the $\delta^{13}C$ values of carbon-bearing ions and molecules that can coexist in equilibrium in a hydrothermal fluid. The isotopic compositions of carbon of solid phases (calcite and graphite) that coexist in chemical and isotopic equilibrium with these ions and molecules will likewise be affected by the pH and fO_2 of the environment. It follows that the $\delta^{13}C$ values of calcite and graphite precipitated from a hydrothermal fluid depend on the pH and fO_2 of the fluid and can vary widely in response to changes in these parameters. This point is illustrated in Figure 20.4 which shows $\delta^{13}C$ contours of calcite and graphite formed at 250°C from a hydrothermal fluid containing 1 mole of C per kilogram of H_2O having $\delta^{13}C_{\Sigma C} = -5\%_0$ and an ionic strength of 1.0. Two distinctive sets of circumstances can be recognized with respect to the isotopic composition of calcite: (1) at high oxygen fugacity ($fO_2 > 10^{-37}$ atmospheres), the $\delta^{13}C$ value of calcite is independent of fO_2 and varies only as a function of increasing pH; (2) at oxygen fugacities less than 10^{-37} atmosphere, $\delta^{13}C$ of calcite decreases with increasing pH and increases with decreasing fO_2. The $\delta^{13}C$ values of graphite depend strongly on fO_2 and become pH-dependent only above pH $\sim$ 7. However, these considerations apply only to calcite and graphite that may form in a closed system in which isotopic equilibrium is maintained among all phases. As soon as solid phases appear, carbon is removed from the hydrothermal fluid, and the isotopic composition of the remaining carbon is thereby changed. This will be reflected by the isotopic composition of solid phases that

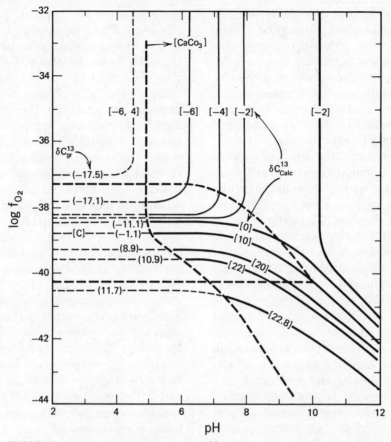

FIGURE 20.4 Possible variation of $\delta^{13}C$ values of calcite and graphite as functions of pH and the fugacity of oxygen. The diagram is based on the assumption that isotopic equilibrium is maintained and that the removal of solid phases does not affect the isotopic composition of the carbon remaining in solution. Temperature = 250°C, ionic strength = 1.0, $\delta^{13}C$ (total carbon) = −5 percent, total carbon concentration = 1 mole/kg H_2O. (Redrawn from H. Ohmoto, Economic Geology, 1972, vol. 67, No. 5, 551–578, Figure 14, with permission for the use of the diagram by H. Ohmoto.)

may form subsequently. It is not clear to what extent hydrothermal systems in nature approach the ideal case shown in Figure 20.4.

Nevertheless, the important point to be made here is that the $\delta^{13}C$ values of hydrothermal calcite may depend not only on the isotopic composition of the total carbon in the system but also on the hydrogen ion activity and fugacity of oxygen. The isotopic composition of hydrothermal calcite by itself does not enable us to specify both pH and fO_2 at the time of deposition. However, the isotopic composition of sulfur of coexisting sulfide minerals is also related to these parameters. Therefore, both pH and fO_2 may be specified from a consideration of the isotopic compositions of carbon and sulfur in cogenetic minerals of hydrothermal ore deposits.

9 SUMMARY

Carbon is one of the most abundant elements in the universe and is the basis of life on Earth. Its isotopes are fractionated by inorganic exchange reactions and during photosynthesis in green plants. Carbonates are generally enriched in ^{13}C, while organic matter in plants and animals is enriched in ^{12}C. Fossil fuels are likewise enriched in ^{12}C consistent with their derivation from biogenic matter. Combustion of fossil fuels during the past century has not only increased the CO_2 content of the atmosphere by about 10 percent but has also increased the abundance of ^{12}C. Reduced carbon in sedimentary rocks of Early Precambrian is enriched in ^{12}C indicative of a biogenic origin. The ^{12}C content of such carbon may be lowered by loss of CH_4 during incipient stages of low-grade regional metamorphism.

Carbon isotopes are fractionated in the system CO_2 (gas)-carbonate ions (aqueous)-$CaCO_3$ (solid) in such a way that calcite is enriched in ^{13}C by about 10‰ at 20°C relative to CO_2 gas. The $\delta^{13}C$ values of marine carbonates are close to zero relative to PDB and do not vary appreciably with age. Lacustrine carbonates are somewhat enriched in ^{12}C due to incorporation of CO_2 derived by decay of plant debris in soil. Carbonates in carbonaceous chondrites are strongly enriched in ^{13}C, while terrestrial carbonates formed by oxidation of bacteriogenic CH_4 are strongly enriched in ^{12}C. Carbonates in sedimentary rocks of Precambrian age are generally similar isotopically to those of Phanerozoic age.

The isotopic composition of carbon in the mantle of the Earth is difficult to determine because isotope fractionation effects are important even at very high temperatures. The $\delta^{13}C$ values of carbonatites and diamonds exhibit significant variations either because of isotopic heterogeneity of carbon in the mantle or because of isotope fractionation, or both.

Oxidized carbon in igneous rocks has variable $\delta^{13}C$ values and may have been deposited to a large extent by circulating aqueous solutions after crystallization. The isotopic composition of reduced carbon in igneous rocks overlaps that of biogenic carbon, but an inorganic origin for the reduced carbon is possible. Volcanic gases contain CO_2 and lesser concentrations of CH_4 which appear to be in isotopic equilibrium and permit temperature determinations. The methane may be produced by reduction of CO_2 with hydrogen. The CO_2 is probably derived from crustal rocks including limestones in subsurface. Geochemical balance calculations indicate a $\delta^{13}C$ value of $-7.0‰$ for average crustal carbon.

Contact metamorphism of carbonate rocks leads to a decrease in $\delta^{13}C$ values and implies that CO_2 gas liberated by decarbonation reactions is enriched in ^{13}C. However, $\delta^{13}C$ values of marbles generally do not vary systematically with increasing grade of regional metamorphism. Graphite is variably enriched in ^{12}C compatible with a biogenic source for the carbon. However, some graphites containing "heavy" carbon may be abiogenic in origin. Coexisting calcite and graphite are not in isotopic equilibrium in many cases.

The isotopic compositions of carbonate minerals and CO_2 gas in fluid inclusions in hydrothermal ore deposits are variable but suggest a deep-seated source. The $\delta^{13}C$ values of calcite and graphite deposited in a closed system in isotopic equilibrium may be strongly affected by the pH and the fugacity of oxygen in the hydrothermal fluid at the time of deposition. Both pH and fO_2 can be specified by consideration of the isotopic compositions of carbon and sulfur in cogenetic carbonate and sulfide minerals in hydrothermal ore deposits.

PROBLEMS

1. The carbon isotope fractionation factor at 200°C between CO_2 and diamond is 1.0115 and that between diamond and graphite is 1.0029. What is the value of the fractionation factor between CO_2 and graphite at 200°C and what would be the value of $\delta^{13}C$ for CO_2 in equilibrium with graphite whose $\delta^{13}C$ value is $-15.00\permil$?

2. According to Bottinga (1969b), the carbon isotope fractionation factor between CO_2 and CH_4 is 1.012 at 600°C and 1.025 at 300°C. Use this information to derive a carbon isotope thermometry equation of the form: $\Delta(CO_2-CH_4) = A \times 10^6 T^{-2} + B$, where T is the temperature in degrees Kelvin.

3. If the $\delta^{13}C$ of atmospheric CO_2 were to change from $-7.00\permil$ to $-10\permil$ due to increased combustion of fossil fuel, what effect would this have on $\delta^{13}C$ values of calcite precipitated at 20°C in isotopic equilibrium with atmospheric CO_2 given that α(calcite—CO_2) is 1.0102?

4. According to Hulston and McCabe (1962), the isotopic compositions of carbon of gases emitted by the Seven Dwarfs fumarole on White Island, New Zealand, are as follows: $CO_2 : \delta^{13}C = -4.5\permil$; $CH_4 : \delta^{13}C = -23.3\permil$. Assume that the carbon is in isotopic equilibrium and calculate the equilibration temperature. Make use of the fractionation factors for the system CH_4-CO_2 (Craig, 1953): $T = 298.1°K$, $\alpha(CH_4-CO_2) = 0.943$; $T = 500°K$, $\alpha(CH_4-CO_2) = 0.975$. (**ANSWER:** 250°C.)

5. The average $\delta^{13}C$ value of calcite in cap rock of salt domes from the Gulf Coast region is $-36.2\permil$ (Cheney and Jensen, 1967). Contrast its isotopic composition with that of marine limestones and suggest an explanation for the enrichment in ^{12}C of this calcite.

REFERENCES

Abelson, P. H., and T. C. Hoering (1961) Carbon isotope fractionation in formation of amino acids by photosynthetic organisms. Proc. Nat. Acad. Sci., *47*, No. 5, 623–632.

Barghoorn, E. S., and J. W. Schopf (1966) Microorganisms three billion years old from the Precambrian of South Africa. Science, *152*, 758–763.

Barker, F., and I. Friedman (1969) Carbon isotopes in pelites of the Precambrian Uncompahgre formation, Needle Mountains, Colorado. Geol. Soc. Amer. Bull., *80*, 1403–1408.

Becker, R. H., and R. N. Clayton (1972) Carbon isotopic evidence for the origin of a banded iron-formation in Western Australia. Geochim. Cosmochim. Acta, *36*, 577–595.

Belsky, T., and I. R. Kaplan (1970) Light hydrocarbon gases, C^{13}, and origin of organic matter in carbonaceous chondrites. Geochim. Cosmochim. Acta, *34*, 257–278.

Bottinga, Y. (1969a) Carbon isotope fractionation between graphite, diamond and carbon dioxide. Earth Planet. Sci. Letters, *5*, 301–307.

Bottinga, Y. (1969b) Calculated fractionation factors for carbon and hydrogen isotope exchange in the system calcite-carbon dioxide-graphite, methane-hydrogen-water vapor. Geochim. Cosmochim. Acta, *33*, 49–64.

Broecker, W. S. (1975) Climatic change: Are we on the brink of a pronounced global warming? Science, *189*, 460–463.

Cheney, E. S., and M. L. Jensen (1965) Stable carbon isotopic composition of biogenic carbonates. Geochim. Cosmochim. Acta, *29*, 1331–1346.

Cheney, E. S., and M. L. Jensen (1967) Corrections to carbon isotopic data of Gulf Coast salt-dome cap rock. Geochim. Cosmochim. Acta, *31*, 1345–1346.

Choquette, P. W. (1968) Marine diagenesis of shallow marine lime-mud sediments: Insights from δO^{18} and δC^{13} data. Science, *161*, 1130–1132.

Clayton, R. N. (1963) Carbon isotope abundance in meteoritic carbonates. Science, *140*, 192–193.

Colombo, U., F. Gazzarrini, R. Gonfiantini, G. Sironi, and E. Tongiorgi (1966) Measurements of C^{13}/C^{12} isotope ratios on Italian natural gases and their geochemical interpretation. In Advances in Organic Geochemistry pp. 279–292. Pergamon Press, Oxford.

Conway, C. M., and H. P. Taylor, Jr. (1969) O^{18}/O^{16} and C^{13}/C^{12} ratios of coexisting minerals in the Oka and Magnet Cove carbonatite bodies. J. Geol., *77*, 618–626.

Craig, H. (1953) The geochemistry of the stable carbon isotopes. Geochim. Cosmochim. Acta, *3*, 53–92.

Craig, H. (1954) Geochemical implications of the isotopic composition of carbon in ancient rocks. Geochim. Cosmochim. Acta., *6*, 186.

Craig, H. (1957) Isotopic standards for carbon and oxygen and correction factors for mass spectrometric analysis of carbon dioxide. Geochim. Cosmochim. Acta, *12*, 133–149.

Craig, H. (1963) The isotopic geochemistry of water and carbon in geothermal areas. In Nuclear Geology of Geothermal Areas, pp. 17–53. Spoleto, Sept. 9–13, 1963.

Degens, E. T., M. Behrendt, B. Gotthardt, and E. Reppmann (1968) Metabolic fractionation of carbon isotopes in marine plankton-II. Data on samples collected off the coasts of Peru and Ecuador. Deep Sea Res., *15*, 11–20.

Degens, E. T. (1969) Biogeochemistry of stable carbon isotopes. In Organic Geochemistry; Methods and Results. pp. 304–329, G. Eglinton and M. T. J. Murphy, eds. Springer-Verlag, New York.

Deines, P., and D. P. Gold (1969) The change in carbon and oxygen isotopic composition during contact metamorphism of Trenton limestone by the Mount Royal pluton. Geochim. Cosmochim. Acta, *33*, 421–424.

Deines, P., and D. P. Gold (1973) The isotopic composition of carbonatite and kimberlite carbonates and their bearing on the isotopic composition of deep-seated carbon. Geochim. Cosmochim. Acta, *37*, 1709–1733.

Deines, P., D. Langmuir, and R. S. Harmon (1974) Stable carbon isotope ratios and the existence of a gas phase in the evolution of carbonate ground water. Geochim. Cosmochim. Acta, *38*, 1147–1164.

Deuser, W. G., and E. T. Degens (1967) Carbon isotope fractionation in the system $CO_2(gas)—CO_2(aq.)—HCO_3^-$ (aq.). Nature, *215*, 1033.

Eckelman, W. R., W. S. Broecker, D. W. Whitlock, and J. R. Allsup (1962) Implications of carbon isotopic composition of total organic carbon of some recent sediments and ancient oils. Bull. Amer. Ass. Petrol. Geol., *46*, 699–704.

Emrich, K., D. H. Ehhalt, and J. C. Vogel (1970) Carbon isotope fractionation during the precipitation of calcium carbonate. Earth Planet. Sci. Letters, *8*, 363–371.

Engel, A. E. J., B. Nagy, L. A. Nagy, C. G. Engel, G. O. W. Kemp, and C. M. Drew (1968) Alga-like forms in Onverwacht Series, South Africa: Oldest recognized life-like forms on Earth. Science, *183*, 1005–1008.

Erlenkeuser, H., E. Suess, and H. Willkomm (1974) Industrialization affects heavy metal and carbon isotope concentrations in recent Baltic Sea sediments. Geochim. Cosmochim. Acta, *38*, 823–842.

Farmer, J. G., and M. S. Baxter (1974) Atmospheric carbon dioxide levels as indicated by the stable isotope record in wood. Nature, *247*, 273–274.

Ferrara, G. C., G. Ferrara, and R. Gonfiantini (1963) Carbon isotopic composition of carbon dioxide and methane from steam jets of Tuscany. In Nuclear Geology of Geothermal Areas, pp. 275–282. Spoleto, Sept. 9–13, 1963.

Fritz, P. (1969) The oxygen and carbon isotopic composition of carbonates from the Pine Point lead-zinc ore deposits. Econ. Geol., *64*, 733–742.

Fuex, A. N., and D. R. Baker (1973) Stable carbon isotopes in selected granitic, mafic and ultramafic rocks. Geochim. Cosmochim. Acta, *37*, 2509–2521.

Galimov, E. M., V. A. Kononova, and V. S. Prokhorov (1975) Carbon isotopic composition in carbonatites and carbonatite-like rocks (in connection with the source of carbonatite material). Geochemistry International, *11*, 503–510.

Giardini, A. A., and C. A. Salotti (1969) Kinetics and relations in the calcite-hydrogen reaction and relations in the dolomite-hydrogen and siderite-hydrogen systems. Amer. Mineral., *54*, 1151–1172.

Hahn-Weinheimer, P. (1966) Die isotopische Verteilung von Kohlenstoff und Schwefel in Marmor und anderen Metamorphiten. Geol. Rundschau, *55*, 197–208.

Hathaway, J. C., and E. T. Degens (1969) Methane-derived marine carbonates of Pleistocene age. Science, *165*, 609–692.

Hoefs, J., and M. Schidlowski (1967) Carbon isotope composition of carbonaceous matter from the Precambrian of the Witwatersrand System. Science, *155*, 1096–1097.

Hoefs, J. (1973a) Stable Isotope Geochemistry. Springer-Verlag, New York, 140 p.

Hoefs, J. (1973b) Ein Beitrag zur Isotopengeochemie des Kohlenstoffs in magmatischen Gesteinen. Contrib. Mineral. Petrol., *41*, 277–300.

Hoering, T. C. (1967) The organic geochemistry of Precambrian rocks. In Researches in Geochemistry, vol. 2, pp. 89–111, P. H. Abelson, ed., John Wiley, New York.

Hulston, J. R., and W. J. McCabe (1962) Mass spectrometer measurements in the thermal areas of New Zealand. Part 2. Carbon isotopic ratios. Geochim. Cosmochim. Acta, *26*, 399–410.

Junge, C., M. Schidlowski, R. Eichmann, and H. Pietrek (1975) Model calculations for the terrestrial carbon cycle: Carbon isotope geochemistry and evolution of photosynthetic oxygen. J. Geophys. Res., *80*, 4542–4552.

Keeling, C. D. (1958) The concentration and isotopic abundance of carbon dioxide in rural areas. Geochim. Cosmochim. Acta, *13*, 322–334.

Keeling, C. D. (1961) The concentration and isotopic abundances of carbon dioxide in rural and marine air. Geochim. Cosmochim. Acta, *24*, 277–298.

Keith, M. L., G. M. Anderson, and R. Eichler (1964) Carbon and oxygen isotopic composition of mollusk shells from marine and fresh-water environments. Geochim. Cosmochim. Acta, *28*, 1757–1786.

Keith, M. L., and J. N. Weber (1964) Isotopic composition and environmental classification of selected limestones and fossils. Geochim. Cosmochim. Acta, *28*, 1787–1816.

Koval'skiy, V. V., and N. V. Cherskiy (1973) Possible sources and isotopic composition of carbon in diamonds. Int. Geol. Rev., *15*, 1224–1228.

Lancet, M. S., and E. Anders (1970) Carbon isotope fractionation in the Fischer-Tropsch synthesis and in meteorites. Science, *170*, 980–982.

Landergren, S. (1961) The content of ^{13}C in the graphite-bearing magnetite ores and associated carbonate rocks in the Norberg mining district, Central Sweden. Geol. Foren. Stockholm Forh., *83*, 151–156.

Lowdon, J. A., and W. Dyck (1974) Seasonal variations in the isotope ratios of carbon in maple leaves and other plants. Can. J. Earth Sci., *11*, 79–88.

McKirdy, D. M., and T. G. Powell (1974) Metamorphic alteration of carbon isotopic composition in ancient sedimentary organic matter: New evidence from Australia and South Africa. Geol., *2*, 591–596.

Mitchell, R. H., and J. H. Crocket (1971) Diamond genesis—a synthesis of opposing views. Mineral. Deposits, *6*, 392–403.

Nagy, B. (1970) Porosity and permeability of the early Precambrian Onverwacht chert: Origin of the hydrocarbon content. Geochim. Cosmochim. Acta, *34*, 525–527.

Oehler, D. Z., J. W. Schopf, and K. A. Kvenvolden (1972) Carbon isotopic studies of organic matter in Precambrian rocks. Science, *175*, 1246–1248.

Ohmoto, H. (1972) Systematics of sulfur and carbon isotopes in hydrothermal ore deposits. Econ. Geol., *67*, 551–578.

Oró, J., D. W. Nooner, and A. Zlatkis (1965) Hydrocarbons of biological origin in sediments about two billion years old. Science, *148*, 77–79.

Park, R., and S. Epstein (1960) Carbon isotope fractionation during photosynthesis. Geochim. Cosmochim. Acta, *21*, 110–126.

Parker, P. L. (1964) The biogeochemistry of the stable isotopes of carbon in a marine Bay. Geochim. Cosmochim. Acta, *28*, 1155–1164.

Pineau, F., M. Javoy, and C. J. Allegre (1973) Etude systematique des isotopes de l' oxygene, du carbone et du strontium dans les carbonatites. Geochim. Cosmochim. Acta, *37*, 2363–2377.

Rankama, K. (1963) Progress in Isotope Geology. Interscience, New York, 705 p.

Rosenfeld, W. D., and S. R. Silverman (1959) Carbon isotope fractionation in bacterial production of methane. Science, *130*, 1658–1658.

Rye, R. O., and H. Ohmoto (1974) Sulfur and carbon isotopes and ore genesis. A review. Econ. Geol., *69*, 826–842.

Sackett, W. M. (1968) Carbon isotope composition of natural methane occurrences. Amer. Assoc. Petrol. Geol. Bull., *52*, 853–857.

Sackett, W. M., and R. R. Thompson (1963) Isotopic organic carbon composition of recent continental derived clastic sediments of eastern gulf coast, Gulf of Mexico. Bull. Amer. Ass. Petrol. Geol., *47*, 525–531.

Schidlowski, M., R. Eichmann, and C. Junge (1975) Precambrian sedimentary carbonates: Carbon and oxygen isotope geochemistry and implications for the terrestrial oxygen budget. Precambrian Res., *2*, 1–69.

Schidlowski, M., R. Eichmann, and C. E. Junge (1976) Carbon isotope geochemistry of the Precambrian Lomagundi carbonate province, Rhodesia. Geochim. Cosmochim. Acta, *40*, 449–455.

Schopf, J. W. (1968) Microflora of the Bitter Springs Formation, Late Precambrian, Central Australia. J. Paleontology, *42*, 651–688.

Schopf, J. W. (1970) Precambrian micro-organisms and evolutionary events prior to the origin of vascular plants. Biol. Rev., *45*, 319–352.

Schwarcz, H. P. (1966) Oxygen and carbon isotopic fractionation between coexisting metamorphic calcite and dolomite. J. Geol., *74*, 38–48.

Schwarcz, H. P. (1969) The stable isotopes of carbon. In Handbook of Geochemistry, vol II/1, K. H. Wedepohl, ed. Springer-Verlag, New York.

Sheppard, S. M. F., and H. P. Schwarcz (1970) Fractionation of carbon and oxygen isotopes and magnesium between coexisting metamorphic calcite and dolomite. Contr. Mineral. Petrol., *26*, 161–198.

Shieh, Y. N., and H. P. Taylor (1969) Oxygen and carbon isotope studies of contact metamorphism of carbonate rocks. J. Petrol., *10*, 307–331.

Shultz, D. J., and J. A. Calder (1976) Organic carbon $^{13}C/^{12}C$ variations in estuarine sediments. Geochim. Cosmochim. Acta, *40*, 381–386.

Silverman, S. R. (1964) Investigations of petroleum origin and evolution mechanisms by carbon isotope studies. In Isotopic and Cosmic Chemistry, pp. 92–102, H. Craig, S. L. Miller, and G. J. Wasserburg, eds. North-Holland, Amsterdam.

Smith, B. N., and S. Epstein (1971) Two categories of $^{13}C/^{12}C$ ratios for higher plants. Plant Physiol., *47*, 380–384.

Vogel, J. C., P. M. Grootes, and W. G. Mook (1970) Isotopic fractionation between gaseous and dissolved carbon dioxide. Z. Phys., *230*, 225–238.

Wasserburg, G. J., E. Mazor, and R. E. Zartman (1963) Isotopic and chemical composition of some terrestrial natural gases. In Earth Science and Meteoritics, pp. 219–240, J. Geiss and E. D. Goldberg, eds. North-Holland, Amsterdam.

Wendt, I. (1968) Fractionation of carbon isotopes and its temperature dependence in the system CO_2—gas—CO_2 in solution and HCO_3^-—CO_2 in solution. Earth Planet. Sci. Letters, *4*, 64–68.

Whelan, T., W. M. Sackett, and C. R. Benedict (1973) Enzymatic fractionation of carbon isotopes by phosphoenopyruvate carboxylase from C_4 plants. Plant Physiol., *51*, 1051–1054.

21 SULFUR

Sulfur ($Z = 16$) has four stable isotopes whose approximate abundances are: $^{32}S = 95.02$ percent, $^{33}S = 0.75$ percent, $^{34}S = 4.21$ percent, $^{36}S = 0.02$ percent. Its isotopic composition is expressed in terms of $\delta^{34}S$, which is defined as

$$\delta^{34}S = \left[\frac{\left(\frac{^{34}S}{^{32}S}\right)_{spl} - \left(\frac{^{34}S}{^{32}S}\right)_{std}}{\left(\frac{^{34}S}{^{32}S}\right)_{std}} \right] \times 10^3\text{\textperthousand} \quad (21.1)$$

The standard is the sulfur in troilite (FeS) of the iron meteorite Canyon Diablo whose $^{32}S/^{34}S$ ratio is 22.22. This is an appropriate standard because the isotopic composition of sulfur in mafic igneous rocks is very similar to that of meteorites (Smitheringale and Jensen, 1963; Schneider, 1970). Consequently, the $\delta^{34}S$ value of a given sample of terrestrial sulfur can be taken as a measure of the change that has taken place in its isotopic composition since its initial introduction into the crust of the Earth.

Variations in the isotopic composition of sulfur were first reported by Thode et al. (1949) and are caused by two kinds of processes: (1) reduction of sulfate ions to hydrogen sulfide by certain anaerobic bacteria which results in the enrichment of hydrogen sulfide in ^{32}S; (2) various isotopic exchange reactions between sulfur-bearing ions, molecules, and solids by which ^{34}S is generally concentrated in compounds having the highest oxidation state of sulfur, or greatest bond strength (Bachinski, 1969).

Sulfur is widely distributed in the lithosphere, biosphere, hydrosphere, and atmosphere of the Earth. It occurs in the oxidized form as sulfate in the oceans and in evaporite rocks. It is found in the native state in the cap rock of salt domes and in the rocks of certain volcanic regions. Sulfur also occurs in the reduced form as sulfide in metallic mineral deposits associated with igneous, sedimentary and metamorphic rocks. For this reason the isotopic composition of sulfur is especially useful in the study of sulfide ore deposits.

The isotopic composition of sulfur is commonly measured on SO_2 gas using mass spectrometers equipped with double collectors. Sulfides are converted to SO_2 by reactions with CuO, V_2O_5, or O_2 at temperatures of up to 1000°C. Sulfates are first reduced to sulfide which is precipitated as CdS and then converted to SO_2 gas. Details of these procedures have been discussed by Ricke (1964), Holt and Engelkemeir (1970), and Puchelt et al. (1971). The isotope geochemistry of sulfur has been reviewed by Ault and Kulp (1959), Thode et al. (1961), Thode (1963, 1970), Jensen (1967), and Hoefs (1973).

1
Biogenic Fractionation

The most important cause for variations in the isotopic composition of sulfur in nature is the reduction of sulfate ions by anaerobic bacteria such as *Desulfovibrio desulfuricans* which live in sediment deposited in the oceans and in lakes. These bacteria split oxygen from sulfate ions and excrete H_2S which is enriched in ^{32}S relative to the sulfate. The extent of the fractionation is variable and depends on certain rate-controlling steps in the reactions by which the sulfur is metabolized. In inorganic systems, the extent of isotope fractionation during the reduction of sulfate ion to hydrogen sulfide is due to the different rates at which S—O bonds are broken. Harrison and Thode (1957) demonstrated experimentally that ^{32}S—O bonds are broken more easily than ^{34}S—O bonds. As a result, the first H$_2$S produced by inorganic reduction of SO$_4^=$ is enriched in ^{32}S by about 22‰ compared to the sulfate. Harrison and Thode (1958a) therefore explained the fractionation associated with bacterial reduction of sulfate in terms of a two-stage process consisting of (1) entrance of sulfate into the cell yielding a small isotope shift; and (2) breaking of S—O bonds causing a large change (up to 22‰) in the isotope composition and controlling the rate of the process. This model appeared to be capable of explaining fractionation effects of up to 27‰ obtained in the laboratory with bacterial cultures.

However, Kaplan and Rittenberg (1964) obtained ^{32}S enrichments in H$_2$S of up to 46‰ relative to sulfate. Moreover, sulfide minerals in recently deposited sediment may be enriched in ^{32}S by about 50‰ compared to associated marine sulfate. Evidently, isotope fractionation in nature is more extensive than suggested by most laboratory experiments and is not limited by the fractionation associated with chemical reduction of sulfate ions to hydrogen sulfide. In this connection it is interesting to note that Tudge and Thode (1950) calculated a fractionation factor $\alpha = 1.075$ at 25°C for the chemical equilibrium:

$$^{32}SO_{4(aq)}^= + H_2{}^{34}S_{(g)} \rightleftarrows {}^{34}SO_{4(aq)}^= + H_2{}^{32}S_{(g)}$$
$$(21.2)$$

This equilibrium is unattainable by chemical systems at low temperature but may be approached by enzyme-catalyzed sulfate reduction in bacteria (Trudinger and Chambers, 1973). This exchange equilibrium is capable of causing enrichment of H$_2$S in ^{32}S of up to 75‰ at 25°C relative to the sulfate.

The degree of isotope fractionation by bacteria is known to be inversely proportional to the rate of sulfate reduction, which in turn is controlled by either temperature or sulfate concentration or both. However, Kaplan and Rittenberg (1964) found that the rate of reduction and the magnitude of the isotopic fractionation also depend on the nature of the electron donor. They observed the largest fractionation effect with ethanol as electron donor. Kemp and Thode (1968) were unable to obtain fractionation factors greater than 1.025 under similar conditions, and therefore attributed Kaplan and Rittenberg's results to some unknown alteration in the pathway of sulfate reduction during bacterial metabolism.

Pathways for metabolism of inorganic sulfur by microorganisms have been reviewed by McCready et al. (1974). In general, their conclusion was that the degree of isotope fractionation is variable because the total observed effect consists of several additive steps. Breaking of S—O bonds plays an important role as does the exchange reaction between sulfite and bisulfide ion. The total observed effect depends on the extent to which the individual steps are controlled by the physiology of the cell which in turn depends on the substrate. Rees (1973) developed a mathematical model based on the recognition that the overall effect results from the transfer of sulfur among several reservoirs within the cell and that the extent of fractionation is controlled by the rates of reactions by which these reservoirs interact. The isotopic composition of H$_2$S liberated by bacteria depends also on the magnitude of the sulfate reservoir. If it is essentially infinite, the δ^{34}S values of H$_2$S will remain constant, assuming no change in the bacterial metabolism. On the other hand, if the size of the sulfate reservoir is limited, its isotopic composition will change due to the preferential removal of ^{32}S as H$_2$S. As a consequence, the $\delta^{34}S$ value of the sulfide will also change as a function of time because the enrichment in ^{32}S relative to the standard will be diminished even though the fractionation factor with respect to sulfate remains constant.

2
Sulfur in Recent Sediment

Laboratory experiments with resting suspensions of *Desulfovibrio desulfuricans* generally indicate that metabolic H_2S produced by reduction of sulfate is enriched in ^{32}S up to about 25‰ relative to the sulfate. These experimental results provide an explanation for differences in the isotopic composition of sulfur observed in modern sedimentary basins. However, there are important differences between laboratory cultures and bacteria in the natural environment. Bacteria in nature are not resting, but grow and die. They may live in open or closed systems with respect to the availability of sulfate, the environmental temperatures may be lower than those of laboratory experiments, and the food supply may fluctuate. As a consequence, modern euxinic environments commonly contain H_2S which is enriched in ^{32}S by 50‰ or more compared to coexisting sulfate, even though such extreme fractionation has been difficult to achieve in the laboratory. For example, sulfate in solution in the Black Sea has a $\delta^{34}S$ value of $+19.2‰$ relative to troilite in Canyon Diablo, while dissolved H_2S has a value of about $-31.9‰$ (Vinogradov et al., 1962). Similar results were obtained by Hartmann and Nielsen (1969) for sulfur in pore water of sediment collected in the Bay of Kiel in the Baltic Sea. Their data show a difference of about 50 to 60‰ between the $\delta^{34}S$ values of dissolved sulfate and sulfide. The difference is primarily due to the action of sulfur-reducing bacteria. However, the $\delta^{34}S$ of

dissolved sulfide is further modified by precipitation of pyrite which concentrates ^{34}S.

Hartmann and Nielsen (1969) found that the $\delta^{34}S$ values of both sulfate and sulfide dissolved in pore water increase as a function of depth in sediment cores, that is, both become enriched in ^{34}S. The strongest ^{34}S enrichment was observed in their core 2092 shown in Figure 21.1 in which $\delta^{34}S$ for dissolved sulfate increases from $+20.0‰$ at the top of the core to $+60.7‰$ at a depth of about 30 cm. The $\delta^{34}S$ value of dissolved sulfide in this core changes from $-24.6‰$ at about 4 cm to $+2.0‰$ at 30 cm. At the same time the sulfate *concentration* in pore water of this core decreases with depth, while that of dissolved sulfide increase. However, the *total amount* of sulfur in the pore water decreases sharply, which indicates that sulfur is being precipitated, primarily as sulfides.

The principal carrier of sulfur in the sediments studied by Hartmann and Nielsen (1969) is pyrite (FeS_2). In addition, minor amounts of sulfur may occur as FeS (hydrotroilite), native sulfur, organically bound sulfur, and sulfate (Kaplan et al., 1963). The total sulfur content of sediment in core 2092 (Fig. 21.1) from the Baltic Sea increases from the top of the core to a depth of about 6 cm and then fluctuates irregularly in the deeper portions. The sulfur content of the sediment is much *greater* than that of sulfate-sulfur in the pore water at the top of the core. This indicates that a large fraction of the sulfur in the deeper sediment is derived from sulfate withdrawn from the overlying water column. Apparently the sediment becomes closed to sulfate only after burial under a layer of about 6 cm in thickness.

The $\delta^{34}S$ values of the water-insoluble sulfur in core 2092 from the Baltic Sea *decrease* in the upper 6 cm of the core from -20.2 to $-23.4‰$ and then remain fairly constant in the rest of the core. These values include a small contribution (5 to 10 percent)

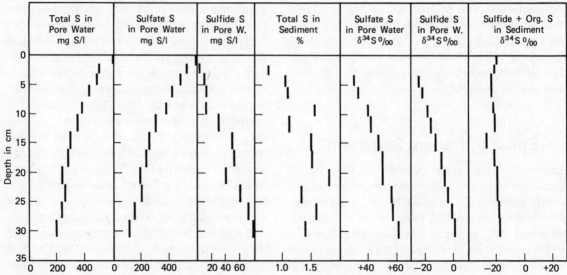

FIGURE 21.1 Variation of sulfur concentration and $\delta^{34}S$ values of pore water and recently deposited sediment in core 2092 from the Bay of Kiel in the Baltic Sea. (Hartmann and Nielsen, 1969.)

of organically bound sulfur which is enriched in ^{34}S relative to troilite of Canyon Diablo. Kaplan et al. (1963) reported average values of $+15.3\%_0$ for marine plants and $+19.6\%_0$ for marine animals. Zooplankton from the Bay of Kiel analyzed by Hartmann and Nielsen (1969) yielded a value of $+18.1\%_0$. Subtraction of the organically bound sulfur from the water-insoluble sulfur of the sediment therefore yields even more negative $\delta^{34}S$ values for the inorganic sulfide phases. The general conclusion is that sulfide phases in recently deposited sediment in the Baltic Sea are enriched in ^{32}S by 35 to $55\%_0$ compared to marine sulfate, and that bacterial fractionation of sulfur in nature is greater than that achieved by most laboratory cultures.

Similar studies by Thode et al. (1960), Vinogradov et al. (1962), and Kaplan et al. (1963) all confirm that sulfide minerals associated with recently deposited marine sediment are enriched in ^{32}S compared to marine sulfate and commonly have negative $\delta^{34}S$ values. Sulfide minerals in lacustrine

sediment are similarly enriched in ^{32}S relative to associated sulfates due to the action of anaerobic bacteria (Deevey et al., 1963; Nakai and Jensen, 1964).

3
Fossil Fuels

a. PETROLEUM

The *concentrations* of organically bound sulfur in petroleum vary within wide limits from 0.1 to 10 percent. The $\delta^{34}S$ values of petroleum also span a wide range from -8 to $+32\%_0$. The isotopic composition of a particular sample of petroleum depends on the source of the sulfur and on the isotope fractionation associated with its incorporation into the petroleum. We have already discussed the evidence based on the $\delta^{13}C$ values that petroleum is derived from biogenic material in marine sediment (Chapter 20). For this reason, marine sulfate is the most likely source of sulfur found in petroleum. Moreover, the reduction of the sulfur is undoubtedly due to the action of anaerobic bacteria which enrich the organically bound sulfur in ^{32}S.

The isotopic composition of sulfur in petroleum samples has been studied by Thode et al. (1958), and Harrison and Thode (1958b), among others. Their conclusions were summarized by Thode and Rees (1970) as follows: (1) although the isotopic composition of sulfur in petroleum varies by more than 40‰, oil from a given pool has the same δ^{34}S value; (2) oil pools in the same reservoir rock within a sedimentary basin have similar δ^{34}S values; (3) oil pools in reservoir rocks of different geologic ages may have significantly different δ^{34}S values; (4) the isotopic composition of sulfur in hydrogen sulfide gas is very similar to that of petroleum from which it was derived. This fourth observation suggests that sulfur isotopes are not fractionated during the splitting off of H_2S gas in the course of maturation of petroleum.

The wide range of δ^{34}S values for petroleum may be caused partly by the fact that the isotopic composition of sulfur in marine sulfate has varied systematically from Precambrian through Phanerozoic time. Thode and Monster (1965) compared δ^{34}S values of petroleum and contemporaneous anhydrite from marine evaporite deposits ranging in age from Silurian to Permian and found that sulfur in petroleum is enriched in ^{32}S by about 15‰. They attributed this enrichment to the action of sulfur-reducing bacteria at the time of deposition of the sediment and during subsequent formation of petroleum. The work of Thode and his colleagues suggests therefore that the sulfur in petroleum of marine origin was derived from marine sulfate, was enriched in ^{32}S by about 15‰ by bacterial action prior to incorporation into the petroleum, and was not modified isotopically during subsequent

maturation of the petroleum. Because δ^{34}S values of marine sulfate are known to have varied systematically in the geologic past, it follows that the δ^{34}S value of a petroleum sample may depend on the age of the source rocks in which it was formed. Consequently, the δ^{34}S value may be useful in identifying the source rocks from which a particular petroleum sample was derived and thus to trace its migration into the reservoir rock in which it accumulated.

These principles were used by Thode and Rees (1970) to study the source and migration of oil in oil fields of northern Iraq. In this area oil pools occur in reservoir rocks of Lower Cretaceous to Tertiary age and of Upper Triassic age. They found that oil in Cretaceous and Tertiary rocks has very similar δ^{34}S values ranging from -5.2 to -5.6‰. The oil in Upper Triassic rocks has a distinctly different average δ^{34}S value of $+2.35$‰. They concluded that all of the oil in Cretaceous to Tertiary reservoir rocks originated from a single source located in rocks of Jurassic to Lower Cretaceous age. After deposition of the Tertiary reservoir rocks, the oil migrated upward through thousands of feet of rocks and accumulated in pools in which it is now found. The difference between δ^{34}S values of oil in the Upper Triassic rocks and in the Cretaceous to Tertiary reservoir rocks rules out the possibility that the oil originated from the same sources. The source of the petroleum in Triassic reservoir rocks is not definitely known, but is presumably to be sought in underlying older rocks.

b. COAL

The *sulfur content* of coal is variable and may range up to 20 percent. It is present in several chemical forms, including (1) sulfides of iron such as pyrite and marcasite (sphalerite, galena and chalcopyrite are found only rarely); (2) sulfates, including gypsum, barite, and sometimes sulfates of iron that result from the oxidation of pyrite or marcasite; (3) elemental sulfur in trace concentrations; (4) organically bound sulfur whose mode of combination in the coal is not well understood.

The *isotopic composition* of sulfur in coal varies widely from $\delta^{34}S = +24$ to $-30‰$ (Rees and Thode, 1970). Smith and Batts (1974) have enlarged the range to $+32.3‰$ on the basis of a detailed study of sulfur in Australian coals ranging in age from Permian to Tertiary. They found that the $\delta^{34}S$ values of organic sulfur in coals containing *less than 1 percent total sulfur* vary within a narrow range from $+4.6$ to $+7.3‰$, whereas the $\delta^{34}S$ values of organic sulfur in *sulfur-rich* coals ($>1\%$ total sulfur) vary more widely from $+2.9$ to $+24.4‰$. Smith and Batts suggested that the $\delta^{34}S$ values of organic sulfur in low-sulfur coal is representative of the isotopic composition of sulfate in the freshwater environment in which the coal was deposited. They therefore concluded from the uniformity of $\delta^{34}S$ values of organic sulfur in low-sulfur coals that the *nonmarine* sulfate in the coal basins of Australia remained fairly constant from the Permian to the Tertiary periods. This is in marked contrast to the $\delta^{34}S$ values of *marine* sulfate which changed from about $+10$ to $+20‰$ during the same period (Holser and Kaplan, 1966).

The greater variability of $\delta^{34}S$ values of the various sulfur compounds in high-sulfur coals may result from postdepositional infiltration of marine or nonmarine sulfate into the coal, accompanied by partial or complete reduction of this sulfate to sulfide by bacteria. Smith and Batts (1974) demonstrated systematic stratigraphic variations of $\delta^{34}S$ values in coal seams overlain by marine sediment and interpreted them in terms of models based on different rates of diffusion of sulfate and/or rates of reduction. Such detailed studies of sulfur isotope compositions of suites of coal samples provide much more useful information than single determinations of $\delta^{34}S$ values of total sulfur in isolated samples.

4
Native Sulfur Deposits

Large deposits of native sulfur occur in the cap rocks of salt domes such as those of the Gulf Coast area of the United States and in certain evaporite sequences in the Mediterranean basin as exemplified by the deposits in Sicily. The explanation of the origin of such native sulfur deposits is based primarily on the isotopic compositions of sulfur and carbon in these deposits, thus illustrating the value of such studies in the solution of certain geological problems.

The salt domes of the Gulf Coast region have been important producers of petroleum and native sulfur. The petroleum occurs in traps along the flanks of the domes while the native sulfur is localized within the calcite and anhydrite rocks which cap the domes. The caps are formed by leaching of rock salt by groundwater and consist largely of anhydrite that was originally disseminated in the rock salt. The presence of calcite in the cap rock is related to the origin of the native sulfur as we shall see later on. The isotopic compositions of sulfur of various sulfur-bearing compounds associated with salt domes in the Gulf Coast area of the United States were determined by Feely and Kulp (1957). Table 21.1 contains a summary of their data for the Boling salt dome in Texas which will serve as a model for the explanation of the occurrence of native sulfur associated with salt domes.

Unaltered samples of rock salt from the salt domes of the Gulf Coast contain from

Table 21.1   Isotopic Composition of Sulfur in the Boling Salt Dome, Texas (Feely and Kulp, 1957)

MATERIAL	$\delta^{34}S^a$ ‰
Anhydrite, disseminated in halite	+16.2
Anhydrite, anhydrite cap rock	+17.9
Anhydrite, calcite cap rock	+26.3
Native sulfur, anhydrite cap rock	+3.0
Native sulfur, calcite cap rock	+2.6
Marcasite and pyrite in calcite rock	+5.8
Native sulfur with above sulfides	+4.2
Sulfate, bleedwater	+25.2
Hydrogen sulfide, bleedwater	+0.8

^{a} Recalculated from $^{32}S/^{34}S$ ratios assuming a value of 22.22 for the $^{34}S/^{32}S$ ratio of Canyon Diablo.

5 to 10 percent of disseminated anhydrite. The $\delta^{34}S$ values of this anhydrite in the salt domes of Louisiana and Texas are very similar, which suggests that they originated from a common source bed believed to be the Louann salt of Jurassic age. In the Boling dome (Table 21.1), the $\delta^{34}S$ value of the disseminated anhydrite is +16.2 percent. The anhydrite of the cap rock is significantly enriched in ^{34}S and has variable $\delta^{34}S$ values averaging +17.9‰ (in the anhydrite cap rock) and +26.3‰ (in calcite rock). In marked contrast, the native sulfur is strongly enriched in ^{32}S compared to the anhydrite and has average $\delta^{34}S$ values of +3.0‰ (in the anhydrite rock) and +2.6‰ (in the calcite rock). Feely and Kulp (1957) attributed this difference to isotope fractionation of sulfur by bacteria which reduced sulfate ions to hydrogen sulfide gas and suggested that the bacteria lived on petroleum that saturated the cap rocks. The preferential incorporation of ^{32}S into hydrogen sulfide left the remaining sulfate variably enriched in ^{34}S, depending on the fraction of sulfate that was converted to sulfide in a given region of the cap rock and on the rate of metabolic activity of the bacteria. The hydrogen sulfide produced by the bacteria was subsequently oxidized by sulfate according to the reaction

$$SO_4^= + 3H_2S \rightarrow 4S + 2H_2O + 2OH^- \quad (21.3)$$

Feely and Kulp (1957) demonstrated experimentally that this reaction is capable of converting about 20 mg of sulfide per liter to native sulfur in one day. A relatively small fraction of H_2S may have been precipitated locally as pyrite or marcasite having $\delta^{34}S$ = +5.8‰. However, salt domes have very low iron contents, which accounts for the low abundance of these minerals in the cap rock.

The calcite in the upper parts of the cap rock formed by precipitation of CO_2 gas liberated by bacteria during metabolic oxidation of petroleum. The $\delta^{13}C$ value of petroleum from Boling dome is −26.2‰ (relative to "sedimentary limestone"), while that of calcite from the cap rock is −32.0‰. Therefore, the calcite is strongly enriched in ^{12}C compared to normal marine limestones and is similar in isotopic composition to the associated petroleum. In fact, the carbon in the calcite is enriched in ^{12}C even when compared to the petroleum. Feely and Kulp (1957) attributed this enrichment to fractionation during metabolic oxidation of the petroleum by the bacteria. The calcium of the calcite was undoubtedly derived from the anhydrite that had yielded sulfate ions for bacterial reduction to H_2S. Thus the isotopic composition of carbon of caprock calcite is consistent with the hypothesis that the native sulfur was produced by bacterial reduction of sulfate to hydrogen sulfide followed by oxidation of native sulfur. Subsequent investigations of other native sulfur deposits associated with marine sedimentary rocks have generally confirmed these conclusions (Dessau et al., 1962; Schneider and Nielsen, 1965; Davis and Kirkland, 1970).

5
Isotopic Evolution of Marine Sulfate

The oceans of the Earth contain a large amount of sulfur (1.3×10^{15} metric tons) in the form of sulfate ions in solution. The isotopic composition of sulfur in modern marine sulfate is constant within narrow limits and is represented by a $\delta^{34}S$ value of about $+20\%_0$. However, the $\delta^{34}S$ values of sulfate minerals from marine evaporite rocks of Phanerozoic age vary systematically as a function of time in the past. This variation implies that significant changes have occurred in the sulfur cycle on the surface of the Earth.

A large number of $\delta^{34}S$ values of marine sulfate minerals of Phanerozoic age have been used to reconstruct the history of the isotopic composition of marine sulfate (Nielsen and Ricke, 1964; Thode and Monster, 1965; Holser and Kaplan, 1966). The results suggest that the $\delta^{34}S$ values of marine sulfate minerals decreased from about $+30\%_0$ during the Cambrian period to about $+10\%_0$ during the Permian and then increased irregularly during the Mesozoic era toward the present value of $+20\%_0$. Thode and Monster (1965) and others have shown that sulfur isotopes are not appreciably fractionated by the precipitation of gypsum from sulfate-bearing brines ($\Delta^{34}S$ gypsum-brine $= +1.65 \pm 0.12\%_0$). Therefore, we may assume that the $\delta^{34}S$ values of marine sulfate minerals are representative of the isotopic composition of sulfur in the brines from which they were precipitated. Nevertheless, the $\delta^{34}S$ values of sulfate minerals deposited during a given period of geologic time generally vary within certain limits, thus complicating the selection of the most representative value of $\delta^{34}S$ of the oceans during a particular time interval. The variability of $\delta^{34}S$ values of contemporaneously deposited marine sulfate

minerals may result from several causes: (1) variable enrichment of ^{34}S in the residual sulfate of isolated evaporative basins due to bacterial reduction to sulfide; (2) input of isotopically light sulfate by rivers; (3) isotopic fractionation in the course of precipitation of minerals from brines. As a result, the $\delta^{34}S$ values of the oceans in the geologic past have been determined primarily on the basis of the preponderance of the evidence and by rejecting aberrant values. Such aberrant values may also result from incorrect age assignments or a nonmarine origin of the deposit. In spite of these difficulties, the general outline of the evolutionary history of the isotopic composition of marine sulfate in post-Precambrian time is now quite well established, as shown in Figure 21.2.

The isotopic composition of sulfates of Precambrian age is not well known. Perry et al. (1971) analyzed barites of presumed sedimentary origin from the Swaziland System of the Barberton Mountain Land, South Africa, known to be more than three billion years old. They reported an average $\delta^{34}S$ value of $+3.4\%_0$ for barites and $+0.42\%_0$ for approximately contemporaneous sulfides. They attributed the small enrichment of the sulfate in ^{34}S to the oxidation of volcanic H_2S to sulfate by green or purple sulfur bacteria.

Sulfur in the form of sulfate ions in aqueous solution enters the ocean primarily by the discharge of fresh water and originates as a weathering product of several sources: (1) sulfide-bearing sedimentary rocks (black shales and carbonate rocks); (2) evaporite rocks of marine origin; and (3) volcanic and primary igneous rocks. On the other hand, sulfur is removed from the oceans by formation of evaporite rocks and by bacterial reduction of sulfate to sulfide followed by precipitation as pyrite, marcasite or hydrotroilite. The concentration and $\delta^{34}S$ value of sulfate in the oceans at the present time

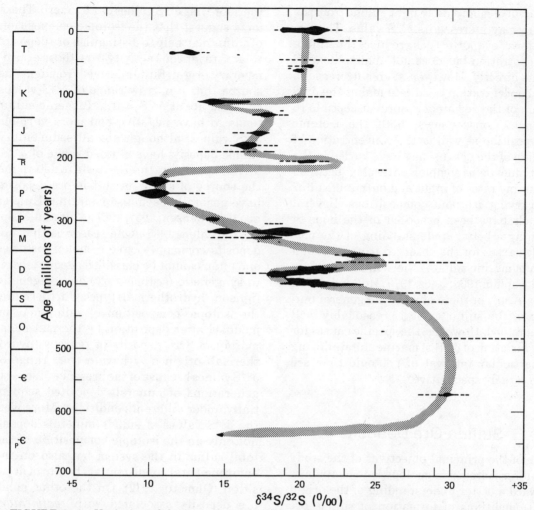

FIGURE 21.2   Variation of the isotopic composition of marine sulfate during Phanerozoic time. (Reproduced from Figure 5 (p. 120) of Holser, W. T. and I. R. Kaplan, Chemical Geology, vol. 1, No. 2, 93–135, 1966, by permission of Elsevier Scientific Publishing Co.)

reflect a particular state of the sulfur cycle in terms of quantities and isotopic compositions of sulfur that enter and leave the oceans in unit time. The fact that the isotopic composition of sulfur in the oceans has varied systematically throughout geologic time indicates that significant changes have occurred in the sulfur cycle in the past.

Both the rate of input of sulfur and its isotopic composition may vary as a function of time depending on the rates of weathering and erosion of different kinds of sedimentary rocks, or on the intensity of igneous activity on a worldwide basis. In general, increased erosion of black shale containing isotopically light sulfides may lower the $\delta^{34}S$ value of oceanic sulfate. On the other hand, increased reduction of sulfate by bacteria and removal of isotopically light sulfide from the ocean may increase its $\delta^{34}S$ value. The formation of evaporite rocks reduces the sulfate concentration but does not change its $\delta^{34}S$ value directly. However, such a decrease in the concentration of sulfate makes the $\delta^{34}S$ value of the remainder more susceptible to change. Consequently, both the isotopic composition as well as the concentration of sulfate of the oceans may have varied in the past due to a complex interplay between changing rates of input and output of sulfur of varying isotopic compositions. Several models have been proposed in the hope of gaining a better understanding of the possible causes for the observed isotopic evolution of marine sulfate (Nielsen, 1966; Holser and Kaplan, 1966; Rees, 1970; Holland, 1973). As a result of these efforts, the general outline of the sulfur cycle is reasonably well understood. However, the specific causes for the variation of $\delta^{34}S$ of marine sulfate during a particular interval of geologic time are still largely speculative.

6
Sulfide Ore Deposits

One of the principal objectives of the study of sulfur isotopes in geology is to contribute toward a better understanding of the origin and conditions of formation of sulfide ore deposits. It is important, both for theoretical and practical reasons, to distinguish between ore deposits formed as a result of igneous activity and those that are of sedimentary origin. We know that bacteria living in freshly deposited sediment reduce sulfate and enrich the resulting H_2S in ^{32}S. Consequently, sulfur that has been subjected to bacterial reduction becomes enriched in ^{32}S compared to marine sulfate. On the other hand, the sulfur associated with igneous rocks derived from the upper mantle is isotopically similar to that of meteorites and

therefore has $\delta^{34}S$ values close to zero. These facts suggest that the isotopic composition of sulfur may help to distinguish ore deposits related to igneous activity from those of sedimentary origin. Sulfur derived from igneous sources has a narrow range of $\delta^{34}S$ values centered about $\delta^{34}S = 0\%_0$. Biogenic sulfur tends to have negative and more variable $\delta^{34}S$ values, although not all sedimentary sulfide deposits have a broad range of $\delta^{34}S$.

Attempts to use these criteria to determine the source of sulfur in sulfide ore deposits have generally been disappointing (Kulp et al., 1956; Jensen, 1957, 1967; Ault and Kulp, 1960; Gehlen, 1965). One reason is that ore deposits commonly have complex histories and often cannot be classified adequately into syngenetic (sedimentary) and epigenetic (igneous-hydrothermal) types. Furthermore, the isotopic composition of sulfur may be modified after deposition by thermal metamorphism. Ore deposits of *igneous* (hydrothermal) origin may have a *wide* range of $\delta^{34}S$ values because of the presence of several generations of minerals deposited sequentially under different conditions. Moreover, the $\delta^{34}S$ values of sulfide minerals depend not only on the isotopic composition of the total sulfur in the system but also on the environmental conditions at the site of deposition (Ohmoto, 1972). On the other hand, ore deposits associated with *sedimentary* rocks may have a *narrow* range of $\delta^{34}S$ values because they formed from an infinite sulfate reservoir under constant environmental conditions.

Nevertheless, the isotopic composition of sulfur in stratabound sulfide deposits associated with sedimentary and volcanic rocks may be related to that of marine sulfate at the time of deposition. Sangster (1968) compared the average $\delta^{34}S$ values of 66 stratabound sulfide deposits to the $\delta^{34}S$ values of contemporaneous marine sulfate and demonstrated that the sulfur in sedimentary types is enriched in ^{32}S by 11.7‰ on the average, while volcanic types are enriched by 17.5‰ compared to contemporaneous marine sulfate. Such average differences are consistent with the assumption that the sulfur in these stratabound ore deposits was derived from marine sulfate by bacterial reduction. Consequently, the $\delta^{34}S$ values of stratabound sulfide deposits of different ages may also reflect the time-dependent variation of the isotope composition of marine sulfate (Fig. 21.2).

7
Isotope Fractionation Among Sulfide Minerals

When sulfide minerals are precipitated from aqueous solutions or crystallize from sulfide liquids, small differences may occur in the $\delta^{34}S$ values of cogenetic minerals due to isotopic equilibration among the solids and between the solids and the liquid. The existence of these differences was first pointed out by Sakai (1957). Subsequent work by other investigators, summarized by Thode (1970), established the general validity of this phenomenon and led to the suggestion that such differences in the isotopic composition of sulfur in cogenetic sulfide minerals may reflect the temperature of isotope equilibration (Tatsumi, 1965). Theoretical considerations by Sakai (1968) and Bachinski (1969) based on bond strengths indicate that the ^{34}S-enrichment of some common sulfide minerals should *decrease* in the order: pyrite > sphalerite > chalcopyrite > galena. Analytical data for coexisting suites of these minerals generally confirm this prediction and indicate thereby that isotopic equilibrium may be closely approached by these minerals in nature. Consequently, the $\delta^{34}S$ values of cogenetic sulfide minerals may be used to determine the temperature of equilibration, provided the variation of $\delta^{34}S$

values as a function of temperature is either calculated theoretically or determined experimentally.

The $\Delta^{34}S$ values of pyrite-galena, sphalerite-galena, and pyrite-sphalerite were calculated by Sakai (1968) for the temperature range from 27° to 527°C, shown in Figure 21.3, as a function $10^6/T^2$, where T is the absolute temperature. It is apparent that the $\Delta^{34}S$ values for these mineral pairs decrease linearly with increasing temperature above about 150°C. The pyrite-galena pair is the most sensitive geothermometer, but is rarely formed in equilibrium. For that reason, the pair sphalerite-galena is found to be most useful. Calculations by Hulston, quoted by Groves et al. (1970), suggest a slightly lower

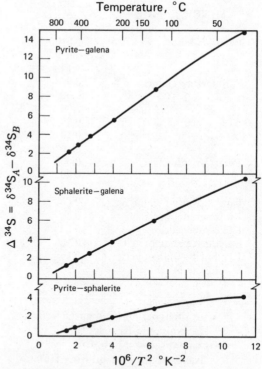

FIGURE 21.3 Fractionation of sulfur isotopes among cogenetic mineral pairs as a function of temperature, based on calculated values of isotope equilibrium constants by Sakai (1968).

slope for the sphalerite-galena fractionation line than that of Sakai (1968).

The fractionation of sulfur isotopes by sulfide minerals has also been studied experimentally (Grootenboer and Schwarcz, 1969; Kajiwara and Krouse, 1971; Czamanske and Rye, 1974). The results confirm that $\delta^{34}S$ values of mineral pairs are linearly related to the equilibration temperatures expressed as $10^6/T^2$. The experimental results are therefore presented as equations of the form

$$\Delta^{34}S = \frac{A \times 10^6}{T^2} \qquad (21.4)$$

where A is a constant equal to the slope of straight lines through the origin. Table 21.2 contains a compilation of calculated and experimentally determined values of A for different mineral pairs. The isotopic temperatures derivable from these equations are still somewhat uncertain but, in general, they do conform with temperature estimates based on other criteria.

Rye (1974) has made a detailed comparison of the filling temperatures of fluid inclusions in sphalerite mainly from Providencia, Mexico, and the isotopic temperatures in-

Table 21.2 **Isotope Fractionation of Sulfur Among Coexisting Sulfide Minerals in Isotopic Equilibrium with an External Sulfur Reservoir**[a]

MINERAL PAIR	A	TEMPERATURE RANGE, °C	REFERENCE
Pyrite-galena	1.319	27–527	1
	~0.9	340–690	2
	1.10	250–600	3
Sphalerite-galena	0.963	27–527	1
	~0.62	340–690	2
	~$0.66 \times 10^6 T^{-2} - 0.1$		2 by 4
	0.80	250–600	3
	0.70	275–600	4
	~0.78	0–1000	5
Pyrite-sphalerite	0.356	27–527	1
	~0.26	340–690	2
	0.30	250–600	3
Chalcopyrite-galena	0.65	250–600	3
Pyrite-chalcopyrite	0.45	250–600	3
Pyrite-pyrrhotite	0.30	250–600	3
Sphalerite-chalcopyrite	0.15	250–600	3
Pyrrhotite-chalcopyrite	0.15	250–600	3
Sphalerite-pyrrhotite	~0	250–600	3

1. Sakai, 1968
2. Grootenboer and Schwarcz, 1969
3. Kajiwara and Krouse, 1971
4. Czamanske and Rye, 1974
5. Groves et al., 1970, quoting Hulston

[a] The temperature dependence of $\Delta^{34}S$ values of mineral pairs is expressed as:

$$\Delta^{34}S = A \times 10^6/T^2$$

where $\Delta^{34}S$ is the difference between $\delta^{34}S$ values of two coexisting minerals, A is a constant and T is the absolute temperature.

dicated by the calibration equations for sphalerite and galena given in Table 21.2. The results generally support the equation of Czamanske and Rye (1974). The isotopic temperatures based on the equation of Kajiwara and Krouse (1971) are about 40°C higher and those based on Grootenboer and Schwarcz (1969) are 20°C lower on the average than the filling temperatures. For example, the $\Delta^{34}S$ value (sphalerite-galena) for sample 60-H-65 from the Zinc West Level 14 of Providencia, Mexico, is 1.88‰. The filling temperature is 320° to 340°C. The isotopic equilibration temperature is 335°C according to Czamanske and Rye (1974), 380°C according to Kajiwara and Krouse (1971), and 300°C using the equation of Grootenboer and Schwarcz (1969). Despite this uncertainty, the use of sulfur isotopes to determine equilibration temperatures of coexisting sulfide minerals is an important advance in the study of ore deposits.

8
Isotope Fractionation in Ore-forming Fluids

The preceding discussion (sections 6 and 7) was based on the assumption that the $\delta^{34}S$ values of sulfide minerals depend only on the isotopic composition of sulfur in the fluid from which they formed and on the temperature of isotopic equilibration. However, Sakai (1968) originally pointed out that the $\delta^{34}S$ value of aqueous sulfide ions depends on the relative proportions of H_2S, HS^-, and S^{-2}, all of which fractionate sulfur isotopes. The abundances of these ionic and molecular species are controlled by chemical equilibria involving hydrogen ions. Therefore, the $\delta^{34}S$ value of the aqueous sulfide ion, and of sulfide minerals in equilibrium with it, depends on the pH of the ore-forming fluid. This idea has been extended by Ohmoto (1972) to include the

effects of other physical and chemical parameters on the $\delta^{34}S$ values of sulfide minerals that can precipitate from a given ore-forming fluid. His treatment of this complex problem indicates that the $\delta^{34}S$ values of sulfide minerals reflect not only the temperature and isotopic composition of sulfur in the fluid but also the pH, the fugacities of oxygen and sulfur, the total sulfur content, and the ionic strength of the ore-forming fluid.

The sulfur-bearing species that may be important in ore-forming fluids below about 500°C include: H_2S, HS^-, S^{-2}, SO_4^{-2}, HSO_4^-, KSO_4^-, and $NaSO_4^-$. These species are related to each other by the following chemical equilibria:

$$
\begin{aligned}
H_2S(aq) &\rightleftarrows H^+ + HS^- \\
HS^- &\rightleftarrows H^+ + S^= \\
2H^+ + SO_4^= &\rightleftarrows H_2S(aq) + 2O_2 \\
HSO_4^- &\rightleftarrows H^+ + SO_4^= \\
KSO_4^- &\rightleftarrows K^+ + SO_4^= \\
NaSO_4^- &\rightleftarrows Na^+ + SO_4^=
\end{aligned} \qquad (21.5)
$$

It is apparent that the activity of hydrogen ion controls the relative proportions of H_2S, HS^-, and S^{-2} that can coexist, while the fugacity of oxygen controls the abundance of SO_4^{-2} relative to aqueous H_2S. The distribution of ^{34}S among these species is illustrated in Figure 21.4 by the calculated fractionation factors of Sakai (1968) relative to S^{-2}. The diagram shows that SO_4^{-2} is strongly enriched in ^{34}S relative to S^{-2} and that the enrichment increases with decreasing temperature. Likewise, $H_2S(aq)$ and HS^- concentrate ^{34}S, but do so to a considerably lesser extent.

The data in Figure 21.4 can be used to demonstrate qualitatively the effect of the fugacity of oxygen and the pH on the $\delta^{34}S$ value of sulfide ion in an ore-forming fluid at a particular temperature. A high oxygen fugacity favors SO_4^{-2} which concentrates

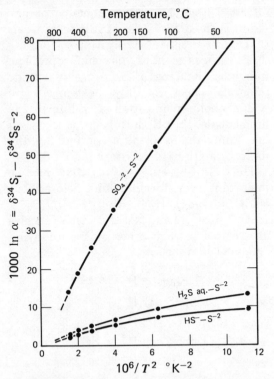

FIGURE 21.4 Fractionation of sulfur isotopes among SO_4^{-2}, H_2S (aq), HS^-, and S^{-2} as a function of the temperature. Note that SO_4^{-2} is strongly enriched in ^{34}S relative to S^{-2} and that the enrichment increases with decreasing temperature. Fractionation of sulfur isotopes among H_2S (aq) and HS^- is less pronounced but these ions clearly prefer ^{34}S over ^{32}S compared to the sulfide ion. These fractionation factors were calculated by Sakai (1968) from ratios of reduced partition coefficients.

^{34}S and thereby enriches the sulfide ion in ^{32}S. Therefore, sulfide minerals precipitating at high oxygen fugacity from a given ore-bearing fluid are enriched in ^{32}S compared to the same mineral forming at lower oxygen fugacity. The pH-dependence of the $\delta^{34}S$ value of a sulfide mineral can be demonstrated similarly. Let us assume that the oxygen fugacity is sufficiently low so that SO_4^{-2} is unimportant and H_2S(aq),

HS^-, and S^{-2} are dominant. A decrease in pH (i.e., increase in the activity of hydrogen ion) favors the formation of H_2S(aq) and HS^-, both of which prefer ^{34}S. The result is that the sulfide ion becomes enriched in ^{32}S and minerals forming at low pH are enriched in ^{32}S compared to the same mineral forming at higher pH from the same fluid.

Ohmoto (1972) has worked out these relationships quantitatively and has plotted $\delta^{34}S$ contours on mineral stability fields in the system $Fe{-}Ba{-}O_2{-}S_2$. Figure 21.5 is a simplified example of such a diagram for the case $\sum S = 0.1$ moles/kg H_2O, $T = 250°C$, ionic strength $= 1.0$, and $\delta^{34}S_{\Sigma s} = 0‰$. The total sulfur content of the system affects the size of the stability fields while the ionic strength determines the values of the activity coefficients of ions in aqueous solution. The temperature not only controls the magnitude of the isotope fractionation factors (Fig. 21.4) but also determines the values of the chemical equilibrium constants. Figure 21.5 indicates that pyrite precipitating from a fluid whose $\delta^{34}S$ value is 0‰ can have $\delta^{34}S$ ranging from about $+5$ to $-27‰$, depending on the fugacity of oxygen and the pH. Evidently, these parameters play an important role in determining the $\delta^{34}S$ values of sulfide minerals that may precipitate from a given ore-bearing fluid. However, these effects do not prevent us from interpreting the *differences* in $\delta^{34}S$ values of cogenetic sulfide minerals in terms of equilibration temperatures as outlined in section 7 of this chapter. The treatment by Ohmoto does not include the effect of removal of sulfur on the $\delta^{34}S$ of the sulfur remaining in the system.

Ohmoto's contribution is particularly important because it may enable us to use $\delta^{34}S$ and $\delta^{13}C$ values of coexisting sulfur and carbon-bearing minerals to determine the pH and the oxygen fugacity of the

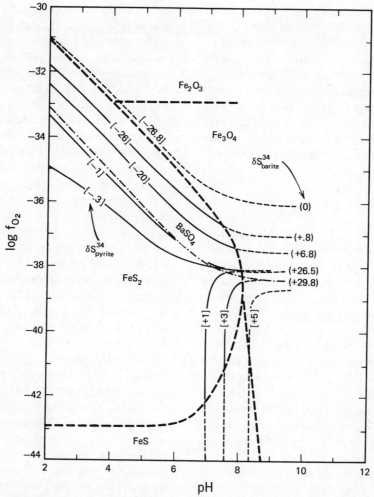

FIGURE 21.5 Isotopic composition of sulfur in pyrite and barite in the system Fe-S-O as a function of pH and the fugacity of oxygen. Square brackets indicate $\delta^{34}S$ of pyrite; round brackets denote $\delta^{34}S$ of barite. Note that $\delta^{34}S$ of pyrite varies widely depending on the fugacity of oxygen and the pH of the solution. The $\delta^{34}S$ of the total sulfur in the system is equal to zero per mil. The Fe-S-O mineral boundaries represent a total sulfur content of 0.1 moles S/kg H_2O, $T = 250°C$, ionic strength = 1.0. The dash-dot line is the barite soluble/insoluble boundary ($m_{Ba+2} \times m_{\Sigma S} = 10^{-4}$). (Adapted from Figure 5 (p. 559) of Ohmoto, H., Economic Geology, vol. 67, No. 5, 551–578, 1972 with permission for the use of the diagram from H. Ohmoto.)

ore-bearing fluid during the formation of an ore body. This idea was elaborated by Rye and Ohmoto (1974) by means of a composite diagram on which they traced the likely evolution of pH and oxygen fugacity of several ore deposits. The isotope geology of sulfur as applied to ore deposits has clearly experienced a renaissance and promises to contribute significantly to our understanding of ore-forming processes.

9   SUMMARY

The isotopes of sulfur are fractionated during the reduction of sulfate to sulfide by bacteria (*Desulfovibrio desulfuricans*) and by isotope exchange reactions among sulfur-bearing compounds, ions, and molecules. Bacterial reduction of sulfate in nature enriches hydrogen sulfide in ^{32}S by 50‰ or more compared to the sulfate. Laboratory cultures have generally yielded enrichments of only 27‰. The magnitude of the enrichment depends on a variety of factors, including the rate of reduction, the temperature, the nature and availability of the food supply, and the size of the sulfate reservoir.

Hydrogen sulfide and sulfide minerals in Recent marine sediment are variously enriched in ^{32}S compared to contemporary marine sulfate. The effect is attributable to the action of bacteria. Enrichment in ^{32}S and variability of $\delta^{34}S$ are believed to be characteristic of sulfur that has passed through the sedimentary cycle.

The $\delta^{34}S$ values of petroleum range from -8 to $+32$‰ and appear to be enriched in ^{32}S by about 15‰ compared to marine sulfate at the time of formation of the petroleum. On the other hand, the $\delta^{34}S$ values of organically bound sulfur in low-sulfur coal are fairly constant, which has been taken as evidence that the isotope composition of sulfate in fresh water remained constant, while that of marine sulfate varied systematically.

The origin of native sulfur deposits in the cap rock of salt domes and in certain evaporite sequences has been explained on the basis of isotopic compositions of carbon and sulfur. Native sulfur was produced as a by-product of the reduction of sulfate by bacteria feeding on petroleum. The calcite cap rock is strongly enriched in ^{12}C because it formed by precipitation of metabolic CO_2 derived from petroleum.

Systematic variations of the $\delta^{34}S$ values of marine sulfate minerals of Phanerozoic age indicate significant changes in the sulfur cycle operating on the surface of the Earth. These changes involve the rates of total sulfur input and output, as well as the isotopic composition of the sulfur entering and leaving the oceans. It is possible that both the concentration and the isotopic composition of sulfate in the oceans have varied in the past.

The isotopic composition of sulfur in sulfide ore deposits has received much attention. Attempts to develop criteria for distinguishing between ore deposits of igneous-hydrothermal origin from those of sedimentary-syngenetic origin have failed because of extensive overlap of $\delta^{34}S$ values of such deposits. Theoretical and experimental studies indicate that isotope exchange reactions among cogenetic sulfide minerals and ore-forming fluids may achieve equilibrium and that the temperature of equilibration can be estimated from $\Delta^{34}S$ values of coexisting sulfides. The $\delta^{34}S$ values of individual minerals do not necessarily reflect the isotopic composition of sulfur in the fluid from which they formed but may be controlled by changes in the pH and the fugacity of oxygen. By combining isotopic data for carbon and sulfur in coexisting carbon-and sulfur-bearing minerals, both pH and oxygen fugacity may be determined.

PROBLEMS

1. The $\delta^{34}S$ value of sphalerite from Providencia, Mexico, is $+1.38\%_0$ and that of cogenetic galena is $-0.71\%_0$ (Rye, 1974). What is the value of the isotope-fractionation factor (α) for the

sphalerite-galena pair? (**ANSWER:** 1.00209.)

2. The $\delta^{34}S$ value of sphalerite from Providencia, Mexico, is $+1.21\%_0$ and that of cogenetic galena is $-1.16\%_0$ (Rye, 1974). Calculate the sulfur isotope equilibration temperature using the equation of Czamanske and Rye (1974). (**ANSWER:** 270°C.)

3. Using the equation of Czamanske and Rye (1974) for the sphalerite-galena pair, plot a graph showing the fractionation factor (α) as a function of temperature from 100° to 600°C for this mineral pair.

4. The following filling temperatures and $\delta^{34}S$ values were reported by Rye (1974):

LOCALITY	SPHALERITE $\delta^{34}S\%_0$	GALENA $\delta^{34}S\%_0$	FILLING TEMP. °C (SPHALERITE)
Flat Gap, Tenn.	+28.70	+24.89	125
Hill Mine, Ill.	+7.48	+3.83	142
Creede, Colo.	-1.62	-4.10	225
Providencia, Mex.	+1.08	-1.62	245
Providencia, Mex.	+1.03	-1.42	250
Providencia, Mex.	+0.62	-1.53	311
Casapalca, Peru	+0.09	-1.85	324
Providencia, Mex.	+1.02	-0.86	330
Providencia, Mex.	+0.61	-1.1	345

Use these data to derive an equation relating $\Delta^{34}S$ (sphalerite-galena) versus $10^6 T^{-2}$, where T is the absolute temperature.

5. At 135°C, $\Delta^{34}S$ (sphalerite-galena) = $+5.8\%_0$ and $\Delta^{34}S$ (pyrite-sphalerite) = $+2.7\%_0$. Calculate the isotopic fractionation factor between pyrite and galena at that temperature. (**ANSWER:** 1.0085.)

REFERENCES

Ault, W. U., and J. L. Kulp (1959) Isotopic geochemistry of sulphur. Geochim. Cosmochim. Acta, *16*, 201–235.

Ault, W. U., and J. L. Kulp (1960) Sulfur isotopes and ore deposits. Econ. Geol., *55*, 73–100.

Bachinski, D. J. (1969) Bond strength and sulfur isotopic fractionation in coexisting sulfides. Econ. Geol., *64*, 56–65.

Czamanske, G. K., and R. O. Rye (1974) Experimentally determined sulfur isotope fractionation between sphalerite and galena in the temperature range 600° to 275°C. Econ. Geol., *69*, 17–25.

Davis, J. B., and D. W. Kirkland (1970) Native sulfur deposition in the Castile formation, Gulberson County, Texas. Econ. Geol., *65*, 107.

Deevey, E. S., Jr., N. Nakai, and M. Stuiver (1963) Fractionation of sulfur and carbon isotopes in a meromictic lake. Science, *139*, 407–408.

Dessau, G., M. L. Jensen, and N. Nakai (1962) Geology and isotopic studies of Sicilian sulfur deposits. Econ. Geol., *57*, 410–438.

Feely, H. W., and J. L. Kulp (1957) The origin of the Gulf Coast salt dome sulphur deposits. Bull. Amer. Assoc. Petrol. Geologists, *41*, 1802–1853.

Gehlen, K. V. (1965) Schwefel-Isotope und die Genese von Erzlagerstätten. Geol. Rundschau, *55*, 178–197.

Grootenboer, J., and H. P. Schwarcz (1969) Experimentally determined sulfur isotope fractionations between sulfide minerals. Earth Planet. Sci. Letters, *7*, 162–166.

Groves, D. I., M. Solomon, and T. A. Rafter (1970) Sulfur isotope fractionation and fluid inclusion studies at the Rex Hill Mine, Tasmania. Econ. Geol., *65*, 459–469.

Harrison, A. G., and H. G. Thode (1957) The kinetic isotope effect in the chemical reduction of sulphate. Trans. Faraday Soc., *53*, 1–4.

Harrison, A. G., and H. G. Thode (1958a) Mechanism of the bacterial reduction of sulfate from isotope fractionation studies. Trans. Faraday Soc., *54*, 84–92.

Harrison, A. G., and H. G. Thode (1958b) Sulphur isotope abundances in hydrocarbons and source rocks of the Uinta Basin, Utah, Bull. Amer. Assoc. Petrol. Geologists, *42*, 2642–2649.

Hartmann, M., and H. Nielsen (1969) δ^{34}S-Werte in rezenten Meeressedimenten und ihre Deutung am Beispiel einiger Sedimentprofile aus der westlichen Ostsee. Geol. Rundschau, *58*, 621–655.

Holland, H. D. (1973) Systematics of the isotopic composition of sulfur in the oceans during the Phanerozoic and its implications for atmospheric oxygen. Geochim. Cosmochim. Acta, *37*, 2605–2616.

Holser, W. T., and I. R. Kaplan (1966) Isotope geochemistry of sedimentary sulfates. Chem. Geol., *1*, 93–135.

Holt, B. D. and A. G. Engelkemeir (1970) The thermal decomposition of barium sulfate to sulfur dioxide for mass spectrometric analysis. Anal. Chem., *42*, 1451–1453.

Jensen, M. L. (1957) Sulphur isotopes and mineral paragenesis. Econ. Geol., *52*, 269–281.

Jensen, M. L. (1967) Sulfur isotopes and mineral genesis. In Geochemistry of hydrothermal Ore Deposits, H. L. Barnes, ed. Holt, Rinehart and Winston, New York and London.

Kajiwara, Y., and H. R. Krouse (1971) Sulfur isotope partitioning in metallic sulfide systems. Can. J. Earth Sci., *8*, 1397–1408.

Kaplan, I. R., K. O. Emery, and S. C. Rittenberg (1963) The distribution and isotopic abundance of sulfur in recent marine sediments off southern California. Geochim. Cosmochim. Acta, *27*, 297–331.

Kaplan, I. R., and S. C. Rittenberg (1964) Microbiological fractionation of sulfur isotopes. J. Gen. Microbiol., *34*, 195–212.

Kemp, A. L. W., and H. G. Thode (1968) The mechanism of the bacterial reduction of sulfate and of sulfite from isotope fractionation studies. Geochim. Cosmochim. Acta, *32*, 71–91.

Kulp, J. L., W. U. Ault, and H. W. Feely (1956) Sulfur isotope abundances in sulfide minerals. Econ. Geol., *51*, 139–149.

McCready, R. G. L., I. R. Kaplan, and G. A. Din (1974) Fractionation of sulfur isotopes by the yeast *Saccharomyces cerevisiae*. Geochim. Cosmochim. Acta, *38*, 1239–1253.

Nakai, N., and M. L. Jensen (1964) The kinetic isotope effect in the bacterial reduction and oxidation of sulfur. Geochim. Cosmochim. Acta, *28*, 1893–1912.

Nielsen, H., and W. Ricke (1964) S-Isotopenverhältnisse von Evaporiten aus Deutschland. Ein Beitrag zur Kenntnis von δ^{34}S im Meerwasser-Sulfat. Geochim. Cosmochim. Acta, *28*, 577–591.

Nielsen, H. (1966) Schwefelisotope im marinen Kreislauf und das S^{34} der früheren Meere. Geol. Rundschau, *55*, 160–172.

Ohmoto, H. (1972) Systematics of sulfur and carbon isotopes in hydrothermal ore deposits. Econ. Geol., *67*, 551–578.

Perry, E. C., Jr., J. Monster, and T. Reimer (1971) Sulfur isotopes in Swaziland system barites and the evolution of the Earth's atmosphere. Science, *171*, 1105–1016.

Puchelt, H., B. R. Sabels, and T. C. Hoering (1971) Preparation of sulfur hexafluoride for isotope geochemical analysis. Geochim. Cosmochim. Acta, *35*, 625–628.

Rees, C. E. (1970) The sulphur isotope balance of the ocean: An improved model. Earth Planet. Sci. Letters, *7*, 336–370.

Rees, C. E. (1973) A steady state model for sulfur isotope fractionation in bacterial reduction processes. Geochim. Cosmochim. Acta, *37*, 1141–1162.

Ricke, W. (1964) Präparation von Schwefeldioxid für massenspektrometrische Bestimmungen des Schwefel-Isotopen-Verhältnisses S^{34}/S^{32} in natürlichen Schwefelverbindungen. Z. Anal. Chem., *199*, 401–413.

Rye, R. O. (1974) A comparison of sphalerite-galena sulfur isotope temperatures with filling temperatures of fluid inclusions. Econ. Geol., *69*, 26–32.

Rye, R. O., and H. Ohmoto (1974) Sulfur and carbon isotopes and ore genesis: A review. Econ. Geol., *69*, 826–842.

Sakai, H. (1957) Fractionation of sulfur isotopes in nature. Geochim. Cosmochim. Acta, *12*, 150–169.

Sakai, H. (1968) Isotopic properties of sulfur compounds in hydrothermal processes. Geochem. J., *2*, 29–49.

Sangster, D. F. (1968) Relative sulfur isotope abundances of ancient seas and stratabound sulphide deposits. Geol. Assoc. Canada Proc., *19*, 79–91.

Schneider, A., and H. Nielsen (1965) Zur Genese des elementaren Schwefels im Gips von Weenzen. Contrib. Mineral. Petrol., *11*, 705–717.

Schneider, A. (1970) The sulfur isotope composition of basaltic rocks. Contrib. Mineral. Petrol., *24*, 95.

Smith, J. W., and B. D. Batts (1974) The distribution and isotopic composition of sulfur in coal. Geochim. Cosmochim. Acta, *38*, 121–133.

Smitheringale, W. G., and M. L. Jensen (1963) Sulfur isotopic composition of the Triassic igneous rocks of eastern United States. Geochim. Cosmochim. Acta, *27*, 1183–1207.

Tatsumi, T. (1965) Sulfur isotopic fractionation between coexisting sulfide minerals from some Japanese ore deposits. Econ. Geol., *60*, 1645–1659.

Thode, H. G., J. MacNamara, and C. B. Collins (1949) Natural variations in the isotopic content of sulphur and their significance. Can. J. Res., *B-27*, 361–373.

Thode, H. G., J. Monster, and H. B. Dunford (1958) Sulphur isotope abundances in petroleum and associated materials. Bull. Amer. Assoc. Petrol. Geologists, *42*, 2619–2641.

Thode, H. G., A. G. Harrison, and J. Monster (1960) Sulphur isotope fractionation in early diagenesis of recent sediments of north-east Venezuela. Bull. Amer. Assoc. Petrol. Geologists, *44*, 1809–1817.

Thode, H. G., J. Monster, and H. B. Dunford (1961) Sulphur isotope geochemistry. Geochim. Cosmochim. Acta, *25*, 150–174.

Thode, H. G. (1963) Sulphur isotope geochemistry. In Studies in analytical geochemistry, pp. 25–41, D. M. Shaw, ed. Roy. Soc. Can. Spec. Pub. No. 6, University of Toronto Press, Toronto.

Thode, H. G., and J. Monster (1965) Sulfur-isotope geochemistry of petroleum, evaporites and ancient seas. In Fluids in Subsurface Environments, pp. 367–377, A. Young and J. E. Galley, eds. Amer. Assoc. Petrol. Geologists, Mem. 4.

Thode, H. G., and C. E. Rees (1970) Sulfur isotope geochemistry and Middle East oil studies. Endeavour, *29*, 24–28.

Thode, H. G. (1970) Sulfur isotope geochemistry and fractionation between coexisting minerals. Mineral. Soc. Amer. Spec. Paper, *3*, 133–144.

Trudinger, P. A., and L. A. Chambers (1973) Reversibility of bacterial sulfate reduction and its relevance to isotopic fractionation. Geochim. Cosmochim. Acta, *37*, 1775–1778.

Tudge, A. P., and H. G. Thode (1950) Thermodynamic properties of isotopic compounds of sulfur. Can. J. Res., *B28*, 567–578.

Vinogradov, A. P., V. A. Grinenko, and V. I. Ustinov (1962) Isotopic composition of sulfur compounds in the Black Sea. Geokhimiya, *10*, 973–997.

APPENDIX I
FITTING OF ISOCHRONS FOR DATING BY THE Rb-Sr METHOD

The line to be fitted to the data points on an isochron diagram should take into account the reliability of the observations. To do this, a set of coefficients is defined to weight each observation according to its relative precision of measurement, such that the most reliably determined values receive most weight and most strongly influence the estimate of the slope and intercept of the line. There are a number of ways to derive this set of weight coefficients. The simplest method is to assign all observations a weight of 1. However, this method is unsatisfactory for several reasons which we shall not elaborate. A more appropriate set of weight coefficients for small numbers of samples can be formed from the experimental errors. If each observation is weighted by the reciprocal of the square of its standard deviation, the resulting line reflects the known reliability of the observations. Although the reliability of an observation can be determined from replicate analyses, if is generally not well known, and weighting with experimental errors may, therefore, lead to poor estimates of the slope and intercept values of the isochron. A third set of weight coefficients can be formed by using the reciprocals of squares of residuals derived during the iterative procedure of fitting the isochron. (See commentary for Model 3 in Fig. A.1.) The three sets of weight coefficients described above yield identical estimates of well-determined isochrons, if the analytical data and the

estimates of their errors are both good. For this reason, all three have been used in the accompanying computer program to let the user see the influence of differently chosen weights.

The program listed in Figure A.1 was prepared by M. J. McSaveney, based on equation 6 of York (1969). It is written in Fortran IV and uses 126K of storage in the IBM 370/165 computer. The program listing is preceded by a commentary that provides additional information.

A data deck for the program must have a leading card specifying the number of sets of data to be fitted with isochrons (from 0 to 999). This number is punched, right justified, in I3 format in the first three columns. *Each set* of data following this card *must* have the following cards:

1. A title card identifying the set of samples,
2. A card of isochron information,
3. A series of cards containing the observations, one sample per card,
4. A card of precision parameters.

Explanations of these cards and their formats are given in the program. Blank cards are read as zeros, and the card count for each set of data must be exact. The program can handle up to 30 samples per isochron as dimensioned, but can handle any number of samples, if the dimension statement is changed. For more than 100 samples, the tolerance level must be specified (see program and later section). The correlation coefficient between errors for each sample can be omitted from the data cards if it is unknown or if it is a constant. Analytical errors must be specified for each observation and must *not* be given as fractional or percentage errors.

The card for precision parameters (item 4 above) requires some further explanation. In Model 3, where residuals are used for weighting, very large or even infinite weights can arise when the residuals are very small or zero. Therefore, a variable cutoff is used in which errors are assumed to vary linearly with the magnitudes of the coordinates at rates equal to the *percent* standard deviations derived from replicate analyses of a selected sample (e.g., 0.9 percent for $^{87}Rb/^{86}Sr$ and 0.05 percent for $^{87}Sr/^{86}Sr$). These values provide the PRESX and PRESY for the precision parameters. In addition, a small constant term (CONX) has been added in order to provide a finite estimated error for samples whose rubidium concentrations are so small that their $^{87}Rb/^{86}Sr$ ratios approach zero (e.g., CONX = 0.0001). Since the $^{87}Sr/^{86}Sr$ ratio in geological samples cannot be zero, the corresponding fixed error term CONY = 0. The cutoff actually used in Model 3 is 0.32 of the estimated errors so that only 25 percent of any given set of samples can have the maximum weight.

The constants discussed above are used only in Model 3. The other parameters to be entered in card 4 of each data set are used in all three models to calculate weighting factors. (See commentary in Fig. A.1.) They control the magnitudes of residuals that will be tolerated in the solution and include (1) the tolerance level; (2) the desired level of confidence; and (3) the proportion of a normal distribution that is included in the

calculation (Beyer, 1966). These may be left blank to use the internal table, or they may be specified to override the table. A very large tolerance level (e.g., 99.99) blocks the rejection, whereas a low tolerance level gives a trimmed estimate. The last number on this card is the correlation coefficient. If it is given separately on each observation card, this number should be set at -2.0 on card 4. If the correlation coefficient is the same for all data, it may be specified here as a number between -1 and $+1$ (but not including $+1$ which causes overflow in the weighting vector **W** in Model 1). If the space is left blank, the program assumes no correlation, even if values are punched on the observation cards.

An example of the program output is given in Figure A.2 based on test data derived from a line with a slope of 0.005 and an intercept of 0.7200, with random and normally distributed errors of up to 1 percent added to the $^{87}Rb/^{86}Sr$ values and up to 0.05 percent to the $^{87}Sr/^{86}Sr$ values (Brooks et al., 1972). The actual isochron has an age of 358.82 m.y., but because of the small sample size and because the residuals are not normally distributed about this isochron, this theoretical age is not recoverable from the data. (With errors proportional to the magnitude of the values, the residuals are lognormally distributed.)

With real data, well-determined isochrons yield identical model dates through the three weighting models. The third model usually gives a smaller value of the error estimate. In general, the model rejecting the fewest points should be selected. Data rejection, however, is a reliable trouble flag. When no data are rejected in any model, the three model dates should never be averaged because each is the best estimate derivable from a particular model. The choice of model is left to the user but should be specified in the presentation of the results.

FIGURE A-1   Listing of the computer program for weighted regression analysis of data points to a straight line. Based on York (1969) with modifications by M. J. McSaveney. (Pg. 427–433.)

```
C
C        PROGRAM FOR ANALYSIS OF CORRELATED AND UNCORRELATED ISOCHRON DATA
C        (WILL ALSO FIT LINES TO ANY SET OF CORRELATED OR UNCORRELATED DATA
C        PAIRS)
C
C        REQUIRES:
C                  NUMBER OF SETS OF DATA FOR WHICH ISOCHRONS ARE TO BE
C                  CALCULATED
C
C                  TWO SETS OF MEASURED VALUES AND THEIR ANALYTICAL ERRORS
C                  (FOR EXAMPLE,  A SET OF RB87/SR86 VALUES, AND A SET OF
C                  SR87/SR86 VALUES)
C                  CORRELATION BETWEEN ERRORS IN THE TWO SETS OF DATA
C                  ( IF ANY )
C
C                  ESTIMATED SLOPE FROM A PLOT OF THE DATA
C
C                  NUMBER OF SAMPLES TO BE FITTED TO EACH ISOCHRON
C
C                  NAME OF DECAYING ELEMENT
C                  (IF THE FIRST 4 LETTERS OF THIS NAME READ 'OMIT'
C                  THE PROGRAM WILL FIT A LINE BUT WILL NOT CALCULATE
C                  AN AGE).
C
C                  DECAY CONSTANT TO BE USED
C
C                  LABELS FOR THE TWO SETS OF MEASURED VALUES
C
C                  TOLERANCE LEVEL FOR TESTING FIT OF DATA TO THE
C                  CALCULATED ISOCHRON
C                  THE CHOSEN LEVEL OF CONFIDENCE AND THE PROPORTION OF THE
C                  EXPECTED POPULATION TO BE INCLUDED IN THE CALCULATION
C                  (TOLERANCE LEVELS ARE OBTAINED FROM TABLES, BASED ON THE
C                  NUMBER OF DEGREES OF FREEDOM, NUMBER OF SAMPLES MINUS
C                  TWO, AND AN ARBITRARILY CHOSEN LEVEL OF CONFIDENCE,
C                  THE 95% LEVEL IS RECOMMENDED)
C                  (IF THE TOLERANCE LEVEL IS NOT SPECIFIED, THE PROGRAM USES
C                  INTERNAL TABLES)
C
C                  EXPECTED STANDARD ERRORS FOR SAMPLES CONTAINING NONE OF
C                  THE MEASURED ELEMENTS
C                  PERCENTAGE ANALYTICAL ERRORS FOR STANDARD LABORATORY
C                  SAMPLES
C
C                  CORRELATION SIGNAL TO TELL PROGRAM THAT THE DATA IS
C                  CORRELATED
C                  RELATE =-2  CORRELATION COEFFICIENTS ARE GIVEN FOR EACH
C                  POINT
C                  RELATE=0     DATA ARE NOT CORRELATED
C                  RELATE=SOME SPECIFIED VALUE NOT EQUAL TO -2 OR 0,
C                  THIS VALUE IS USED FOR ALL DATA
C
C
C        PROGRAM USES EQUATION 6 OF YORK, D.J. (1969): LEAST SQUARES FITTING
C        OF A STRAIGHT LINE WITH CORRELATED ERRORS: EARTH & PLANETARY
C        SCIENCE LETTERS 5(1969)320-324, TO ESTIMATE THE WEIGHTED MINIMUM
C        VARIANCE STRAIGHT LINE FOR CORRELATED OR UNCORRELATED DATA
C        (THE MINIMUM VARIANCE SOLUTION IS ALWAYS OBTAINED BY THIS
C        METHOD.   ONLY IF THE RESIDUALS OF THE FIT OF THE DATA TO THE
C        ISOCHRON ARE NORMALLY DISTRIBUTED, WILL THE CALCULATED ISOCHRON
C        ALSO BE THE MAXIMUM LIKELIHOOD SOLUTION)
C
C        THREE WEIGHTING MODELS ARE USED:
C            1)   EQUAL WEIGHT TO ALL DATA
C                 (THIS REQUIRES WEIGHTING THE X VALUES WITH 1 AND THE Y
C                 VALUES WITH THE INVERSE SQUARE OF THE ESTIMATED
C                 GRADIENT )
C
C            2)   INVERSE SQUARES OF ANALYTICAL ERRORS FOR WEIGHTS
C
C            3)   INVERSE SQUARES OF RESIDUALS TO CALCULATED LINE
C                 FOR WEIGHTS (THIS DROPS THE NUMBER OF DEGREES OF FREEDOM
C                 BY 1)
```

```
C     THE TOLERANCE LEVEL IS USED TO CHECK THE FIT OF THE DATA IN
C     ALL WEIGHTING MODELS.  ANY POINT WHOSE RESIDUALS ARE ONE TOLERANCE
C     LEVEL OR GREATER THAN THE ANALYTICAL ERROR FOR THAT POINT IS
C     REMOVED FROM THE SOLUTION FOR THE LINE BY MOVING ALL DATA FOR THAT
C     POINT TO THE APPARENT END OF THE ARRAYS AND REDUCING THE APPARENT
C     NUMBER OF SAMPLES BY 1 FOR EACH POINT REJECTED.
C     IF TOO MANY POINTS ARE REJECTED, THE PROGRAM TAKES THE FIRST HALF OF
C     THE DATA, ESTIMATES A NEW APPROXIMATE SLOPE, AND TRYS AGAIN.
C     AT EACH ITERATION, ALL POINTS ARE CHECKED FOR FIT, AND WILL BE
C     BROUGHT BACK INTO THE SOLUTION IF THEY SHOULD HAPPEN TO FIT
C     IF PROGRAM CANNOT REACH A SATISFACTORY SOLUTION AFTER 9 CYCLES, EACH
C     WITH UP TO 9 ITERATIONS, THE CYCLING IS STOPPED, THE SOLUTION IS PRINTED
C     WITH ERROR MESSAGES (UP TO 3), AND THE NEXT WEIGHT MODEL IS ATTEMPTED.
C
C
C               THE PROGRAM FOLLOWS:
C
C
      REAL*4 SPEC1,SPEC2,EL,EM,ENT,OMIT,LA,BE,L1,LAB,BEL,L2
      DIMENSION X(030),Y(030),U(030),V(030),P(030),Q(030),W(030),SQW(030
     1),RESX(030),RESY(030),SPEC1(030),SPEC2(030),R(030),S(030),IBAD(030
     2),C(030),TOLTBL(31)
      DATA OMIT,TOLTBL/'OMIT',999.0,32.019,8.380,5.369,4.275,3.712,3.369
     1,3.136,3.532,3.379,3.259,3.162,3.081,3.012,2.954,2.903,2.858,2.819
     2,3.656,3.615,3.457,3.350,3.272,3.213,3.165,3.126,3.066,3.021,2.986,
     3,2.958,2.233/
C     READ NUMBER OF SETS OF DATA TO BE WORKED OVER ( I3 FORMAT)
C     BEGINING IN COLUMN 1, EG. 009 FOR 9 SETS, 011 FOR 11 SETS
      READ(5,11)L
      DO 100 J=1,L
      ITRASH=0
      INSIDE=0
      WRITE(6,99)
C     READ A TITLE FOR THE DATA SET. THE FIRST COLUMN MUST BE BLANK,
C     THE REMAINING 79 COLUMNS MAY BE ANYTHING THAT IDENTIFIES THE DATA
      READ(5,12)
      WRITE(6,12)
C     READ ESTIMATED SLOPE, NUMBER OF DATA PAIRS, NAME OF DECAYING ELEMENT
C     DECAY CONSTANT, AND LABELS FOR THE TWO SETS OF MEASURED VALUES
C     AN EXAMPLE OF THE FORMAT FOLLOWS
C.0050      10        RUBIDIUM 87        1.3900E-11 RB87/SR86    SR87/SR86
C
      READ(5,1)B,N,EL,EM,ENT,DECAY,LA,BE,L1,LAB,BEL,L2
      NSTORE=N
      BSTORE=B
      DECA=DECAY*1.0E+06
C     READ IN THE DATA SET, FIRST THE SAMPLE IDENTIFICATION, THEN IN
C     SUCCESSION, THE X VALUE, ITS ANALYTICAL ERROR, THE Y VALUE, ITS ERROR,
CGS12      AND LASTLY THE CORRELATION BETWEEN X AND Y ERRORS
CGS12      10.187      0.1356         0.8543          0.0005          -0.0053
      READ(5,10)(SPEC1(I),SPEC2(I),X(I),R(I),Y(I),S(I),C(I),I=1,N)
      WRITE(6,34)B,N
C
C     READ PRECISION PARAMETERS: THE EXPECTED STANDARD ERRORS FOR SAMPLES
C     WITH NONE OF THE MEASURED ELEMENTS, AND THE % STANDARD ERRORS OF
C     AVERAGE LABORATORY ANALYSES OF STANDARD SAMPLES.  THESE ARE USED TO
C     LIMIT THE MAXIMUM PERMISSIBLE WEIGHT IN WEIGHTING MODEL 3.
C     THE SAME CARD ALSO CARRIES THE TOLERANCE VALUE FOR REJECTION
C     THE CHOSEN LEVEL OF CONFIDENCE, THE PROPORTION OF THE EXPECTED
C     DISTRIBUTION TO BE USED FOR CALCULATION OF THE LINE, AND A
C     CORRELATION SIGNAL.  AN 8F10.5 FORMAT IS USED
C     THE NUMBERS BEGIN IN COLUMNS 1, 11, 21, 31, 41, 51, 61, & 71
C        (IF THE TOLERANCE LEVEL IS NOT SPECIFIED, THE PROGRAM USES AN INTERNAL
C         TABLE AND SETS A LEVEL SUCH THAT LESS THAN HALF A SAMPLE SHOULD BE
```

```
C          REJECTED AT EITHER END OF A NORMAL DISTRIBUTION OF ERRORS AT 95%
C          CONFIDENCE  -  THIS TABLE IS SET FOR 3 TO 100 SAMPLES, WITH
C          TOLERANCE LEVELS AT HIGHER DEGREES OF FREEDOM BEING SET BY LINEAR
C          INTERPOLATION BETWEEN TABULATED VALUES)

           READ(5,888) CONX,PRESX,CONY,PRESY,TOL,CONFID, PROP, RELATE
           IF(TOL.NE.0.0)INSIDE=1
           PRESX=PRESX/100
           PRESY=PRESY/100
           WRITE(6,901)
           IT=-2
C          SET WEIGHTS FOR MODEL 1
           DO 999 I=1,N
           P(I)=1.
   999     Q(I)=1/(B*B)
           GO TO 208
   445     WRITE(6,902)
C          WEIGHT MODEL 1 CAN LEAD TO POOR ESTIMATES OF SLOPE AND INTERCEPT,
C          MODEL 2 IS RESET TO ORIGINAL ESTIMATE OF SLOPE
           N=NSTORE
           B=BSTORE
C          SET WEIGHTS FOR MODEL 2
           DO 446 I=1,N
           P(I)=1/(R(I)*R(I))
   446     Q(I)=1/(S(I)*S(I))
   208     ITER=0
C          SET TOLERANCE LEVEL
           IF(IT.EQ.-1)GO TO 25
           IF(INSIDE.EQ.1)GO TO 25
           IDF=NSTORE-2
           IF(IT.EQ.0)IDF=IDF-1
           PROP=95.0
           IF(IDF.LT.9)PROP=90.0
           IF(IDF.GT.18)PROP=99.0
           CONFID=95.0
           IF(IDF.GT.20)GO TO 26
           IF(IDF.LT.1)IDF=1
           TOL=TOLTBL(IDF)
           GO TO 25
    26     IF(IDF.GT.50)GO TO 27
           IN=(IDF-20)/5
           INN=IDF-20-5*IN
           IN=IN+20
           TOL=TOLTBL(IN)+INN*(TOLTBL(IN+1)-TOLTBL(IN))/5.0
           GO TO 25
    27     IN=(IDF-50)/10
           INN=IDF-50-10*IN
           IN=IN+26
           TOL=TOLTBL(IN)+INN*(TOLTBL(IN+1)-TOLTBL(IN))/10.
    25     CONTINUE
   205     ITERAT=0
   200     CONTINUE
           ITERAT=ITERAT+1
           XBAR=0
           YBAR=0
           SUMA=0
           SUMB=0
           SUMD=0
           SUMP=0
           SUMQ=0
           SUMS=0
           SUMT=0
           SUMW=0
C          CHECK CORRELATION PARAMETER, AND SET UP W ARRAY ACCORDINGLY
           IF(RELATE.LT.-1) GO TO 6
           IF(RELATE)8,7,8
C          CORRELATION THE SAME FOR ALL POINTS
     8     DO 9 I=1,N
           TEMP=P(I)*Q(I)
           TEMP=TEMP/(B*B*Q(I)+P(I)-2.*B*RELATE*SQRT(TEMP))
           W(I)=TEMP
           SQW(I)=TEMP*TEMP
           SUMW=SUMW+TEMP
           C(I)=RELATE
     9     CONTINUE
           GO TO 161
     7     CONTINUE
C          NO CORRELATION
           DO 2 I=1,N
           C(I)=0.
           W(I) = P(I)*Q(I)/(B*B*Q(I) + P(I))
           SQW(I) = W(I)**2
     2     SUMW = SUMW + W(I)
           GO TO 161
     6     CONTINUE
C          CORRELATION AS GIVEN FOR EACH POINT
           DO 161 I=1,N
           TEMP=P(I)*Q(I)
           TEMP=TEMP/(B*B*Q(I)+P(I)-2.*B*C(I)*SQRT(TEMP))
           W(I)=TEMP
           SUMW=SUMW+TEMP
           SQW(I)=TEMP*TEMP
```

```
  161 CONTINUE
C         CALCULATE THE CENTER OF MASS OF THE POINTS TO BE FITTED TO THE LINE
C         (ALSO KNOWN AS THE CENTROID OF REVOLUTION)
          DO 3 I=1,N
          XBAR=XBAR+W(I)*X(I)
    3     YBAR=YBAR+W(I)*Y(I)
          XBAR=XBAR/SUMW
          YBAR=YBAR/SUMW
C         CALCULATE SLOPE OF THE ISOCHRON
          DO 4 I=1,N
          U(I)=X(I)-XBAR
          V(I)=Y(I)-YBAR
          T=SQRT(P(I)*Q(I))
          SUMA=SUMA+SQW(I)*V(I)*(U(I)/Q(I)+B*V(I)/P(I)-C(I)*V(I)/T)
          SUMB=SUMB+SQW(I)*U(I)*(U(I)/Q(I)+B*V(I)/P(I)-B*C(I)*U(I)/T)
    4     CONTINUE
          SLOPEC=SUMA/SUMB
          CINT=YBAR-SLOPEC*XBAR
          TEST1=ABS(B-SLOPEC)
          TEST2=ABS(B*0.0005)
          B=SLOPEC
C         TEST SLOPE ESTIMATE FOR RECYCLING
          IF(ITERAT.EQ.9)GO TO 38
          IF(TEST1.LE.TEST2) GO TO 38
          IF(IT+1)213,200,201
  213     DO 215 I=1,N
  215     Q(I)=1/(B*B)
          GO TO 200
C         RESET WEIGHT MATRIX FOR RECYCLING IN MODEL 3
  201     DO 202 I=1,N
          T=P(I)*Q(I)
          SEE=C(I)*SQRT(T)
          TEMP=(CINT+SLOPEC*X(I)-Y(I))*W(I)/T
          T=TEMP*(SEE-SLOPEC*Q(I))
          SEE=TEMP*(P(I)-SLOPEC*SEE)
          TEMP=T*T
C         25% OF DISTRIBUTION SHOULD LIE WITHIN 0.32 SIGMA(THEORETICAL).
C         THESE ARE GIVEN SIMILAR MAXIMUM WEIGHTS TO AVOID OVERWEIGHTING.
          ERROR=(CONX+PRESX*X(I))*0.32
          ERROR=ERROR*ERROR
          IF(TEMP.LT.ERROR)TEMP=ERROR
          P(I)=1/TEMP
          ERROR=(CONY+PRESY*Y(I))*0.32
          ERROR=ERROR*ERROR
          TEMP=SEE*SEE
          IF(TEMP.LT.ERROR)TEMP=ERROR
          Q(I)=1/TEMP
  202     CONTINUE
          GO TO 200
   38     CONTINUE
          ITER=ITER+1
          SUMRX = 0.0
          SUMRY = 0.0
          KOUNT=0
C         TEST FIT OF THE DATA, KEEPING TRACK OF ILL-FITTING DATA
          DO 31 I=1,N
          T=P(I)*Q(I)
          SEE=C(I)*SQRT(T)
          TEMP=(CINT+SLOPEC*X(I)-Y(I))*W(I)/T
          RESX(I)=TEMP*(SEE-SLOPEC*Q(I))
          RESY(I)=TEMP*(P(I)-SLOPEC*SEE)
          IF(ABS(RESX(I)).LE.TOL*R(I))GO TO 777
          KOUNT=KOUNT+1
          IBAD(KOUNT)=I
          GO TO 31
  777     CONTINUE
          IF(ABS(RESY(I)).LE.TOL*S(I))GO TO 776
          KOUNT=KOUNT+1
          IBAD(KOUNT)=I
          GO TO 31
  776     CONTINUE
          SUMD=SUMD+W(I)*U(I)*U(I)
          SUMS = SUMS + W(I)*(SLOPEC*U(I) - V(I))**2
          SUMT=SUMT+W(I)*X(I)*X(I)
          SUMRX=SUMRX+P(I)*RESX(I)*RESX(I)
          SUMRY=SUMRY+Q(I)*RESY(I)*RESY(I)
          SUMP=SUMP+P(I)
          SUMQ=SUMQ+Q(I)
   31     CONTINUE
          ISET=1
          IF(KOUNT.EQ.0)GO TO 5
C         TEST FOR REJECTION OF TOO MUCH DATA
C         IF LESS THAN TWO SAMPLES ARE LEFT, TRY FITTING TO FIRST HALF OF THE
C         DATA, WITH A RECALCULATED ESTIMATE OF THE SLOPE
C         THIS IS TRIED ONCE
          IF(KOUNT.LT.N-1)GO TO 15
          IF(ITRASH.EQ.1)GO TO 996
          IF(KOUNT.GT.N)GO TO 996
          N=NSTORE/2+1
          B=(Y(1)-Y(N))/(X(1)-X(N))
          ITRASH=ITRASH+1
          GO TO 205
```

430

```
   15 CONTINUE
C     SHIFT BAD DATA TO THE END OF THE ARRAY AND DECREASE THE ARRAY
C     LENGTH BY 1 FOR EACH POINT REJECTED.
      DO 5 I=1,KOUNT
      KK=KOUNT-I+1
      K=IBAD(KK)
   14 ST1=SPEC1(K)
      ST2=SPEC2(K)
      ST3=X(K)
      ST4=R(K)
      ST5=Y(K)
      ST6=S(K)
      ST7=C(K)
      ST8=P(K)
      ST9=Q(K)
      SPEC1(K)=SPEC1(N)
      SPEC2(K)=SPEC2(N)
      X(K)=X(N)
      R(K)=R(N)
      Y(K)=Y(N)
      S(K)=S(N)
      C(K)=C(N)
      P(K)=P(N)
      Q(K)=Q(N)
      SPEC1(N)=ST1
      SPEC2(N)=ST2
      X(N)=ST3
      R(N)=ST4
      Y(N)=ST5
      S(N)=ST6
      C(N)=ST7
      P(N)=ST8
      Q(N)=ST9
      IF(ISET.EQ.0)GO TO 13
      N=N-1
    5 CONTINUE
      NBAD=N+1
      IF(N.EQ.NSTORE)GO TO 13
      IF(NBAD.GT.NSTORE)GO TO 996
      IF(NBAD.LT.1)GO TO 996
C     CHECK THE OLD BAD DATA FOR POINTS THAT NOW FIT THE NEW LINE
      DO 13 I=NBAD,NSTORE
      T=P(I)*Q(I)
      SEE=C(I)*SQRT(T)
      WWW=T/(SLOPEC*SLOPEC*Q(I)+P(I)-2*SLOPEC*C(I)* SQRT(T))
      TEMP=(CINT+SLOPEC*X(I)-Y(I))*WWW/T
      RESX(I)=TEMP*(SEE-SLOPEC*Q(I))
      RESY(I)=TEMP*(P(I)-SLOPEC*SEE)
      IF(ABS(RESX(I)).GT.TOL*R(I))GO TO 13
      IF(ABS(RESY(I)).GT.TOL*S(I))GO TO 13
      K=I
      N=N+1
      ISET=0
      GO TO 14
   13 CONTINUE
C     TEST DATA REJECTION FOR RECYCLING
      IF(ITER.EQ.9)GO TO 206
      IF(KOUNT.NE.0)GO TO 205
      IF(ISET.EQ.0)GO TO 205
      IF(TEST1.GT.TEST2)GO TO 206
      GO TO 209
C     PROGRAM GETS TO HERE IF IT CAN NOT REACH A SATISFACTORY SOLUTION
C     FOR A GIVEN WEIGHT MODEL AFTER 9 ITERATIONS, IT STATES THIS, PRINTS
C     OUT THE UNSATISFACTORY SOLUTION, AND MOVES TO THE NEXT WEIGHT
C     MODEL OR SET OF DATA.
  206 WRITE(6,207)
      IF(TEST1.GT.TEST2)WRITE(6,210)
      IF(KOUNT.NE.0)WRITE(6,211)
      IF(ISET.EQ.0)WRITE(6,212)
  209 CONTINUE
      AN=N
      IF(NSTORE-N)996,17,19
   19 NBAD=N+1
C     SET WEIGHTS OF REJECTED POINTS EQUAL TO 0.0 FOR EASY IDENTIFICATION
      DO 17 I=NBAD,NSTORE
      P(I)=0
      Q(I)=0
   17 CONTINUE
      IF(N.GT.2)GO TO 998
C     IF TOO MANY POINTS HAVE BEEN REJECTED, SET ERROR LIMITS OF SLOPE
C     AND INTERCEPT VERY HIGH
      WRITE(6,990)
      SIGMAB=0.01
      SIGMAA=1.
      GO TO 997
  998 CONTINUE
      IF(IT+1)602,602,601
  601 AN=AN-1
      IF(AN.GT.2.)GO TO 602
      SIGMAB=0.01
      SIGMAA=1.
      GO TO 997
```

```
      602 CONTINUE
C         CALCULATE STANDARD ERROR OF SLOPE AND INTERCEPT
          SIGMAB=SQRT(SUMS/((AN-2.)*SUMD))
          SIGMAA=SIGMAB*SQRT(SUMT/SUMW)
C         CALCULATE VARIANCE OF UNIT WEIGHT OF AN OBSERVATION
          CHI2=SUMRX+SUMRY
          SUMRX=SQRT(SUMRX/SUMP/N)
          SUMRY=SQRT(SUMRY/SUMQ/N)
      997 CONTINUE
C         BEGIN TO WRITE OUT DATA, AND RESULTS
          WRITE(6,32)LA,BE,L1,LAB,BEL,L2
          WRITE (6,33) (SPEC1(I),SPEC2(I),X(I),R(I),P(I),RESX(I),Y(I),S(I),Q
         1(I),RESY(I),C(I),I=1,NSTORE)
          WRITE(6,43)SUMRX,SUMRY
          WRITE (6,37) SLOPEC, SIGMAB, CINT, SIGMAA
          XYSUM=0
          XSUM=0
          YSUM=0
          SUMX2=0
          SUMY2=0
C         CALCULATE CORRELATION COEFFICIENT FOR THE DATA THAT FIT THE LINE
          DO 40 I=1,N
          XYSUM=XYSUM+X(I)*Y(I)
          XSUM=XSUM+X(I)
          SUMX2=SUMX2+X(I)*X(I)
          SUMY2=SUMY2+Y(I)*Y(I)
          YSUM=YSUM+Y(I)
       40 CONTINUE
          AN=N
          XYSUM=XYSUM*SLOPEC
          XSUM=XSUM*SLOPEC
          SUMX2=SUMX2*SLOPEC*SLOPEC
          ANUMER = (AN*XYSUM) - (XSUM*YSUM)
          DENOM=SQRT((AN*SUMX2-XSUM*XSUM)*(AN*SUMY2-YSUM*YSUM))
          CORCOE = ANUMER/DENOM
          WRITE(6,42)CORCOE,CHI2
C         CALCULATE AGE ESTIMATE, AND ITS STANDARD ERROR, IF AN ISOCHRON
C         IS BEING FITTED, OTHERWISE TEST FOR THE WORD  OMIT AND SKIP THE SECTION
          IF(EL.EQ.OMIT)GO TO 556
          DATE=ALOG(SLOPEC+1.)/DECA
          ERROR=SIGMAB/(SLOPEC+1)/DECA
          WRITE(6,41)DATE,ERROR,EL,EM,ENT,DECAY
      556 CONTINUE
C         WRITE OUT THE NUMBER AND NAMES OF THE BAD DATA, IF ANY
          KOUNT=NSTORE-N
          IF(KOUNT.EQ.0)GO TO 850
          WRITE(6,346)KOUNT
          NN=N+1
          WRITE(6,164)(SPEC1(I),SPEC2(I),I=NN,NSTORE)
          WRITE(6,444)N
      850 IT=IT+1
C         WRITE OUT THE STATEMENT OF PROBABILITY USED TO TEST THE FIT OF
C         EACH DATA POINT TO THE LINE
          SAMPL=(1-PROP/100.)*NSTORE/2.
          ICHANC=1.+1./(1.-CONFID/100.)
          WRITE(6,160)TOL,PROP,CONFID,ICHANC,SAMPL
          IF(IT)445,347,100
      347 CONTINUE
          WRITE(6,331)
          N=NSTORE
C         INITIALIZE WEIGHTS FOR MODEL 3
C         MODEL 3 REQUIRES VERY GOOD ESTIMATES OF SLOPE AND INTERCEPT,
C         INITIALIZE WEIGHTS FOR MODEL 3, USING SLOPE AND INTERCEPT ESTIMATES
C         FROM MODEL 2
          A=CINT
          DO 348 I=1,N
          TEMP=A+B*X(I)-Y(I)
          TEMP=TEMP*TEMP*(1+B*B)
          T=(CONY+PRESY*Y(I))*0.32
          T=T*T
          IF(TEMP.LT.T)TEMP=T
          Q(I)=1/TEMP
          T=(CONX+PRESX*X(I))*0.32
          T=T*T
          TEMP=TEMP/(B*B)
          IF(TEMP.LT.T)TEMP=T
          P(I)=1/TEMP
      348 CONTINUE
          GO TO 208
      996 WRITE(6,995)
      100 CONTINUE
          STOP
        1 FORMAT(F10.6,I4,6X,3A4,8X,E10.4,6A4)
       10 FORMAT(2A4,2X,F14.8,F15.10,F15.8,F15.10,F10.7)
       11 FORMAT(I3)
       12 FORMAT(1H0,'DATE,SAMPLE NOS. & LOCATION, ETC.
         1                                     ')
       32 FORMAT(     '  SAMPLE',24X,3A4,38X,3A4,//,13X,'VALUE',7X,'SIGMA',6X,'
         1WEIGHT',6X,'RESIDUAL',11X,'VALUE',7X,'SIGMA',6X,'WEIGHT',6X,'RESID
         2UAL',6X,'CORRELATION',/)
       33 FORMAT (1X, 2A4,F10.4,F11.4,E13.2,F12.5,F18.4,F11.4,E13.2,F12.5,9X
         1,F7.4)
```

```
   34 FORMAT (1H0,'ESTIMATED SLOPE =',F15.8,'       NO. OF SAMPLES =',I10)
   37 FORMAT( /,8X,'THE BEST SLOPE   =',F13.8,'+/-',F11.8,5X,'THE BEST IN
     1TERCEPT  =',F15.8,'+/-',F11.8)
   41 FORMAT(      /,' ESTIMATED AGE OF THE SUITE =',F7.1,'  +/- ',F5.1,
     1' MILLIONS OF YEARS, FOR A ',3A4,' DECAY CONSTANT OF ',E10.4,/)
   42 FORMAT(1H0,25X,'CORRELATION COEFFICIENT =',F10.6,10X,'GOODNESS OF
     1FIT (CHI SQUARED) =',F6.2)
   43 FORMAT(10X,'WEIGHTED ROOT MEAN SQUARE RESIDUAL =',F9.5,0X,'WEIGHTE
     1D ROOT MEAN SQUARE RESIDUAL =',F9.5)
   99 FORMAT(1H1)
  160 FORMAT(      /,' A TOLERANCE LEVEL OF ',F5.2,' WAS USED TO TEST THE
     1FIT OF THE DATA, AT THIS LEVEL ',F4.1,'% OF A NORMAL POPULATION WO
     2ULD BE',/,' INCLUDED IN THE CALCULATION AT THE ',F4.1,'% CONFIDENC
     3E LEVEL (THAT IS, THERE IS ONLY 1 CHANCE IN ',I4,' THAT MORE THAN
     4',F5.2,/,' SAMPLES WOULD BE REJECTED FROM EACH END OF A NORMAL DIS
     5TRIBUTION OF ERRORS ABOUT THE LINE).',/)
  164 FORMAT(10X,13(2A4,',')/,13(2A4,',')/,13(2A4,',')/,13(2A4,',')/
     1)
  207 FORMAT(' THE FOLLOWING SOLUTION MAY BE TECHNICALLY UNSATISFACTORY
     1BECAUSE:')
  210 FORMAT(' LAST ADJUSTMENT OF SLOPE STILL EXCEEDED 0.05% OF THE SLOPE
     1E ESTIMATE ON THE 9 TH ITERATION')
  211 FORMAT(' AN OBSERVATION(S) WAS REJECTED IN THE LAST (9TH) ITERATIO
     1N')
  212 FORMAT(' AN OBSERVATION(S) WAS REINTRODUCED AT THE END OF THE LAST
     1 (9TH) ITERATION')
  331 FORMAT(1H0,25X,'LINE FITTED WITH RECIPROCALS OF SQUARES OF RESIDUA
     1LS FOR WEIGHTS',/,26X,'---- ------ ---- ----------- -- ------- --
     2--------- ---------')
  346 FORMAT(10X,'THE FOLLOWING ',I3,' SAMPLES DO NOT FIT THE CALCULATED
     1 ISOCHRON:',/)
  444 FORMAT(10X,'THEY HAVE NOT BEEN USED TO ESTIMATE THE ISOCHRON, OR T
     1HE CORRELATION COEFFICIENT',/,10X,'THE CALCULATED LINE IS THE WEIG
     2HTED BEST ESTIMATE OF THE REMAINING ',I3,' SAMPLES')
  888 FORMAT(8F10.5)
  901 FORMAT(/,50X,'ALL POINTS WEIGHTED EQUALLY',/,50X,'--- ------- -----
     1--- -------')
  902 FORMAT(///,25X,'RECIPROCALS OF SQUARES OF ANALYTICAL ERRORS USED F
     1OR WEIGHTS',/,25X,'----------- -- ------- -- ---------- ------- ----
     2- --- -------')
  990 FORMAT(' ****************************LINE FITTED TO 2 POINTS OR LESS
     1***********************************')
  995 FORMAT(' THESE DATA DO NOT RESEMBLE A STRAIGHT LINE, AND CANNOT BE
     1 ADJUSTED BY THIS PROGRAM.')
      END
/*
```

FIGURE A-2 Sample output, showing the three different weighting schemes used in the program listed in Figure A-1. The test data are from Brooks et al. (1972).

THEORETICAL TEST DATA SET 1 FROM BROOKS, HART, AND WENDT (1972)

ESTIMATED SLOPE =   0.00500000   NO. OF SAMPLES =   10

ALL POINTS WEIGHTED EQUALLY

SAMPLE		RB87/SR86				SR87/SR86			
	VALUE	SIGMA	WEIGHT	RESIDUAL	VALUE	SIGMA	WEIGHT	RESIDUAL	CORRELATION
1	10.1870	0.1019	0.10E+01	0.04343	0.7707	0.0004	0.40E+05	-0.00022	0.0
2	20.2820	0.2028	0.10E+01	-0.09816	0.8200	0.0004	0.40E+05	0.00049	0.0
3	29.8890	0.2990	0.10E+01	0.13160	0.8706	0.0004	0.40E+05	-0.00066	0.0
4	39.9000	0.3990	0.10E+01	0.00119	0.9196	0.0005	0.40E+05	-0.00001	0.0
5	49.8750	0.4990	0.10E+01	0.07579	0.9704	0.0005	0.40E+05	-0.00038	0.0
6	60.3420	0.6034	0.10E+01	-0.30353	1.0192	0.0005	0.40E+05	0.00153	0.0
10	99.6200	0.9962	0.10E+01	0.05332	1.2202	0.0006	0.40E+05	-0.00027	0.0
8	79.5280	0.7953	0.10E+01	0.19688	1.1207	0.0006	0.40E+05	-0.00099	0.0
9	89.9820	0.8998	0.10E+01	-0.10029	1.1702	0.0006	0.40E+05	0.00050	0.0
7	70.8890	0.7089	0.0	-0.48992	1.0704	0.0005	0.0	0.00245	0.0
	WEIGHTED ROOT MEAN SQUARE RESIDUAL = 0.04693				WEIGHTED ROOT MEAN SQUARE RESIDUAL = 0.00024				

THE BEST SLOPE = 0.00502557+/- 0.00001803   THE BEST INTERCEPT =   0.71902812+/- 0.00109974

CORRELATION COEFFICIENT = 0.999947   GOODNESS OF FIT (CHI SQUARED) = 0.36

ESTIMATED AGE OF THE SUITE = 360.6 +/- 1.3 MILLIONS OF YEARS, FOR A RUBIDIUM 87 DECAY CONSTANT OF 0.1390E-10

THE FOLLOWING 1 SAMPLES DO NOT FIT THE CALCULATED ISOCHRON:

7
THEY HAVE NOT BEEN USED TO ESTIMATE THE ISOCHRON, OR THE CORRELATION COEFFICIENT
THE CALCULATED LINE IS THE WEIGHTED BEST ESTIMATE OF THE REMAINING 9 SAMPLES

A TOLERANCE LEVEL OF 3.14 WAS USED TO TEST THE FIT OF THE DATA, AT THIS LEVEL 90.0% OF A NORMAL POPULATION WOULD BE INCLUDED IN THE CALCULATION AT THE 95.0% CONFIDENCE LEVEL (THAT IS, THERE IS ONLY 1 CHANCE IN 20 THAT MORE THAN 0.50 SAMPLES WOULD BE REJECTED FROM EACH END OF A NORMAL DISTRIBUTION OF ERRORS ABOUT THE LINE).

RECIPROCALS OF SQUARES OF ANALYTICAL ERRORS USED FOR WEIGHTS

SAMPLE		RB87/SR86				SR87/SR86			
	VALUE	SIGMA	WEIGHT	RESIDUAL	VALUE	SIGMA	WEIGHT	RESIDUAL	CORRELATION
1	10.1870	0.1019	0.96E+02	0.02025	0.7707	0.0004	0.67E+07	-0.00006	0.0
2	20.2820	0.2028	0.24E+02	-0.18577	0.8200	0.0004	0.59E+07	0.00015	0.0
3	29.8890	0.2990	0.11E+02	0.25795	0.8706	0.0004	0.53E+07	-0.00011	0.0
4	39.9000	0.3990	0.63E+01	0.05129	0.9196	0.0005	0.47E+07	-0.00001	0.0
5	49.8750	0.4990	0.40E+01	0.23098	0.9704	0.0005	0.43E+07	-0.00004	0.0
6	60.3420	0.6034	0.27E+01	-0.47052	1.0192	0.0005	0.38E+07	0.00007	0.0
10	99.6200	0.9962	0.10E+01	0.36815	1.2202	0.0006	0.27E+07	-0.00003	0.0
8	79.5280	0.7953	0.16E+01	0.57819	1.1207	0.0006	0.32E+07	-0.00006	0.0
9	89.9820	0.8998	0.12E+01	0.03015	1.1702	0.0006	0.29E+07	-0.00000	0.0
7	70.8890	0.7089	0.20E+01	-0.80018	1.0704	0.0005	0.35E+07	0.00009	0.0
	WEIGHTED ROOT MEAN SQUARE RESIDUAL = 0.05405				WEIGHTED ROOT MEAN SQUARE RESIDUAL = 0.00003				

THE BEST SLOPE = 0.00500753+/- 0.00001930   THE BEST INTERCEPT =   0.71948916+/- 0.00055702

CORRELATION COEFFICIENT = 0.999905   GOODNESS OF FIT (CHI SQUARED) = 4.69

ESTIMATED AGE OF THE SUITE = 359.3 +/- 1.4 MILLIONS OF YEARS, FOR A RUBIDIUM 87 DECAY CONSTANT OF 0.1390E-10

A TOLERANCE LEVEL OF 3.14 WAS USED TO TEST THE FIT OF THE DATA, AT THIS LEVEL 90.0% OF A NORMAL POPULATION WOULD BE INCLUDED IN THE CALCULATION AT THE 95.0% CONFIDENCE LEVEL (THAT IS, THERE IS ONLY 1 CHANCE IN 20 THAT MORE THAN 0.50 SAMPLES WOULD BE REJECTED FROM EACH END OF A NORMAL DISTRIBUTION OF ERRORS ABOUT THE LINE).

LINE FITTED WITH RECIPROCALS OF SQUARES OF RESIDUALS FOR WEIGHTS

SAMPLE		RB87/SR86				SR87/SR86			
	VALUE	SIGMA	WEIGHT	RESIDUAL	VALUE	SIGMA	WEIGHT	RESIDUAL	CORRELATION
1	10.1870	0.1019	0.94E+03	0.00160	0.7707	0.0004	0.62E+08	-0.00000	0.0
2	20.2820	0.2028	0.53E+02	-0.14119	0.8200	0.0004	0.22E+07	0.00069	0.0
3	29.8890	0.2990	0.10E+03	0.09330	0.8706	0.0004	0.42E+07	-0.00046	0.0
4	39.9000	0.3990	0.61E+02	-0.07076	0.9196	0.0005	0.44E+08	0.00002	0.0
5	49.8750	0.4990	0.39E+02	0.07429	0.9704	0.0005	0.40E+08	-0.00001	0.0
6	60.3420	0.6034	0.94E+01	-0.34460	1.0192	0.0005	0.38E+06	0.00168	0.0
10	99.6200	0.9962	0.98E+01	0.04378	1.2202	0.0006	0.25E+08	-0.00000	0.0
8	79.5280	0.7953	0.15E+02	0.25365	1.1207	0.0006	0.21E+07	-0.00037	0.0
9	89.9820	0.8998	0.12E+02	-0.26083	1.1702	0.0006	0.28E+08	0.00002	0.0
7	70.8890	0.7089	0.0	-0.53150	1.0704	0.0005	0.0	0.00259	0.0
	WEIGHTED ROOT MEAN SQUARE RESIDUAL = 0.02203				WEIGHTED ROOT MEAN SQUARE RESIDUAL = 0.00004				

THE BEST SLOPE = 0.00502432+/- 0.00001129   THE BEST INTERCEPT =   0.71946442+/- 0.00030635

CORRELATION COEFFICIENT = 0.999949   GOODNESS OF FIT (CHI SQUARED) = 8.73

ESTIMATED AGE OF THE SUITE = 360.5 +/- 0.8 MILLIONS OF YEARS, FOR A RUBIDIUM 87 DECAY CONSTANT OF 0.1390E-10

THE FOLLOWING 1 SAMPLES DO NOT FIT THE CALCULATED ISOCHRON:

7
THEY HAVE NOT BEEN USED TO ESTIMATE THE ISOCHRON, OR THE CORRELATION COEFFICIENT
THE CALCULATED LINE IS THE WEIGHTED BEST ESTIMATE OF THE REMAINING 9 SAMPLES

A TOLERANCE LEVEL OF 3.37 WAS USED TO TEST THE FIT OF THE DATA, AT THIS LEVEL 90.0% OF A NORMAL POPULATION WOULD BE INCLUDED IN THE CALCULATION AT THE 95.0% CONFIDENCE LEVEL (THAT IS, THERE IS ONLY 1 CHANCE IN 20 THAT MORE THAN 0.50 SAMPLES WOULD BE REJECTED FROM EACH END OF A NORMAL DISTRIBUTION OF ERRORS ABOUT THE LINE).

REFERENCES

Beyer, W. H., ed. (1966) Handbook of Tables for Probability and Statistics. The Chem. Rubber Co., Cleveland, Ohio, 502 p.

Brooks, C., S. R. Hart, and I. Wendt (1972) Realistic use of two-error regression treatments as applied to rubidium strontium data. Rev. Geophys. Space Phys., *10*, 551–557.

York, D. (1969) Least squares fitting of a straight line with correlated errors. Earth Planet. Sci. Letters, *5*, 320–324.

APPENDIX II
THE GEOLOGICAL TIME SCALE
FOR THE PHANEROZOIC

ERA	PERIOD	EPOCH	AGE (EUROPE)	AGES IN MILLIONS OF YEARS				
				1	2	3	4	5
Cenozoic	Quaternary	Holocene Pleistocene	not listed					
				1	1	1.5–2		
	Neogene	Pliocene Miocene	not listed	11	13	7		
				25	25	26		
	Paleogene	Oligocene Eocene Paleocene	not listed	40	36	37–38		
				60	58	53–54		
				70 ± 2	63	65	64	64–65
Mesozoic	Cretaceous	Late	Mastrichtian	· · · ·	· · · ·	70	· · · ·	70–71
			Campanian	· · · ·	· · · ·	76	· · · ·	82
			Santonian	· · · ·	· · · ·	82	· · · ·	86
			Coniacian	· · · ·	· · · ·	88	· · · ·	87
			Turonian			94		89–90
			Cenomanian	· · · ·	· · · ·	100	102	94
		Early	Albian	· · · ·	· · · ·	106		
			Aptian	· · · ·	· · · ·	112		
			Barremian	· · · ·	· · · ·	118		
			Hauterivian	· · · ·	· · · ·	124		
			Valanginian	· · · ·	· · · ·	130		
			Berriasian	135 ± 5	135	136	140	
	Jurassic	Malm (Late)	Portlandian	· · · ·	· · · ·	146		
			Kimmeridgian	· · · ·	· · · ·	151		
			Oxfordian	· · · ·	· · · ·	157		
		Dogger (Middle)	Callovian	· · · ·	· · · ·	162		
			Bathonian	· · · ·	· · · ·	167		
			Bajocian	· · · ·	· · · ·	172		
			Aalenian	· · · ·	· · · ·	· · · ·		
		Lias (Early)	Toarcian	· · · ·	· · · ·	178		
			Pliensbachian	· · · ·	· · · ·	183		
			Sinemurian	· · · ·	· · · ·	188		
			Hettangian	180 ± 5	181	190–195	208	

ERA	PERIOD	EPOCH	AGE (EUROPE)	AGES IN MILLIONS OF YEARS				
				1	2	3	4	5
	Triassic	Late	Rhaetian Norian Carnian	....	....	....		
				....	....	....		
				....	....	205?		
		Middle	Ladinian Anisian	....	....	....		
				....	....	215?		
		Early	Scythian					
Paleozoic	Permian			225 ± 5	230	225	> 242	
		Late	Tatarian Kazanian	....	....	230		
				....	....	240		
		Early	Kungurian Sakmarian	....	....	255 – 258		
	Carboniferous (Pennsylvanian) (Mississippian)			270 ± 5	280	280	284	
		Late	Stephanian Westphalian Namurian	....	....	290 – 295		
				....	....	310 – 315		
		Early	Visean Tournaisian	....	310	325	335	
				....	....	335 – 340		
	Devonian			350 ± 5	345	345	360	
		Late	Famennian Frasnian	....	....	353		
				....	....	359		
		Middle	Givetian Couvinian	....	....			
				....	....	370		
		Early	Emsian Siegenian Gedinnian	....	....	374		
				....	....	390		
	Silurian			400 ± 10	405	395	409	
		Late	Ludlovian Wenlockian	....	....	....		
				....	....	....		
		Early	Llandoverian Valentian	....	....	430 – 440		
	Ordovician			440 ± 10	425?	....	> 436	
		Late	Ashgillian Caradocian	....		....		
				....	....	445		

ERA	PERIOD	EPOCH	AGE (EUROPE)	AGES IN MILLIONS OF YEARS				
				1	2	3	4	5
		Early	Llandeilian	....	....	....		
			Llanvirnian	....	....	....		
			Arenigian	....	....	~500		
			Skiddavian	....	....	....		
			Tremadocian					
				500 ± 15	500	....	~500	
	Cambrian	Late	Shidertinian	....	....	....		
			Tuorian	....	....	515		
		Middle	Mayan	....	....	....		
			Amgan	....	....	....		
			Lenan	....	....	540		
		Early	Aldanian					
				600 ± 20	600?	570	~564	

1. Holmes (1959)
2. Kulp (1961)
3. Harland et al. (1964)
4. Armstrong and McDowall (1974)
5. Obradovich and Cobban (1975)

REFERENCES

Armstrong, R. L., and W. G. McDowall (1974) Proposed refinement of the Phanerozoic time scale. Int. Meeting Geochron., Cosmochron., Isotope Geol., Aug. 26–31, Un. Paris, France.

Harland, W. B., A. Gilbert Smith, and B. Wilcock, eds. (1964) The Phanerozoic time scale. Quart. J. Geol. Soc. London, *1205*, 458 p.

Holmes, A, (1959) A revised geological time scale. Trans. Edinb. Geol. Soc., *17*, 183–216.

Kulp, J. L. (1961) Geologic time scale. Science, *133*, 1105–1114.

Obradovich, J. D., and W. A. Cobban (1975) A time-scale for the Late Cretaceous of the western interior of North America. Geol. Assoc. Canada, Spec. Paper, no 13, 31–54.

AUTHOR INDEX

Bofinger, V. M., 88, 93
Bogard, D. D., 109, 139
Boger, P. D., 100, 106, 133, 135, 136, 139
Bohr, N., 5, 41, 43
Bolinger, J., 224
Boltwood, B. B., 6, 7, 10
Bothe, W., 5
Bottinga, Y., 354, 355, 356, 370, 371, 373, 375, 387, 388, 389, 390, 391, 392, 396, 397
Boudin A., 186, 188, 193
Bowen, N. L., 362
Bowman, H. R., 100, 106
Bowman, J. R., 106, 122
Bracken, N. T. J., 191, 195
Brereton, N. R., 167, 168, 171, 173, 179
Brickwedde, F. G., 8, 11
Bridgwater, D., 94
Brinkman, G. A., 186, 193
Brock, M. R., 245
Broecker, W. S., 291, 292, 293, 300, 301, 302, 303, 304, 317, 319, 320, 323, 327, 346, 382, 397, 398
Brookins, D. G., 203, 222, 274, 280
Brooks, C., 90, 93, 94, 103, 106, 121, 139, 141, 434, 436
Brooks, C. K., 188, 194
Brown, E. W., 10
Brown, H., 266
Brown, J. F., 176, 179
Bruland, K. W., 303
Brunfelt, A. O., 54, 63
Brunhes, B., 153
Bryhni, I., 171, 177, 179
Buchsbaum, R., 347
Buford, P., 8, 279
Burcham, W. E., 305, 319
Burchart, J. R., 268, 271, 278
Burchfield, J. D., 2, 10
Burger, A. J., 205, 223
Burke, W. H., Jr., 314, 319
Burnett, D. S., 108, 139, 143, 144
Butler, W. A., 218, 222

Caelles, J. C., 142
Calder J. A., 381, 401
Cameron, A. G. W., 27, 43
Campbell, N. R., 75, 93, 147, 162
Carmichael, C. M., 176, 179
Carpenter, B. S., 269, 278, 280
Carter, J. A., 304
Catanzaro, E. J., 75, 93, 210, 211, 214, 215, 222, 261, 264
Chadwick, J., 5, 10, 26

Chamberlin, T. C., 2, 10
Chambers, L. A., 404, 422
Chan, L. H., 278
Chen, J. H., 230, 245
Cheney, E. S., 386, 396, 397
Cherdyntsev, V. V., 8, 288, 302
Cherskiy, N. V., 387, 399
Choquette, P. W., 385, 397
Christensen, R. J., 375
Christner, M., 106
Church, S. E., 114, 139, 253, 254, 255, 262, 264, 274, 277, 278
Clague, D. A., 162
Clark, A. H., 142
Clarke, S. E., Jr., 225
Clarke, W. B., 224
Clauer, N., 89, 93
Clausen, H. B., 332, 346, 347
Clayton, R. N., 325, 342, 346, 348, 351, 353, 354, 357, 358, 368, 371, 375, 376, 377, 386, 397
Cliff, R. A., 110, 139
Cobban, W. A., 439, 440
Coleman, M. L., 190, 194
Coleman, R. C., 143
Coleman, R. G., 371, 378
Collerson, K. D., 94
Collins, C. B., 422
Colombo, U., 381, 397
Compston, W., 68, 72, 88, 93, 94, 358, 375
Condie, K. C., 100, 106, 122, 144
Conway, C. M., 368, 375, 386, 397
Cooper, J. A., 149, 162, 253, 264
Corless, J. T., 190, 194, 195
Cortecci, G., 336, 346
Coscio, M. R., Jr., 143
Cotlovskay, F. I., 164
Cowan, C. L., 26
Cowart, J. B., 303
Cox, A., 154, 155, 162
Craig, H., 8, 10, 312, 319, 325, 326, 327, 328, 329, 333, 334, 335, 336, 337, 339, 340, 341, 342, 343, 344, 346, 376, 379, 388, 389, 390, 392, 396, 397, 398, 401
Crawford, A. R., 274, 278
Cristy, S. S., 281
Crocket, J. H., 142, 184, 193, 387, 400
Crookes, 5, 45
Crozaz, G., 295, 296, 301, 302, 303
Cumming, G. L., 266
Cunningham, B. B., 222, 245
Curie, I., 5
Curie, M. S., 3, 4, 5, 10

White, R. W., 179
Whiteford, D. J., 117, 145
Whitlock, D. W., 398
Wien, W., 65
Wilcock, B., 9, 141, 440
Wilgain, S., 200, 223, 246, 284, 285, 303
Willkomm, H., 398
Wilson, A. F., 371, 378
Wilson, E. E., 73
Winchester, J. W., 190, 194

Windom, H., 295, 304
Wollenberg, H. A., 126, 145
Wood, A., 75, 93, 147, 162
Woods, R. T., 279

Yachenko, M. L., 9
Yamaguchi, M., 206, 223
Yanagi, T., 119, 145
Yaniv, A., 303
Yardley, D. H., 141

SUBJECT INDEX

Baltimore gneiss, 92, 95
Barbados, 292, 301
Barberton Mountain Land, South Africa, 410
Barite, 284, 408, 410, 417
Barium, 41, 43, 67, 75
Barn, 58
Basalt, 76, 112, 115, 117, 118, 122, 123, 152, 160, 174, 176, 187, 199, 252-256, 257, 354, 359
Basaltic achondrites, 108, 109, 110, 143, 178, 180, 359
Base Roi Baudouin, Antarctica, 295, 296
Basin and Range province, 142
Batholith, 125, 126, 137, 141
Bathurst, New Brunswick, 236
Bay of Kiel, Germany, 405, 406
Beacon Supergroup, Antarctica, 129, 140
Bear province, Canada, 159
Beardmore glacier, Antarctica, 141
Belt Supergroup, Montana, 247, 375
Benioff zone, 114, 115, 117, 119, 126
Benzene, 313
Beryl, 153
Beryllium, 17, 20, 48, 75
Beta decay, 25-36, 75, 147, 183, 186, 189, 199-202, 307
Beta particle, 4, 25-36, 54, 75, 148, 183, 199-202
Binding energy, 16, 33
Biotite, 75, 82, 84, 85, 86, 151, 153, 155, 156, 157, 160, 171, 172, 177, 187, 197, 211, 267, 275, 356, 358, 370, 371
Birunga volcanic area, East Africa, 103
Bismuth, 67, 295
Bitter Springs Formation, Australia, 400
Biwabik Iron Formation, Minnesota, 370
Bjurbole (meteorite), 373
Black Hills, South Dakota, 246
Black hole, 13
Black Rock Desert, Utah, 106, 144
Black Sea, 405
Blanchard Springs cavern, U.S.A., 304
Bleiberg, Austria, 234
Blocking temperature, 155, 157, 272
Bluebell mine, British Columbia, 376
Bohemia mining district, Oregon, 363
Boling salt dome, Texas, 408, 409
Boltzman's constant, 272
Bone necrosis, 54
Boron, 20
Boulder Creek batholith, Colorado, 206, 210
Branched beta decay, 33-36, 186, 189
Brice Creek stock, Oregon, 363
Brine, 88, 121, 140, 340-344, 363, 367, 410

British Columbia, 235, 236, 240, 241, 246, 376
Brockman Iron Formation, Australia, 386
Broken Hill, Australia, 223, 236
Bronzite chondrite, 109, 142
Brunderheim (meteorite), 181, 191, 373
Brunhes epoch, 154, 338
Butte, Montana, 377
Byrd Station, Antarctica, 331, 332

Calcite, 130, 132, 152, 191, 271, 336, 353, 354, 355, 358, 368, 385, 386, 391, 393, 394, 408, 409
Calcium, 67, 75, 76, 101, 102, 105, 147, 148, 165, 167, 188-191, 323, 392
Calcium carbonate, 75, 76, 128-132, 191, 288, 289, 316, 336, 367, 384-386
Caledonian orogeny, 85, 86, 178
California, 125, 126, 127, 142, 171, 172, 224, 278, 279, 288, 296, 297, 298, 317, 319, 341, 354, 382, 390, 391
Calutron, 68
Cambrian period, 87, 241, 367, 383, 385, 410
Cambridge University, 4, 5
Camp Century, Greenland, 301, 330, 331, 332
Canada, 101, 102, 156, 157, 160, 235, 236, 240, 244, 245, 246, 259, 264, 320, 360, 361, 378, 386, 392
Canadian Shield, 112, 156, 157, 158, 159, 161, 164
Canary Islands, 253, 265
Cancer, 5, 54
Canyon Diablo (meteorite), 230, 403, 405, 406, 409
Cape Kidnappers, New Zealand, 280
Captain's Flat, N.S.W., Australia, 236
Capture cross-section, 58
Carbon, 8, 67, 68, 323, 379-395
Carbon-14, 305-317
Carbon dioxide, 65, 149, 306, 307, 313, 325, 379, 380, 382. 387, 389-390, 391
Carbonaceous chondrite, 109, 142, 358, 359, 386
Carbonate rocks, marine, 128-132, 384-386
  nonmarine, 129, 384-386
Carbonatites, 113, 368, 369, 375, 386-388
Caribbean Sea, 294, 303, 338
Carn Chuinneag, Scotland, 85, 86, 94, 95
Carnallite, 75, 88
Casapalca, Peru, 419
Cascade Range, Oregon, 363, 364, 378
Castile Formation, Texas, 420
Cataracts, 54
Catastrophism, 1
Cathode rays, 3, 65

Mt. Ascutney, Vermont, 277
Mt. Vesuvius, Italy, 298
Mull, Scotland, 378
Muntsche Tundra pluton, U.S.S.R., 138, 139
Muscovite, 75, 85, 86, 88, 150, 153, 157, 160, 197, 267, 268, 271, 275, 353, 354, 356, 358, 371
Musgrave Range, Australia, 378
Mushandike granite, Rhodesia, 86
Muskox Intrusive, N.W. T., Canada, 360, 361

Nain province, Canada, 159
Nakhla (meteorite), 108, 143
National Academy of Sciences, U.S.A., 7
National Bureau of Standards, 269, 311, 312, 325
Natural gas, 381-382
Near Islands, Alaska, 162
Needle Mountains, Colorado, 397
Negatron decay, 25-30
Nellore, India, 280
Nelson, British Columbia, 240
Nelson batholith, British Columbia, 241
Neodymium, 183, 196
Neon, 5, 65, 67
Nepheline, 153, 181
Neutrino, 26, 27, 30, 31, 33
Neutron, 5, 13, 16, 25, 30, 41, 54, 55, 165-177, 218, 268, 305, 306, 308, 314
Neutron activation, 54-61, 149, 165-177, 187, 190
Neutron capture cross-section, 58, 165
Neutron flux, 58, 59, 165, 268
Neutron number, 13, 14, 17, 25, 31, 33, 37
Neutron stars, 13
Nevada, 376, 377, 390
New Brunswick, Canada, 236
New Guinea, 143
New Jersey, 198
New Mexico, 245
New South Wales, Australia, 164, 236
New York, 93, 369
New Zealand, 106, 140, 180, 244, 280, 320, 354, 390, 396
Nickel, 80, 184
Niland, California, 341
Nimrod Glacier, Antarctica, 124
Nimrod Group, Antarctica, 124
Nitrogen, 8, 67, 149, 305, 323
Nobel Prize, 4, 5, 8, 306
Noranda, Quebec, 245, 259, 264
Norberg district, Sweden, 399
North America, 120, 125, 128, 140, 162, 212, 317
North Carolina, 124, 125
North Channel, Lake Huron, 101, 102, 105
North Island, New Zealand, 106, 140, 390
Northern Light Lake, Ontario, 138, 141
Norton County (meteorite), 109, 139
Norway, 162, 179, 320, 334
Nuanetsi igneous province, South Africa, 142
Nuclear reactions, 15, 16, 17, 27, 54-56, 167, 306
Nuclear reactor, 19, 41, 55, 56, 59, 165, 203, 268, 309
Nucleon, 13, 17
Nucleosynthesis, 20, 48, 78, 107
Nucleus, 13, 25-41, 54-61
Nuclide, 14
Nuevo Laredo (meteorite), 191, 230
Nunivak event, 154

Oahu, Hawaii, 144
Obsidian, 106
Ocean of Storms, moon, 143
Oceanic islands, 115, 116, 119, 121
Oceans, 128-137
Oddo-Harkins rule, 20
Oil shale, 381
Oka, Quebec, 368, 386
Oklo, Gabon, 203, 223
Olduvai event, 154
Olivenza (meteorite), 373
Olivine, 187, 271, 298, 299, 300, 356, 358, 359, 362
Ontario, 72, 80, 138, 141, 163, 179, 225, 236, 244, 342, 378
Onverwacht Group, South Africa, 86, 93, 94, 224, 383
Opal, 288, 339
Ophiolite, 278
Optical spectrograph, 197
Orbital electron capture, 25, 33
Ordinary lead, 234, 235-237
Ordovician period, 87, 124, 180, 391
Ore deposits, 67, 80, 115, 126, 185, 228, 231, 232, 235-237, 239, 241, 362-365, 392-395, 412-418
Oregon, 363, 364
Orijärvi, Finland, 236
Orogenic belt, 120, 156, 157-160, 161
Orogeny, 157
Orthoclase, 75, 82, 85, 147, 316
Osmiridium, 184, 185
Osmium, 183-185
Oxford University, 85

Oxygen, 8, 67, 68, 149, 305, 323-344, 351-372
Oxygen isotope shift, 341

Pacific ocean, 114, 117, 121, 140, 141, 144, 179, 253, 254, 256, 288, 327, 333, 334, 335, 338, 339
Paleothermometry, 336-340
Paleozoic era, 126, 128, 132, 172
Palisade sill, New Jersey, 198
Papua, New Guinea, 143
Paris, 3, 4
Pasamonte (meteorite), 178, 191, 194
PDB standard, 312, 325, 379
Peedee Formation, South Carolina, 312, 379
Pegmatite, 77, 153, 187, 189, 205, 359
Pennsylvania, 93, 197, 222
Penokean orogeny, 160
Periodic table, 15
Permian period, 131, 132, 407, 408, 410
Peru, 317, 419
Petersburg (meteorite), 180
Petroleum, 381-382, 406-407, 408
Philipsite, 284, 299
Phlogopite, 75, 118, 121, 153, 267, 271, 274, 391
Phosphate, 339, 340
Photosynthesis, 307, 312, 330, 379, 380
Pigeonite, 271
Pine Point, N.W.T., Canada, 392
Piston core, 106
Pitchblende, 205
Plagioclase, 75, 76, 85, 90, 152, 153, 172, 187, 354, 355, 356, 358, 369
Planck's constant, 5, 323
Plate tectonics, 114-121
Platinum, 184, 285
Pleistocene, 27, 198, 222, 283, 332, 337, 338, 339, 342, 343, 385
Pleochroic haloes, 6, 275-276
Pliocene epoch, 103
Poland, 3
Polarity reversals (magnetic), 153-155
Polonium, 4
Pollucite, 271
Polylithionite, 190
Porphyry copper deposits, 142, 377
Position decay, 25, 30-33, 147
Potassium, 48, 67, 75, 76, 121, 147-161, 165-177, 188-190, 323
Precambrian, 87, 95, 101, 120, 124, 126, 129, 151, 155, 172, 176, 241, 256, 375, 382, 383, 386, 410
Precambrian shield, 101, 120, 140, 156, 160, 214
Precambrian time scale, 157-160

Primeval lead, 227, 229, 230, 231, 233, 242, 256
Primordial strontium, 107, 109, 111, 137
Priorite, 186
Production rate, 56-58, 201
Promethium, 15
Protactinium, 292-294
Proterozoic eon, 158, 179
Proton, 4, 13, 25, 30, 33, 41, 54
Proton number, 14, 17, 31, 33
Providencia mining district, Mexico, 377, 414, 415, 419
Pseudoisochron, 103
Pueblito de Allende (meteorite), 109, 110, 358
Pyrite, 405, 407, 408, 409, 410, 413, 414, 417
Pryoxene, 152, 153, 187, 188, 356, 358, 359, 362
Pyrrhotite, 414

Quantum mechanics, 5
Quartz, 271, 285, 298, 299, 300, 353, 354, 355, 356, 358, 359, 362, 370, 391
Quaternary period, 154, 283
Quebec, 368, 391
Queen Alexandra Range, Antarctica, 106, 122
Queensland, Australia, 222, 236
Quilangarssuit, Greenland, 85, 87

Rad, 54
Radiation damage, 151, 197, 208, 213, 275, 289
Radiation safety, 53-54
Radioactive equilibrium, 52
Radioactivity, 2, 4, 6, 7, 17, 25-40, 45, 75, 77, 121, 147, 197, 213
Radiogenic daughter, 46, 75
Radiogenic lead, 237-243
Radiography, 33
Radiolarians, 339
Radiometer, 79
Radionuclide, 17
Radium, 4, 6, 75
Radium Institute, Paris, 4
Rajasthan, India, 278
Rare earth elements, 61, 183, 185, 186
Rayleigh distillation, 327
Rb/Sr ratio, 76, 79, 103, 108, 109, 112, 119, 121, 137
Reading Prong, New Jersey, 179
Recoil energy, 37-38, 275, 289
Recrystallization, 81, 88
Red Sea, 100, 133, 135, 136, 139, 140, 333, 337, 343, 344
Redwood Falls, Minnesota, 215
Regression methods, 80, 89-90, 425
Rem, 54

Storm Peak, Antarctica, 122, 123
Straight line, 79, 82, 97-104, 135
Straits of Bab el Mandeb, Red Sea, 344
Strontianite, 76
Strontium, 66, 67, 72, 75-95, 101, 105, 107-137, 337
Subduction zone, 114, 115, 118, 119
Substitution, ionic, 75
Sudbury, Ontario, 80, 236
Suess effect, 309, 310
Sulfate, 335, 403, 407, 409, 410-412
Sulfur, 8, 67, 68, 323, 403-418
Sullivan mine, B.C., Canada, 236, 241, 242, 245
Sumatra, 117
Sumbawa, Indonesia, 117
Sun, 26, 27, 48, 108, 308, 309, 311
Sunda arc, 117, 145
Superior province, Canada, 158
Survey meter, 54
Swaziland sequence, South Africa, 93, 94, 410
Sweden, 311, 400
Swimming pool reactor, 55
Switzerland, 274
Sybella granite, Australia, 222
Syenite, 76, 113, 199, 359, 368
Sylvite, 75, 88, 147, 151, 189, 190

Tantalite, 183
Tantalum, 66
Tanzanite, 279
Tar sands, 381
Target nucleus, 55-56
Tasmania, 420
Taurus-Littrow valley, moon, 174-175
Tachnetium, 15, 183
Tektite, 267, 269, 271
Telluride, 184
Tenerife, Canary Islands, 253, 265
Tennessee, 419
Terbium, 41
Terskey Ala-Tau, U.S.S.R., 224, 266
Tertiary period, 27, 103, 151, 198, 259, 349, 363, 366, 385, 407, 408
Tethys Sea, 129, 144
Texas, 225, 408, 409, 420
Thallium, 67
The Geysers, California, 341, 342, 390
Thermochron, 157
Tholeiite, oceanic, 114, 115, 116, 119, 126
Thorianite, 198
Thorite, 197, 198, 205
Thorium, 3, 5, 6, 40, 45, 121, 197-219, 228, 229, 249-261, 274, 284, 285

Th/U ratio, 240, 242
Thucholite, 383
Thule, Greenland, 331
Time scale, geologic, 7, 9, 157-160, 332, 437
Tin, 19, 20
Tintina Trench, Yukon Territory, Canada, 244, 245
Titanium, 67, 188
Tonga, Pacific Ocean, 265
Toro-Ankole volcanic area, East Africa, 103
Toronto, Ontario, 235, 237
Torridonian system, Scotland, 88, 89, 94
Tourmaline, 153, 275, 363
Trail, British Columbia, 240, 241
Transantarctic Mountains, 87, 122, 123, 124, 129, 140, 150
Transcendental function, 204, 232
Trans-uranium elements, 41
Transvaal, South Africa, 223
Travertine, 191
Tremolite, 391
Trench, deep sea, 114, 115, 117
Triassic period, 125, 126, 129, 132, 172, 407
T.R.I.G.A. reactor, 167
Trinidad, 265
Tristan da Cunha, 298
Tritium, 328, 330
Troilite, 229, 403, 405
Trondhjemite, 128
Tudor gabbro, Ontario, 179
Tungsten, 66

Uganda, Africa, 103
Uinta basin, Utah, 420
Uivak gneiss, Labrador, 138
Ukrainian Precambrian shield, 160, 164
Ultramafic rocks, 76, 113, 126, 176, 180, 184, 198, 199, 260, 359, 368
Uncompahgre Formation, Colorado, 397
Uniformitarianism, 1, 2
United States of America, 41, 141, 241, 243, 247, 255, 288, 408
University College, London, 4
University of British Columbia, 235
University of California, 305
University of Chicago, 65, 312, 379
University of Toronto, 235
Ural Mountains, U.S.S.R., 185
Uraninite, 3, 6, 7, 77, 198, 205, 383
Uranium, 3, 6, 40, 41, 45, 55, 73, 121, 197-219, 227, 228, 229, 230, 249-261, 267, 268, 269, 271, 274, 283-300, 337, 386
Uranyl ion, 198, 284, 289

10